MAZDA TRUCKS 1987-93 REPAIR MANUAL

CHILTON'S

Covers all U.S. and Canadian models of Mazda B2200, B2600, MPV and Navajo

by **Matthew E. Frederick,** A.S.E., S.A.E.

PUBLISHED BY HAYNES NORTH AMERICA, Inc.

Manufactured in USA
© 1995 Haynes North America, Inc.
ISBN 0-8019-8964-7
Library of Congress Catalog Card No. 98-74836
4567890123 9876543210

Haynes Publishing Group
Sparkford Nr Yeovil
Somerset BA22 7JJ England

Haynes North America, Inc
861 Lawrence Drive
Newbury Park
California 91320 USA

ABCDE
FGHIJ
KLM

Chilton is a registered trademark of W.G. Nichols, Inc., and has been licensed to Haynes North America, Inc.

Contents

1. GENERAL INFORMATION AND MAINTENANCE

- 1-2 HOW TO USE THIS BOOK
- 1-2 TOOLS AND EQUIPMENT
- 1-4 SERVICING YOUR VEHICLE SAFELY
- 1-6 FASTENERS, MEASUREMENTS AND CONVERSIONS
- 1-9 SERIAL NUMBER IDENTIFICATION
- 1-12 ROUTINE MAINTENANCE
- 1-38 FLUIDS AND LUBRICANTS
- 1-50 TRAILER TOWING
- 1-50 TOWING THE VEHICLE
- 1-51 JUMP STARTING A DEAD BATTERY
- 1-52 JACKING

2. ENGINE ELECTRICAL

- 2-2 DISTRIBUTOR IGNITION SYSTEM
- 2-6 DISTRIBUTORLESS IGNITION SYSTEM
- 2-7 FIRING ORDERS
- 2-8 CHARGING SYSTEM
- 2-10 STARTING SYSTEM
- 2-13 SENDING UNITS AND SENSORS

3. ENGINE AND ENGINE OVERHAUL

- 3-2 ENGINE MECHANICAL
- 3-41 EXHAUST SYSTEM
- 3-44 ENGINE RECONDITIONING

4. DRIVEABILITY AND EMISSION CONTROLS

- 4-2 EMISSION CONTROLS
- 4-9 ELECTRONIC ENGINE CONTROLS
- 4-20 TROUBLE CODES
- 4-23 COMPONENT LOCATIONS
- 4-30 VACUUM DIAGRAMS

5. FUEL SYSTEM

- 5-2 BASIC FUEL SYSTEM DIAGNOSIS
- 5-2 FUEL LINES AND FITTINGS
- 5-3 CARBURETED FUEL SYSTEM
- 5-7 FUEL INJECTION SYSTEM
- 5-15 FUEL TANK

6. CHASSIS ELECTRICAL

- 6-2 UNDERSTANDING AND TROUBLESHOOTING ELECTRICAL SYSTEMS
- 6-8 BATTERY CABLES
- 6-8 HEATING AND AIR CONDITIONING
- 6-14 CRUISE CONTROL
- 6-16 ENTERTAINMENT SYSTEMS
- 6-20 WINDSHIELD WIPERS
- 6-24 INSTRUMENTS AND SWITCHES
- 6-27 LIGHTING
- 6-34 TRAILER WIRING
- 6-35 CIRCUIT PROTECTION
- 6-39 WIRING DIAGRAMS

Contents

7 DRIVE TRAIN

- **7-2** MANUAL TRANSMISSION
- **7-4** CLUTCH
- **7-7** AUTOMATIC TRANSMISSION
- **7-12** TRANSFER CASE
- **7-13** DRIVELINE
- **7-18** FRONT DRIVE AXLE
- **7-24** REAR AXLE

8 SUSPENSION AND STEERING

- **8-2** FRONT SUSPENSION
- **8-24** REAR SUSPENSION
- **8-29** STEERING

9 BRAKES

- **9-2** BRAKE OPERATING SYSTEM
- **9-7** DISC BRAKES
- **9-12** REAR DRUM BRAKES
- **9-18** PARKING BRAKE
- **9-20** REAR WHEEL ANTI-LOCK BRAKE SYSTEM
- **9-25** 4-WHEEL ANTI-LOCK BRAKE SYSTEM (4WABS)

10 BODY AND TRIM

- **10-2** EXTERIOR
- **10-9** INTERIOR

GLOSSARY

- **10-19** GLOSSARY

MASTER INDEX

- **10-23** MASTER INDEX

SAFETY NOTICE

Proper service and repair procedures are vital to the safe, reliable operation of all motor vehicles, as well as the personal safety of those performing repairs. This manual outlines procedures for servicing and repairing vehicles using safe, effective methods. The procedures contain many NOTES, CAUTIONS and WARNINGS which should be followed, along with standard procedures to eliminate the possibility of personal injury or improper service which could damage the vehicle or compromise its safety.

It is important to note that repair procedures and techniques, tools and parts for servicing motor vehicles, as well as the skill and experience of the individual performing the work vary widely. It is not possible to anticipate all of the conceivable ways or conditions under which vehicles may be serviced, or to provide cautions as to all possible hazards that may result. Standard and accepted safety precautions and equipment should be used when handling toxic or flammable fluids, and safety goggles or other protection should be used during cutting, grinding, chiseling, prying, or any other process that can cause material removal or projectiles.

Some procedures require the use of tools specially designed for a specific purpose. Before substituting another tool or procedure, you must be completely satisfied that neither your personal safety, nor the performance of the vehicle will be endangered.

Although information in this manual is based on industry sources and is complete as possible at the time of publication, the possibility exists that some car manufacturers made later changes which could not be included here. While striving for total accuracy, the authors or publishers cannot assume responsibility for any errors, changes or omissions that may occur in the compilation of this data.

PART NUMBERS

Part numbers listed in this reference are not recommendations by Haynes North America, Inc. for any product brand name. They are references that can be used with interchange manuals and aftermarket supplier catalogs to locate each brand supplier's discrete part number.

SPECIAL TOOLS

Special tools are recommended by the vehicle manufacturer to perform their specific job. Use has been kept to a minimum, but where absolutely necessary, they are referred to in the text by the part number of the tool manufacturer. These tools can be purchased, under the appropriate part number, from your local dealer or regional distributor, or an equivalent tool can be purchased locally from a tool supplier or parts outlet. Before substituting any tool for the one recommended, read the SAFETY NOTICE at the top of this page.

ACKNOWLEDGMENTS

The publisher expresses appreciation to Mazda Motor Corporation for their generous assistance.

All rights reserved. No part of this book may be reproduced or transmitted in any form or by any means, electronic or mechanical, including photocopying, recording or by any information storage or retrieval system, without permission in writing from the copyright holder.

While every attempt is made to ensure that the information in this manual is correct, no liability can be accepted by the authors or publishers for loss, damage or injury caused by any errors in, or omissions from, the information given.

1. GENERAL INFORMATION AND MAINTENANCE

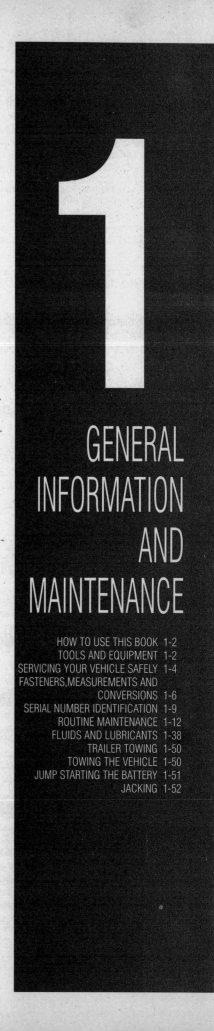

HOW TO USE THIS BOOK 1-2
WHERE TO BEGIN 1-2
AVOIDING TROUBLE 1-2
MAINTENANCE OR REPAIR? 1-2
AVOIDING THE MOST COMMON MISTAKES 1-2
TOOLS AND EQUIPMENT 1-2
SPECIAL TOOLS 1-4
SERVICING YOUR VEHICLE SAFELY 1-4
DO'S 1-4
DON'TS 1-5
FASTENERS, MEASUREMENTS AND CONVERSIONS 1-6
BOLTS, NUTS AND OTHER THREADED RETAINERS 1-6
TORQUE 1-6
 TORQUE WRENCHES 1-6
 TORQUE ANGLE METERS 1-7
STANDARD AND METRIC MEASUREMENTS 1-7
SERIAL NUMBER IDENTIFICATION 1-9
VEHICLE 1-9
 PICK-UP & MPV 1-9
 NAVAJO 1-9
ENGINE 1-10
 PICK-UP & MPV 1-10
 NAVAJO 1-10
TRANSMISSION 1-12
 MANUAL TRANSMISSION 1-12
 AUTOMATIC TRANSMISSION 1-12
DRIVE AXLE 1-12
 FRONT 1-12
 REAR 1-12
TRANSFER CASE 1-12
ROUTINE MAINTENANCE AND TUNE-UP 1-12
AIR CLEANER 1-12
 REMOVAL & INSTALLATION 1-12
FUEL FILTER 1-14
 REMOVAL & INSTALLATION 1-14
PCV VALVE 1-15
 REMOVAL & INSTALLATION 1-15
EVAPORATIVE EMISSION CANISTER 1-15
 SERVICING 1-15
BATTERY 1-15
 PRECAUTIONS 1-15
 GENERAL MAINTENANCE 1-15
 BATTERY FLUID 1-16
 CABLES 1-16
 CHARGING 1-17
 REPLACEMENT 1-17
BELTS 1-17
 INSPECTION 1-17
 ADJUSTMENT 1-18
 REMOVAL & INSTALLATION 1-18
TIMING BELTS 1-21
 INSPECTION 1-21
HOSES 1-21
 INSPECTION 1-21
 REMOVAL & INSTALLATION 1-22
CV-BOOTS 1-25
 INSPECTION 1-25
SPARK PLUGS 1-25
 SPARK PLUG HEAT RANGE 1-25
 REMOVAL & INSTALLATION 1-26
 INSPECTION & GAPPING 1-26
SPARK PLUG WIRES 1-28
 TESTING 1-28
 REMOVAL & INSTALLATION 1-28
DISTRIBUTOR CAP AND ROTOR 1-29
 REMOVAL & INSTALLATION 1-29
 INSPECTION 1-29
IGNITION TIMING 1-29
 GENERAL INFORMATION 1-29
 INSPECTION & ADJUSTMENT 1-30
VALVE LASH 1-30
 ADJUSTMENT 1-30
IDLE SPEED AND MIXTURE ADJUSTMENTS 1-31
 IDLE SPEED 1-31
 MIXTURE ADJUSTMENT 1-32
AIR CONDITIONING 1-32
 SYSTEM SERVICE & REPAIR 1-32
 PREVENTIVE MAINTENANCE 1-34
 SYSTEM INSPECTION 1-34
WINDSHIELD WIPERS 1-35
 ELEMENT (REFILL) CARE & REPLACEMENT 1-35
TIRES AND WHEELS 1-35
 TIRE ROTATION 1-35
 TIRE DESIGN 1-36
 TIRE STORAGE 1-36
 INFLATION & INSPECTION 1-36
FLUIDS AND LUBRICANTS 1-38
FLUID DISPOSAL 1-38
FUEL AND ENGINE OIL RECOMMENDATIONS 1-38
 ENGINE OIL 1-38
 FUEL 1-39
 OPERATION IN FOREIGN COUNTRIES 1-39
ENGINE OIL 1-39
 OIL LEVEL CHECK 1-39
 OIL & FILTER CHANGE 1-39
MANUAL TRANSMISSION 1-40
 FLUID RECOMMENDATIONS 1-40
 LEVEL CHECK 1-40
 DRAIN & REFILL 1-41
AUTOMATIC TRANSMISSION 1-41
 FLUID RECOMMENDATION 1-41
 LEVEL CHECK 1-41
 DRAIN AND REFILL 1-42
TRANSFER CASE 1-43
 FLUID RECOMMENDATION 1-43
 LEVEL CHECK 1-43
 DRAIN AND REFILL 1-43
FRONT AND REAR DRIVE AXLE 1-43
 FLUID RECOMMENDATIONS 1-43
 LEVEL CHECK 1-43
 DRAIN AND REFILL 1-43
COOLING SYSTEM 1-44
 FLUID RECOMMENDATIONS 1-45
 LEVEL CHECK 1-45
 CHECK THE RADIATOR CAP 1-46
 CLEAN RADIATOR OF DEBRIS 1-46
 DRAIN, REFILL & FLUSHING 1-46
BRAKE MASTER CYLINDER 1-47
 FLUID RECOMMENDATIONS 1-47
 LEVEL CHECK 1-47
CLUTCH MASTER CYLINDER 1-47
 FLUID RECOMMENDATIONS 1-47
 LEVEL CHECK 1-47
POWER STEERING PUMP 1-48
 FLUID RECOMMENDATIONS 1-48
 LEVEL CHECK 1-48
MANUAL STEERING GEAR 1-48
 FLUID RECOMMENDATIONS 1-48
 LEVEL CHECK 1-48
CHASSIS MAINTENANCE 1-48
BODY LUBRICATION AND MAINTENANCE 1-48
 LOCK CYLINDERS 1-48
 DOOR HINGES AND HINGE CHECKS 1-48
 TAILGATE 1-48
 BODY DRAIN HOLES 1-48
WHEEL BEARINGS 1-48
 REPACKING 1-48
TRAILER TOWING 1-50
GENERAL RECOMMENDATIONS 1-50
TRAILER WEIGHT 1-50
HITCH (TONGUE) WEIGHT 1-50
ENGINE 1-50
TRANSMISSION 1-50
HANDLING A TRAILER 1-50
TOWING THE VEHICLE 1-50
JUMP STARTING A DEAD BATTERY 1-51
JUMP STARTING PRECAUTIONS 1-51
JUMP STARTING PROCEDURE 1-51
JACKING 1-52
JACKING PRECAUTIONS 1-53
SPECIFICATIONS CHARTS
VEHICLE IDENTIFICATION CHART 1-9
ENGINE IDENTIFICATION 1-10
GENERAL ENGINE SPECIFICATIONS 1-11
GASOLINE ENGINE TUNE-UP SPECIFICATIONS 1-33
MAINTENANCE INTERVALS 1-53
CAPACITIES 1-59

1-2 GENERAL INFORMATION AND MAINTENANCE

HOW TO USE THIS BOOK

This Chilton's Total Car Care manual for 1987–93 Mazda Trucks is intended to help you learn more about the inner workings of your vehicle while saving you money on its upkeep and operation.

The beginning of the book will likely be referred to the most, since that is where you will find information for maintenance and tune-up. The other sections deal with the more complex systems of your vehicle. Systems (from engine through brakes) are covered to the extent that the average do-it-yourselfer can attempt. This book will not explain such things as rebuilding a differential because the expertise required and the special tools necessary make this uneconomical. It will, however, give you detailed instructions to help you change your own brake pads and shoes, replace spark plugs, and perform many more jobs that can save you money and help avoid expensive problems.

A secondary purpose of this book is a reference for owners who want to understand their vehicle and/or their mechanics better.

Where to Begin

Before removing any bolts, read through the entire procedure. This will give you the overall view of what tools and supplies will be required. So read ahead and plan ahead. Each operation should be approached logically and all procedures thoroughly understood before attempting any work.

If repair of a component is not considered practical, we tell you how to remove the part and then how to install the new or rebuilt replacement. In this way, you at least save labor costs.

Avoiding Trouble

Many procedures in this book require you to "label and disconnect . . ." a group of lines, hoses or wires. Don't be think you can remember where everything goes—you won't. If you hook up vacuum or fuel lines incorrectly, the vehicle may run poorly, if at all. If you hook up electrical wiring incorrectly, you may instantly learn a very expensive lesson.

You don't need to know the proper name for each hose or line. A piece of masking tape on the hose and a piece on its fitting will allow you to assign your own label. As long as you remember your own code, the lines can be reconnected by matching your tags. Remember that tape will dissolve in gasoline or solvents; if a part is to be washed or cleaned, use another method of identification. A permanent felt-tipped marker or a metal scribe can be very handy for marking metal parts. Remove any tape or paper labels after assembly.

Maintenance or Repair?

Maintenance includes routine inspections, adjustments, and replacement of parts which show signs of normal wear. Maintenance compensates for wear or deterioration. Repair implies that something has broken or is not working. A need for a repair is often caused by lack of maintenance. for example: draining and refilling automatic transmission fluid is maintenance recommended at specific intervals. Failure to do this can shorten the life of the transmission/transaxle, requiring very expensive repairs. While no maintenance program can prevent items from eventually breaking or wearing out, a general rule is true: MAINTENANCE IS CHEAPER THAN REPAIR.

TOOLS AND EQUIPMENT

♦ See Figures 1 thru 15

Without the proper tools and equipment it is impossible to properly service your vehicle. It would be virtually impossible to catalog every tool that you would need to perform all of the operations in this book. It would be unwise for the amateur to rush out and buy an expensive set of tools on the theory that he/she may need one or more of them at some time.

The best approach is to proceed slowly, gathering a good quality set of those tools that are used most frequently. Don't be misled by the low cost of bargain tools. It is far better to spend a little more for better quality. Forged wrenches, 6 or 12-point sockets and fine tooth ratchets are by far preferable to their less expensive counterparts. As any good mechanic can tell you, there are few worse experiences than trying to work on a vehicle with bad tools. Your monetary savings will be far outweighed by frustration and mangled knuckles.

Two basic mechanic's rules should be mentioned here. First, whenever the left side of the vehicle or engine is referred to, it means the driver's side. Conversely, the right side of the vehicle means the passenger's side. Second, screws and bolts are removed by turning counterclockwise, and tightened by turning clockwise unless specifically noted.

Safety is always the most important rule. Constantly be aware of the dangers involved in working on an automobile and take the proper precautions. Please refer to the information in this section regarding SERVICING YOUR VEHICLE SAFELY and the SAFETY NOTICE on the acknowledgment page.

Avoiding the Most Common Mistakes

Pay attention to the instructions provided. There are 3 common mistakes in mechanical work:

1. Incorrect order of assembly, disassembly or adjustment. When taking something apart or putting it together, performing steps in the wrong order usually just costs you extra time; however, it CAN break something. Read the entire procedure before beginning. Perform everything in the order in which the instructions say you should, even if you can't see a reason for it. When you're taking apart something that is very intricate, you might want to draw a picture of how it looks when assembled in order to make sure you get everything back in its proper position. When making adjustments, perform them in the proper order. One adjustment possibly will affect another.

2. Overtorquing (or undertorquing). While it is more common for overtorquing to cause damage, undertorquing may allow a fastener to vibrate loose causing serious damage. Especially when dealing with aluminum parts, pay attention to torque specifications and utilize a torque wrench in assembly. If a torque figure is not available, remember that if you are using the right tool to perform the job, you will probably not have to strain yourself to get a fastener tight enough. The pitch of most threads is so slight that the tension you put on the wrench will be multiplied many times in actual force on what you are tightening.

There are many commercial products available for ensuring that fasteners won't come loose, even if they are not torqued just right (a very common brand is Loctite®). If you're worried about getting something together tight enough to hold, but loose enough to avoid mechanical damage during assembly, one of these products might offer substantial insurance. Before choosing a threadlocking compound, read the label on the package and make sure the product is compatible with the materials, fluids, etc. involved.

3. Crossthreading. This occurs when a part such as a bolt is screwed into a nut or casting at the wrong angle and forced. Crossthreading is more likely to occur if access is difficult. It helps to clean and lubricate fasteners, then to start threading the bolt, spark plug, etc. with your fingers. If you encounter resistance, unscrew the part and start over again at a different angle until it can be inserted and turned several times without much effort. Keep in mind that many parts have tapered threads, so that gentle turning will automatically bring the part you're threading to the proper angle. Don't put a wrench on the part until it's been tightened a couple of turns by hand. If you suddenly encounter resistance, and the part has not seated fully, don't force it. Pull it back out to make sure it's clean and threading properly.

Be sure to take your time and be patient, and always plan ahead. Allow yourself ample time to perform repairs and maintenance.

Begin accumulating those tools that are used most frequently: those associated with routine maintenance and tune-up. In addition to the normal assortment of screwdrivers and pliers, you should have the following tools:

• Wrenches/sockets and combination open end/box end wrenches in sizes ⅛–¾ in. and/or 3mm–19mm ¹³⁄₁₆ in. or ⅝ in. spark plug socket (depending on plug type).

➡ **If possible, buy various length socket drive extensions. Universal-joint and wobble extensions can be extremely useful, but be careful when using them, as they can change the amount of torque applied to the socket.**

GENERAL INFORMATION AND MAINTENANCE

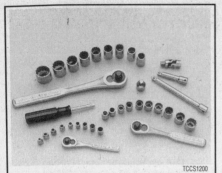
Fig. 1 All but the most basic procedures will require an assortment of ratchets and sockets

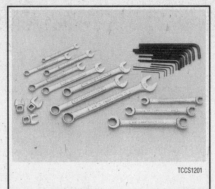
Fig. 2 In addition to ratchets, a good set of wrenches and hex keys will be necessary

Fig. 3 A hydraulic floor jack and a set of jackstands are essential for lifting and supporting the vehicle

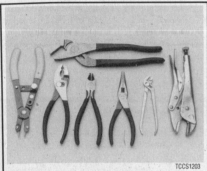
Fig. 4 An assortment of pliers, grippers and cutters will be handy for old rusted parts and stripped bolt heads

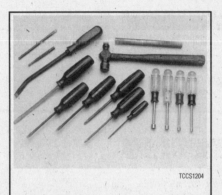
Fig. 5 Various drivers, chisels and prybars are great tools to have in your toolbox

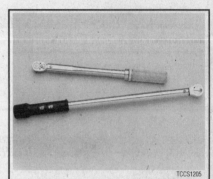
Fig. 6 Many repairs will require the use of a torque wrench to assure the components are properly fastened

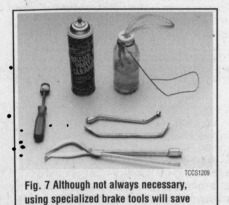
Fig. 7 Although not always necessary, using specialized brake tools will save time

Fig. 8 A few inexpensive lubrication tools will make maintenance easier

Fig. 9 Various pullers, clamps and separator tools are needed for many larger, more complicated repairs

- Jackstands for support.
- Oil filter wrench.
- Spout or funnel for pouring fluids.
- Grease gun for chassis lubrication (unless your vehicle is not equipped with any grease fittings)
- Hydrometer for checking the battery (unless equipped with a sealed, maintenance-free battery).
- A container for draining oil and other fluids.
- Rags for wiping up the inevitable mess.

In addition to the above items there are several others that are not absolutely necessary, but handy to have around. These include an equivalent oil absorbent gravel, like cat litter, and the usual supply of lubricants, antifreeze and fluids. This is a basic list for routine maintenance, but only your personal needs and desire can accurately determine your list of tools.

After performing a few projects on the vehicle, you'll be amazed at the other tools and non-tools on your workbench. Some useful household items are: a large turkey baster or siphon, empty coffee cans and ice trays (to store parts), a ball of twine, electrical tape for wiring, small rolls of colored tape for tagging lines or hoses, markers and pens, a note pad, golf tees (for plugging vacuum lines), metal coat hangers or a roll of mechanic's wire (to hold things out of the way), dental pick or similar long, pointed probe, a strong magnet, and a small mirror (to see into recesses and under manifolds).

A more advanced set of tools, suitable for tune-up work, can be drawn up easily. While the tools are slightly more sophisticated, they need not be outrageously expensive. There are several inexpensive tach/dwell meters on the market that are every bit as good for the average mechanic as a professional model. Just be sure that it goes to a least 1200–1500 rpm on the tach scale and that it works on 4, 6 and 8-cylinder engines. The key to these purchases is to make them with an eye towards adaptability and wide range. A basic list of tune-up tools could include:

- Tach/dwell meter.
- Spark plug wrench and gapping tool.

1-4 GENERAL INFORMATION AND MAINTENANCE

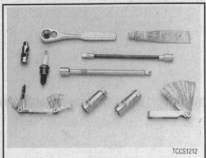

Fig. 10 A variety of tools and gauges should be used for spark plug gapping and installation

Fig. 11 Inductive type timing light

Fig. 12 A screw-in type compression gauge is recommended for compression testing

Fig. 13 A vacuum/pressure tester is necessary for many testing procedures

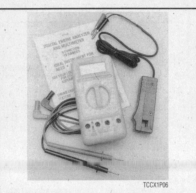

Fig. 14 Most modern automotive multimeters incorporate many helpful features

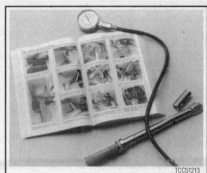

Fig. 15 Proper information is vital, so always have a Chilton Total Car Care manual handy

- Feeler gauges for valve adjustment.
- Timing light.

The choice of a timing light should be made carefully. A light which works on the DC current supplied by the vehicle's battery is the best choice; it should have a xenon tube for brightness. On any vehicle with an electronic ignition system, a timing light with an inductive pickup that clamps around the No. 1 spark plug cable is preferred.

In addition to these basic tools, there are several other tools and gauges you may find useful. These include:

- Compression gauge. The screw-in type is slower to use, but eliminates the possibility of a faulty reading due to escaping pressure.
- Manifold vacuum gauge.
- 12V test light.
- A combination volt/ohmmeter
- Induction Ammeter. This is used for determining whether or not there is current in a wire. These are handy for use if a wire is broken somewhere in a wiring harness.

As a final note, you will probably find a torque wrench necessary for all but the most basic work. The beam type models are perfectly adequate, although the newer click types (breakaway) are easier to use. The click type torque wrenches tend to be more expensive. Also keep in mind that all types of torque wrenches should be periodically checked and/or recalibrated. You will have to decide for yourself which better fits your pocketbook, and purpose.

Special Tools

Normally, the use of special factory tools is avoided for repair procedures, since these are not readily available for the do-it-yourself mechanic. When it is possible to perform the job with more commonly available tools, it will be pointed out, but occasionally, a special tool was designed to perform a specific function and should be used. Before substituting another tool, you should be convinced that neither your safety nor the performance of the vehicle will be compromised.

Special tools can usually be purchased from an automotive parts store or from your dealer. In some cases special tools may be available directly from the tool manufacturer.

SERVICING YOUR VEHICLE SAFELY

▶ See Figures 16, 17 and 18

It is virtually impossible to anticipate all of the hazards involved with automotive maintenance and service, but care and common sense will prevent most accidents.

The rules of safety for mechanics range from "don't smoke around gasoline," to "use the proper tool(s) for the job." The trick to avoiding injuries is to develop safe work habits and to take every possible precaution.

Do's

- Do keep a fire extinguisher and first aid kit handy.
- Do wear safety glasses or goggles when cutting, drilling, grinding or prying, even if you have 20–20 vision. If you wear glasses for the sake of vision, wear safety goggles over your regular glasses.

GENERAL INFORMATION AND MAINTENANCE 1-5

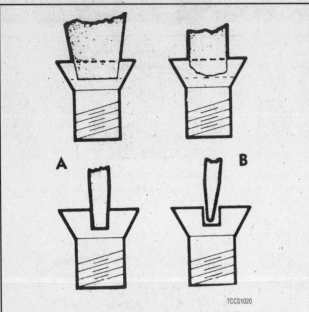

Fig. 16 Screwdrivers should be kept in good condition to prevent injury or damage which could result if the blade slips from the screw

Fig. 17 Using the correct size wrench will help prevent the possibility of rounding off a nut

Fig. 18 NEVER work under a vehicle unless it is supported using safety stands (jackstands)

- Do shield your eyes whenever you work around the battery. Batteries contain sulfuric acid. In case of contact with, flush the area with water or a mixture of water and baking soda, then seek immediate medical attention.
- Do use safety stands (jackstands) for any undervehicle service. Jacks are for raising vehicles; jackstands are for making sure the vehicle stays raised until you want it to come down.
- Do use adequate ventilation when working with any chemicals or hazardous materials. Like carbon monoxide, the asbestos dust resulting from some brake lining wear can be hazardous in sufficient quantities.
- Do disconnect the negative battery cable when working on the electrical system. The secondary ignition system contains EXTREMELY HIGH VOLTAGE. In some cases it can even exceed 50,000 volts.
- Do follow manufacturer's directions whenever working with potentially hazardous materials. Most chemicals and fluids are poisonous.
- Do properly maintain your tools. Loose hammerheads, mushroomed punches and chisels, frayed or poorly grounded electrical cords, excessively worn screwdrivers, spread wrenches (open end), cracked sockets, slipping ratchets, or faulty droplight sockets can cause accidents.
- Likewise, keep your tools clean; a greasy wrench can slip off a bolt head, ruining the bolt and often harming your knuckles in the process.
- Do use the proper size and type of tool for the job at hand. Do select a wrench or socket that fits the nut or bolt. The wrench or socket should sit straight, not cocked.
- Do, when possible, pull on a wrench handle rather than push on it, and adjust your stance to prevent a fall.
- Do be sure that adjustable wrenches are tightly closed on the nut or bolt and pulled so that the force is on the side of the fixed jaw.
- Do strike squarely with a hammer; avoid glancing blows.
- Do set the parking brake and block the drive wheels if the work requires a running engine.

Don'ts

- Don't run the engine in a garage or anywhere else without proper ventilation—EVER! Carbon monoxide is poisonous; it takes a long time to leave the human body and you can build up a deadly supply of it in your system by simply breathing in a little at a time. You may not realize you are slowly poisoning yourself. Always use power vents, windows, fans and/or open the garage door.
- Don't work around moving parts while wearing loose clothing. Short sleeves are much safer than long, loose sleeves. Hard-toed shoes with neoprene soles protect your toes and give a better grip on slippery surfaces. Watches and jewelry is not safe working around a vehicle. Long hair should be tied back under a hat or cap.
- Don't use pockets for toolboxes. A fall or bump can drive a screwdriver deep into your body. Even a rag hanging from your back pocket can wrap around a spinning shaft or fan.
- Don't smoke when working around gasoline, cleaning solvent or other flammable material.
- Don't smoke when working around the battery. When the battery is being charged, it gives off explosive hydrogen gas.
- Don't use gasoline to wash your hands; there are excellent soaps available. Gasoline contains dangerous additives which can enter the body through a cut or through your pores. Gasoline also removes all the natural oils from the skin so that bone dry hands will suck up oil and grease.
- Don't service the air conditioning system unless you are equipped with the necessary tools and training. When liquid or compressed gas refrigerant is released to atmospheric pressure it will absorb heat from whatever it contacts. This will chill or freeze anything it touches.
- Don't use screwdrivers for anything other than driving screws! A screwdriver used as an prying tool can snap when you least expect it, causing injuries. At the very least, you'll ruin a good screwdriver.
- Don't use an emergency jack (that little ratchet, scissors, or pantograph jack supplied with the vehicle) for anything other than changing a flat! These jacks are only intended for emergency use out on the road; they are NOT designed as a maintenance tool. If you are serious about maintaining your vehicle yourself, invest in a hydraulic floor jack of at least a 1½ ton capacity, and at least two sturdy jackstands.

1-6 GENERAL INFORMATION AND MAINTENANCE

FASTENERS, MEASUREMENTS AND CONVERSIONS

Bolts, Nuts and Other Threaded Retainers

▶ See Figures 19 and 20

Although there are a great variety of fasteners found in the modern car or truck, the most commonly used retainer is the threaded fastener (nuts, bolts, screws, studs, etc.). Most threaded retainers may be reused, provided that they are not damaged in use or during the repair. Some retainers (such as stretch bolts or torque prevailing nuts) are designed to deform when tightened or in use and should not be reinstalled.

Whenever possible, we will note any special retainers which should be replaced during a procedure. But you should always inspect the condition of a retainer when it is removed and replace any that show signs of damage. Check all threads for rust or corrosion which can increase the torque necessary to achieve the desired clamp load for which that fastener was originally selected. Additionally, be sure that the driver surface of the fastener has not been compromised by rounding or other damage. In some cases a driver surface may become only partially rounded, allowing the driver to catch in only one direction. In many of these occurrences, a fastener may be installed and tightened, but the driver would not be able to grip and loosen the fastener again.

If you must replace a fastener, whether due to design or damage, you must ALWAYS be sure to use the proper replacement. In all cases, a retainer of the same design, material and strength should be used. Markings on the heads of most bolts will help determine the proper strength of the fastener. The same material, thread and pitch must be selected to assure proper installation and safe operation of the vehicle afterwards.

Thread gauges are available to help measure a bolt or stud's thread. Most automotive and hardware stores keep gauges available to help you select the proper size. In a pinch, you can use another nut or bolt for a thread gauge. If the bolt you are replacing is not too badly damaged, you can select a match by finding another bolt which will thread in its place. If you find a nut which threads properly onto the damaged bolt, then use that nut to help select the replacement bolt.

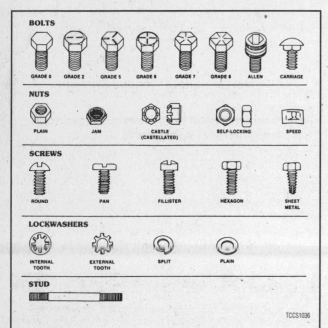

Fig. 19 There are many different types of threaded retainers found on vehicles

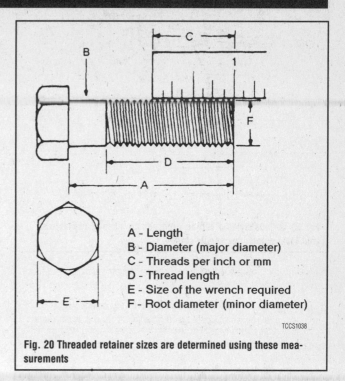

A - Length
B - Diameter (major diameter)
C - Threads per inch or mm
D - Thread length
E - Size of the wrench required
F - Root diameter (minor diameter)

Fig. 20 Threaded retainer sizes are determined using these measurements

✻✻ WARNING

Be aware that when you find a bolt with damaged threads, you may also find the nut or drilled hole it was threaded into has also been damaged. If this is the case, you may have to drill and tap the hole, replace the nut or otherwise repair the threads. NEVER try to force a replacement bolt to fit into the damaged threads.

Torque

Torque is defined as the measurement of resistance to turning or rotating. It tends to twist a body about an axis of rotation. A common example of this would be tightening a threaded retainer such as a nut, bolt or screw. Measuring torque is one of the most common ways to help assure that a threaded retainer has been properly fastened.

When tightening a threaded fastener, torque is applied in three distinct areas, the head, the bearing surface and the clamp load. About 50 percent of the measured torque is used in overcoming bearing friction. This is the friction between the bearing surface of the bolt head, screw head or nut face and the base material or washer (the surface on which the fastener is rotating). Approximately 40 percent of the applied torque is used in overcoming thread friction. This leaves only about 10 percent of the applied torque to develop a useful clamp load (the force which holds a joint together). This means that friction can account for as much as 90 percent of the applied torque on a fastener.

TORQUE WRENCHES

▶ See Figure 21

In most applications, a torque wrench can be used to assure proper installation of a fastener. Torque wrenches come in various designs and most automotive supply stores will carry a variety to suit your needs. A torque wrench should be used any time we supply a specific torque value for a fastener. Again, the

GENERAL INFORMATION AND MAINTENANCE

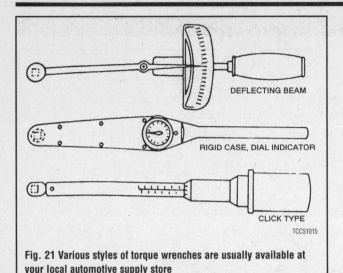

Fig. 21 Various styles of torque wrenches are usually available at your local automotive supply store

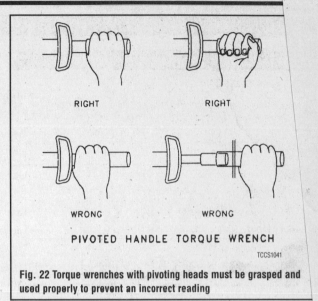

Fig. 22 Torque wrenches with pivoting heads must be grasped and used properly to prevent an incorrect reading

general rule of "if you are using the right tool for the job, you should not have to strain to tighten a fastener" applies here.

Beam Type

The beam type torque wrench is one of the most popular types. It consists of a pointer attached to the head that runs the length of the flexible beam (shaft) to a scale located near the handle. As the wrench is pulled, the beam bends and the pointer indicates the torque using the scale.

Click (Breakaway) Type

Another popular design of torque wrench is the click type. To use the click type wrench you pre-adjust it to a torque setting. Once the torque is reached, the wrench has a reflex signaling feature that causes a momentary breakaway of the torque wrench body, sending an impulse to the operator's hand.

Pivot Head Type

♦ See Figure 22

Some torque wrenches (usually of the click type) may be equipped with a pivot head which can allow it to be used in areas of limited access. BUT, it must be used properly. To hold a pivot head wrench, grasp the handle lightly, and as you pull on the handle, it should be floated on the pivot point. If the handle comes in contact with the yoke extension during the process of pulling, there is a very good chance the torque readings will be inaccurate because this could alter the wrench loading point. The design of the handle is usually such as to make it inconvenient to deliberately misuse the wrench.

➥ **It should be mentioned that the use of any U-joint, wobble or extension will have an effect on the torque readings, no matter what type of wrench you are using. For the most accurate readings, install the socket directly on the wrench driver. If necessary, straight extensions (which hold a socket directly under the wrench driver) will have the least effect on the torque reading. Avoid any extension that alters the length of the wrench from the handle to the head/driving point (such as a crow's foot). U-joint or wobble extensions can greatly affect the readings; avoid their use at all times.**

Rigid Case (Direct Reading)

A rigid case or direct reading torque wrench is equipped with a dial indicator to show torque values. One advantage of these wrenches is that they can be held at any position on the wrench without affecting accuracy. These wrenches are often preferred because they tend to be compact, easy to read and have a great degree of accuracy.

TORQUE ANGLE METERS

Because the frictional characteristics of each fastener or threaded hole will vary, clamp loads which are based strictly on torque will vary as well. In most applications, this variance is not significant enough to cause worry. But, in certain applications, a manufacturer's engineers may determine that more precise clamp loads are necessary (such is the case with many aluminum cylinder heads). In these cases, a torque angle method of installation would be specified. When installing fasteners which are torque angle tightened, a predetermined seating torque and standard torque wrench are usually used first to remove any compliance from the joint. The fastener is then tightened the specified additional portion of a turn measured in degrees. A torque angle gauge (mechanical protractor) is used for these applications.

Standard and Metric Measurements

♦ See Figure 23

Throughout this manual, specifications are given to help you determine the condition of various components on your vehicle, or to assist you in their installation. Some of the most common measurements include length (in. or cm/mm), torque (ft. lbs., inch lbs. or Nm) and pressure (psi, in. Hg, kPa or mm Hg). In most cases, we strive to provide the proper measurement as determined by the manufacturer's engineers.

Though, in some cases, that value may not be conveniently measured with what is available in your toolbox. Luckily, many of the measuring devices which are available today will have two scales so the Standard or Metric measurements may easily be taken. If any of the various measuring tools which are available to you do not contain the same scale as listed in the specifications, use the accompanying conversion factors to determine the proper value.

The conversion factor chart is used by taking the given specification and multiplying it by the necessary conversion factor. For instance, looking at the first line, if you have a measurement in inches such as "free-play should be 2 in." but your ruler reads only in millimeters, multiply 2 in. by the conversion factor of 25.4 to get the metric equivalent of 50.8mm. Likewise, if the specification was given only in a Metric measurement, for example in Newton Meters (Nm), then look at the center column first. If the measurement is 100 Nm, multiply it by the conversion factor of 0.738 to get 73.8 ft. lbs.

1-8 GENERAL INFORMATION AND MAINTENANCE

CONVERSION FACTORS

LENGTH-DISTANCE

Inches (in.)	x 25.4	= Millimeters (mm)	x .0394	= Inches
Feet (ft.)	x .305	= Meters (m)	x 3.281	= Feet
Miles	x 1.609	= Kilometers (km)	x .0621	= Miles

VOLUME

Cubic Inches (in3)	x 16.387	= Cubic Centimeters	x .061	= in3
IMP Pints (IMP pt.)	x .568	= Liters (L)	x 1.76	= IMP pt.
IMP Quarts (IMP qt.)	x 1.137	= Liters (L)	x .88	= IMP qt.
IMP Gallons (IMP gal.)	x 4.546	= Liters (L)	x .22	= IMP gal.
IMP Quarts (IMP qt.)	x 1.201	= US Quarts (US qt.)	x .833	= IMP qt.
IMP Gallons (IMP gal.)	x 1.201	= US Gallons (US gal.)	x .833	= IMP gal.
Fl. Ounces	x 29.573	= Milliliters	x .034	= Ounces
US Pints (US pt.)	x .473	= Liters (L)	x 2.113	= Pints
US Quarts (US qt.)	x .946	= Liters (L)	x 1.057	= Quarts
US Gallons (US gal.)	x 3.785	= Liters (L)	x .264	= Gallons

MASS-WEIGHT

Ounces (oz.)	x 28.35	= Grams (g)	x .035	= Ounces
Pounds (lb.)	x .454	= Kilograms (kg)	x 2.205	= Pounds

PRESSURE

Pounds Per Sq. In. (psi)	x 6.895	= Kilopascals (kPa)	x .145	= psi
Inches of Mercury (Hg)	x .4912	= psi	x 2.036	= Hg
Inches of Mercury (Hg)	x 3.377	= Kilopascals (kPa)	x .2961	= Hg
Inches of Water (H_2O)	x .07355	= Inches of Mercury	x 13.783	= H_2O
Inches of Water (H_2O)	x .03613	= psi	x 27.684	= H_2O
Inches of Water (H_2O)	x .248	= Kilopascals (kPa)	x 4.026	= H_2O

TORQUE

Pounds-Force Inches (in-lb)	x .113	= Newton Meters (N·m)	x 8.85	= in-lb
Pounds-Force Feet (ft-lb)	x 1.356	= Newton Meters (N·m)	x .738	= ft-lb

VELOCITY

Miles Per Hour (MPH)	x 1.609	= Kilometers Per Hour (KPH)	x .621	= MPH

POWER

Horsepower (Hp)	x .745	= Kilowatts	x 1.34	= Horsepower

FUEL CONSUMPTION*

Miles Per Gallon IMP (MPG)	x .354	= Kilometers Per Liter (Km/L)
Kilometers Per Liter (Km/L)	x 2.352	= IMP MPG
Miles Per Gallon US (MPG)	x .425	= Kilometers Per Liter (Km/L)
Kilometers Per Liter (Km/L)	x 2.352	= US MPG

*It is common to covert from miles per gallon (mpg) to liters/100 kilometers (1/100 km), where mpg (IMP) x 1/100 km = 282 and mpg (US) x 1/100 km = 235.

TEMPERATURE

Degree Fahrenheit (°F) = (°C x 1.8) + 32
Degree Celsius (°C) = (°F − 32) x .56

Fig. 23 Standard and metric conversion factors chart

GENERAL INFORMATION AND MAINTENANCE 1-11

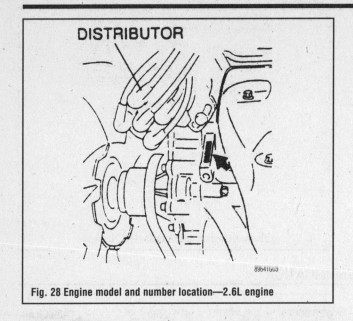

Fig. 28 Engine model and number location—2.6L engine

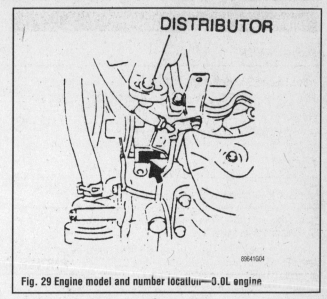

Fig. 29 Engine model and number location—3.0L engine

GENERAL ENGINE SPECIFICATIONS

Year	Engine Code	Engine Displacement Liters (cc)	Fuel System Type	Net Horsepower @ rpm	Net Torque @ rpm (ft. lbs.)	Bore x Stroke (in.)	Compression Ratio	Oil Pressure (psi) @ idle rpm
1987	F2	2.2 (2184)	2 bbl	85 @ 4500	118 @ 2500	3.395 x 3.700	8.6:1	60 @ 2000
	G54B	2.6 (2555)	2 bbl	105 @ 5000	139 @ 2500	3.586 x 3.858	8.7:1	57 @ 2000
1988	F2	2.2 (2184)	2 bbl	85 @ 4500	118 @ 2500	3.395 x 3.700	8.6:1	43-57 @ 3000
	G54B	2.6 (2555)	2 bbl	105 @ 5000	139 @ 2500	3.586 x 3.858	8.7:1	78 @ 3000
1989	F2	2.2 (2184)	2 bbl	85 @ 4500	118 @ 2500	3.395 x 3.700	8.6:1	43-57 @ 3000
	G6	2.6 (2606)	MFI	121 @ 4600	149 @ 3500	3.622 x 3.858	8.4:1	44-58 @ 3000
	JE	3.0 (2954)	MFI	150 @ 5000	165 @ 4000	3.540 x 3.050	8.5:1	53-75 @ 3000
1990	F2	2.2 (2184)	①	②	③	3.395 x 3.700	8.6:1	43-57 @ 3000
	G6	2.6 (2606)	MFI	121 @ 4600	149 @ 3500	3.622 x 3.858	8.4:1	44-58 @ 3000
	JE	3.0 (2954)	MFI	150 @ 5000	165 @ 4000	3.540 x 3.050	8.5:1	53-75 @ 3000
1991	F2	2.2 (2184)	①	②	③	3.395 x 3.700	8.6:1	43-57 @ 3000
	G6	2.6 (2606)	MFI	121 @ 4600	149 @ 3500	3.622 x 3.858	8.4:1	44-58 @ 3000
	JE	3.0 (2954)	MFI	150 @ 5000	165 @ 4000	3.540 x 3.050	8.5:1	53-75 @ 3000
	X	4.0 (4016)	MFI	155 @ 4200	220 @ 2400	3.950 x 3.320	9.0:1	40-60 @ 2000
1992	F2	2.2 (2184)	①	②	③	3.395 x 3.700	8.6:1	43-57 @ 3000
	G6	2.6 (2606)	MFI	121 @ 4600	149 @ 3500	3.622 x 3.858	8.4:1	44-58 @ 3000
	JE	3.0 (2954)	MFI	150 @ 5000	165 @ 4000	3.540 x 3.050	8.5:1	53-75 @ 3000
	X	4.0 (4016)	MFI	155 @ 4200	220 @ 2400	3.950 x 3.320	9.0:1	40-60 @ 2000
1993	F2	2.2 (2184)	①	②	③	3.395 x 3.700	8.6:1	43-57 @ 3000
	G6	2.6 (2606)	MFI	121 @ 4600	149 @ 3500	3.622 x 3.858	8.4:1	44-58 @ 3000
	JE	3.0 (2954)	MFI	150 @ 5000	165 @ 4000	3.540 x 3.050	8.5:1	53-75 @ 3000
	X	4.0 (4016)	MFI	155 @ 4200	220 @ 2400	3.950 x 3.320	9.0:1	40-60 @ 2000

MFI - Multiport Fuel Injection

NOTE: Horsepower and torque are SAE net figures. They are measured at the rear of the transmission with all accessories installed and operating. Since the figures vary when a given engine is installed in different models, some are representative rather than exact.

① Except Ca. - 2 bbl
   Ca - MFI

② Except Ca. - 85 @ 4500
   Ca - 91 @ 4500

③ Except Ca. - 188 @ 2500
   Ca - 118 @ 2000

Transmission

MANUAL TRANSMISSION

On manual transmissions, the identification number is located on a plate attached to the main transmission case. On the plate you find the assigned part number, serial number and bar code used for inventory purposes.

B2200 Pick-up models use the M4M-D 4-speed and M5M-D 5-speed transmissions. They are essentially the same unit geared differently.

B2600 Pick-up and MPV models use the R5M-D (R5MX-D w/ 4x4) 5-speed transmission.

The Navajo uses the M5OD 5-speed transmission.

AUTOMATIC TRANSMISSION

Pick-up Models

In 1987, a 3-speed lock-up torque converter transmission was introduced. Its designation is L3N71B. It was offered only on the B2200. This unit continued on the B2200 through 1988.

Also in 1987 a 4-speed unit was offered as an option on the B2200 and as standard on the B2600. It is designated L4N71B. This unit continues through 1988. In 1989, only the L4N71B was offered on pick-ups. From 1990, the N4A-HL 4-speed is used on all pickups.

The identification number is stamped on a plate that is bolted to the side of the transmission.

The tag identifies when the transmission was built, it's code letter and model number.

MPV Models

♦ See Figure 30

The MPV uses a 4-speed overdrive transmission called the N4A-HL. This unit is a conventional, hydraulically controlled transmission.

As an option, there is an electronically controlled 4-speed automatic, designated R4A-EL. The transmission is computer controlled and has no owner serviceable or adjustable components other than the neutral start switch.

The identification number is stamped on a plate that is bolted to the side of the transmission.

The tag identifies when the transmission was built, it's code letter and model number.

4-Wheel Drive MPVs use the R4AX-EL, which is essentially the same unit as the R4A-EL.

Navajo Models

The identification number is stamped on a plate that hangs from the lower left extension housing bolt. The tag identifies when the transmission was built, it's code letter and model number.

The Navajo uses a Ford A4LD 4-speed transmission.

ROUTINE MAINTENANCE AND TUNE-UP

Proper maintenance and tune-up is the key to long and trouble-free vehicle life, and the work can yield its own rewards. Studies have shown that a properly tuned and maintained vehicle can achieve better gas mileage than an out-of-tune vehicle. As a conscientious owner and driver, set aside a Saturday morning, say once a month, to check or replace items which could cause major problems later. Keep your own personal log to jot down which services you performed, how much the parts cost you, the date, and the exact odometer reading at the time. Keep all receipts for such items as engine oil and filters, so that they may be referred to in case of related problems or to determine operating expenses. As a do-it-yourselfer, these receipts are the only proof you have that the required maintenance was performed. In the event of a warranty problem, these receipts will be invaluable.

The literature provided with your vehicle when it was originally delivered includes the factory recommended maintenance schedule. If you no longer have this literature, replacement copies are usually available from the dealer. A maintenance schedule is provided later in this section, in case you do not have the factory literature.

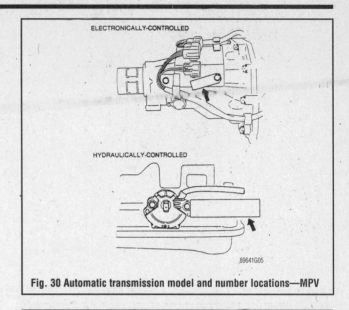

Fig. 30 Automatic transmission model and number locations—MPV

Drive Axle

FRONT

The identification number is stamped on a plate on the differential housing.

REAR

♦ See Figure 31

The rear axle identification code is stamped on a metal tag hanging from one of the axle cover-to-carrier bolts in the cover bolt circle.

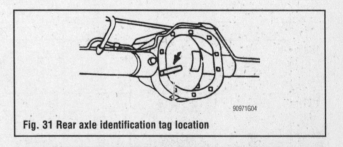

Fig. 31 Rear axle identification tag location

Transfer Case

All vehicles can be equipped with a mechanical or electronic shift transfer case. The identification number is stamped on a plate on the side of the case.

Air Cleaner

REMOVAL & INSTALLATION

Replacing the air cleaner element is a simple, routine maintenance operation. You should be careful, however, to keep dust and dirt out of the air cleaner housing, as they accelerate engine wear. If the outside of the air cleaner housing is dusty, wipe it with a clean rag before beginning work.

Carbureted Pickups

♦ See Figure 32

1. Remove the wing nut at the center of the housing and then unclip the retaining clips situated around the sides.

GENERAL INFORMATION AND MAINTENANCE 1-13

UNDERHOOD COMPONENT LOCATIONS—MPV

1. Vehicle Identification Number (VIN) (stamped into the firewall)
2. EVAP canister
3. Brake master cylinder
4. Automatic transmission fluid dipstick
5. Engine oil dipstick
6. Distributor cap and rotor
7. Windshield washer fluid reservoir
8. Battery
9. Power steering pump
10. Engine oil fill cap
11. Air filter housing
12. Coolant recovery tank

1-14 GENERAL INFORMATION AND MAINTENANCE

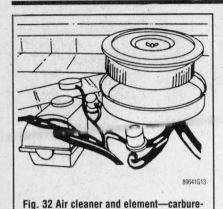

Fig. 32 Air cleaner and element—carburetor equipped

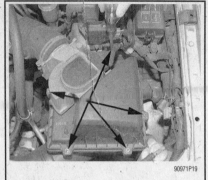

Fig. 33 Loosen the 5 air cleaner housing cover retaining bolts

Fig. 34 While holding the air cleaner housing cover up, remove the filter element out of the housing

2. Pull the top cover off and remove the air cleaner element.
3. When installing the new element, make sure it seats squarely around the bulge in the center of the lower air cleaner housing. Install the housing top, turning it until the wing nut mounting stud lines up with the hole in the top (it's usually off center). Note that the top cover should seat tightly all around. Install the wing nut and reclip the clips.

Fuel Injected Pickups and MPV

♦ See Figures 33 and 34

1. Loosen the clamp on the air intake hose and pull the hose off the housing.
2. Disconnect the airflow sensor electrical connector.
3. Unbolt and remove the housing. Note the direction in which the element is installed and install the new element in the same way (it may be marked TOP). Install the top of the housing and connect the other components.

Navajo

1. Loosen the clamps that secure the hose assembly to the air cleaner.
2. Remove the screws that attach the air cleaner to the bracket.
3. Disconnect the hose and inlet tube from the air cleaner.
4. Remove the screws attaching the air cleaner cover.
5. Remove the air filter and tubes.
6. Install the new element. Place the cover into position and install the retaining screws. Connect the other components. Don't overtighten the hose clamps! A torque of 12–15 inch lbs. is sufficient.

Fuel Filter

On the B2200 and 1987 B2600 models, it is located in the engine compartment, on the intake manifold side, next to the carbon canister.
On the 1988-93 B2600 and the Navajo, it is located on the right rear frame member, near the fuel tank.
On the MPV, it is located on the right fender well just below the battery.

REMOVAL & INSTALLATION

Carbureted Engines

♦ See Figure 35

To replace the filter, loosen the clamps at both ends of the filter and pull off the hoses. Pop the filter from its clamp.

Fuel Injected Engines

PICK-UPS AND MPV

♦ See Figure 36

1. Relieve the pressure from the fuel system.
2. Disconnect the battery negative cable.
3. Slide the fuel line clamps back off the connections on the filter. Then, slowly pull one connection off just until fuel begins to seep out. Place a small plastic container or stuff a couple of rags under the filter to absorb any excess fuel.
4. Pull both fuel hoses off the connectors. Remove the filter from the clamp. On some models, it will be necessary to unbolt the fuel filter from the mounting bracket.
5. Place the new filter in the clamp or attach to the bracket if so equipped.
6. Connect the hoses to the filter connections and secure the hoses with the clamps.
7. Connect the negative battery cable.
8. Start the engine and check the fuel filter connections for leaks.

NAVAJO

♦ See Figure 37

1. Relieve the fuel system pressure.
2. Raise and support the vehicle safely.

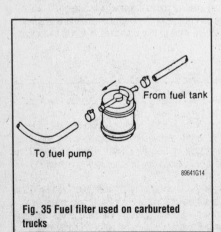

Fig. 35 Fuel filter used on carbureted trucks

Fig. 36 Location of fuel filter on the right side engine compartment fender well—MPV shown

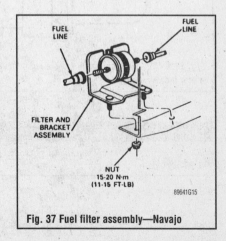

Fig. 37 Fuel filter assembly—Navajo

GENERAL INFORMATION AND MAINTENANCE

3. Remove the push connect fittings at both ends of fuel filter. Install new retainer clips in each push connect fitting.
4. Remove the fuel filter bracket mounting nuts.
5. Remove the filter from the bracket by loosening the filter retaining clamp enough to allow the filter to pass through.

To install:

→ The flow arrow direction should be positioned as installed in the bracket to ensure proper flow of fuel through the replacement filter.

6. Install the filter in the bracket, ensuring proper direction of flow as noted by arrow. Tighten clamp to 15–25 inch lbs. (2–3 Nm).
7. Install the bracket and filter onto the vehicle. Tighten the mounting nuts to 11–15 ft. lbs. (15–20 Nm).
8. Install the push connect fittings at both ends of the filter.
9. Lower the vehicle.
10. Start the engine and check for leaks.

Clean all dirt and/or grease from the fuel filter fittings. The fuel filter uses a "hairpin" clip retainer. Spread the two hairpin clip legs about 1/8 in. (3mm) each to disengage it from the fitting, then pull the clip outward. Use finger pressure only, do not use any tools. Push the quick connect fittings onto the filter ends. Mazda recommends that the retaining clips be replaced whenever removed. The fuel tubes used on these fuel systems are manufactured in 5/16 in. and 3/8 in. diameters. Each fuel tube takes a different size hairpin clip, so keep this in mind when purchasing new clips. A click will be heard when the hairpin clip snaps into its proper position. Pull on the lines with moderate pressure to ensure proper connection. Start the engine and check for fuel leaks. If the inertia switch (reset switch) was disconnected to relieve the fuel system pressure, cycle the ignition switch from the **OFF** to **ON** position several times to re-charge the fuel system before attempting to start the engine.

PCV Valve

The Positive Crankcase Ventilation (PCV) valve should be inspected for blockage periodically. It is located in the valve cover, connected to the intake manifold by a vacuum hose. To inspect the PCV valve, refer to Section 4.

REMOVAL & INSTALLATION

♦ See Figure 38

Remove the PCV valve by simply disconnecting the vacuum hose from the valve, then pulling the valve from the rocker cover grommet.

→ If your engine exhibits lower than normal gas mileage and poor idle characteristics for no apparent reason, suspect the PCV valve. It is probably blocked and should be replaced.

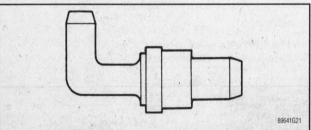

Fig. 38 A PCV valve is a small simple component used to divert and meter blowby gases

Evaporative Emission Canister

SERVICING

♦ See Figure 39

The evaporative canister is located under the hood in the engine compartment and is designed to store fuel vapors and prevent their escaping into the atmosphere. On early models, when the engine is not running, fuel that has

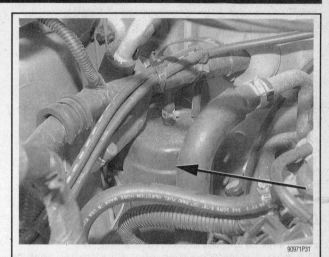

Fig. 39 Location of the evaporative canister near the firewall on the right side fender well—MPV shown

evaporated into the condenser tank is returned to the fuel tank as the ambient temperatures rise and the vapors are condensed. Later models do not have a condenser tank. During periods when the engine is running, fuel vapor that has not condensed in the condenser tank moves to the carbon canister. The stored vapors are removed by fresh air through the bottom of the inlet hole and passed through the air cleaner to the combustion chamber. No maintenance or service should be required, except in the case of damage or malfunction. The canister should be inspected for damage or leaks at the hose fittings. Repair or replace any old or cracked hoses. If either condition should exist, simply replace that component.

Battery

PRECAUTIONS

Always use caution when working on or near the battery. Never allow a tool to bridge the gap between the negative and positive battery terminals. Also, be careful not to allow a tool to provide a ground between the positive cable/terminal and any metal component on the vehicle. Either of these conditions will cause a short circuit, leading to sparks and possible personal injury.

Do not smoke or all open flames/sparks near a battery; the gases contained in the battery are very explosive and, if ignited, could cause severe injury or death.

All batteries, regardless of type, should be carefully secured by a battery hold-down device. If not, the terminals or casing may crack from stress during vehicle operation. A battery which is not secured may allow acid to leak, making it discharge faster. The acid can also eat away at components under the hood.

Always inspect the battery case for cracks, leakage and corrosion. A white corrosive substance on the battery case or on nearby components would indicate a leaking or cracked battery. If the battery is cracked, it should be replaced immediately.

GENERAL MAINTENANCE

Always keep the battery cables and terminals free of corrosion. Check and clean these components about once a year.

Keep the top of the battery clean, as a film of dirt can help discharge a battery that is not used for long periods. A solution of baking soda and water may be used for cleaning, but be careful to flush this off with clear water. DO NOT let any of the solution into the filler holes. Baking soda neutralizes battery acid and will de-activate a battery cell.

Batteries in vehicles which are not operated on a regular basis can fall victim to parasitic loads (small current drains which are constantly drawing current from the battery). Normal parasitic loads may drain a battery on a vehicle that is in storage and not used for 6–8 weeks. Vehicles that have additional accessories such as a phone or an alarm system may discharge a battery sooner. If

1-16 GENERAL INFORMATION AND MAINTENANCE

the vehicle is to be stored for longer periods in a secure area and the alarm system is not necessary, the negative battery cable should be disconnected to protect the battery.

Remember that constantly deep cycling a battery (completely discharging and recharging it) will shorten battery life.

BATTERY FLUID

♦ See Figure 40

Check the battery electrolyte level at least once a month, or more often in hot weather or during periods of extended vehicle operation. On non-sealed batteries, the level can be checked either through the case (if translucent) or by removing the cell caps. The electrolyte level in each cell should be kept filled to the split ring inside each cell, or the line marked on the outside of the case.

If the level is low, add only distilled water through the opening until the level is correct. Each cell must be checked and filled individually. Distilled water should be used, because the chemicals and minerals found in most drinking water are harmful to the battery and could significantly shorten its life.

If water is added in freezing weather, the vehicle should be driven several miles to allow the water to mix with the electrolyte. Otherwise, the battery could freeze.

Although some maintenance-free batteries have removable cell caps, the electrolyte condition and level on all sealed maintenance-free batteries must be checked using the built-in hydrometer "eye." The exact type of eye will vary. But, most battery manufacturers, apply a sticker to the battery itself explaining the readings.

➡ **Although the readings from built-in hydrometers will vary, a green eye usually indicates a properly charged battery with sufficient fluid level. A dark eye is normally an indicator of a battery with sufficient fluid, but which is low in charge. A light or yellow eye usually indicates that elec-** trolyte has dropped below the necessary level. In this last case, sealed batteries with an insufficient electrolyte must usually be discarded.

Checking the Specific Gravity

♦ See Figures 41, 42 and 43

A hydrometer is required to check the specific gravity on all batteries that are not maintenance-free. On batteries that are maintenance-free, the specific gravity is checked by observing the built-in hydrometer "eye" on the top of the battery case.

✴✴ CAUTION

Battery electrolyte contains sulfuric acid. If you should splash any on your skin or in your eyes, flush the affected area with plenty of clear water. If it lands in your eyes, get medical help immediately.

The fluid (sulfuric acid solution) contained in the battery cells will tell you many things about the condition of the battery. Because the cell plates must be kept submerged below the fluid level in order to operate, the fluid level is extremely important. And, because the specific gravity of the acid is an indication of electrical charge, testing the fluid can be an aid in determining if the battery must be replaced. A battery in a vehicle with a properly operating charging system should require little maintenance, but careful, periodic inspection should reveal problems before they leave you stranded.

At least once a year, check the specific gravity of the battery. It should be between 1.20 and 1.26 on the gravity scale. Most auto stores carry a variety of inexpensive battery hydrometers. These can be used on any non-sealed battery to test the specific gravity in each cell.

The battery testing hydrometer has a squeeze bulb at one end and a nozzle at the other. Battery electrolyte is sucked into the hydrometer until the float is lifted from its seat. The specific gravity is then read by noting the position of the float. If gravity is low in one or more cells, the battery should be slowly charged and checked again to see if the gravity has come up. Generally, if after charging, the specific gravity between any two cells varies more than 50 points (0.50), the battery should be replaced, as it can no longer produce sufficient voltage to guarantee proper operation.

CABLES

♦ See Figures 44, 45, 46 and 47

Once a year (or as necessary), the battery terminals and the cable clamps should be cleaned. Loosen the clamps and remove the cables, negative cable first. On top post batteries, the use of a puller specially made for this purpose is recommended. These are inexpensive and available in most parts stores. Side terminal battery cables are secured with a small bolt.

Clean the cable clamps and the battery terminal with a wire brush, until all corrosion, grease, etc., is removed and the metal is shiny. It is especially important to clean the inside of the clamp thoroughly (an old knife is useful here), since a small deposit of oxidation there will prevent a sound connection and inhibit starting or charging. Special tools are available for cleaning these parts, one type for conventional top post batteries and another type for side terminal batteries. It is also a good idea to apply some dielectric grease to the terminal, as this will aid in the prevention of corrosion.

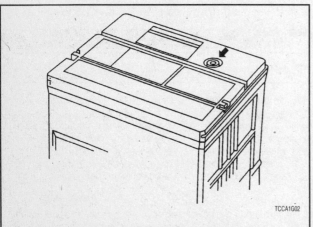

Fig. 40 Maintenance-free batteries usually contain a built-in hydrometer to check fluid level

Fig. 41 On non-sealed batteries, the fluid level can be checked by removing the cell caps

Fig. 42 If the fluid level is low, add only distilled water until the level is correct

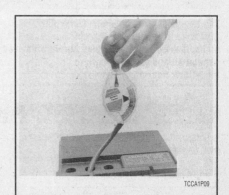

Fig. 43 Check the specific gravity of the battery's electrolyte with a hydrometer

GENERAL INFORMATION AND MAINTENANCE 1-17

Fig. 44 The underside of this special battery tool has a wire brush to clean post terminals

Fig. 45 Place the tool over the battery posts and twist to clean until the metal is shiny

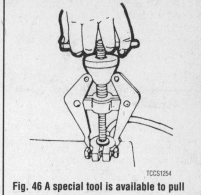
Fig. 46 A special tool is available to pull the clamp from the post

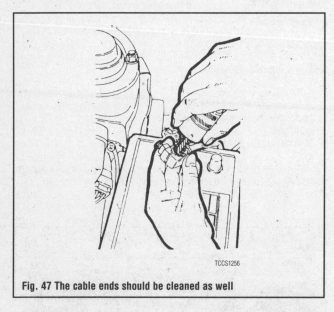
Fig. 47 The cable ends should be cleaned as well

After the clamps and terminals are clean, reinstall the cables, negative cable last; DO NOT hammer the clamps onto battery posts. Tighten the clamps securely, but do not distort them. Give the clamps and terminals a thin external coating of grease after installation, to retard corrosion.

Check the cables at the same time that the terminals are cleaned. If the cable insulation is cracked or broken, or if the ends are frayed, the cable should be replaced with a new cable of the same length and gauge.

CHARGING

> **CAUTION**
>
> The chemical reaction which takes place in all batteries generates explosive hydrogen gas. A spark can cause the battery to explode and splash acid. To avoid personal injury, be sure there is proper ventilation and take appropriate fire safety precautions when working with or near a battery.

A battery should be charged at a slow rate to keep the plates inside from getting too hot. However, if some maintenance-free batteries are allowed to discharge until they are almost "dead," they may have to be charged at a high rate to bring them back to "life." Always follow the charger manufacturer's instructions on charging the battery.

REPLACEMENT

When it becomes necessary to replace the battery, select one with an amperage rating equal to or greater than the battery originally installed. Deterioration and just plain aging of the battery cables, starter motor, and associated wires makes the battery's job harder in successive years. This makes it prudent to install a new battery with a greater capacity than the old.

Belts

INSPECTION

♦ See Figures 48, 49, 50, 51 and 52

Inspect the belts for signs of glazing or cracking. A glazed belt will be perfectly smooth from slippage, while a good belt will have a slight texture of fabric visible. Cracks will usually start at the inner edge of the belt and run outward. All worn or damaged drive belts should be replaced immediately. It is best to replace all drive belts at one time, as a preventive maintenance measure, during this service operation.

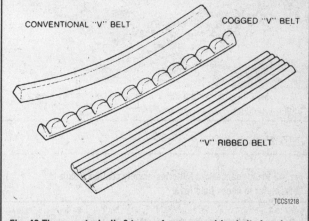
Fig. 48 There are typically 3 types of accessory drive belts found on vehicles today

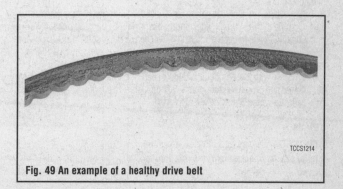
Fig. 49 An example of a healthy drive belt

1-18 GENERAL INFORMATION AND MAINTENANCE

Fig. 50 Deep cracks in this belt will cause flex, building up heat that will eventually lead to belt failure

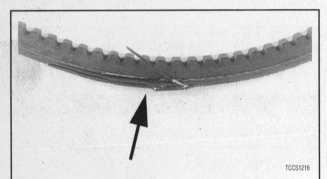

Fig. 51 The cover of this belt is worn, exposing the critical reinforcing cords to excessive wear

Fig. 52 Installing too wide a belt can result in serious belt wear and/or breakage

ADJUSTMENT

The drive belts should be checked for wear and tension as indicted in the Maintenance Interval charts. If the belt is worn, cracked or frayed, replace it with a new one. To check the belt tension:

Some models may use an adjusting bolt on either the idler pulley (a pulley which is not directly associated with any of the accessories) or on the power steering pump itself. This bolt makes adjustment much easier because you don't have to hold the accessory under a great deal of tension while tightening mounting bolts.

On models where the idler pulley has a lockbolt at its center, simply loosen the lockbolt and then turn the adjusting bolt clockwise to increase belt tension or counterclockwise to decrease it or remove the belt. Just don't forget to retighten the lockbolt when tension is correct, or vibration may cause it be lost. Recheck the tension with the lockbolt tightened and readjust if necessary.

On models with the 4 ridge or 5 ridge ribbed type V-belt driving both the air conditioner and power steering pump, belt tension is much greater. With a used belt, deflection should only be about ¼ in.

V-Belts

1. Apply thumb pressure (about 22 lbs.) to the fan belt midway between the pulleys and check the deflection. I should be approximately ⅜ in. for new belts and ½ in. for used belts.

2. To adjust the tension, loosen the alternator mounting bolt and adjusting bolt.
3. Move the alternator in the direction necessary to loosen or tighten the tension.
4. Tighten the mounting and adjusting bolts and recheck the tension.

Navajo

The Navajo uses a single accessory drive belt that is tensioned by a spring loaded tensioner, no adjustment is necessary. While there is no adjusting with the automatic tensioner and Poly-V belt used, it may be necessary to check the belt tension if squealing or excessive belt wear is noticed. If deflection is greater than ¼ in., the belt probably should be replaced. Before replacement, however, check the accessory component mountings for looseness and check the condition of the automatic tensioner. Repair or replace parts as necessary.

REMOVAL & INSTALLATION

Pick-up and MPV
♦ See Figure 53

POWER STEERING PUMP BELT
♦ See Figures 54, 55 and 56

1. Turn the ignition **OFF** and remove the key. Allow the engine to cool.
2. Loosen the idler pulley locknut to release the drive belt tension.
3. Remove the power steering belt.

To install:

4. Install the power steering bolt and make sure it is correctly seated on the pulleys.
5. Adjust the power steering belt tension/deflection by turning the adjusting bolt. A new belt should deflect 0.26–0.28 in. (6.6–7.2mm) and a used belt should deflect 0.28–0.31 in. (7–8mm).

Fig. 53 (A) Alternator belt, (B) Power steering pump belt, (C) A/C compressor belt—MPV with 3.0L engine shown

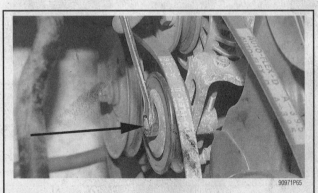

Fig. 54 Using a box wrench, loosen the power steering pump belt idler pulley lockbolt

GENERAL INFORMATION AND MAINTENANCE

6. Tighten the idler pulley locknut to 27–38 ft. lbs. (37–52 Nm).
7. Run the engine for 5 minutes and then recheck the belt deflection.

ALTERNATOR BELT

◆ See Figures 57, 58, 59 and 60

1. Turn the ignition **OFF** and remove the key. Allow the engine to cool.
2. Remove the power steering pump belt and A/C compressor belt, if necessary.
3. Loosen the alternator upper bracket adjusting bolt.
4. Loosen the lower through-bolt.
5. Remove the alternator belt.

To install:

6. Install the alternator belt and make sure it is correctly seated on the pulleys.
7. Turn the adjusting bolt to adjust the belt deflection. New belts should deflect 0.39–0.47 in. (10–12mm). Used belts should deflect 0.43–0.51 in. (11–13mm).
8. Once the correct deflection has been reached, tighten the upper bracket lock bolt to 14–19 ft. lbs. (19–25 Nm).
9. Tighten the lower through-bolt to 28–38 ft. lbs. (38–51 Nm).
10. Install and tension the power steering pump belt and A/C compressor belt, if necessary.
11. Run the engine for 5 minutes and then recheck the belt deflection.

A/C COMPRESSOR BELT

◆ See Figures 61, 62 and 63

1. Turn the ignition **OFF** and remove the key. Allow the engine to cool.
2. Remove the power steering pump belt and alternator belt, if necessary.

Fig. 55 Loosen the power steering pump belt by turning the idler pulley adjusting bolt

Fig. 56 Remove the power steering pump belt

Fig. 57 To access the alternator pivot bolt, you must first remove the power steering pump pulley

Fig. 58 Loosen the alternator pivot bolt located just below the alternator assembly

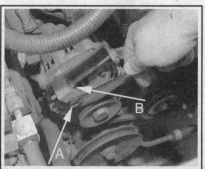

Fig. 59 Loosen the alternator adjuster lockbolt (A), then turn the adjustment nut (B) to loosen the slack on the belt . . .

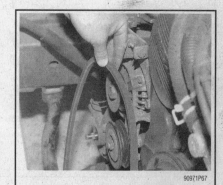

Fig. 60 . . . then remove the alternator belt

Fig. 61 Loosen the A/C compressor belt idler pulley lockbolt using an open end wrench

Fig. 62 Loosen the A/C compressor belt idler pulley adjusting bolt just enough to be able to remove the belt

Fig. 63 Remove the A/C compressor belt

1-20 GENERAL INFORMATION AND MAINTENANCE

3. Loosen the tensioner bracket adjusting bolt and tensioner pulley center locknut to release the drive belt tension.
4. Remove the A/C belt.

To install:

5. Install the A/C belt, and make sure it is correctly seated on the pulleys.
6. Adjust the A/C belt tension/deflection by turning the tensioner bracket adjusting bolt. A new belt should deflect 0.33–0.39 in. (8.5–10mm), and a used belt should deflect 0.39–0.45 in. (10–11.5mm).
7. Tighten the tensioner pulley center locknut to 27–38 ft. lbs. (37–52 Nm).
8. Install and tension the alternator belt, if necessary, and the power steering pump belt.
9. Run the engine for 5 minutes and then recheck the belt deflection.

Navajo

♦ See Figure 64

All Navajo and B Series Pick-up engines utilize one wide-ribbed V-belt to drive the engine accessories such as the water pump, alternator, air conditioner

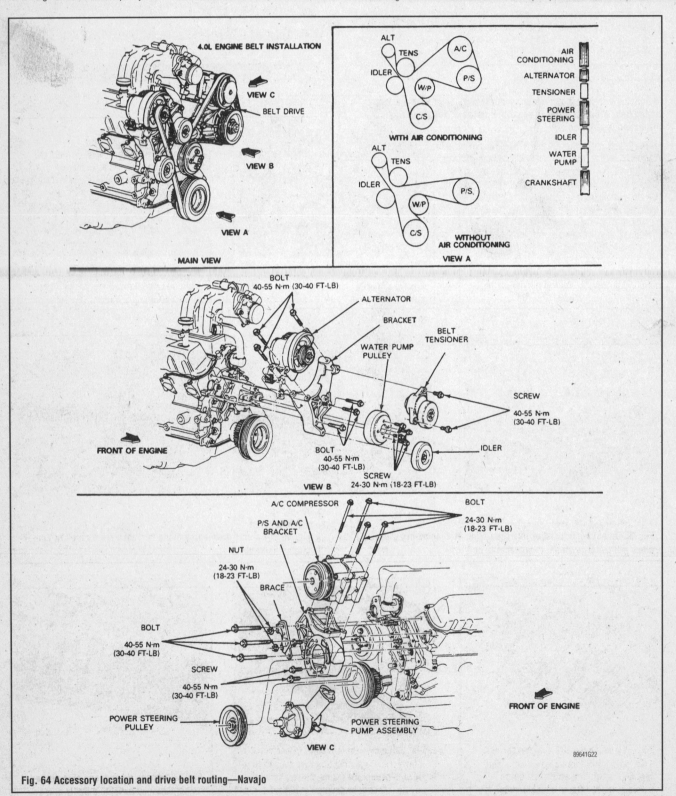

Fig. 64 Accessory location and drive belt routing—Navajo

GENERAL INFORMATION AND MAINTENANCE

compressor, air pump, etc. Because this belt uses a spring loaded tensioner for adjustment, belt replacement tends to be somewhat easier than on engines where accessories are pivoted and bolted in place for tension adjustment, such as the MPV. Basically, all belt replacement involves is to pivot the tensioner to loosen the belt, then slide the belt off of the pulleys. The two most important points are to pay CLOSE attention to the proper belt routing (since serpentine belts tend to be "snaked" all different ways through the pulleys) and to make sure the V-ribs are properly seated in all the pulleys.

Timing Belts

INSPECTION

♦ See Figures 65, 66, 67 and 68

➡ Only the 2.2L B Series Pick-up and 3.0L MPV engines use a rubber timing belt. All other engines use a timing chain, and no periodic inspection is required.

The 2.2L B Series Pick-up and 3.0L MPV engines utilizes a timing belt to drive the camshaft from the crankshaft's turning motion and to maintain proper valve timing. Some manufacturer's schedule periodic timing belt replacement to assure optimum engine performance, to make sure the motorist is never stranded should the belt break (as the engine will stop instantly) and for some (manufacturer's with interference motors) to prevent the possibility of severe internal engine damage should the belt break.

Although these engines are not listed as interference motors (a motor whose valves might contact the pistons if the camshaft was rotated separately from the crankshaft) the first 2 reasons for periodic replacement still apply. Mazda schedules the replacement mileage at 30,000 miles (48,000 km) for the 2.2L engine and 60,000 miles (96,000 km) for the 3.0L engine, however, most belt manufacturers recommend intervals anywhere from 45,000 miles (72,500 km) to 90,000 miles (145,000 km). You will have to decide for yourself if the peace of mind offered by a new belt is worth it on higher mileage engines.

But whether or not you decide to replace it, you would be wise to check it periodically to make sure it has not become damaged or worn. Generally speaking, a severely damaged belt will show as engine performance would drop dramatically, but a damaged belt (which could give out suddenly) may not give as much warning. In general, any time the engine timing cover(s) is(are) removed you should inspect the belt for premature parting, severe cracks or missing teeth. Also, an access plug is provided in the upper portion of the timing cover so that camshaft timing can be checked without cover removal. If timing is found to be off, cover removal and further belt inspection or replacement is necessary.

➡ For timing belt service, refer to section 3.

Hoses

INSPECTION

♦ See Figures 69, 70, 71 and 72

Upper and lower radiator hoses along with the heater hoses should be checked for deterioration, leaks and loose hose clamps at least every 15,000 miles (24,000 km). It is also wise to check the hoses periodically in early spring and at the beginning of the fall or winter when you are performing other maintenance. A quick visual inspection could discover a weakened hose which might have left you stranded if it had remained unrepaired.

Whenever you are checking the hoses, make sure the engine and cooling system are cold. Visually inspect for cracking, rotting or collapsed hoses, and replace as necessary. Run your hand along the length of the hose. If a weak or swollen spot is noted when squeezing the hose wall, the hose should be replaced.

Fig. 69 The cracks developing along this hose are a result of age-related hardening

Fig. 65 Do not bend, twist or turn the timing belt inside out. Never allow oil, water or steam to contact the belt

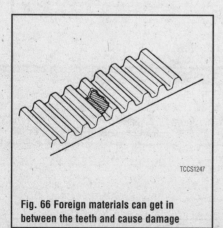
Fig. 66 Foreign materials can get in between the teeth and cause damage

Fig. 67 Inspect the timing belt for cracks, fraying, glazing or damage of any kind

Fig. 68 Damage on only one side of the timing belt may indicate a faulty guide

1-22 GENERAL INFORMATION AND MAINTENANCE

Fig. 70 A hose clamp that is too tight can cause older hoses to separate and tear on either side of the clamp

Fig. 71 A soft spongy hose (identifiable by the swollen section) will eventually burst and should be replaced

Fig. 72 Hoses are likely to deteriorate from the inside if the cooling system is not periodically flushed

REMOVAL & INSTALLATION

▶ See Figures 73, 74, 75, 76 and 77

1. Remove the radiator pressure cap.

✴✴ CAUTION

Never remove the pressure cap while the engine is running, or personal injury from scalding hot coolant or steam may result. If possible, wait until the engine has cooled to remove the pressure cap. If this is not possible, wrap a thick cloth around the pressure cap and turn it slowly to the stop. Step back while the pressure is released from the cooling system. When you are sure all the pressure has been released, use the cloth to turn and remove the cap.

2. Position a clean container under the radiator and/or engine draincock or plug, then open the drain and allow the cooling system to drain to an appropriate level. For some upper hoses, only a little coolant must be drained. To

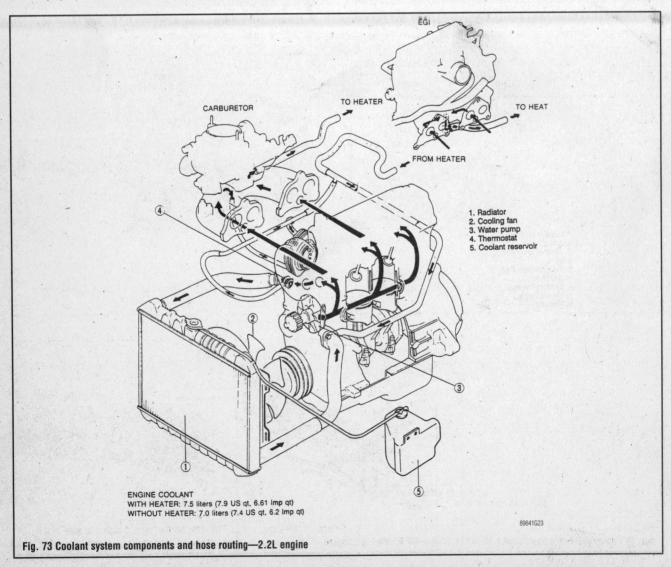

Fig. 73 Coolant system components and hose routing—2.2L engine

GENERAL INFORMATION AND MAINTENANCE

remove hoses positioned lower on the engine, such as a lower radiator hose, the entire cooling system must be emptied.

> ✱✱ **CAUTION**
>
> When draining coolant, keep in mind that cats and dogs are attracted by ethylene glycol antifreeze, and are quite likely to drink any that is left in an uncovered container or in puddles on the ground. This will prove fatal in sufficient quantity. Always drain coolant into a sealable container. Coolant may be reused unless it is contaminated or several years old.

3. Loosen the hose clamps at each end of the hose requiring replacement. Clamps are usually either of the spring tension type (which require pliers to squeeze the tabs and loosen) or of the screw tension type (which require screw or hex drivers to loosen). Pull the clamps back on the hose away from the connection.

4. Twist, pull and slide the hose off the fitting, taking care not to damage the neck of the component from which the hose is being removed.

➡ If the hose is stuck at the connection, do not try to insert a screwdriver or other sharp tool under the hose end in an effort to free it, as the connection and/or hose may become damaged. Heater connections especially may be easily damaged by such a procedure. If the hose is to be replaced, use a single-edged razor blade to make a slice along the portion of the hose which is stuck on the connection, perpendicular to the end of the hose. Do not cut deep so as to prevent damaging the connection. The hose can then be peeled from the connection and discarded.

5. Clean both hose mounting connections. Inspect the condition of the hose clamps and replace them, if necessary.

To install:

6. Dip the ends of the new hose into clean engine coolant to ease installation.
7. Slide the clamps over the replacement hose, then slide the hose ends over the connections into position.
8. Position and secure the clamps at least ¼ in. (6.35mm) from the ends of the hose. Make sure they are located beyond the raised bead of the connector.

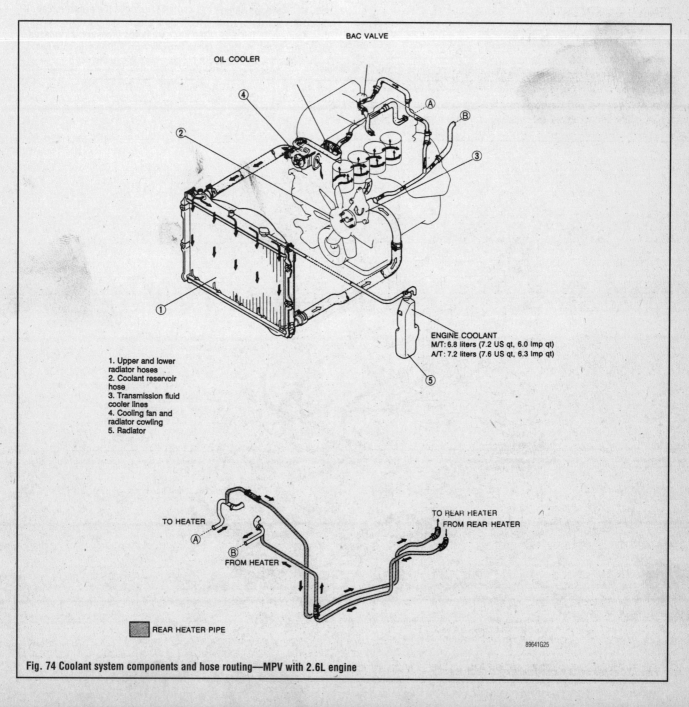

Fig. 74 Coolant system components and hose routing—MPV with 2.6L engine

1-24 GENERAL INFORMATION AND MAINTENANCE

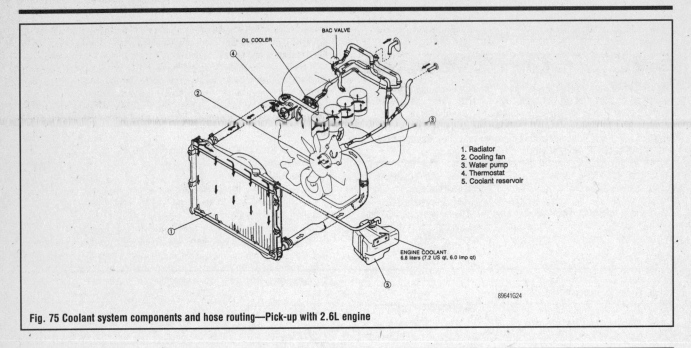

Fig. 75 Coolant system components and hose routing—Pick-up with 2.6L engine

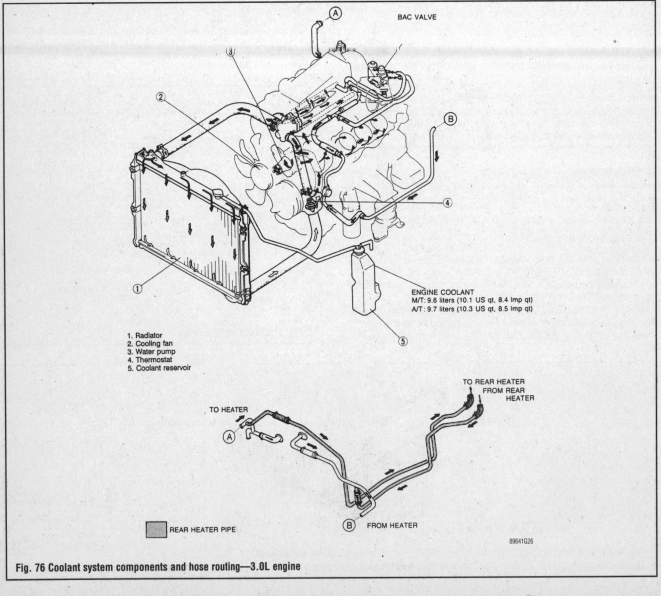

Fig. 76 Coolant system components and hose routing—3.0L engine

GENERAL INFORMATION AND MAINTENANCE 1-25

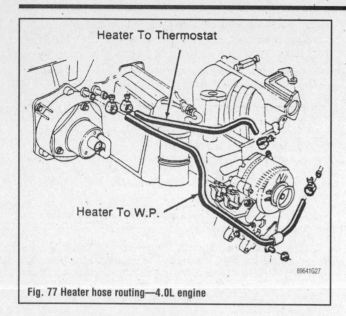

Fig. 77 Heater hose routing—4.0L engine

9. Close the radiator or engine drains and properly refill the cooling system with the clean drained engine coolant or a suitable mixture of ethylene glycol coolant and water.

10. If available, install a pressure tester and check for leaks. If a pressure tester is not available, run the engine until normal operating temperature is reached (allowing the system to naturally pressurize), then check for leaks.

✲✲ CAUTION

If you are checking for leaks with the system at normal operating temperature, BE EXTREMELY CAREFUL not to touch any moving or hot engine parts. Once temperature has been reached, shut the engine OFF, and check for leaks around the hose fittings and connections which were removed earlier.

CV-Boots

INSPECTION

♦ See Figures 78 and 79

The Pick-up and MPV models equipped with 4-wheel drive use an Independent Front Suspension (IFS) drive axle which utilizes Constant Velocity (CV) joint equipped axle half-shafts. These shafts utilize rubber boots to protect the joints and retain the grease.

The CV (Constant Velocity) boots should be checked for damage each time the oil is changed and any other time the vehicle is raised for service. These boots keep water, grime, dirt and other damaging matter from entering the CV-joints. Any of these could cause early CV-joint failure which can be expensive to repair. Heavy grease thrown around the inside of the front wheel(s) and on the brake caliper(s) can be an indication of a torn boot. Thoroughly check the boots for missing clamps and tears. If the boot is damaged, it should be replaced immediately. Please refer to Section 7 for procedures.

Spark Plugs

♦ See Figure 80

A typical spark plug consists of a metal shell surrounding a ceramic insulator. A metal electrode extends downward through the center of the insulator and protrudes a small distance. Located at the end of the plug and attached to the side of the outer metal shell is the side electrode. The side electrode bends in at a 90° angle so that its tip is just past and parallel to the tip of the center electrode. The distance between these two electrodes (measured in thousandths of an inch or hundredths of a millimeter) is called the spark plug gap.

The spark plug does not produce a spark but instead provides a gap across which the current can arc. The coil produces anywhere from 20,000 to 50,000 volts (depending on the type and application) which travels through the wires to the spark plugs. The current passes along the center electrode and jumps the gap to the side electrode, and in doing so, ignites the air/fuel mixture in the combustion chamber.

SPARK PLUG HEAT RANGE

♦ See Figure 81

Spark plug heat range is the ability of the plug to dissipate heat. The longer the insulator (or the farther it extends into the engine), the hotter the plug will operate; the shorter the insulator (the closer the electrode is to the block's cooling pas-

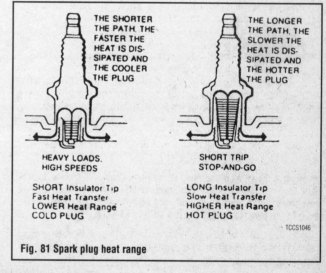

Fig. 81 Spark plug heat range

Fig. 78 CV-boots must be inspected periodically for damage

Fig. 79 A torn boot should be replaced immediately

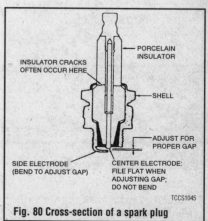

Fig. 80 Cross-section of a spark plug

sages) the cooler it will operate. A plug that absorbs little heat and remains too cool will quickly accumulate deposits of oil and carbon since it is not hot enough to burn them off. This leads to plug fouling and consequently to misfiring. A plug that absorbs too much heat will have no deposits but, due to the excessive heat, the electrodes will burn away quickly and might possibly lead to preignition or other ignition problems. Preignition takes place when plug tips get so hot that they glow sufficiently to ignite the air/fuel mixture before the actual spark occurs. This early ignition will usually cause a pinging during low speeds and heavy loads.

The general rule of thumb for choosing the correct heat range when picking a spark plug is: if most of your driving is long distance, high speed travel, use a colder plug; if most of your driving is stop and go, use a hotter plug. Original equipment plugs are generally a good compromise between the 2 styles and most people never have the need to change their plugs from the factory-recommended heat range.

REMOVAL & INSTALLATION

▶ See Figures 82, 83 and 84

➡ Mazda recommends replacing the spark plugs every 30,000 miles (48,000km).

A set of spark plugs usually requires replacement after about 20,000–30,000 miles (32,000–48,000km), depending on your style of driving. In normal operation plug gap increases about 0.001 in. (0.025mm) for every 2500 miles (4000km). As the gap increases, the plug's voltage requirement also increases. It requires a greater voltage to jump the wider gap and about two to three times as much voltage to fire the plug at high speeds than at idle. The improved air/fuel ratio control of modern fuel injection combined with the higher voltage output of modern ignition systems will often allow an engine to run significantly longer on a set of standard spark plugs, but keep in mind that efficiency will drop as the gap widens (along with fuel economy and power).

When you're removing spark plugs, work on one at a time. Don't start by removing the plug wires all at once, because, unless you number them, they may become mixed up. Take a minute before you begin and number the wires with tape. Also, an anti-seize compound should be used before installing the plugs into the cylinder head.

1. Disconnect the negative battery cable, and if the vehicle has been run recently, allow the engine to thoroughly cool.
2. Carefully twist the spark plug wire boot to loosen it, then pull upward and remove the boot from the plug. Be sure to pull on the boot and not on the wire, otherwise the connector located inside the boot may become separated.
3. Using compressed air, blow any water or debris from the spark plug well to assure that no harmful contaminants are allowed to enter the combustion chamber when the spark plug is removed. If compressed air is not available, use a rag or a brush to clean the area.

➡ Remove the spark plugs when the engine is cold, if possible, to prevent damage to the threads. If removal of the plugs is difficult, apply a few drops of penetrating oil or silicone spray to the area around the base of the plug, and allow it a few minutes to work.

4. Using a spark plug socket that is equipped with a rubber insert to properly hold the plug, turn the spark plug counterclockwise to loosen and remove the spark plug from the bore.

✱✱ WARNING

Be sure not to use a flexible extension on the socket. Use of a flexible extension may allow a shear force to be applied to the plug. A shear force could break the plug off in the cylinder head, leading to costly and frustrating repairs.

To install:

5. Inspect the spark plug boot for tears or damage. If a damaged boot is found, the spark plug wire must be replaced.
6. Using a wire feeler gauge, check and adjust the spark plug gap. When using a gauge, the proper size should pass between the electrodes with a slight drag. The next larger size should not be able to pass while the next smaller size should pass freely.

➡ Coat the spark plug threads with an anti-seize compound before installing it into the cylinder head.

7. Carefully thread the plug into the bore by hand. If resistance is felt before the plug is almost completely threaded, back the plug out and begin threading again. In small, hard to reach areas, an old spark plug wire and boot could be used as a threading tool. The boot will hold the plug while you twist the end of the wire and the wire is supple enough to twist before it would allow the plug to crossthread.

✱✱ WARNING

Do not use the spark plug socket to thread the plugs. Always carefully thread the plug by hand or using an old plug wire to prevent the possibility of crossthreading and damaging the cylinder head bore.

8. Carefully tighten the spark plug. If the plug you are installing is equipped with a crush washer, seat the plug, then tighten about ¼ turn to crush the washer. If you are installing a tapered seat plug, tighten the plug to specifications provided by the vehicle or plug manufacturer.
9. Apply a small amount of silicone dielectric compound to the end of the spark plug lead or inside the spark plug boot to prevent sticking, then install the boot to the spark plug and push until it clicks into place. The click may be felt or heard, then gently pull back on the boot to assure proper contact.

INSPECTION & GAPPING

▶ See Figures 85, 86, 87 and 88

Check the plugs for deposits and wear. If they are not going to be replaced, clean the plugs thoroughly. Remember that any kind of deposit will decrease the efficiency of the plug. Plugs can be cleaned on a spark plug cleaning machine, which can sometimes be found in service stations, or you can do an acceptable job of cleaning with a stiff brush. If the plugs are cleaned, the electrodes must be filed flat. Use an ignition points file, not an emery board or the like, which will leave deposits. The electrodes must be filed perfectly flat with sharp edges; rounded edges reduce the spark plug voltage by as much as 50%.

Check spark plug gap before installation. The ground electrode (the L-shaped one connected to the body of the plug) must be parallel to the center

Fig. 82 Remove the spark plug wire boot from the spark plug

Fig. 83 Using a spark plug socket, an extension and a ratchet, unscrew the spark plug from the cylinder head . . .

Fig. 84 . . . then pull the socket and spark plug straight out from the cylinder head

GENERAL INFORMATION AND MAINTENANCE

A normally worn spark plug should have light tan or gray deposits on the firing tip.

A carbon fouled plug, identified by soft, sooty, black deposits, may indicate an improperly tuned vehicle. Check the air cleaner, ignition components and engine control system.

This spark plug has been **left in the engine too long,** as evidenced by the extreme gap- Plugs with such an extreme gap can cause misfiring and stumbling accompanied by a noticeable lack of power.

An oil fouled spark plug indicates an engine with worn poston rings and/or bad valve seals allowing excessive oil to enter the chamber.

A physically damaged spark plug may be evidence of severe detonation in that cylinder. Watch that cylinder carefully between services, as a continued detonation will not only damage the plug, but could also damage the engine.

A bridged or almost bridged spark plug, identified by a build-up between the electrodes caused by excessive carbon or oil build-up on the plug.

Fig. 85 Inspect the spark plug to determine engine running conditions

1-28 GENERAL INFORMATION AND MAINTENANCE

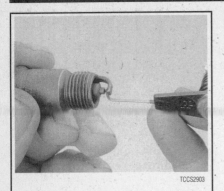

Fig. 86 Checking the spark plug gap with a feeler gauge

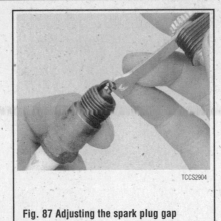

Fig. 87 Adjusting the spark plug gap

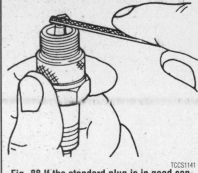

Fig. 88 If the standard plug is in good condition, the electrode may be filed flat—WARNING: do not file platinum plugs

electrode and the specified size wire gauge (please refer to the Tune-Up Specifications chart for details) must pass between the electrodes with a slight drag.

➡NEVER adjust the gap on a used platinum type spark plug.

Always check the gap on new plugs as they are not always set correctly at the factory. Do not use a flat feeler gauge when measuring the gap on a used plug, because the reading may be inaccurate. A round-wire type gapping tool is the best way to check the gap. The correct gauge should pass through the electrode gap with a slight drag. If you're in doubt, try one size smaller and one larger. The smaller gauge should go through easily, while the larger one shouldn't go through at all. Wire gapping tools usually have a bending tool attached. Use that to adjust the side electrode until the proper distance is obtained. Absolutely never attempt to bend the center electrode. Also, be careful not to bend the side electrode too far or too often as it may weaken and break off within the engine, requiring removal of the cylinder head to retrieve it.

Spark Plug Wires

TESTING

◆ See Figure 89

At every tune-up/inspection, visually check the spark plug cables for burns, cuts, or breaks in the insulation. Check the boots and the nipples on the distributor cap and/or coil. Replace any damaged wiring.

➡Whenever a high tension wire is removed for any reason from a spark plug, coil or distributor terminal housing, silicone grease should be applied to the boot before it is reconnected.

Every 50,000 miles (80,000 Km) or 60 months, the resistance of the wires should be checked using an ohmmeter. Wires with excessive resistance will cause misfiring, and may make the engine difficult to start in damp weather. Resistance should measure as follows:

- Pickups and MPV—16k ohms per 3.28 feet (1m) of wire
- Navajo—7k ohms per foot (30cm) of wire

It's a good idea to replace the wires in sets, rather than individually. The wires should also be tested for resistance with an ohmmeter.

➡Only test one spark plug wire at a time. When the check is complete return the plug wire to its original location. If the wire is defective and more wires are to be checked, mark the wire as such, return it to its original location, then inspect the other wires. Once all of the wires are checked, replace the defective wires one at a time. This will avoid any mix-ups.

REMOVAL & INSTALLATION

◆ See Figures 90 and 91

When removing spark plug wires, use great care. Grasp and twist the insulator back and forth on the spark plug to free the insulator. Do not pull on the wire directly as it may become separated from the connector inside the insulator.

To install:

➡Whenever a high tension wire is removed for any reason form a spark plug, coil or distributor terminal housing, silicone grease must be applied to the boot before it is reconnected. Using a small clean tool, coat the entire interior surface of the boot with silicone grease or an equivalent.

1. Install each wire in or on the proper terminal of the coil pack or distributor cap. Be sure the terminal connector inside the insulator is fully seated. The No. 1 terminal is identified on the cap.
2. Remove wire separators from old wire set and install them on new set in approximately same position.
3. Connect wires to proper spark plugs. Be certain all wires are fully seated on terminals.

Fig. 89 Checking individual plug wire resistance with a digital ohmmeter

Fig. 90 Note that the spark plug wires and distributor cap are numbered for correct wiring placement at the factory (C refers to ignition coil wire)

Fig. 91 Be sure to number all the spark plug wires to the correct terminal on the distributor cap before disconnecting them

GENERAL INFORMATION AND MAINTENANCE

Distributor Cap and Rotor

➡ The 2.2L, 2.6L and 3.0L engines are equipped with a distributor ignition system. Only the 4.0L engine in the Navajo utilizes a distributorless ignition.

REMOVAL & INSTALLATION

Distributor Cap

♦ See Figures 92 and 93

1. As a precaution, label all of the spark plug wires with their perspective cylinder number.
2. Loosen the cap hold-down screws.
3. Remove the cap by lifting straight up to prevent damage to the rotor blade and spring.

➡ If the plug wires do not have enough slack to allow cap removal, confirm that they are properly labeled and remove them from the cap.

4. Installation is the reverse of the removal procedure. Note the position of the square alignment locator, or keyway, and tighten the hold-down screws.

Distributor Rotor

♦ See Figure 94

1. Remove the distributor cap.
2. Pull straight up on the rotor to disengage it from the shaft and armature.
3. Installation is the reverse of the removal procedure. Align the locating boss on the rotor with the hole on the armature, or the flat on the shaft, then insure that it is fully seated on the shaft.

INSPECTION

Distributor Cap

1. Wash the inside and outside surfaces of the cap with soap and water then dry it with compressed air.
2. Inspect the cap for cracks, broken or worn carbon button, or carbon tracks. Also inspect the cap terminals for dirt and corrosion.
3. Replace the cap if any of the above conditions are observed.

Distributor Rotor

1. Wash the rotor with soap and water then dry with it compressed air.
2. Inspect the rotor for cracks, carbon tracks, burns or damage to the blade or spring.
3. Replace the rotor if any of the above conditions are observed.

Ignition Timing

GENERAL INFORMATION

♦ See Figures 95 and 96

Ignition timing is an important part of the tune-up. It should be checked to compensate for timing belt or gear wear at the interval specified in the Maintenance Chart. An inductive type DC light, one that can be used on electronic ignition, and powered by the vehicle's battery, is the most frequently used by professional tuners. The bright flash put out by the DC light makes the timing marks stand out on even the brightest of days. The DC light attaches to the spark plug and the wire with an adapter and two clips attached to the battery posts for power.

Fig. 92 Loosen the three distributor cap retaining screws . . .

Fig. 93 . . . then remove the cap from the distributor

Fig. 94 After removing the distributor cap, pull the rotor straight up and off of the distributor shaft

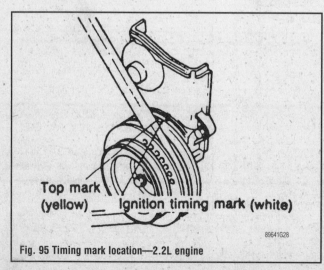

Fig. 95 Timing mark location—2.2L engine

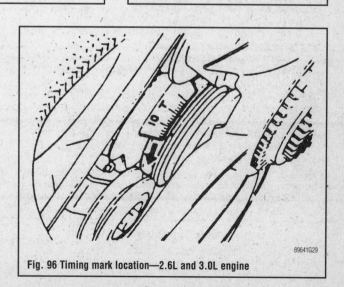

Fig. 96 Timing mark location—2.6L and 3.0L engine

1-30 GENERAL INFORMATION AND MAINTENANCE

> ※※ **CAUTION**
>
> When performing this or any other operation with the engine running, be very careful of the alternator belt and pulleys. Make sure that your timing light wires don't interfere with the belt.

Ignition timing is the measurement, in degrees of crankshaft rotation, of the point at which the spark plugs fire in each of the cylinders. It is measured in degrees before or after Top Dead Center (TDC) of the compression stroke. Ignition timing is adjusted by turning the distributor body in the engine. Ideally, the air/fuel mixture in the cylinder will be ignited by the spark plug just before the piston passes TDC of the compression stroke. If this happens, the piston will be beginning its downward motion of the power stroke just as the compressed and ignited air/fuel mixture begins to develop a considerable amount of pressure. The expansion of the air/fuel mixture then forces the piston down on the power stroke and turns the crankshaft. Because it takes time for the mixture to burn, the spark plug must fire a little before the piston reaches TDC. Otherwise, the mixture will not be burned completely early enough in the downstroke and the full power of the explosion will not be used by the engine. The timing measurement is given in degrees of crankshaft rotation before the piston reaches TDC (BTDC). If the setting for the ignition timing is 5° BTDC (5B), the spark plug must fire 5° before each piston reaches TDC. This only holds true, however, when the engine is at idle speed. As the engine speed increases, the pistons go faster. The spark plugs have to ignite the fuel even sooner if it is to be completely ignited when the piston reaches TDC. To do this, the distributor has a means to advance the timing of the spark as the engine speed increases. This is accomplished by centrifugal weights within the distributor and a vacuum diaphragm, mounted on the side of the distributor. It is necessary to disconnect the vacuum line from the diaphragm when the ignition timing is being set. If the ignition is set too far advanced (BTDC), the ignition and expansion of the fuel in the cylinder will occur too soon and there will be excessive temperature and pressure. This causes engine ping. If the ignition spark is set too far retarded, after TDC (ATDC), the piston will have already passed TDC and started on its way down when the fuel is ignited. This will cause the piston to be forced down for only a portion of its travel and creates less pressure in the cylinder, resulting in poor engine performance and lack of power. The timing is best checked with a timing light. This device is connected in series (or through induction) with the No. 1 spark plug. The current which fires the spark plug also causes the timing light to flash. The timing marks are located at the front crankshaft pulley and consist of a notch on the crankshaft pulley and a scale of degrees of crankshaft rotation attached to the front cover. When the engine is running, the timing light is aimed at the marks on the flywheel pulley and the pointer.

INSPECTION & ADJUSTMENT

1987–88

1. Raise the hood and clean and mark the timing marks. Chalk or fluorescent paint makes a good, visible mark.
2. Disconnect the vacuum line to the distributor and plug the disconnected line. Disconnect the line at the vacuum source, not at the distributor.
3. Connect a timing light to the front (no.1) cylinder, a power source and ground. Follow the manufacturer's instructions.
4. Connect a tachometer to the engine.
5. Start the engine and reduce the idle to 700–750 rpm to be sure that the centrifugal advance mechanism is not working.
6. With the engine running, shine the timing light at the timing pointer and observe the position of the pointer in relation to the timing mark on the crankshaft pulley. All models have two notches on the pulley. Looking straight down on the marks, the one on the left is TDC, the one on the right is BTDC
7. If the timing is not as specified, adjust the timing by loosening the distributor holddown bolt and rotating the distributor in the proper direction. When the proper ignition timing is obtained, tighten the holddown bolt on the distributor.
8. Check the centrifugal advance mechanism by accelerating the engine to about 2,000 rpm. If the ignition timing advances, the mechanism is working properly.
9. Stop the engine and remove the timing light.
10. Reset the idle to specifications.
11. Remove the tachometer.

1989–93 Pickup and MPV

1. Run the engine to normal operating temperature.
2. Turn all electrical accessories OFF.
3. Disconnect the vacuum hoses from the vacuum control and plug it.
4. On fuel injected models, connect a jumper wire between the green, 1-pin test connector and ground.
5. Check the idle speed, and, if necessary, adjust it.
6. Mark the correct timing mark with white paint.
7. Connect a timing light according to the manufacturer's instructions.
8. With the engine running at idle, aim the timing light at the timing marks. If the correct timing is not indicated, loosen the distributor locknut and turn the distributor as needed to align the marks.
9. Tighten the locknut and recheck the timing.
10. Reconnect the vacuum hose and remove the jumper wire.

Navajo

With the EEC-IV EDIS systems, no ignition timing adjustment is possible (preset at the factory to 10° BTDC and is computer controlled) and none should be attempted.

Valve Lash

Valve adjustment determines how far the valves enter the cylinder and how long they stay open and closed. If the valve clearance is too large, part of the lift of the camshaft will be used in removing the excessive clearance. Consequently, the valve will not be opening as far as it should, it will start to open too late and will close too early. This condition has two effects: the valve train components will emit a tapping sound as they take up the excessive clearance and as the valves slam shut, and the engine will perform poorly because the valves don't open fully and allow the proper amount of gases to flow into and out of the engine. If the valve clearance is too small, the intake valves and the exhaust valves will open too far and they will not fully seat on the cylinder head when they close. When a valve seats itself on the cylinder head, it does two things: it seals the combustion chamber so that none of the gasses in the cylinder escape and it cools itself by transferring some of the heat it absorbs from the combustion in the cylinder to the cylinder head and to the engine's cooling system. If the valve clearance is too small, the engine will run poorly because of the gases escaping from the combustion chamber. The valves will also become overheated and will warp, since they cannot transfer heat unless they are touching the valve seat in the cylinder head.

➥**While all valve adjustments must be made as accurately as possible, it is better to have the valve adjustment slightly loose than slightly tight, as a burned valve may result from overly tight adjustments. This holds true for valve adjustments on most engines.**

ADJUSTMENT

1987–88 Engines

♦ See Figure 97

These engines use hydraulic lash adjusters for the intake and exhaust valves, which require no routine adjustment. However, these engines also incorporate a 3rd valve per cylinder called the jet valve which should be periodically checked and adjusted.

1. Run the engine to normal operating temperature. Then, shut it off.
2. Remove the valve cover.
3. Rotate the engine by hand until the No.1 cylinder is at TDC compression. Both the intake and exhaust valves will feel loose and the timing mark for 0 will be aligned with the pointer.
4. Check the jet valve clearance with a flat feeler gauge. Clearance for all engines should be 0.010 inch (0.25mm). If not, loosen the locknut on the rocker arm and turn the adjusting screw until a slight drag is felt on the feeler gauge. The jet valve spring is relatively weak, so don't press down on the valve stem or you'll get an erroneous reading.
5. Turn the engine by hand, in the normal direction of rotation, until each cylinder in the firing order reaches TDC of its compression stroke and adjust each jet valve in turn.
6. Install the valve cover.

GENERAL INFORMATION AND MAINTENANCE 1-31

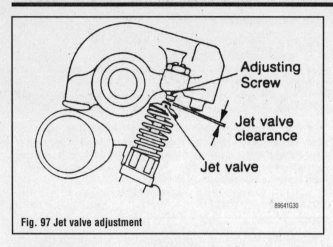

Fig. 97 Jet valve adjustment

1989–93 Engines

No routine adjustment is necessary or possible.

Idle Speed and Mixture Adjustments

IDLE SPEED

▶ See Figure 98

1987 Engines

EXCEPT B2200 W/AUTOMATIC TRANSMISSION

1. Connect a tachometer to the engine.
2. Run the engine to normal operating temperature.
3. Check the idle speed in NEUTRAL or PARK. If it is not 800–850 rpm, turn the Throttle Adjusting Screw until it is.

B2200 W/AUTOMATIC TRANSMISSION AND WO/AIR CONDITIONING

1. Connect a tachometer to the engine.
2. Run the engine to normal operating temperature.
3. Disconnect the vacuum hose from the servo vacuum unit.
4. Connect direct intake vacuum to the servo vacuum unit.
5. Set the parking brake and block the wheels.
6. Check the idle speed in DRIVE. If it is not 920–970 rpm, turn the Throttle Adjusting Screw until it is.
7. Reconnect the hose properly.

B2200 W/AUTOMATIC TRANSMISSION AND W/AIR CONDITIONING

1. Connect a tachometer to the engine.
2. Run the engine to normal operating temperature.
3. Disconnect the vacuum hose from the servo vacuum unit lower nipple.
4. Connect direct intake vacuum to the servo vacuum unit lower nipple.
5. Set the parking brake and block the wheels.
6. Check the idle speed in DRIVE. If it is not 1,300–1,500 rpm, turn the Throttle Adjusting Screw until it is.
7. Reconnect the hose properly.

1988–93 CARBURETED ENGINES

1. Connect a tachometer to the engine.
2. Run the engine to normal operating temperature, make sure the choke is fully opened. The cooling fan motor must not be operating during this operation.
3. Check the idle speed in NEUTRAL or PARK. If it is not 800–850 rpm, turn the Throttle Adjusting Screw until it is.

1989–93 FUEL INJECTED ENGINES (EXCEPT NAVAJO)

▶ See Figures 98 thru 103

1. Place manual transmission in neutral or automatic transmission in park.
2. Make sure all accessories are Off.
3. Connect a tachometer and timing light to the engine.
4. Warm up the engine to normal operating temperature.
5. Check the ignition timing and adjust, if necessary.
6. Ground the green 1 pin test connector to the body with a jumper wire.
7. Check the idle speed. Specifications are as follows: 2.2L and 2.6L engines—Manual transmission: 730–770 rpm. Automatic transmission: 750–790 rpm. 3.0L engine—Manual and automatic transmission: 780–820 rpm.
8. If the idle speed is not within specification, adjust by turning the air adjusting screw.
9. After adjustment, disconnect the jumper wire from the test connector. Recheck the ignition timing.

NAVAJO

1. Place manual transmission in neutral or automatic transmission in Park. Apply the parking brake.

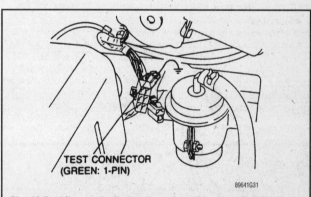

Fig. 99 For idle speed adjustment, connect a jumper wire between the test connector green 1-pin and a good ground—Pickups equipped with fuel injection

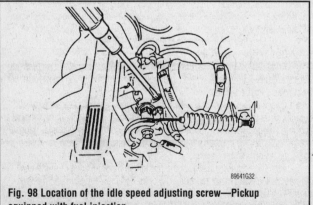

Fig. 98 Location of the idle speed adjusting screw—Pickup equipped with fuel injection

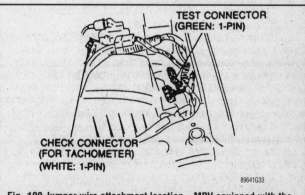

Fig. 100 Jumper wire attachment location—MPV equipped with the 2.6L engine

GENERAL INFORMATION AND MAINTENANCE

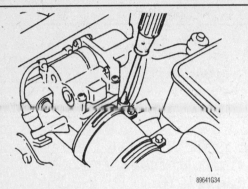

Fig. 101 Location of the idle speed adjustment screw—MPV equipped with the 2.6L engine

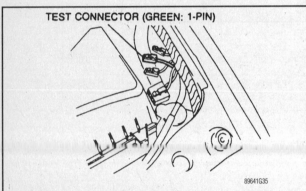

Fig. 102 Jumper wire attachment location—MPV equipped with the 3.0L engine

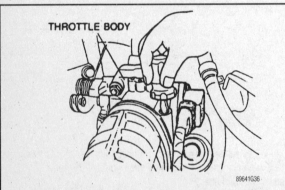

Fig. 103 Location of the idle speed adjustment screw—MPV equipped with the 3.0L engine

2. Make sure the heater and accessories are Off.
3. Start the engine and bring to normal operating temperature. Make sure the throttle lever is resting on the throttle plate stop screw.
4. Check the ignition timing and adjust, if necessary.
5. Shut off the engine and disconnect the negative battery cable for 5 minutes minimum. Reconnect the negative battery cable.
6. Start the engine and let it stabilize for 2 minutes. Rev the engine and let it return to idle, lightly depress and release the accelerator and let the engine idle.
7. If the engine does not idle properly, shut off the engine and disconnect the idle speed control-air bypass solenoid.
8. Run the engine at 2500 rpm for 30 seconds, then let it idle for 2 minutes.
9. Check/adjust the idle rpm to 675 rpm by turning the throttle plate stop screw.

➡ If the screw must be turned in, shut the engine off and make the estimated adjustment, then start the engine and repeat Steps 8 and 9.

10. Shut the engine off and repeat Steps 8 and 9.
11. Shut the engine off and disconnect the negative battery cable for 5 minutes minimum.
12. With the engine off, reconnect the idle speed control-air bypass solenoid. Make sure the throttle is not stuck in the bore and the linkage is not preventing the throttle from closing.
13. Start the engine and let it stabilize for 2 minutes. Rev the engine and let it return to idle, lightly depress and release the accelerator and let the engine idle.

➡ A condition may occur where the engine rpm will oscillate. This can be caused by the throttle plates being open enough to allow purge flow. To make sure of this condition, disconnect the carbon canister purge line and plug it. If purge is present, the throttle plates must be closed until the purge flow induced idle oscillations stop.

MIXTURE ADJUSTMENT

1987–88 Engines

➡ You'll need a dwell meter with a 90° scale.

1. Run the engine to normal operating temperature.
2. On the 2.6L engine, connect the dwell meter (+) positive lead to the yellow/green wire in the check connector, located near the No.3 air control valve and reed valve case assembly. Connect the (-) lead to ground.
3. On the 2.2L engine, connect the dwell meter (+) positive lead to the brown/yellow wire in the check connector, located near the No.3 air control valve and reed valve case assembly. Connect the (-) lead to ground.
4. With the engine at the normal idle speed, the dwell meter should read 27–45° for the 2.6L engine, or 20–70° for the 2.2L engine.
5. If not, check the oxygen sensor, the wiring between the oxygen sensor and the feedback control unit, and the control unit itself. See Section 5 for procedures. If all components check out satisfactorily, adjust the mixture with the adjusting screw.

1989–93 Engines

The idle mixture is controlled by the Electronic Control Unit (ECU) and is not adjustable.

Air Conditioning

SYSTEM SERVICE & REPAIR

➡ It is recommended that the A/C system be serviced by an EPA Section 609 certified automotive technician utilizing a refrigerant recovery/recycling machine.

The do-it-yourselfer should not service his/her own vehicle's A/C system for many reasons, including legal concerns, personal injury, environmental damage and cost. The following are some of the reasons why you may decide not to service your own vehicle's A/C system.

According to the U.S. Clean Air Act, it is a federal crime to service or repair (involving the refrigerant) a Motor Vehicle Air Conditioning (MVAC) system for money without being EPA certified. It is also illegal to vent R-12 and R-134a refrigerants into the atmosphere. Selling or distributing A/C system refrigerant (in a container which contains less than 20 pounds of refrigerant) to any person who is not EPA 609 certified is also not allowed by law.

State and/or local laws may be more strict than the federal regulations, so be sure to check with your state and/or local authorities for further information. For further federal information on the legality of servicing your A/C system, call the EPA Stratospheric Ozone Hotline.

➡ Federal law dictates that a fine of up to $25,000 may be levied on people convicted of venting refrigerant into the atmosphere. Additionally, the EPA may pay up to $10,000 for information or services leading to a criminal conviction of the violation of these laws.

When servicing an A/C system you run the risk of handling or coming in contact with refrigerant, which may result in skin or eye irritation or frostbite. Although low in toxicity (due to chemical stability), inhalation of concentrated refrigerant fumes is dangerous and can result in death; cases of fatal cardiac

GENERAL INFORMATION AND MAINTENANCE 1-33

GASOLINE ENGINE TUNE-UP SPECIFICATIONS

Year	Engine CODE	Engine Displacement Liters (cc)	Spark Plugs Gap (in.)	Ignition Timing (deg.) MT	Ignition Timing (deg.) AT	Fuel Pump (psi)	Idle Speed (rpm) MT	Idle Speed (rpm) AT	Valve Clearance In.	Valve Clearance Ex.
1987	F2	2.2 (2184)	0.030	6B	6B	③	825	825	HYD.	HYD.
	G54B	2.6 (2555)	0.041	7B	7B	2.8-3.6	825	825	HYD.	HYD.
1988	F2	2.2 (2184)	0.030	6B	6B	③	825	825	HYD.	HYD.
	G54B	2.6 (2555)	0.041	7B	7B	2.8-3.6	825	825	HYD.	HYD.
1989	F2	2.2 (2184)	0.030	6B ②	6B ②	③	825	825	HYD.	HYD.
	G6	2.6 (2606)	0.041	5B ②	5B ②	64-85	750 ②	750 ②	HYD.	HYD.
	JE	3.0 (2954)	0.041	11B ②	11B ②	64-85	800 ②	800 ②	HYD.	HYD.
1990	F2	2.2 (2184)	①	6B ②	6B ②	③	④②	⑥②	HYD.	HYD.
	G6	2.6 (2606)	0.041	5B ②	5B ②	64-85	750 ②	750 ②	HYD.	HYD.
	JE	3.0 (2954)	0.041	11B ②	11B ②	64-85	800 ②	800 ②	HYD.	HYD.
1991	F2	2.2 (2184)	①	6B ②	6B ②	③	④②	⑥②	HYD.	HYD.
	G6	2.6 (2606)	0.041	5B ②	5B ②	64-85	750	750	HYD.	HYD.
	JE	3.0 (2954)	0.041	11B ②	11B ②	64-85	800	800	HYD.	HYD.
	X	4.0 (4016)	0.054	10B	10B	35-45	⑤	⑤	HYD.	HYD.
1992	F2	2.2 (2184)	①	6B ②	6B ②	③	④②	⑥②	HYD.	HYD.
	G6	2.6 (2606)	0.041	5B ②	5B ②	64-85	750	750	HYD.	HYD.
	JE	3.0 (2954)	0.041	11B ②	11B ②	64-85	800	800	HYD.	HYD.
	X	4.0 (4016)	0.054	10B ②	10B ②	35-45	⑤	⑤	HYD.	HYD.
1993	F2	2.2 (2184)	①	6B ②	6B ②	③	④②	⑥②	HYD.	HYD.
	G6	2.6 (2606)	0.041	5B ②	5B ②	64-85	750	750	HYD.	HYD.
	JE	3.0 (2954)	0.041	11B ②	11B ②	64-85	800	800	HYD.	HYD.
	X	4.0 (4016)	0.054	10B	10B	35-45	⑤	⑤	HYD.	HYD.

NOTE: The underhood specifications sticker often reflects tune-up specification changes in production. Sticker figures must be used if they disagree with those in this chart.

HYD.-Hydraulic valve lifters

① Carbureted engine: 0.030
   Fuel injected engine: 0.041
② With test connector grounded
③ Carbureted engine
   Manual transmission-mechanical pump: 3.7-4.7
   Automatic transmission-electric pump: 2.8-3.6
   Fuel injected engine: 64-85
④ Carbureted engine: 800
   Fuel injected engine: 750
⑤ See the underhood Vehicle Emission Control Information (VECI) label
⑥ Carbureted engine: 800
   Fuel injected engine: 770

arrhythmia have been reported in people accidentally subjected to high levels of refrigerant. Some early symptoms include loss of concentration and drowsiness.

➡ **Generally, the limit for exposure is lower for R-134a than it is for R-12. Exceptional care must be practiced when handling R-134a.**

Also, refrigerants can decompose at high temperatures (near gas heaters or open flame), which may result in hydrofluoric acid, hydrochloric acid and phosgene (a fatal nerve gas).

R-12 refrigerant can damage the environment because it is a Chlorofluorocarbon (CFC), which has been proven to add to ozone layer depletion, leading to increasing levels of UV radiation. UV radiation has been linked with an increase in skin cancer, suppression of the human immune system, an increase in cataracts, damage to crops, damage to aquatic organisms, an increase in ground-level ozone, and increased global warming.

R-134a refrigerant is a greenhouse gas which, if allowed to vent into the atmosphere, will contribute to global warming (the Greenhouse Effect).

It is usually more economically feasible to have a certified MVAC automotive technician perform A/C system service on your vehicle. Some possible reasons for this are as follows:

- While it is illegal to service an A/C system without the proper equipment, the home mechanic would have to purchase an expensive refrigerant recovery/recycling machine to service his/her own vehicle.
- Since only a certified person may purchase refrigerant—according to the Clean Air Act, there are specific restrictions on selling or distributing A/C sys-

1-34 GENERAL INFORMATION AND MAINTENANCE

tem refrigerant—it is legally impossible (unless certified) for the home mechanic to service his/her own vehicle. Procuring refrigerant in an illegal fashion exposes one to the risk of paying a $25,000 fine to the EPA.

R-12 Refrigerant Conversion

If your vehicle still uses R-12 refrigerant, one way to save A/C system costs down the road is to investigate the possibility of having your system converted to R-134a. The older R-12 systems can be easily converted to R-134a refrigerant by a certified automotive technician by installing a few new components and changing the system oil.

The cost of R-12 is steadily rising and will continue to increase, because it is no longer imported or manufactured in the United States. Therefore, it is often possible to have an R-12 system converted to R-134a and recharged for less than it would cost to just charge the system with R-12.

If you are interested in having your system converted, contact local automotive service stations for more details and information.

PREVENTIVE MAINTENANCE

▶ See Figures 104 and 105

Although the A/C system should not be serviced by the do-it-yourselfer, preventive maintenance can be practiced and A/C system inspections can be performed to help maintain the efficiency of the vehicle's A/C system. For preventive maintenance, perform the following:

- The easiest and most important preventive maintenance for your A/C system is to be sure that it is used on a regular basis. Running the system for five minutes each month (no matter what the season) will help ensure that the seals and all internal components remain lubricated.

➡ Some newer vehicles automatically operate the A/C system compressor whenever the windshield defroster is activated. When running, the compressor lubricates the A/C system components; therefore, the A/C system would not need to be operated each month.

- In order to prevent heater core freeze-up during A/C operation, it is necessary to maintain proper antifreeze protection. Use a hand-held coolant tester (hydrometer) to periodically check the condition of the antifreeze in your engine's cooling system.

➡ Antifreeze should not be used longer than the manufacturer specifies.

- For efficient operation of an air conditioned vehicle's cooling system, the radiator cap should have a holding pressure which meets manufacturer's specifications. A cap which fails to hold these pressures should be replaced.
- Any obstruction of or damage to the condenser configuration will restrict air flow which is essential to its efficient operation. It is, therefore, a good rule to keep this unit clean and in proper physical shape.

➡ Bug screens which are mounted in front of the condenser (unless they are original equipment) are regarded as obstructions.

- The condensation drain tube expels any water which accumulates on the bottom of the evaporator housing into the engine compartment. If this tube is obstructed, the air conditioning performance can be restricted and condensation buildup can spill over onto the vehicle's floor.

SYSTEM INSPECTION

▶ See Figure 106

Although the A/C system should not be serviced by the do-it-yourselfer, preventive maintenance can be practiced and A/C system inspections can be performed to help maintain the efficiency of the vehicle's A/C system. For A/C system inspection, perform the following:

The easiest and often most important check for the air conditioning system consists of a visual inspection of the system components. Visually inspect the air conditioning system for refrigerant leaks, damaged compressor clutch, abnormal compressor drive belt tension and/or condition, plugged evaporator drain tube, blocked condenser fins, disconnected or broken wires, blown fuses, corroded connections and poor insulation.

A refrigerant leak will usually appear as an oily residue at the leakage point in the system. The oily residue soon picks up dust or dirt particles from the sur-

Fig. 104 A coolant tester can be used to determine the freezing and boiling levels of the coolant in your vehicle

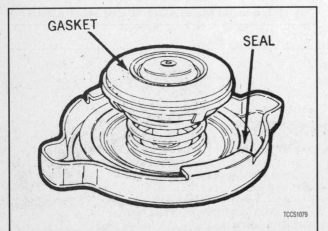

Fig. 105 To ensure efficient cooling system operation, inspect the radiator cap gasket and seal

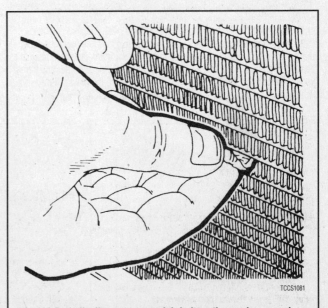

Fig. 106 Periodically remove any debris from the condenser and radiator fins

GENERAL INFORMATION AND MAINTENANCE

rounding air and appears greasy. Through time, this will build up and appear to be a heavy dirt impregnated grease.

For a thorough visual and operational inspection, check the following:
- Check the surface of the radiator and condenser for dirt, leaves or other material which might block air flow.
- Check for kinks in hoses and lines. Check the system for leaks.
- Make sure the drive belt is properly tensioned. When the air conditioning is operating, make sure the drive belt is free of noise or slippage.
- Make sure the blower motor operates at all appropriate positions, then check for distribution of the air from all outlets with the blower on **HIGH** or **MAX**.

➡ **Keep in mind that under conditions of high humidity, air discharged from the A/C vents may not feel as cold as expected, even if the system is working properly. This is because vaporized moisture in humid air retains heat more effectively than dry air, thereby making humid air more difficult to cool.**

- Make sure the air passage selection lever is operating correctly. Start the engine and warm it to normal operating temperature, then make sure the temperature selection lever is operating correctly.

Windshield Wipers

ELEMENT (REFILL) CARE & REPLACEMENT

▶ See Figures 107, 108 and 109

For maximum effectiveness and longest element life, the windshield and wiper blades should be kept clean. Dirt, tree sap, road tar and so on will cause streaking, smearing and blade deterioration if left on the glass. It is advisable to wash the windshield carefully with a commercial glass cleaner at least once a month. Wipe off the rubber blades with the wet rag afterwards. Do not attempt to move wipers across the windshield by hand; damage to the motor and drive mechanism will result.

To inspect and/or replace the wiper blade elements, place the wiper switch in the **LOW** speed position and the ignition switch in the **ACC** position. When the wiper blades are approximately vertical on the windshield, turn the ignition switch to **OFF**.

Examine the wiper blade elements. If they are found to be cracked, broken or torn, they should be replaced immediately. Replacement intervals will vary with usage, although ozone deterioration usually limits element life to about one year. If the wiper pattern is smeared or streaked, or if the blade chatters across the glass, the elements should be replaced. It is easiest and most sensible to replace the elements in pairs.

If your vehicle is equipped with aftermarket blades, there are several different types of refills and your vehicle might have any kind. Aftermarket blades and arms rarely use the exact same type blade or refill as the original equipment.

Regardless of the type of refill used, be sure to follow the part manufacturer's instructions closely. Make sure that all of the frame jaws are engaged as the refill is pushed into place and locked. If the metal blade holder and frame are allowed to touch the glass during wiper operation, the glass will be scratched.

Tires and Wheels

Common sense and good driving habits will afford maximum tire life. Make sure that you don't overload the vehicle or run with incorrect pressure in the tires. Either of these will increase tread wear. Fast starts, sudden stops and sharp cornering are hard on tires and will shorten their useful life span.

➡ **For optimum tire life, keep the tires properly inflated, rotate them often and have the wheel alignment checked periodically.**

Inspect your tires frequently. Be especially careful to watch for bubbles in the tread or sidewall, deep cuts or underinflation. Replace any tires with bubbles in the sidewall. If cuts are so deep that they penetrate to the cords, discard the tire. Any cut in the sidewall of a radial tire renders it unsafe. Also look for uneven tread wear patterns that may indicate the front end is out of alignment or that the tires are out of balance.

TIRE ROTATION

▶ See Figure 110

Tires must be rotated periodically to equalize wear patterns that vary with a tire's position on the vehicle. Tires will also wear in an uneven way as the front steering/suspension system wears to the point where the alignment should be reset.

Rotating the tires will ensure maximum life for the tires as a set, so you will not have to discard a tire early due to wear on only part of the tread. Regular rotation is required to equalize wear.

When rotating "unidirectional tires," make sure that they always roll in the same direction. This means that a tire used on the left side of the vehicle must not be switched to the right side and vice-versa. Such tires should only be rotated front-to-rear or rear-to-front, while always remaining on the same side of the vehicle. These tires are marked on the sidewall as to the direction of rotation; observe the marks when reinstalling the tire(s).

Some styled or "mag" wheels may have different offsets front to rear. In these cases, the rear wheels must not be used up front and vice-versa. Furthermore, if these wheels are equipped with unidirectional tires, they cannot be rotated unless the tire is remounted for the proper direction of rotation.

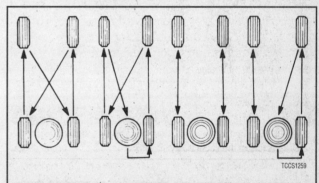

Fig. 110 Common tire rotation patterns for 4 and 5-wheel rotations

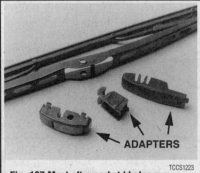

Fig. 107 Most aftermarket blades are available with multiple adapters to fit different vehicles

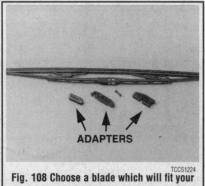

Fig. 108 Choose a blade which will fit your vehicle, and that will be readily available next time you need blades

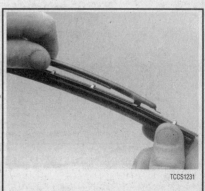

Fig. 109 When installed, be certain the blade is fully inserted into the backing

1-36 GENERAL INFORMATION AND MAINTENANCE

➥The compact or space-saver spare is strictly for emergency use. It must never be included in the tire rotation or placed on the vehicle for everyday use.

TIRE DESIGN

▶ See Figure 111

For maximum satisfaction, tires should be used in sets of four. Mixing of different brands or types (radial, bias-belted, fiberglass belted) should be avoided. In most cases, the vehicle manufacturer has designated a type of tire on which the vehicle will perform best. Your first choice when replacing tires should be to use the same type of tire that the manufacturer recommends.

When radial tires are used, tire sizes and wheel diameters should be selected to maintain ground clearance and tire load capacity equivalent to the original specified tire. Radial tires should always be used in sets of four.

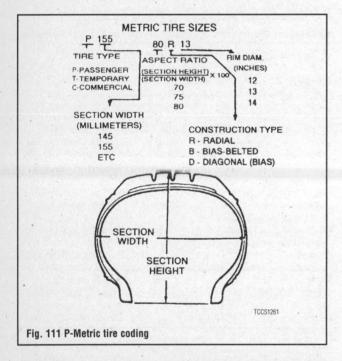

Fig. 111 P-Metric tire coding

✱✱ CAUTION

Radial tires should never be used on only the front axle.

When selecting tires, pay attention to the original size as marked on the tire. Most tires are described using an industry size code sometimes referred to as P-Metric. This allows the exact identification of the tire specifications, regardless of the manufacturer. If selecting a different tire size or brand, remember to check the installed tire for any sign of interference with the body or suspension while the vehicle is stopping, turning sharply or heavily loaded.

Snow Tires

Good radial tires can produce a big advantage in slippery weather, but in snow, a street radial tire does not have sufficient tread to provide traction and control. The small grooves of a street tire quickly pack with snow and the tire behaves like a billiard ball on a marble floor. The more open, chunky tread of a snow tire will self-clean as the tire turns, providing much better grip on snowy surfaces.

To satisfy municipalities requiring snow tires during weather emergencies, most snow tires carry either an M + S designation after the tire size stamped on the sidewall, or the designation "all-season." In general, no change in tire size is necessary when buying snow tires.

Most manufacturers strongly recommend the use of 4 snow tires on their vehicles for reasons of stability. If snow tires are fitted only to the drive wheels, the opposite end of the vehicle may become very unstable when braking or turning on slippery surfaces. This instability can lead to unpleasant endings if the driver can't counteract the slide in time.

Note that snow tires, whether 2 or 4, will affect vehicle handling in all non-snow situations. The stiffer, heavier snow tires will noticeably change the turning and braking characteristics of the vehicle. Once the snow tires are installed, you must re-learn the behavior of the vehicle and drive accordingly.

➥Consider buying extra wheels on which to mount the snow tires. Once done, the "snow wheels" can be installed and removed as needed. This eliminates the potential damage to tires or wheels from seasonal removal and installation. Even if your vehicle has styled wheels, see if inexpensive steel wheels are available. Although the look of the vehicle will change, the expensive wheels will be protected from salt, curb hits and pothole damage.

TIRE STORAGE

If they are mounted on wheels, store the tires at proper inflation pressure. All tires should be kept in a cool, dry place. If they are stored in the garage or basement, do not let them stand on a concrete floor; set them on strips of wood, a mat or a large stack of newspaper. Keeping them away from direct moisture is of paramount importance. Tires should not be stored upright, but in a flat position.

INFLATION & INSPECTION

▶ See Figures 112 thru 117

The importance of proper tire inflation cannot be overemphasized. A tire employs air as part of its structure. It is designed around the supporting strength of the air at a specified pressure. For this reason, improper inflation drastically reduces the tire's ability to perform as intended. A tire will lose some

Fig. 112 Tires with deep cuts, or cuts which bulge, should be replaced immediately

Fig. 113 Radial tires have a characteristic sidewall bulge; don't try to measure pressure by looking at the tire. Use a quality air pressure gauge

GENERAL INFORMATION AND MAINTENANCE 1-37

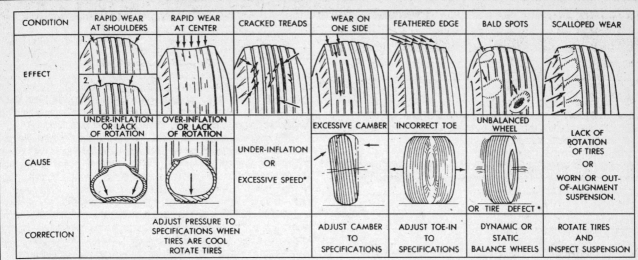

Fig. 114 Common tire wear patterns and causes

air in day-to-day use; having to add a few pounds of air periodically is not necessarily a sign of a leaking tire.

Two items should be a permanent fixture in every glove compartment: an accurate tire pressure gauge and a tread depth gauge. Check the tire pressure (including the spare) regularly with a pocket type gauge. Too often, the gauge on the end of the air hose at your corner garage is not accurate because it suffers too much abuse. Always check tire pressure when the tires are cold, as pressure increases with temperature. If you must move the vehicle to check the tire inflation, do not drive more than a mile before checking. A cold tire is generally one that has not been driven for more than three hours.

A plate or sticker is normally provided somewhere in the vehicle (door post, hood, tailgate or trunk lid) which shows the proper pressure for the tires. Never counteract excessive pressure build-up by bleeding off air pressure (letting some air out). This will cause the tire to run hotter and wear quicker.

✲✲ CAUTION

Never exceed the maximum tire pressure embossed on the tire! This is the pressure to be used when the tire is at maximum loading, but it is rarely the correct pressure for everyday driving. Consult the owner's manual or the tire pressure sticker for the correct tire pressure.

Once you've maintained the correct tire pressures for several weeks, you'll be familiar with the vehicle's braking and handling personality. Slight adjustments in tire pressures can fine-tune these characteristics, but never change the cold pressure specification by more than 2 psi. A slightly softer tire pressure will give a softer ride but also yield lower fuel mileage. A slightly harder tire will give crisper dry road handling but can cause skidding on wet surfaces. Unless you're fully attuned to the vehicle, stick to the recommended inflation pressures.

All automotive tires have built-in tread wear indicator bars that show up as ½ in. (13mm) wide smooth bands across the tire when 1/16 in. (1.5mm) of tread remains. The appearance of tread wear indicators means that the tires should be replaced. In fact, many states have laws prohibiting the use of tires with less than this amount of tread.

You can check your own tread depth with an inexpensive gauge or by using a Lincoln head penny. Slip the Lincoln penny (with Lincoln's head upside-down) into several tread grooves. If you can see the top of Lincoln's head in 2 adjacent grooves, the tire has less than 1/16 in. (1.5mm) tread left and should be replaced. You can measure snow tires in the same manner by using the "tails" side of the Lincoln penny. If you can see the top of the Lincoln memorial, it's time to replace the snow tire(s).

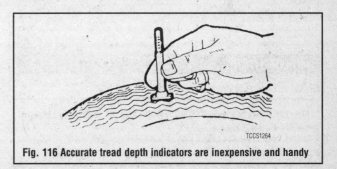

Fig. 116 Accurate tread depth indicators are inexpensive and handy

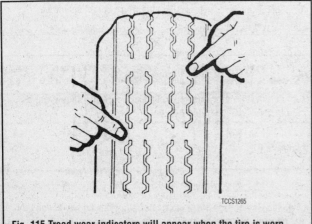

Fig. 115 Tread wear indicators will appear when the tire is worn

Fig. 117 A penny works well for a quick check of tread depth

1-38 GENERAL INFORMATION AND MAINTENANCE

FLUIDS AND LUBRICANTS

Fluid Disposal

Used fluids such as engine oil, transmission fluid, antifreeze and brake fluid are hazardous wastes and must be disposed of properly. Before draining any fluids, consult with your local authorities; in many areas, waste oil, antifreeze, etc. is being accepted as a part of recycling programs. A number of service stations and auto parts stores are also accepting waste fluids for recycling.

Be sure of the recycling center's policies before draining any fluids, as many will not accept different fluids that have been mixed together.

Fuel and Engine Oil Recommendations

ENGINE OIL

♦ See Figures 118, 119 and 120

The recommended oil viscosities for sustained temperatures ranging from below 0°F (–20°C) to above 32°F (0°C) are listed in this Section. They are broken down into multi–viscosity and single viscosities. Multi–viscosity oils are recommended because of their wider range of acceptable temperatures and driving conditions.

➥ **Mazda recommends that SAE 5W–30 viscosity engine oil should be used for all climate conditions, however, SAE 10W–30 is acceptable for vehicles operated in moderate to hot climates.**

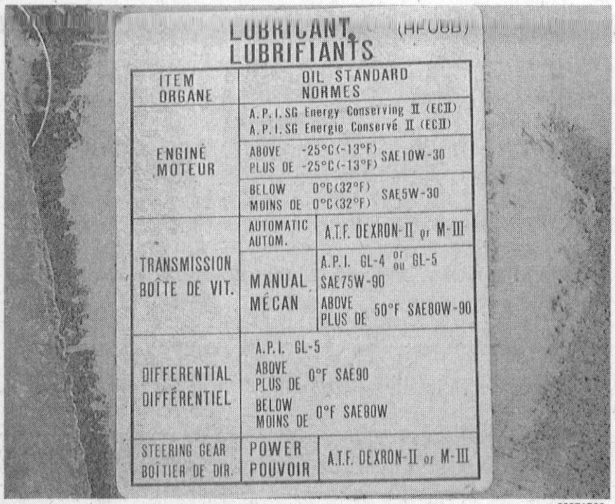

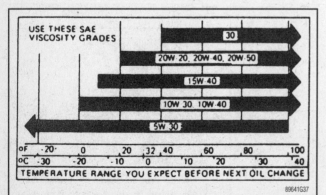

Fig. 118 Lubricant specifications sticker located on the underside of the hood—MPV shown

Fig. 119 Engine oil viscosity chart

When adding oil to the crankcase or changing the oil or filter, it is important that oil of an equal quality to original equipment be used in your truck. The use of inferior oils may void the warranty, damage your engine, or both.

The SAE (Society of Automotive Engineers) grade number of oil indicates the viscosity of the oil (its ability to lubricate at a given temperature). The lower the SAE number, the lighter the oil; the lower the viscosity, the easier it is to crank the engine in cold weather but the less the oil will lubricate and protect the engine in high temperatures. This number is marked on every oil container.

Oil viscosities should be chosen from those oils recommended for the lowest anticipated temperatures during the oil change interval. Due to the need for an oil that embodies both good lubrication at high temperatures and easy cranking in cold weather, multigrade oils have been developed. Basically, a multigrade oil is thinner at low temperatures and thicker at high temperatures. For example, a 10W–40 oil (the W stands for winter) exhibits the characteristics of a 10 weight (SAE 10) oil when the truck is first started and the oil is cold. Its lighter weight allows it to travel to the lubricating surfaces quicker and offer less resistance to starter motor cranking than, say, a straight 30 weight (SAE 30) oil. But after the engine reaches operating temperature, the 10W–40 oil begins acting like straight 40 weight (SAE 40) oil, its heavier weight providing greater lubrication with less chance of foaming than a straight 30 weight oil:

Fig. 120 Look for the API oil identification label when choosing your engine oil

The API (American Petroleum Institute) designations, also found on the oil container, indicates the classification of engine oil used under certain given operating conditions. Only oils designated for use Service SG heavy duty detergent should be used in your truck. Oils of the SG type perform may functions inside the engine besides their basic lubrication. Through a balanced system of metallic detergents and polymeric dispersants, the oil prevents high and low temperature deposits and also keeps sludge and dirt particles in suspension. Acids, particularly sulfuric acid, as well as other by–products of engine combustion are neutralized by the oil. If these acids are allowed to concentrate, they can cause corrosion and rapid wear of the internal engine parts.

✳✳ CAUTION

Non–detergent motor oils or straight mineral oils should not be used in your Ford gasoline engine.

Synthetic Oil

There are many excellent synthetic and fuel–efficient oils currently available that can provide better gas mileage, longer service life, and in some cases better engine protection. These benefits do not come without a few hitches, however; the main one being the price of synthetic oils, which is three or four times the price per quart of conventional oil.

GENERAL INFORMATION AND MAINTENANCE

Synthetic oil is not for every truck and every type of driving, so you should consider your engine's condition and your type of driving. Also, check your truck's warranty conditions regarding the use of synthetic oils.

FUEL

All of these vehicles must use lead–free gasoline. It is recommended that these vehicles avoid the use of premium grade gasoline. This is due to the engine control system being calibrated towards the use of regular grade gasoline. The use of premium grades may actually cause driveability problems. Also, Mazda advises against the use of gasoline with an octane rating lower than 87 which can cause persistant and heavy knocking, and may cause internal engine damage.

OPERATION IN FOREIGN COUNTRIES

If you plan to drive your vehicle outside the United States or Canada, there is a possibility that fuels will be too low in anti–knock quality and could produce engine damage. It is wise to consult with local authorities upon arrival in a foreign country to determine the best fuels available.

Engine Oil

The SAE grade number indicates the viscosity of the engine oil, or its ability to lubricate under a given temperature. The lower the SAE grade number, the lighter the oil; the lower the viscosity, the easier it is to crank the engine in cold weather.

The API (American Petroleum Institute) designation indicates the classification of engine oil for use under given operating conditions. Only oils designated for Service SG should be used to provide maximum engine protection. Both the SAE grade number and the API designation can be found on the container.

➡ **Non-detergent or straight mineral oils should not be used.**

Oil viscosities should be chosen from those oils recommended for the lowest anticipated temperatures during the oil change interval.

OIL LEVEL CHECK

▶ See Figures 121, 122 and 123

When checking the oil level, it is best that the oil be at operating temperature, although checking the level immediately after stopping will give a false reading because all of the oil will not have drained back into the crankcase. Be sure that the truck is on a level surface, allowing tine for all of the oil to drain back into the crankcase.
1. Open the hood and locate the engine oil dipstick.
2. Remove the dipstick and wipe it clean with a rag.
3. Insert the dipstick fully into the tube and remove it again. Hold the dipstick horizontal and read the level on the dipstick. The level should be between the **F** (Full) and **L** (Low) marks. If the oil level is at or below the **L** mark, sufficient oil should be added to restore the level to the proper place. Oil is added through the capped opening in the top of the valve cover. See the section on "Oil and Fuel Recommendations" for the proper viscosity and oil to use.

4. Replace the dipstick and check the level after adding oil. Be careful not to overfill the crankcase.

OIL & FILTER CHANGE

▶ See Figures 123 thru 132

> ※※ **CAUTION**
>
> The EPA warns that prolonged contact with used engine oil may cause a number of skin disorders, including cancer! You should make every effort to minimize your exposure to used engine oil. Protective gloves should be worn when changing the oil. Wash your hands and any other exposed skin areas as soon as possible after exposure to used engine oil. Soap and water, or waterless hand cleaner should be used.

The engine oil and filter should always be changed together. To skip an oil filter change is to leave a quart of contaminated oil in the engine. Engine oil and filter should be changed according to the schedule in the Maintenance Intervals chart. Under conditions such as:
- Driving in dusty conditions
- Continuous trailer pulling or RV use
- Extensive or prolonged idling
- Extensive short trip operation in freezing temperatures (when the engine is not thoroughly warmed up)
- Frequent long runs at high speeds and high ambient temperatures
- Stop-and-go service, such as delivery trucks

the oil change interval and filter replacement interval should be cut in half. Operation of the engine in severe conditions, such as a dust storm, may require an immediate oil and filter change.

To change the engine oil and filter, the truck should be parked on a level surface and the engine should be at operating temperature. This is to ensure that foreign matter will be drained away with the oil and not left behind in the engine to form sludge, which will happen if the engine is drained cold. Oil that is slightly brownish when drained is a good sign that the contaminants are being drained away. You should have available a container that will hold at least five quarts, a wrench to fit the oil drain plug, a spout for pouring in new oil and some rags to clean up the inevitable mess. If the filter is being replaced, you will also need a band wrench to fit the filter.

1. Position the truck on a level surface and set the parking brake or block the wheels. Slide a drain pan under the oil drain plug.
2. From under the truck, loosen, but do not remove the oil drain plug. Cover your hand with a heavy rag or glove and slowly unscrew the drain plug. Push the plug against the threads to prevent oil from leaking past the threads.

> ※※ **CAUTION**
>
> The engine oil will be hot. Keep your arms, face and hands away from the oil as it drains out.

3. As the plug comes to the end of the threads, whisk it away from the hole, letting the oil drain into the pan, which hopefully is still under the drain plug.

Fig. 121 The oil level dipstick on the 3.0L engine is located on the right side of the engine

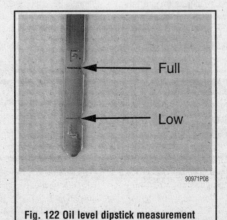

Fig. 122 Oil level dipstick measurement

Fig. 123 Pour the engine oil, using a funnel, through the oil fill hole located on the right side valve cover—MPV

1-40 GENERAL INFORMATION AND MAINTENANCE

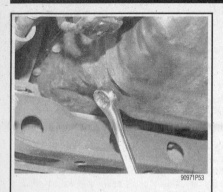

Fig. 124 Using a box wrench, loosen the oil pan drain plug

Fig. 125 Unscrew the plug by hand. Keep an inward pressure on the plug as you unscrew it, so the oil won't escape until you pull the plug away to drain the oil

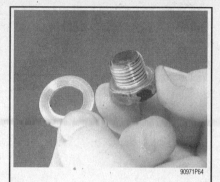

Fig. 126 Clean up the oil pan drain plug and always replace the crush washer

Fig. 127 An oil filter can be removed with a variety of tools. One tool that is good to use is the filter wrench . . .

Fig. 128 . . . another good tool is an end cap filter wrench used along with a ratchet and extension

Fig. 129 Remove the oil filter. Be careful not to make a big mess

Fig. 130 Be sure to clean the gasket mounting surface of any dirt before installing a new oil filter

Fig. 131 Before installing a new oil filter, lightly coat the rubber gasket with clean engine oil

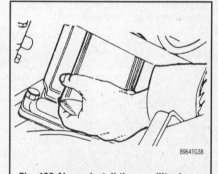

Fig. 132 Always install the new filter by hand (an oil filter wrench will usually lead to overtightening)

This method usually avoids the messy task of reaching into a tub full of hot, dirty oil to retrieve a usually elusive drain plug. Crawl out from under the truck and wait for the oil to drain.

4. When the oil is drained, install the drain plug.
5. Change the engine oil filter as necessary or desired. Loosen the filter with a band wrench and spin the filter off by hand. Be careful of the one quart of hot, dirty oil that inevitably overflows the filter.
6. Coat the rubber gasket on a new filter with engine oil and install the filter. Screw the filter onto the mounting stud and tighten according to the directions on the filter.
7. Refill the engine with the specified amount of clean engine oil. Be sure to use the proper viscosity. Pour the oil in through the capped opening.
8. Run the engine for several minutes, checking for oil pressure and leaks. Check the level of the oil and add if necessary.

➡ Take the drained oil, in a suitable container, to your local service station, or a special designated oil recycling station for disposal.

Manual Transmission

FLUID RECOMMENDATIONS

API GL-4 or GL-5 80W-90 or 75W-90 gear lube is recommended.

LEVEL CHECK

1. Clean the dirt away from the area of the filler plug.
2. Jack the truck if necessary and support it on jackstands, make sure the vehicle is level.

GENERAL INFORMATION AND MAINTENANCE 1-41

3. Remove the filler plug from the case. The filler plug is the one on the side of the case. Do not remove the plug from the bottom of the case unless you wish to drain the transmission.

4. If lubricant flows from the area of the filler plug as it is removed, the level is satisfactory. If lubricant does not flow from the filler hole when the plug is removed, add enough of the specified lubricant to bring the level to the bottom of the filler hole with the truck in a level position.

DRAIN & REFILL

Pickups and MPV

▶ See Figures 133, 134 and 135

The same procedure is used for both 4-speed and 5-speed transmission, but note that the 5-speed transmissions have two filler plugs and two drain plugs. Thus, to drain the 5-speed, both drain plugs must be removed, and to fill it, both filler plugs must be removed. The truck should be parked on a level surface, and the transmission should be at normal operating temperature (oil hot).

1. With the truck parked on a level surface (parking brake applied), place a pan of at least four quarts capacity under the transmission drain plug(s).

2. Remove the filler plug(s) to provide a vent; this will speed the draining process.

3. On the pickup, remove the drain plug(s) and allow the old oil to drain into the pan. On the MPV, remove plugs labeled "A", "B" and "C" in the illustration. Apply upward pressure on the drain plug until you can pull it out and direct the fluid into the drain pan.

4. Clean the drain plug(s) thoroughly and replace. On the pickup, tighten to 15–20 ft. lbs. (20–27 Nm), if you have a torque wrench handy; otherwise, just snug the plug or plugs in. Overtightening will strip the aluminum threads in the case. On the MPV, coat all plug threads with sealant and install them. Tighten plug "A" to 29–43 ft. lbs. (39–58 Nm), plug "B" and "D" to 18–29 ft. lbs. (24–39 Nm).

5. Add lubricant through the filler plug(s) until it comes right up to the edge of the filler hole. Use the recommended gear oil. It usually comes in a squeeze bottle with a nozzle attached to the cap, but you can use a squeeze bulb or suction gun for additions.

6. Install the filler plug(s). Tightening torque is 15–20 ft. lbs. (20–27 Nm) for pickups; 18–29 ft. lbs. (24–39 Nm) for MPV. Check for leaks after the vehicle has been driven for a few miles.

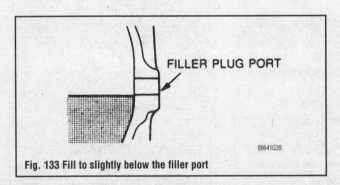

Fig. 133 Fill to slightly below the filler port

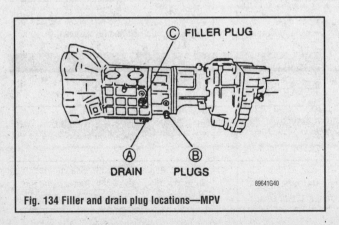

Fig. 134 Filler and drain plug locations—MPV

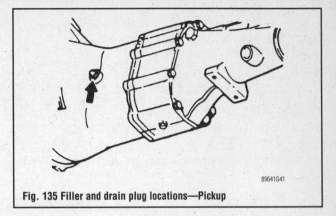

Fig. 135 Filler and drain plug locations—Pickup

Navajo

Drain and refill the transmission daily if the vehicle has been operating in water. All you have to do is remove the drain plug which is located at the bottom of the transmission. Allow all the lubricant to run out before replacing the plug. Replace with the recommended gear oil. If you are experiencing hard shifting and the weather is very cold, use a lighter weight fluid in the transmission. If you don't have a pressure gun to install the oil, use a suction gun.

Automatic Transmission

In 1987, a 3-speed lock-up torque converter version of the 3N71B was introduced. Its designation is L3N71B. It was offered only on the B2200. Aside from the lock-up feature, the main difference lies in the fact that there are no band adjustments. This unit continues on the B2200 through 1988.

Also in 1987 a 4-speed unit was offered as an option on the B2200 and as standard on the B2600. It is designated L4N71B. This unit continues through 1988.

In 1989, only the L4N71B is offered on pick-ups.

From 1990, the N4A-HL 4-speed is used on all pickups.

The MPV is equipped with a 4-speed unit designated N4A-HL. An optional, electronically controlled 4-speed is available. Its designation is R4A-EL.

4-Wheel Drive MPVs use the R4AX-EL, which is essentially the same unit as the R4A-EL.

The Navajo uses a Ford A4LD 4-speed unit.

FLUID RECOMMENDATION

All automatics use Dexron®II ATF.

LEVEL CHECK

▶ See Figures 136, 137 and 138

1. Drive the vehicle for several miles to bring the fluid level to operating temperature.

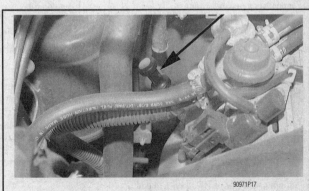

Fig. 136 Automatic transmission fluid level dipstick is located on the right side of the engine in the rear of the compartment—MPV shown

1-42 GENERAL INFORMATION AND MAINTENANCE

2. Park the truck on a level surface.
3. Put the automatic transmission in PARK. Leave the engine running.
4. Remove the dipstick from the tube and wipe it clean.
5. Reinsert the dipstick so that it is fully seated.
6. Remove the dipstick and note the reading. If the fluid is at or below the **Add** mark, add sufficient fluid to bring the level to the **Full** mark. Do not overfill the transmission. Overfilling will lead to fluid aeration.

DRAIN AND REFILL

♦ See Figures 138 thru 146

The automatic transmission fluid is a long lasting type, and Mazda does not specify that it need ever be changed. However, if you have brought the truck used, driven it in water deep enough to reach the transmission, or used the truck for trailer pulling or delivery service, you may want to change the fluid and filter. It is a good idea to measure the amount of fluid drained from the transmission, and to use this as a guide when refilling. Some parts of the transmission, such as the torque converter, will not drain completely, and using the dry refill capacity listed in the Capacities Chart may lead to overfilling.

1. Drive the truck until it is at normal operating temperature.
2. If a hoist is not being used, park the truck on a level surface, block the wheels, and set the parking brake. If you raise the truck on jackstands, check to see that it is reasonably level before draining the transmission.
3. There is no drain plug, so the transmission pan must be removed to drain the fluid. Carefully remove the screws from the pan and lower the pan at the corner. Allow the fluid to drain into a suitable container. After the fluid has drained, remove the pan.

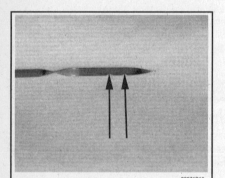

Fig. 137 Automatic transmission fluid level dipstick measurement-Fluid level should be between the two arrows

Fig. 138 Use a funnel (to avoid spills) and add the required type and amount of ATF

Fig. 139 Using a 10 mm socket, remove the transmission oil pan bolts except for one mounting bolt at each corner

Fig. 140 After removing the rear corner pan bolts, carefully lower the pan by hand, making sure that a minimal amount of trans fluid spills out and soaks your arm

Fig. 141 Lower down the transmission oil pan

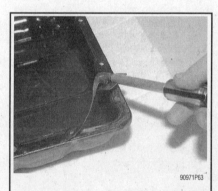

Fig. 142 Once the pan is removed, discard the old gasket and insure that the mating surfaces are clean

Fig. 143 Remove the O-magnet from the bottom of the transmission oil pan and clean it of any dirt or small metallic fragments using a towel

Fig. 144 Remove the four transmission oil filter mounting bolts using a 10mm socket

Fig. 145 Remove the transmission filter from the trans valve body. Be sure that the O-ring seal also comes out with the filter

GENERAL INFORMATION AND MAINTENANCE

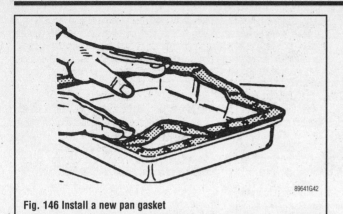

Fig. 146 Install a new pan gasket

4. The filter is bolted to the lower valve body. Remove the filter attaching bolts and remove. Clean it thoroughly in solvent, if it is a screen type, allow it to air dry completely, and replace it. Tightening torque for the attaching bolts is only 24–36 inch lbs. (3–4 Nm), so be careful not to overtighten them.

5. Remove the old gasket and install a new one. The pan may be cleaned with solvent, if desired. After cleaning, allow the pan to air dry thoroughly. Do not use a rag to dry it, or you risk leaving bits of lint in the pan that will clog the transmission fluid passages. When the pan is completely dry, replace it, and tighten the bolts in a circular pattern, working from the center outward. Tighten gently to 40–60 inch lbs. (5–7 Nm).

6. Refill the transmission. Fluid is added through the dipstick tube. This process is considerably easier if you have a funnel and a long tube to pour through. Add three quarts (2.8 liters) of fluid initially.

7. After adding fluid, start the engine and allow it to idle. Shift through all gear positions slowly to allow the fluid to fill all the hydraulic passages, and return the shift lever to Park. Do not race the engine.

8. Run the engine at fast idle to allow the fluid to reach operating temperature. Place the selector lever at **N** or **P** and check the fluid level. It should be above the **L** mark on the hot side of the dipstick. If necessary, add enough fluid to bring the level between the **L** and **F** marks. Do not overfill the transmission. Overfilling will cause foaming, fluid loss, and plate slippage.

9. After a few days of running, check the pan bolts. They will probably have loosened as the gasket has shrunk. Retorque them.

Transfer Case

FLUID RECOMMENDATION

Pickups and MPV

Use API GL-5, SAE 75W-80 gear oil for year-round use when refilling or adding fluid to the transfer case.

Navajo

Use Dexron®II automatic transmission fluid when refilling or adding fluid to the transfer case.

LEVEL CHECK

Position the vehicle on level ground. Remove the transfer case fill plug (the upper plug) located on the rear of the transfer case. The fluid level should be up to the fill hole. If lubricant doesn't run out when the plug is removed, add lubricant until it does run out and then replace the fill plug.

DRAIN AND REFILL

The transfer case is serviced at the same time and in the same manner as the transmission. Clean the area around the filler and drain plugs and remove the filler plug on the side of the transfer case. Remove the drain plug on the bottom of the transfer case and allow the lubricant to drain completely. Clean and install the drain plug. Add the proper lubricant.

Front and Rear Drive Axle

FLUID RECOMMENDATIONS

API GL-4 or GL-5 80W-90 or 75W-90 gear lube is recommended.

LEVEL CHECK

▶ See Figures 147 thru 152

The drive axle fluid level is checked from underneath the truck.
1. Clean the dirt and grease away from the area of the filler (top) plug.
2. Remove the filler plug. The lubricant level should be even with the bottom of the filler plug hole.
3. If lubricant is required, use only the specified type. It will probably have to be pumped in through the filler hole. Hypoid SAE 90 lubricant usually does not pour very well.

➡On Navajo models with the front locking differential, add 2 oz. of friction modifier Ford part No. EST-M2C118-A. On Navajo models with the rear locking differential, use only locking differential fluid Ford part No. ESP-M2C154-A or its equivalent, and add 4 oz. of friction modifier Ford part No. EST-M2C118-A.

DRAIN AND REFILL

Pickups and MPV

▶ See Figures 147, 152 thru 158

These models use a removable carrier axle which has a drain and fill plug.

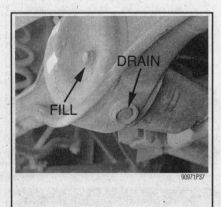

Fig. 147 Location of the drain and fill plugs on the rear differential case

Fig. 148 Wipe the area clean around the differential oil fill plug before opening to prevent any dirt from entering

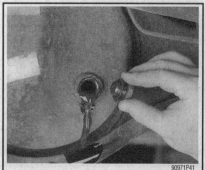

Fig. 149 After loosening the differential oil fill plug using a 24 mm box wrench, remove the plug and allow any excess fluid to drain into a container

1-44 GENERAL INFORMATION AND MAINTENANCE

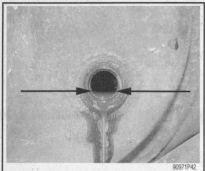

Fig. 150 After removing the fill plug, look inside the hole and inspect the fluid level, if it is overfilled, allow it to drain to the correct level

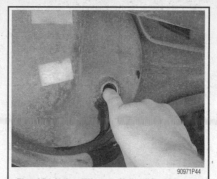

Fig. 151 If the differentail fluid level is not visible, stick your finger into the hole and check to see if the level reaches the bottom level of the filler hole

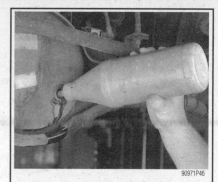

Fig. 152 Fill the differential housing with the correct amount and type of gear oil until fluid starts to trickle out

Fig. 153 Be careful not to lose the fill plug washer, however, inspect the condition of the washer and replace it if necessary

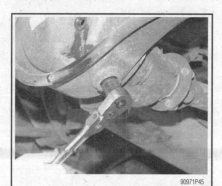

Fig. 154 Using a 24 mm socket, loosen the differential housing fluid drain plug

Fig. 155 Remove the differential drain plug and allow the fluid to drain in an approved container

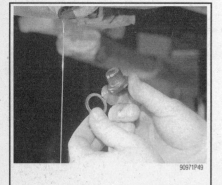

Fig. 156 Be careful not to lose the differential drain plug washer, inspect and replace if necessary

Fig. 157 Most differential drain plugs have a small magnet in the center to catch any small metallic fragments in the fluid

Fig. 158 Always inspect the air vent tube or cap (located on the top of the differential housing) for obstructions and clean if necessary

1. Jack the front and/or rear of the vehicle and support it with jackstands.
2. Position a suitable container under the axle drain plug. Remove the fill plug to provide a vent.
3. Remove the drain plug and allow the lubricant to drain out.

➡ Do not confuse the drain and fill plugs. The drain lug is magnetic to attract fine particles of metal which are inevitably present.

4. Clean the magnetic drain plug.
5. Install the drain plug.
6. Fill the rear axle with the specified amount and type of fluid. Install the filler plug.
7. Lower the truck to the ground and drive the truck, checking for leaks after the fluid is warm.

Navajo

Remove the filler plug. Remove the oil with a suction gun. Refill the axle housings with the SAE 80W/90 gear oil. Be sure and clean the area around the drain plug before removing the plug.

Cooling System

◆ See Figure 159

✱✱ CAUTION

Never remove the radiator cap under any conditions while the engine is running! Failure to follow these instructions could result in dam-

GENERAL INFORMATION AND MAINTENANCE 1-45

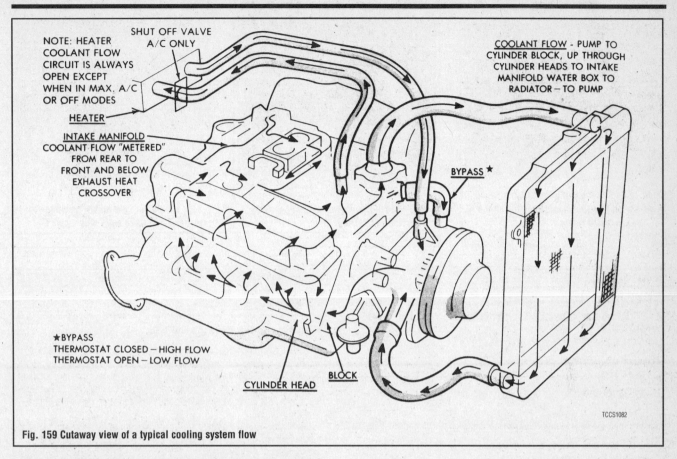

Fig. 159 Cutaway view of a typical cooling system flow

age to the cooling system or engine and/or personal injury. To avoid having scalding hot coolant or steam blow out of the radiator, use extreme care when removing the radiator cap from a hot radiator. Wait until the engine has cooled, then wrap a thick cloth around the radiator cap and turn it slowly to the first stop. Step back while the pressure is released from the cooling system. When you are sure the pressure has been released, press down on the radiator cap (still have the cloth in position) turn and remove the radiator cap.

At least once every 2 years, the engine cooling system should be inspected, flushed, and refilled with fresh coolant. If the coolant is left in the system too long, it loses its ability to prevent rust and corrosion. If the coolant has too much water, it won't protect against freezing.

The pressure cap should be looked at for signs of age or deterioration. Fan belt and other drive belts should be inspected and adjusted to the proper tension. (See checking belt tension).

Hose clamps should be tightened, and soft or cracked hoses replaced. Damp spots, or accumulations of rust or dye near hoses, water pump or other areas, indicate possible leakage, which must be corrected before filling the system with fresh coolant.

FLUID RECOMMENDATIONS

▶ See Figures 160 and 161

A 50/50 mixture of ethylene glycol and water for year round use. Use a good quality antifreeze that is safe for use with aluminum components.

LEVEL CHECK

▶ See Figures 162, 163 and 164

Most vehicles are equipped with a coolant reservoir tank (expansion tank) connected to the radiator by a small hose. When the engine is cold, look through the plastic tank; the fluid level should be between the FULL and Low lines. If the level is too low, remove the reservoir fill cover and add the proper mix of coolant until it reaches the FULL mark.

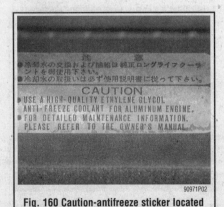

Fig. 160 Caution-antifreeze sticker located on the radiator—MPV

Fig. 161 Sticker showing quantity and coolant type located on the radiator cross-member

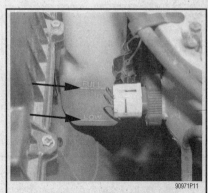

Fig. 162 Coolant overflow tank full and low level measurements

1-46 GENERAL INFORMATION AND MAINTENANCE

Fig. 163 Pull the cap and suction tube out of the coolant overflow tank to top off the cooling system

Fig. 164 Fill the engine coolant overflow tank to the FULL line

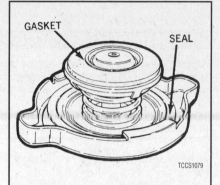

Fig. 165 Be sure the rubber gasket on the radiator cap has a tight seal

CHECK THE RADIATOR CAP

▶ See Figure 165

While you are checking the coolant level, check the radiator cap for a worn or cracked gasket. It the cap doesn't seal properly, fluid will be lost and the engine will overheat.

Worn caps should be replaced with a new one.

CLEAN RADIATOR OF DEBRIS

▶ See Figure 166

Periodically clean any debris — leaves, paper, insects, etc. — from the radiator fins. Pick the large pieces off by hand. The smaller pieces can be washed away with water pressure from a hose.

Carefully straighten any bent radiator fins with a pair of needle nose pliers. Be careful — the fins are very soft. Don't wiggle the fins back and forth too much. Straighten them once and try not to move them again.

DRAIN, REFILL & FLUSHING

▶ See Figures 167, 168, 169 and 170

Completely draining and refilling the cooling system every two years at least will remove accumulated rust, scale and other deposits. Coolant in late model trucks is a 50/50 mixture of ethylene glycol (make sure it is safe for aluminum components) and water for year round use. Use a good quality antifreeze with water pump lubricants, rust inhibitors and other corrosion inhibitors along with acid neutralizers.

1. Drain the existing antifreeze and coolant. It may be necessary to remove a splash shield, on some models, to gain access to the drain at the bottom of

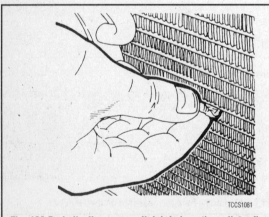

Fig. 166 Periodically remove all debris from the radiator fins

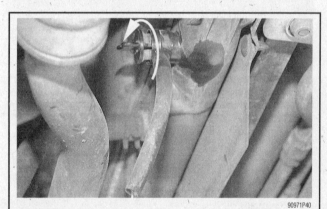

Fig. 167 Turn the radiator draincock counterclockwise to open, draining the engine coolant

Fig. 168 Pouring coolant into the radiator until it reaches just below the filler neck

Fig. 169 To refill the system with coolant, remove the radiator cap. NEVER remove the cap if the system is HOT

Fig. 170 Cooling systems should be pressure tested for leaks periodically

GENERAL INFORMATION AND MAINTENANCE 1-47

the radiator. Open the radiator and engine drain petcocks, or disconnect the bottom radiator hose, at the radiator outlet.

✽✽ CAUTION

When draining the coolant, keep in mind that cats and dogs are attracted by the ethylene glycol antifreeze, and are quite likely to drink any that is left in an uncovered container or in puddles on the ground. This will prove fatal in sufficient quantity. Always drain the coolant into a sealable container. Coolant should be reused unless it is contaminated or several years old.

➥Before opening the radiator petcock, spray it with some penetrating lubricant.

2. Close the petcock or reconnect the lower hose and fill the system with water.
3. Add a can of quality radiator flush.
4. Idle the engine until the upper radiator hose gets hot.
5. Drain the system again.
6. Repeat this process until the drained water is clear and free of scale.
7. Close all petcocks and connect all the hoses.
8. If equipped with a coolant recovery system, flush the reservoir with water and leave empty.
9. Determine the capacity of your coolant system (see capacities specifications). Add a 50/50 mix of quality antifreeze (ethylene glycol) and water to provide the desired protection to the level of the radiator filler port. Fill the coolant reservoir to the correct level. Install the radiator cap.
10. Run the engine to operating temperature.
11. Stop the engine. Allow the engine to cool and check the coolant level in the radiator. Add more coolant if necessary to the radiator.
12. Check the level of protection with an antifreeze tester, replace the cap and check for leaks.

Brake Master Cylinder

FLUID RECOMMENDATIONS

Use only a good quality brake fluid meeting DOT-3 specifications.

LEVEL CHECK

▶ See Figures 171, 172, 173 and 174

Check the level of the fluid in the brake master cylinder at the specified interval or more often.
1. Park the truck on a level surface.
2. If the vehicle is not equipped with a see through reservoir, clean all dirt from the area of the master cylinder reservoir cover.
3. Remove the top from the master cylinder reservoir. Be careful when doing this. Brake fluid that is dripped on painted surfaces will quickly destroy the paint.

4. The level should be maintained approximately $FR1/4-½ in. below the top of the reservoir. On see-through reservoirs, the fluid level should be level with the FULL mark.
5. If brake fluid is needed, use only a good quality brake fluid meeting DOT-3 specifications.
6. If necessary, add fluid to maintain the proper level and replace the top on the master cylinder reservoir securely.

➥If the fluid level is constantly low, it would be a good idea to look into the matter. This is a good indication of problems elsewhere in the system.

Clutch Master Cylinder

FLUID RECOMMENDATIONS

Use only a good quality brake fluid meeting DOT-3 specifications.

LEVEL CHECK

▶ See Figure 175

The clutch master cylinder is located at the left rear corner of the engine compartment. Check the level in the clutch master cylinder in the same manner as the brake master cylinder. The level should be kept at the MAX. mark or approximately 1/4 in. from the top of the cylinder. Use brake fluid in the clutch system. Be sure that the vehicle is on a level surface.

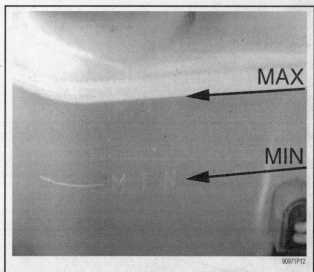

Fig. 171 Level of brake fluid as measured on the reservoir

Fig. 172 Be sure to wipe off the brake fluid reservoir cap and surrounding area before opening to ensure that no contaminants enter the hydraulic system

Fig. 173 Remove the brake fluid reservoir cap being careful not to allow any contaminants to enter the system

Fig. 174 Top off or fill the brake fluid reservoir to the MAX level

1-48 GENERAL INFORMATION AND MAINTENANCE

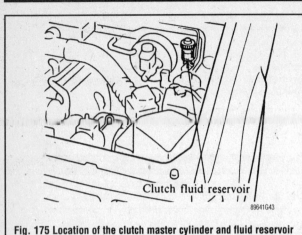

Fig. 175 Location of the clutch master cylinder and fluid reservoir

Power Steering Pump

FLUID RECOMMENDATIONS

All Pickups and MPV models use either DEXRON II Automatic Transmission Fluid (ATF) or ATF Type F. Ensure that the fluid added to the power steering pump reservoir is new and clean.

LEVEL CHECK

▶ See Figures 176, 177 and 178

The power steering reservoir is usually located on the left side of the engine. A dipstick is part of the cap. With the fluid hot—10 minutes of driving—the level should be between the F and L or Max/Min marks. If necessary, add the correct amount and type of fluid.

Manual Steering Gear

FLUID RECOMMENDATIONS

80W-30 gear oil is recommended.

LEVEL CHECK

1. Clean the area around the plug and remove the plug from the top of the gear housing.
2. The oil level should just reach the plug hole.
3. If necessary, add 80W-90 gear oil until the fluid is at the proper level.
4. Install the plug.

Chassis Maintenance

Complete chassis maintenance should include an inspection of all rubber suspension bushings, lubrication of all body hinges, as well inspection of the front suspension components and steering linkage.

Chassis greasing should be performed every 7,500 miles (12,000 km) for most vehicles. Greasing can be performed with a commercial pressurized grease gun or at home by using a hand-operated grease gun. Wipe the grease fittings clean before greasing in order to prevent the possibility of forcing any dirt into the component.

Check the steering, driveshaft and front suspension components for grease fittings. There are far less grease points on the modern vehicle chassis than there were on vehicles of yesteryear.

A water resistant long life grease that meets Mazda's specification should be used for all chassis greasing applications.

Body Lubrication and Maintenance

LOCK CYLINDERS

Apply graphite lubricant sparingly thought the key slot. Insert the key and operate the lock several times to be sure that the lubricant is worked into the lock cylinder.

DOOR HINGES AND HINGE CHECKS

Spray a silicone lubricant on the hinge pivot points to eliminate any binding conditions. Open and close the door several times to be sure that the lubricant is evenly and thoroughly distributed.

TAILGATE

Spray a silicone lubricant on all of the pivot and friction surfaces to eliminate any squeaks or binds. Work the tailgate to distribute the lubricant

BODY DRAIN HOLES

Be sure that the drain holes in the doors and rocker panels are cleared of obstruction. A small screwdriver, or wire hanger, can be used to clear them of any debris.

Wheel Bearings

REPACKING

Rear Wheel Drive

▶ See Figures 179 and 180

➡ The MPV has integral hub/bearing assemblies. No service, other than replacement, is possible.

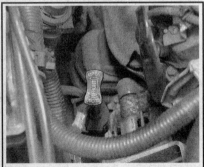

Fig. 176 Power steering fluid level dipstick location in the lower right side of the engine compartment—MPV

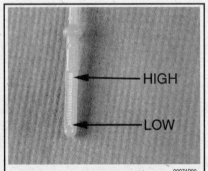

Fig. 177 Power steering fluid level should measure between the HIGH and LOW markings on the dipstick+

Fig. 178 Place long-nosed funnel into the power steering pump dipstick/filler tube and pour in the correct amount of fluid—MPV

GENERAL INFORMATION AND MAINTENANCE   1-49

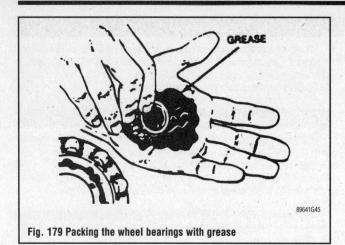

Fig. 179 Packing the wheel bearings with grease

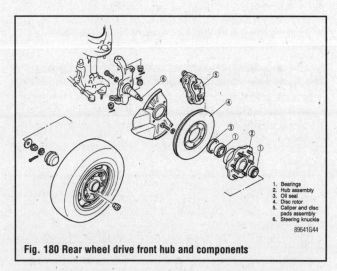

Fig. 180 Rear wheel drive front hub and components

➥Sodium-based grease is not compatible with lithium-based grease. Read the package labels and be careful not to mix the two types. If there is any doubt as to the type of grease used, completely clean the old grease from the bearing and hub before replacing.

Before handling the bearings, there are a few things that you should remember to do and not to do. **Remember to DO the following:**
- Remove all outside dirt from the housing before exposing the bearing.
- Treat a used bearing as gently as you would a new one.
- Work with clean tools in clean surroundings.
- Use clean, dry canvas gloves, or at least clean, dry hands.
- Clean solvents and flushing fluids are a must.
- Use clean paper when laying out the bearings to dry.
- Protect disassembled bearings from rust and dirt. Cover them up.
- Use clean rags to wipe bearings.
- Keep the bearings in oil-proof paper when they are to be stored or are not in use.
- Clean the inside of the housing before replacing the bearing. **Do NOT do the following:**
- Don't work in dirty surroundings.
- Don't use dirty, chipped or damaged tools.
- Try not to work on wooden work benches or use wooden mallets.
- Don't handle bearings with dirty or moist hands.
- Do not use gasoline for cleaning; use a safe solvent.
- Do not spin-dry bearings with compressed air. They will be damaged.

- Do not spin dirty bearings.
- Avoid using cotton waste or dirty cloths to wipe bearings.
- Try not to scratch or nick bearing surfaces.
- Do not allow the bearing to come in contact with dirt or rust at any time.
 1. Remove the wheel hub/bearing assemblies from the vehicle.
 2. Thoroughly clean the bearings and inside of the hub with a nonflammable solvent. Allow them to air dry.
 3. Inspect the bearings for wear, damage, heat discoloration or other signs of fatigue. If they are at all suspect, replace them.
 4. Pack the inside of the hub with clean wheel bearing grease until it is flush packed.
 5. Pack each bearing with clean grease, making sure that it is thoroughly packed. Special devices are sold for packing bearings. They are inexpensive and readily available. If you don't have one, just make certain that the bearing is as full of grease as possible by working it in with your fingers.
 6. Install the wheel hub/bearing assemblies.

4—Wheel Drive

▶ See Figures 179 and 181

The MPV has sealed bearing assembly mounted in the steering knuckle. No service, other than replacement, is possible.

➥For front locking hub removal and service, see Section 7.

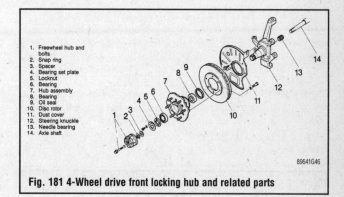

Fig. 181 4-Wheel drive front locking hub and related parts

PICKUPS

 1. Remove the wheel hub and bearing assemblies. See section 7.
 2. Thoroughly clean the bearings and inside of the hub with a nonflammable solvent. Allow them to air dry.
 3. Inspect the bearings for wear, damage, heat discoloration or other signs of fatigue. If they are at all suspect, replace them.
 4. Pack the inside of the hub with clean wheel bearing grease until it is flush packed.
 5. Pack each bearing with clean grease, making sure that it is thoroughly packed. Special devices are sold for packing bearings. They are inexpensive and readily available. If you don't have one, just make certain that the bearing is as full of grease as possible by working it in with your fingers.
 6. Install the wheel hub/bearing assemblies.

NAVAJO

 1. Remove the locking hub assembly.
 2. Remove the wheel bearing assemblies.
 3. Lubricate the bearings with disc brake wheel bearing grease. Clean all old grease from the hub. Pack the cones and rollers. If a bearing packer is not available, work as much lubricant as possible between the rollers and the cages.
 4. Install the wheel bearing assemblies.
 5. Install the locking hub assembly.

1-50 GENERAL INFORMATION AND MAINTENANCE

TRAILER TOWING

General Recommendations

Your vehicle was primarily designed to carry passengers and cargo. It is important to remember that towing a trailer will place additional loads on your vehicles engine, drive train, steering, braking and other systems. However, if you decide to tow a trailer, using the prior equipment is a must.

Local laws may require specific equipment such as trailer brakes or fender mounted mirrors. Check your local laws.

Trailer Weight

The weight of the trailer is the most important factor. A good weight-to-horsepower ratio is about 35:1, 35 lbs. of Gross Combined Weight (GCW) for every horsepower your engine develops. Multiply the engine's rated horsepower by 35 and subtract the weight of the vehicle passengers and luggage. The number remaining is the approximate ideal maximum weight you should tow, although a numerically higher axle ratio can help compensate for heavier weight.

Hitch (Tongue) Weight

♦ See Figure 182

Calculate the hitch weight in order to select a proper hitch. The weight of the hitch is usually 9–11% of the trailer gross weight and should be measured with the trailer loaded. Hitches fall into various categories: those that mount on the frame and rear bumper, the bolt-on type, or the weld-on distribution type used for larger trailers. Axle mounted or clamp-on bumper hitches should never be used.

Check the gross weight rating of your trailer. Tongue weight is usually figured as 10% of gross trailer weight. Therefore, a trailer with a maximum gross weight of 2000 lbs. will have a maximum tongue weight of 200 lbs. Class I trailers fall into this category. Class II trailers are those with a gross weight rating of 2000–3000 lbs., while Class III trailers fall into the 3500–6000 lbs. category. Class IV trailers are those over 6000 lbs. and are for use with fifth wheel trucks, only.

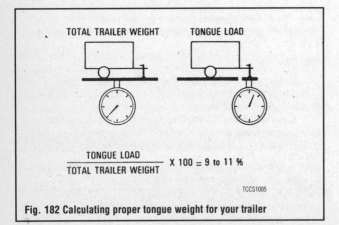

Fig. 182 Calculating proper tongue weight for your trailer

When you've determined the hitch that you'll need, follow the manufacturer's installation instructions, exactly, especially when it comes to fastener torques. The hitch will subjected to a lot of stress and good hitches come with hardened bolts. Never substitute an inferior bolt for a hardened bolt.

Engine

One of the most common, if not THE most common, problems associated with trailer towing is engine overheating. If you have a cooling system without an expansion tank, you'll definitely need to get an aftermarket expansion tank kit, preferably one with at least a 2 quart capacity. These kits are easily installed on the radiator's overflow hose, and come with a pressure cap designed for expansion tanks.

Aftermarket engine oil coolers are helpful for prolonging engine oil life and reducing overall engine temperatures. Both of these factors increase engine life. While not absolutely necessary in towing Class I and some Class II trailers, they are recommended for heavier Class II and all Class III towing. Engine oil cooler systems usually consist of an adapter, screwed on in place of the oil filter, a remote filter mounting and a multi-tube, finned heat exchanger, which is mounted in front of the radiator or air conditioning condenser.

Transmission

An automatic transmission is usually recommended for trailer towing. Modern automatics have proven reliable and, of course, easy to operate, in trailer towing. The increased load of a trailer, however, causes an increase in the temperature of the automatic transmission fluid. Heat is the worst enemy of an automatic transmission. As the temperature of the fluid increases, the life of the fluid decreases.

It is essential, therefore, that you install an automatic transmission cooler. The cooler, which consists of a multi-tube, finned heat exchanger, is usually installed in front of the radiator or air conditioning compressor, and hooked in-line with the transmission cooler tank inlet line. Follow the cooler manufacturer's installation instructions.

Select a cooler of at least adequate capacity, based upon the combined gross weights of the vehicle and trailer.

Cooler manufacturers recommend that you use an aftermarket cooler in addition to, and not instead of, the present cooling tank in your radiator. If you do want to use it in place of the radiator cooling tank, get a cooler at least two sizes larger than normally necessary.

→A transmission cooler can, sometimes, cause slow or harsh shifting in the transmission during cold weather, until the fluid has a chance to come up to normal operating temperature. Some coolers can be purchased with or retrofitted with a temperature bypass valve which will allow fluid flow through the cooler only when the fluid has reached above a certain operating temperature.

Handling A Trailer

Towing a trailer with ease and safety requires a certain amount of experience. It's a good idea to learn the feel of a trailer by practicing turning, stopping and backing in an open area such as an empty parking lot.

TOWING THE VEHICLE

♦ See Figure 183

→Mazda recommends that a flat bed tow service be utilized. If a flat bed tow truck is not available, tow the vehicle with the rear wheels lifted, front hubs unlocked (4WD models) and the steering wheel locked in the straight ahead position using a clamping device designed for towing. DO NOT lock the steering wheel using only the ignition key activated steering lock, as damage to the steering column can occur.

Your vehicle can be towed forward with the driveshaft connected as long as you do not exceed 50 miles in distance and 35 MPH in speed. The transmission must always be placed in the **N** (neutral) position. Severe damage to the transmission can occur if these limits are exceeded. However, if the vehicle must be towed faster than 35 MPH and/or further than 50 miles, disconnect and support the driveshaft. With the driveshaft disconnected, do not exceed 55 MPH. If your vehicle has to be towed backward and is a 4WD model, unlock the front axle driving hubs, to prevent the front differential from rotating and place the transfer case in neutral. Also clamp the steering wheel on all models, in the straight ahead position with a clamping device designed for towing service.

✱✱ WARNING

Do not attach chains to the bumpers or bracketing. All attachments should be made to structural members. Safety chains should also be used. If you are flat towing, remember that the power steering and power brake assists will not work with the engine OFF.

GENERAL INFORMATION AND MAINTENANCE 1-51

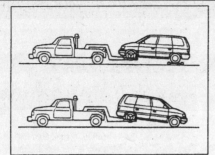
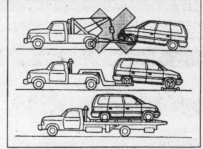

Fig. 183 Proper towing configurations

JUMP STARTING A DEAD BATTERY

♦ See Figure 184

Whenever a vehicle is jump started, precautions must be followed in order to prevent the possibility of personal injury. Remember that batteries contain a small amount of explosive hydrogen gas which is a by-product of battery charging. Sparks should always be avoided when working around batteries, especially when attaching jumper cables. To minimize the possibility of accidental sparks, follow the procedure carefully.

※※ CAUTION

NEVER hook the batteries up in a series circuit or the entire electrical system will go up in smoke, including the starter!

Vehicles equipped with a diesel engine may utilize two 12 volt batteries. If so, the batteries are connected in a parallel circuit (positive terminal to positive terminal, negative terminal to negative terminal). Hooking the batteries up in parallel circuit increases battery cranking power without increasing total battery voltage output. Output remains at 12 volts. On the other hand, hooking two 12 volt batteries up in a series circuit (positive terminal to negative terminal, positive terminal to negative terminal) increases total battery output to 24 volts (12 volts plus 12 volts).

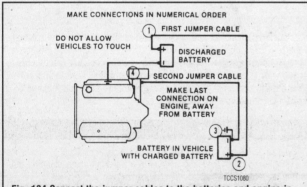

Fig. 184 Connect the jumper cables to the batteries and engine in the order shown

Jump Starting Precautions

- Be sure that both batteries are of the same voltage. Vehicles covered by this manual and most vehicles on the road today utilize a 12 volt charging system.
- Be sure that both batteries are of the same polarity (have the same terminal, in most cases NEGATIVE grounded).
- Be sure that the vehicles are not touching or a short could occur.
- On serviceable batteries, be sure the vent cap holes are not obstructed.
- Do not smoke or allow sparks anywhere near the batteries.
- In cold weather, make sure the battery electrolyte is not frozen. This can occur more readily in a battery that has been in a state of discharge.
- Do not allow electrolyte to contact your skin or clothing.

Jump Starting Procedure

1. Make sure that the voltages of the 2 batteries are the same. Most batteries and charging systems are of the 12 volt variety.
2. Pull the jumping vehicle (with the good battery) into a position so the jumper cables can reach the dead battery and that vehicle's engine. Make sure that the vehicles do NOT touch.
3. Place the transmissions/transaxles of both vehicles in Neutral (MT) or P (AT), as applicable, then firmly set their parking brakes.

➡If necessary for safety reasons, the hazard lights on both vehicles may be operated throughout the entire procedure without significantly increasing the difficulty of jumping the dead battery.

4. Turn all lights and accessories OFF on both vehicles. Make sure the ignition switches on both vehicles are turned to the OFF position.
5. Cover the battery cell caps with a rag, but do not cover the terminals.
6. Make sure the terminals on both batteries are clean and free of corrosion or proper electrical connection will be impeded. If necessary, clean the battery terminals before proceeding.
7. Identify the positive (+) and negative (−) terminals on both batteries.
8. Connect the first jumper cable to the positive (+) terminal of the dead battery, then connect the other end of that cable to the positive (+) terminal of the booster (good) battery.
9. Connect one end of the other jumper cable to the negative (−) terminal on the booster battery and the final cable clamp to an engine bolt head, alternator bracket or other solid, metallic point on the engine with the dead battery. Try to pick a ground on the engine that is positioned away from the battery in order to minimize the possibility of the 2 clamps touching should one loosen during the procedure. DO NOT connect this clamp to the negative (−) terminal of the bad battery.

※※ CAUTION

Be very careful to keep the jumper cables away from moving parts (cooling fan, belts, etc.) on both engines.

10. Check to make sure that the cables are routed away from any moving parts, then start the donor vehicle's engine. Run the engine at moderate speed for several minutes to allow the dead battery a chance to receive some initial charge.
11. With the donor vehicle's engine still running slightly above idle, try to start the vehicle with the dead battery. Crank the engine for no more than 10 seconds at a time and let the starter cool for at least 20 seconds between tries. If the vehicle does not start in 3 tries, it is likely that something else is also wrong or that the battery needs additional time to charge.
12. Once the vehicle is started, allow it to run at idle for a few seconds to make sure that it is operating properly.
13. Turn ON the headlights, heater blower and, if equipped, the rear defroster of both vehicles in order to reduce the severity of voltage spikes and subsequent risk of damage to the vehicles' electrical systems when the cables are disconnected. This step is especially important to any vehicle equipped with computer control modules.
14. Carefully disconnect the cables in the reverse order of connection. Start with the negative cable that is attached to the engine ground, then the negative

1-52 GENERAL INFORMATION AND MAINTENANCE

cable on the donor battery. Disconnect the positive cable from the donor battery and finally, disconnect the positive cable from the formerly dead battery. Be careful when disconnecting the cables from the positive terminals not to allow the alligator clips to touch any metal on either vehicle or a short and sparks will occur.

JACKING

♦ See Figure 185

✳✳ CAUTION

On models equipped with an under chassis mounted spare tire, remove the tire, wheel or tire carrier from the vehicle before it is placed in a high lift position in order to avoid sudden weight release from the chassis.

Your vehicle was supplied with a jack for emergency road repairs. This jack is fine for changing a flat tire or other short term procedures not requiring you to go beneath the vehicle. If it is used in an emergency situation, carefully follow the instructions provided either with the jack or in your owner's manual. Do not attempt to use the jack on any portions of the vehicle other than specified by the vehicle manufacturer. Always block the diagonally opposite wheel when using a jack.

A more convenient way of jacking is the use of a garage or floor jack. You may use the floor jack to jack the vehicle from under the differential/axle, body pinchwelds (MPV only) or frame/crossmember. Be sure and block the diagonally opposite wheel to prevent the vehicle from moving.

Never place the jack under the radiator, engine or transmission components. Severe and expensive damage will result when the jack is raised. Additionally, never jack under the floorpan or bodywork; the metal will deform.

Whenever you plan to work under the vehicle, you must support it on jackstands or ramps. Never use cinder blocks or stacks of wood to support the vehicle, even if you're only going to be under it for a few minutes. Never crawl under the vehicle when it is supported only by the tire-changing jack or other floor jack.

When supporting the vehicle on jackstands, position them under the front pinch welds just behind the front wheels (MPV and Pick-up) or the frame rail/bracket close to the front, behind each front wheel (Navajo). The rear jackstands should be placed under the rear pinch welds just ahead of the rear

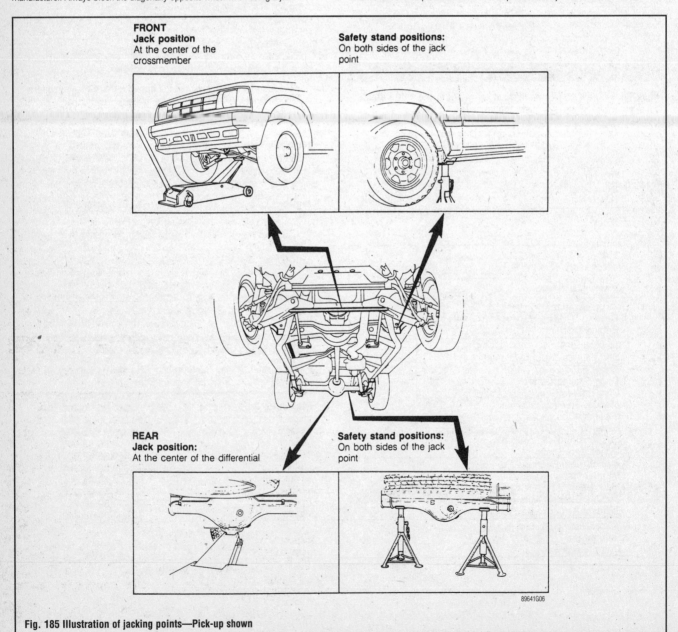

Fig. 185 Illustration of jacking points—Pick-up shown

GENERAL INFORMATION AND MAINTENANCE 1-53

wheels (MPV), both rear spring shackles (Navajo) or under the axle on each side of the differential (Pick-up). Be careful not to touch the rear shock absorber mounting brackets.

➡ **Always position a block of wood or small rubber pad on top of the jack or jackstand to protect the lifting point's finish when lifting or supporting the vehicle.**

Small hydraulic, screw, or scissors jacks are satisfactory for raising the vehicle. Drive-on trestles or ramps are also a handy and safe way to both raise and support the vehicle. Be careful though, some ramps may be too steep to drive your vehicle onto without scraping the front bottom panels. Never support the vehicle on any suspension member (unless specifically instructed to do so by a repair manual) or by an underbody panel.

Jacking Precautions

The following safety points cannot be overemphasized:
- Always block the opposite wheel or wheels to keep the vehicle from rolling off the jack.
- When raising the front of the vehicle, firmly apply the parking brake.
- When the drive wheels are to remain on the ground, leave the vehicle in gear to help prevent it from rolling.
- Always use jackstands to support the vehicle when you are working underneath. Place the stands beneath the vehicle's jacking brackets. Before climbing underneath, rock the vehicle a bit to make sure it is firmly supported.

MAINTENANCE INTERVALS

Component	Serivice	Interval

Intervals are for every: miles/Km or miles/months/Km

Under heavy duty operating conditions, the maintenance intervals should be halved. Service even more frequently if continuous severe operation is experienced.

Component	Serivice	Interval
B2200 (2.2L) 1987-93		
Engine		
Engine Oil and Oil Filter (1)	replace	7,500/7.5/12,000
Choke System (Carb only)	clean	15,000/15/24,000
Idle Switch (Carb only)	inspect	15,000/15/24,000
Drive Belts (1987-88)	inspect	15,000/15/24,000
Drive Belts (1989-93)	inspect	30,000/30/48,000
Air Cleaner Element (2)	replace	30,000/30/48,000
Timing Belt	replace	30,000/48,000
Oxygen Sensor	replace	80,000/128,000
EGR Control Valve	replace	60,000/60/96,000
PCV Valve	inspect	60,000/60/96,000
Emission Hoses & Tubes	replace	60,000/60/96,000
HAC Air Filter	replace	60,000/60/96,000
Ignition System		
Spark Plugs	replace	30,000/30/48,000
Ignition Timing	inspect	60,000/60/96,000
Fuel System		
Idle Speed	inspect	15,000/15/24,000
Fuel Lines	inspect	30,000/30/48,000
Fuel Filter	replace	30,000/30/48,000
Cooling System		
Cooling System	inspect	15,000/15/24,000
Engine Coolant	replace	30,000/30/48,000
Chassis and Body		
Brake Lines & Connections	inspect	30,000/30/48,000
Brake Fluid Level	replace	30,000/30/48,000
Drum Brakes (Rear)	inspect	30,000/30/48,000
Disc Brakes (Front)	inspect	30,000/30/48,000
Manual Steering Gear Oil	inspect	30,000/30/48,000
Steering Operations & Gear Housing	inspect	30,000/30/48,000
Steering Linkage, Tie Rod Ends & arms	inspect	30,000/30/48,000
Suspension Ball Joints (Front)	inspect	30,000/30/48,000
Upper Arm Shafts	lube	30,000/30/48,000
Front Wheel Bearings	lube	30,000/30/48,000
Manual Transmission Oil	replace	60,000/60/96,000
Rear Axle Oil	replace	60,000/60/96,000
Body & Chassis Nuts and Bolts	Tighten	30,000/30/48,000
Exhaust System Heat Shield	inspect	30,000/30/48,000
Air Conditioning System		
Refrigerant	inspect	annually
Compressor	operation	annually

(1) We recommend that the oil filter be replaced with each oil change.
(2) Replace as required under heavy duty driving conditions.

1-54 GENERAL INFORMATION AND MAINTENANCE

MAINTENANCE INTERVALS

Component	Service	Interval

Intervals are for every: miles/Km or miles/months/Km

Under heavy duty operating conditions, the maintenance intervals should be halved. Service even more frequently if continuous severe operation is experienced.

B2600 (2.6L) 1987-88

Component	Service	Interval
Engine		
Engine Oil and Oil Filter (1)	replace	7,500/7.5/12,000
Choke System	clean	30,000/30/48,000
Jet Valve Clearance	Adjust	15,000/15/24,000
Drive Belts	inspect	15,000/15/24,000
Air Cleaner Element (2)	replace	30,000/30/48,000
Oxygen Sensor	replace	50,000/50/80,000
EGR Control Valve	replace	60,000/60/96,000
PCV Valve	inspect	50,000/50/80,000
Emission Hoses & Tubes	replace	60,000/60/96,000
HAC Air Filter	replace	60,000/60/96,000
Intake Temperature Control System	inspect	50,000/50/80,000
Secondary Air System	inspect	50,000/50/80,000
Vacuum Control System Air Filters	replace	50,000/50/80,000
Throttle Position System	inspect	15,000/15/24,000
Ignition System		
Spark Plugs	replace	30,000/30/48,000
Ignition Timing	inspect	50,000/50/80,000
Ignition Cables	replace	50,000/50/80,000
Distributor Spark Advance System	inspect	50,000/50/80,000
Fuel System		
Idle Speed	inspect	15,000/15/24,000
Fuel Lines	inspect	30,000/30/48,000
Fuel Filter	replace	30,000/30/48,000
Cooling System		
Cooling System	inspect	15,000/15/24,000
Engine Coolant	replace	30,000/30/48,000
Chassis and Body		
Brake Lines & Connections	inspect	15,000/15/24,000
Brake Fluid Level	replace	30,000/30/48,000
Clutch Pedal	inspect	15,000/15/24,000
Drum Brakes (Rear)	inspect	30,000/30/48,000
Disc Brakes (Front)	inspect	15,000/15/24,000
Manual Steering Gear Oil	inspect	15,000/15/24,000
Steering Operations & Gear Housing	inspect	30,000/30/48,000
Steering Linkage, Tie Rod Ends & Arms	inspect	15,000/15/24,000
Suspension Ball Joints (Front)	inspect	30,000/30/48,000
Upper Arm Shafts	lube	30,000/30/48,000
Front Wheel Bearings	lube	30,000/30/48,000
Manual Transmission Oil	replace	30,000/30/48,000
Front and/or Rear Axle Oil	replace	30,000/30/48,000
Body & Chassis Nuts and Bolts	Tighten	15,000/15/24,000
Exhaust System Heat Shield	inspect	30,000/30/48,000
Drive Shaft Dust Boots	inspect	30,000/30/48,000
Propeller Shaft Joints	lube	30,000/30/48,000
Automatic Transmission Fluid	replace	30,000/30/48,000
Transfer Case Oil	replace	30,000/30/48,000
Air Conditioning System		
Refrigerant	inspect	annually
Compressor	operation	annually

(1) We recommend that the oil filter be replaced with each oil change.
(2) Replace as required under heavy duty driving conditions.

GENERAL INFORMATION AND MAINTENANCE 1-55

MAINTENANCE INTERVALS

Component	Serivice	Interval
Intervals are for every: miles/Km or miles/months/Km		
Under heavy duty operating conditions, the maintenance intervals should be halved. Service even more frequently if continuous severe operation is experienced.		
B2600i (2.6L) 1989-93		
Engine		
Engine Oil and Oil Filter (1)	replace	7,500/7.5/12,000
Drive Belts	inspect	30,000/30/48,000
Air Cleaner Element	replace	30,000/30/48,000
Oxygen Sensor	replace	80,000/80/128,000
PCV Valve	inspect	50,000/50/80,000
Emission Hoses & Tubes	replace	60,000/60/96,000
Ignition System		
Spark Plugs	replace	30,000/30/48,000
Ignition Timing	inspect	60,000/60/96,000
Fuel System		
Idle Speed	inspect	15,000/15/24,000
Fuel Lines	inspect	30,000/30/48,000
Fuel Filter	replace	60,000/60/96,000
Cooling System		
Cooling System	inspect	15,000/15/24,000
Engine Coolant	replace	30,000/30/48,000
Chassis and Body		
Brake Lines & Connections	inspect	30,000/30/48,000
Brake Fluid Level	replace	30,000/30/48,000
Drum Brakes (Rear)	inspect	60,000/60/96,000
Disc Brakes (Front)	inspect	30,000/30/48,000
Steering Operations & Gear Housing	inspect	30,000/30/48,000
Steering Linkage, Tie Rod Ends & Arms	inspect	30,000/30/48,000
Suspension Ball Joints (Front)	inspect	30,000/30/48,000
Upper Arm Shafts	lube	30,000/30/48,000
Front Wheel Bearings	lube	30,000/30/48,000
Manual Transmission Oil	replace	60,000/60/96,000
Transfer Oil	replace	52,000/52/84,000
Driveshaft Dust Bolts	inspect	30,000/30/48,000
Propeller Shaft Joints	lube	15,000/15/24,000
Rear Axle Oil, Front Axle Oil	replace	60,000/60/96,000
Bolts and Nuts on Chassis and Body	Tighten	30,000/30/48,000
Exhaust System Heat Shield	inspect	30,000/30/48,000
Air Conditioning System		
Refrigerant	inspect	annually
Compressor	operation	annually

(1) We recommend that the oil filter be replaced with each oil change.
(2) Replace as required under heavy duty driving conditions.

GENERAL INFORMATION AND MAINTENANCE

MAINTENANCE INTERVALS

Component	Service	Interval

Intervals are for every: miles/Km or miles/months/Km

Under heavy duty operating conditions, the maintenance intervals should be halved. Service even more frequently if continuous severe operation is experienced.

MPV (2.6L) 1989-93

Component	Service	Interval
Engine		
Engine Oil and Oil Filter (1)	replace	7,500/7.5/12,000
Drive Belts	inspect	30,000/30/48,000
Air Cleaner Element	replace	30,000/30/48,000
Oxygen Sensor	replace	80,000/80/128,000
PCV Valve	inspect	50,000/50/80,000
Emission Hoses & Tubes	replace	80,000/128,000
Ignition System		
Spark Plugs	replace	30,000/30/48,000
Ignition Timing	inspect	60,000/60/96,000
Fuel System		
Idle Speed	inspect	15,000/15/24,000
Fuel Lines	inspect	30,000/30/48,000
Fuel Filter	replace	60,000/60/96,000
Cooling System		
Cooling System	inspect	15,000/15/24,000
Engine Coolant	replace	30,000/30/48,000
Chassis and Body		
Brake Lines & Connections	inspect	30,000/30/48,000
Brake Fluid Level	replace	30,000/30/48,000
Drum Brakes (Rear)	inspect	30,000/30/48,000
Steering Operations & Gear Housing	inspect	30,000/30/48,000
Front Suspension Ball Joints	inspect	30,000/30,48,000
Front Axle Oil (2WD) Rear Axle Oil (4WD, 2WD)	replace	60,000/60/96,000
Manual Transmission Oil	replace	60,000/60,96,000
Transfer Case Oil (4WD)	replace	60,000/60/96,000
Driveshaft Dust Boots (4WD)	inspect	30,000/30/48,000
Propeller Shaft Bolts (4WD)	lube	15,000/15/24,000
Bolts and Nuts on Chassis and Body	Tighten	30,000/30/48,000
Exhaust System Heat Shield	inspect	30,000/30/48,000
All Locks and Hinges	lube	7,500/7/12,000
Air Conditioning System		
Refrigerant	inspect	annually
Compressor	operation	annually

(1) We recommend that the oil filter be replaced with each oil change.
(2) Replace as required under heavy duty driving conditions.

GENERAL INFORMATION AND MAINTENANCE

MAINTENANCE INTERVALS

Component	Serivice	Interval
	Intervals are for every: miles/Km or miles/months/Km	
	Under heavy duty operating conditions, the maintenance intervals should be halved. Service even more frequently if continuous severe operation is experienced.	
	MPV (3.0L) 1989-93	
Engine		
Engine Oil and Oil Filter (1)	replace	7,500/7.5/12,000
Drive Belts	inspect	30,000/30/48,000
Air Cleaner Element	replace	30,000/30/48,000
Oxygen Sensor	replace	80,000/80/128,000
Engine Timing Belt	replace	60,000/96,000
PCV Valve	inspect	50,000/50/80,000
Emission Hoses & Tubes	replace	60,000/60/96,000
Ignition System		
Spark Plugs	replace	30,000/30/48,000
Ignition Timing	inspect	60,000/60/96,000
Fuel System		
Idle Speed	inspect	15,000/15/24,000
Fuel Lines	inspect	30,000/30/48,000
Fuel Filter	replace	60,000/60/96,000
Cooling System		
Cooling System	inspect	15,000/15/24,000
Engine Coolant	replace	30,000/30/48,000
Chassis and Body		
Brake Lines & Connections	inspect	30,000/30/48,000
Brake Fluid Level	replace	30,000/30/48,000
Drum Brakes (Rear)	inspect	60,000/60/96,000
Disc Brakes (Front)	inspect	30,000/30/48,000
Front Suspension Ball Joints	inspect	30,000/30,48,000
Front Axle Oil (2WD) Rear Axle Oil (4WD, 2WD)	replace	60,000/60/96,000
Manual Transmission Oil	replace	60,000/60,96,000
Transfer Case Oil (4WD)	replace	60,000/60/96,000
Driveshaft Dust Boots (4WD)	inspect	30,000/30/48,000
Propeller Shaft Bolts (4WD)	lube	15,000/15/24,000
Bolts and Nuts on Chassis and Body	Tighten	30,000/30/48,000
Exhaust System Heat Shield	inspect	30,000/30/48,000
All Locks and Hinges	lube	7,500/7/12,000
Air Conditioning System		
Refrigerant	inspect	annually
Compressor	operation	annually

(1) We recommend that the oil filter be replaced with each oil change.
(2) Replace as required under heavy duty driving conditions.

1-58 GENERAL INFORMATION AND MAINTENANCE

MAINTENANCE INTERVALS

Component	Service	Interval
Intervals are for every: miles/Km or miles/months/Km		
Under heavy duty operating conditions, the maintenance intervals should be halved. Service even more frequently if continuous severe operation is experienced.		
Navajo (4.0L) 1991-93		
Engine & Emission Control Systems		
Engine Oil and Oil Filter (1)	replace	7,500/12,000
Drive Belts	inspect	30,000/48,000
Air Cleaner Element	replace	30,000/48,000
PCV Valve	replace	60,000/96,000
Ignition System		
Spark Plugs	replace	30,000/48,000
Ignition Wires	replace	60,000/96,000
Cooling System		
Cooling System	inspect	annually
Engine Coolant	replace	30,000/36/48,000
Chassis, Body and Other Systems		
Warning Lights & Gauges	inspect	15,000/24,000
Wheel Lug Nut Torque	inspect	7,500/12,000
Rotate Tires	perform	15,000/24,000
Clutch Reservoir Fluid Level	inspect	7,500/12,000
Shift Linkage Cable (AT)	inspect & lube	7,500/12,000
Front Wheel Bearings	inspect & lube	30,000/48,000
Disc Brake System & Caliper Slide Rails	inspect & lube	15,000/24,000
Drum Brake Linings, Lines and Hoses	inspect	30,000/48,000
Exhaust System	inspect	30,000/48,000
Driveshafts w/Fittings	lube	7,500/12,000
Throttle and Kickdown Linkage	lube	30,000/48,000
Rear Prop Shaft Cardan Joint Centering Ball	lube	7,500/12,000
RH Front Axle Slip Joint (4WD)	lube	30,000/48,000
Front Spindle Bearings & Thrust Bearings	inspect & lube	30,000/48,000
Hub Locks (Inspect) 4WD	lube	30,000/48,000
Transfer Case Oil	replace	60,000/96,000
Steering Linkage (w/Fittings)	lube	7,500/12,000
Air Conditioning System		
Refrigerant	inspect	annually
Compressor	operation	annually

(1) We recommend that the oil filter be replaced with each oil change.
(2) Replace as required under heavy duty driving conditions.

GENERAL INFORMATION AND MAINTENANCE

CAPACITIES

Year	Model	Engine Code	Engine Displacement Liters (cc)	Engine Oil with Filter (qts.)	Transmission (pts.) 4-Spd	Transmission (pts.) 5-Spd	Transmission (pts.) Auto.	Transfer Case (pts.)	Drive Axle Front (pts.)	Drive Axle Rear (pts.)	Fuel Tank (gal.)	Cooling System (qts.)
1987	B2200	F2	2.2 (2184)	4.4	3.6	4.2	15.8	—	—	2.6	①	②
	B2600	G54B	2.6 (2555)	③	—	⑤	15.8	4	2.6	3.6	①	⑥
1988	B2200	F2	2.2 (2184)	4.4	3.6	4.2	15.8	—	—	2.6	①	②
	B2600	G54B	2.6 (2555)	③	—	⑤	15.8	4	2.6	3.6	①	⑥
1989	B2200	F2	2.2 (2184)	4.3	3.6	4.2	15.8	—	—	2.6	①	⑥
	B2600i	G6	2.6 (2606)	5.0	—	6.8	15.8	4.2	3.2	3.6	14.8	7.2
	MPV	G6	2.6 (2606)	5.0	—	5.2	18.2	3.2	3.6	3.2	15.9	⑨
	MPV	JE	3.0 (2954)	5.0	—	5.2	18.2	3.2	3.6	3.2	19.6	⑧
1990	B2200	F2	2.2 (2184)	4.3	3.6	4.2	15.8	—	—	2.6	①	⑥
	B2600i	G6	2.6 (2606)	5.0	—	⑨	⑩	4.2	3.2	3.6	14.8	7.2
	MPV	G6	2.6 (2606)	5.0	—	5.2	⑩	3.2	3.6	3.2	15.9	⑦
	MPV	JE	3.0 (2954)	5.0	—	5.2	⑩	3.2	3.6	3.2	19.6	⑧
1991	B2200	F2	2.2 (2184)	4.3	3.6	4.2	15.8	—	—	2.6	①	②
	B2600i	G6	2.6 (2606)	5.0	—	⑨	⑩	4.2	3.2	3.6	14.8	②
	MPV	G6	2.6 (2606)	5.0	—	⑪	⑩	4.2	3.6	3.2	15.9	⑦
	MPV	JE	3.0 (2954)	5.0	—	⑪	⑩	4.2	3.6	3.2	19.6	⑧
	Navajo	X	4.0 (4016)	5.0	—	5.6	⑫	2.5	3.5	5.3	19.3	⑬
1992	B2200	F2	2.2 (2184)	4.3	—	4.2	15.8	—	—	2.6	①	②
	B2600i	G6	2.6 (2606)	5.0	—	⑨	⑩	4.2	3.2	3.6	14.8	②
	MPV	G6	2.6 (2606)	5.0	—	6.0	⑩	4.2	3.6	3.2	15.9	⑦
	MPV	JE	3.0 (2954)	5.0	—	6.0	⑩	4.2	3.6	3.2	19.6	⑧
	Navajo	X	4.0 (4016)	5.0	—	5.6	⑫	2.5	3.5	5.3	19.3	⑬
1993	B2200	F2	2.2 (2184)	4.3	—	4.2	15.8	—	—	2.6	①	②
	B2600i	G6	2.6 (2606)	5.0	—	⑨	⑩	4.2	3.2	3.6	14.8	②
	MPV	G6	2.6 (2606)	5.0	—	6.0	⑩	4.2	3.6	3.2	15.9	⑦
	MPV	JE	3.0 (2954)	5.0	—	6.0	⑩	4.2	3.6	3.2	19.6	⑧
	Navajo	X	4.0 (4016)	5.0	—	5.6	⑫	2.5	3.5	5.3	19.3	⑬

Note: Capacity specifications are approximate figures. If the capacity specifications differ with the specifications listed in the vehicle owner's manual, use the figures in the owner's manual.

Always pour fluids in slowly, being careful not to overfill the system. Top off only where necessary.

① Short bed, Cab Plus-14.8
　Long bed-17.4
② With heater-7.9
　Without heater-7.3
③ 2WD-4.4
　4WD-5.2
④ 2WD-4.0
　4WD-4.8
⑤ 2WD-6.0
　4WD-6.8
⑥ With heater-7.9
　Without heater-7.4
⑦ With manual transmission-7.2
　With automatic transmission-7.6
⑧ With manual transmission-10.1
　With automatic transmission-10.3
⑨ 2WD-6.0
　4WD-6.8
⑩ Hydraulically controlled-15.8
　Electronically controlled-17.2
⑪ 2WD-5.2
　4WD-6.0
⑫ 2WD-19.4
　4WD-20.0
⑬ Standard cooling system, all transmission and air conditioning-8.1
　Super cooling with automatic transmission and air conditioning-8.5

1-60 GENERAL INFORMATION AND MAINTENANCE

ENGLISH TO METRIC CONVERSION: MASS (WEIGHT)

Current mass measurement is expressed in pounds and ounces (lbs. & ozs.). The metric unit of mass (or weight) is the kilogram (kg). Even although this table does not show conversion of masses (weights) larger than 15 lbs, it is easy to calculate larger units by following the data immediately below.

To convert ounces (oz.) to grams (g): multiply th number of ozs. by 28
To convert grams (g) to ounces (oz.): multiply the number of grams by .035

To convert pounds (lbs.) to kilograms (kg): multiply the number of lbs. by .45
To convert kilograms (kg) to pounds (lbs.): multiply the number of kilograms by 2.2

lbs	kg	lbs	kg	oz	kg	oz	kg
0.1	0.04	0.9	0.41	0.1	0.003	0.9	0.024
0.2	0.09	1	0.4	0.2	0.005	1	0.03
0.3	0.14	2	0.9	0.3	0.008	2	0.06
0.4	0.18	3	1.4	0.4	0.011	3	0.08
0.5	0.23	4	1.8	0.5	0.014	4	0.11
0.6	0.27	5	2.3	0.6	0.017	5	0.14
0.7	0.32	10	4.5	0.7	0.020	10	0.28
0.8	0.36	15	6.8	0.8	0.023	15	0.42

ENGLISH TO METRIC CONVERSION: TEMPERATURE

To convert Fahrenheit (°F) to Celsius (°C): take number of °F and subtract 32; multiply result by 5; divide result by 9
To convert Celsius (°C) to Fahrenheit (°F): take number of °C and multiply by 9; divide result by 5; add 32 to total

Fahrenheit (F)		Celsius (C)		Fahrenheit (F)		Celsius (C)		Fahrenheit (F)		Celsius (C)	
°F	°C	°C	°F	°F	°C	°C	°F	°F	°C	°C	°F
−40	−40	−38	−36.4	80	26.7	18	64.4	215	101.7	80	176
−35	−37.2	−36	−32.8	85	29.4	20	68	220	104.4	85	185
−30	−34.4	−34	−29.2	90	32.2	22	71.6	225	107.2	90	194
−25	−31.7	−32	−25.6	95	35.0	24	75.2	230	110.0	95	202
−20	−28.9	−30	−22	100	37.8	26	78.8	235	112.8	100	212
−15	−26.1	−28	−18.4	105	40.6	28	82.4	240	115.6	105	221
−10	−23.3	−26	−14.8	110	43.3	30	86	245	118.3	110	230
−5	−20.6	−24	−11.2	115	46.1	32	89.6	250	121.1	115	239
0	−17.8	−22	−7.6	120	48.9	34	93.2	255	123.9	120	248
1	−17.2	−20	−4	125	51.7	36	96.8	260	126.6	125	257
2	−16.7	−18	−0.4	130	54.4	38	100.4	265	129.4	130	266
3	−16.1	−16	3.2	135	57.2	40	104	270	132.2	135	275
4	−15.6	−14	6.8	140	60.0	42	107.6	275	135.0	140	284
5	−15.0	−12	10.4	145	62.8	44	112.2	280	137.8	145	293
10	−12.2	−10	14	150	65.6	46	114.8	285	140.6	150	302
15	−9.4	−8	17.6	155	68.3	48	118.4	290	143.3	155	311
20	−6.7	−6	21.2	160	71.1	50	122	295	146.1	160	320
25	−3.9	−4	24.8	165	73.9	52	125.6	300	148.9	165	329
30	−1.1	−2	28.4	170	76.7	54	129.2	305	151.7	170	338
35	1.7	0	32	175	79.4	56	132.8	310	154.4	175	347
40	4.4	2	35.6	180	82.2	58	136.4	315	157.2	180	356
45	7.2	4	39.2	185	85.0	60	140	320	160.0	185	365
50	10.0	6	42.8	190	87.8	62	143.6	325	162.8	190	374
55	12.8	8	46.4	195	90.6	64	147.2	330	165.6	195	383
60	15.6	10	50	200	93.3	66	150.8	335	168.3	200	392
65	18.3	12	53.6	205	96.1	68	154.4	340	171.1	205	401
70	21.1	14	57.2	210	98.9	70	158	345	173.9	210	410
75	23.9	16	60.8	212	100.0	75	167	350	176.7	215	414

**DISTRIBUTOR IGNITION
SYSTEM 2-2**
DESCRIPTION AND OPERATION 2-2
DIAGNOSIS AND TESTING 2-3
   SERVICE PRECAUTIONS 2-3
   TESTING PROCEDURES 2-3
   IGNITION COIL SECONDARY
    VOLTAGE (SPARK) TEST 2-3
   PICKUP COIL RESISTANCE 2-3
   EXTERNAL RESISTOR 2-4
   SPARK ADVANCE CONTROL 2-4
ADJUSTMENTS 2-4
   AIR GAP ADJUSTMENT 2-4
IGNITION COIL 2-4
   TESTING 2-4
   REMOVAL & INSTALLATION 2-5
IGNITION MODULE (IGNITER) 2-5
   REMOVAL & INSTALLATION 2-5
DISTRIBUTOR 2-5
   REMOVAL & INSTALLATION 2-5
**DISTRIBUTORLESS IGNITION
SYSTEM 2-6**
GENERAL INFORMATION 2-6
DIAGNOSIS AND TESTING 2-6
   SERVICE PRECAUTIONS 2-6
   PRELIMINARY CHECKS 2-6
   SECONDARY SPARK TEST 2-6
IGNITION COIL PACK 2-6
   TESTING 2-6
   REMOVAL & INSTALLATION 2-7
IGNITION MODULE 2-7
   REMOVAL & INSTALLATION 2-7
CAMSHAFT POSITION (CMP) AND
  CRANKSHAFT POSITION (CKP)
  SENSORS 2-7
FIRING ORDERS 2-7
CHARGING SYSTEM 2-8
GENERAL INFORMATION 2-8
ALTERNATOR PRECAUTIONS 2-9
ALTERNATOR 2-9
   TESTING 2-9
   REMOVAL & INSTALLATION 2-10
REGULATOR 2-10
   REMOVAL & INSTALLATION 2-10
STARTING SYSTEM 2-10
GENERAL INFORMATION 2-10
STARTER 2-11
   TESTING 2-11
   REMOVAL & INSTALLATION 2-12
   SOLENOID REPLACEMENT 2-13
**SENDING UNITS AND
SENSORS 2-13**
COOLANT TEMPERATURE SENDER 2-13
   TESTING 2-13
   REMOVAL & INSTALLATION 2-13
OIL PRESSURE SENDER SWITCH 2-14
   TESTING 2-14
   REMOVAL & INSTALLATION 2-14
LOW OIL LEVEL SENSOR 2-14
   TESTING 2-14
   REMOVAL & INSTALLATION 2-15
TROUBLESHOOTING CHART
  BASIC STARTING SYSTEM
   PROBLEMS 2-16

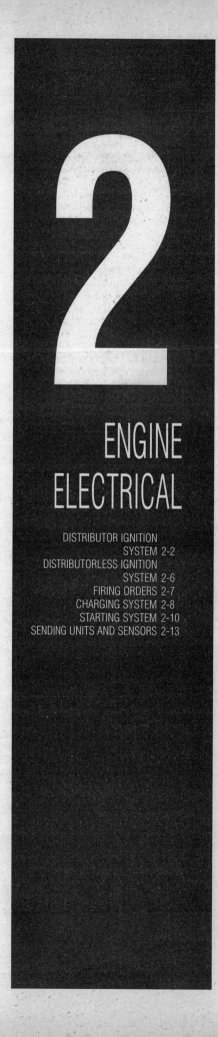

2

ENGINE ELECTRICAL

DISTRIBUTOR IGNITION
SYSTEM 2-2
DISTRIBUTORLESS IGNITION
SYSTEM 2-6
FIRING ORDERS 2-7
CHARGING SYSTEM 2-8
STARTING SYSTEM 2-10
SENDING UNITS AND SENSORS 2-13

2-2 ENGINE ELECTRICAL

DISTRIBUTOR IGNITION SYSTEM

➡ For information on understanding electricity and troubleshooting electrical circuits, please refer to Section 6 of this manual.

Description and Operation

◆ See Figures 1 thru 6

Electronic ignition systems offer many advantages over the old conventional breaker point type ignition system. By eliminating the points, maintenance requirements are greatly reduced. An electronic ignition system is capable of producing much higher voltage which in turn aids in starting, reduces spark plug fouling and provides better emission control. A basic electronic ignition system consists of a distributor with a signal generator, and ignition coil and an electronic igniter. The "signal generator" is used to activate the electronic components of the igniter. It is located in the distributor and consists of three main components; the signal rotor, sometimes call a reluctor (do not confuse with the normal rotor located under the cap), the pickup coil, and the permanent magnet (breaker). The signal rotor revolves with the distributor shaft, while the pickup and the permanent magnet are stationary. As the signal rotor spins, the teeth on it pass a projection leading from the pickup coil. When this happens, voltage is allowed to flow through the system, firing the spark plugs. There is no physical contact and no electrical arcing, hence no need to replace burnt or worn parts. Fuel injected versions of the Pickup, from 1990, and 2.6L MPV models use a distributor mounted crank angle sensor in place of the usual distributor mounted electronic components. The most commonly replaced parts are the rotor and cap, which are still routine maintenance items. Other items, such as the pickup coil and signal rotor, are replaced when they fail. An air gap adjustment is possible only on the 1987-88 2.6L engine. Adjustments are not possible on later models.

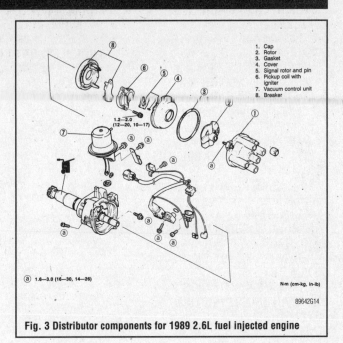

Fig. 3 Distributor components for 1989 2.6L fuel injected engine

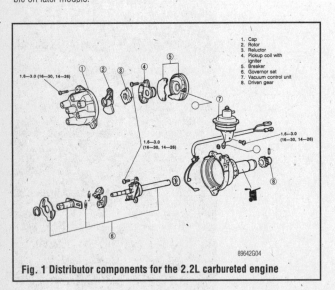

Fig. 1 Distributor components for the 2.2L carbureted engine

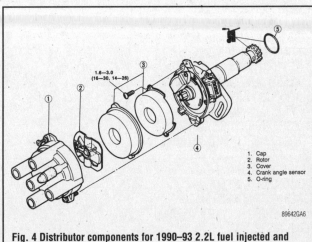

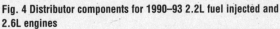

Fig. 4 Distributor components for 1990–93 2.2L fuel injected and 2.6L engines

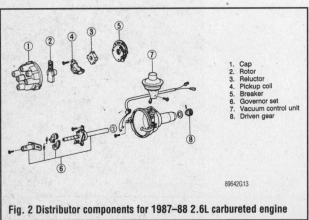

Fig. 2 Distributor components for 1987–88 2.6L carbureted engine

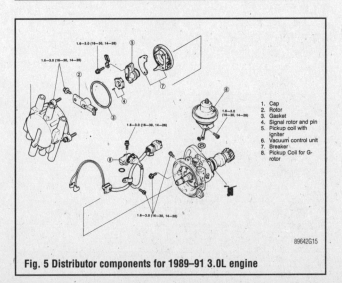

Fig. 5 Distributor components for 1989–91 3.0L engine

ENGINE ELECTRICAL 2-3

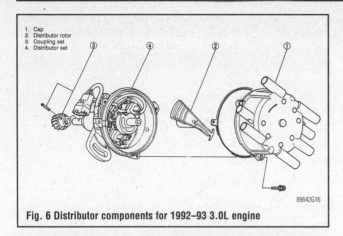

1. Cap
2. Distributor rotor
3. Coupling set
4. Distributor set

Fig. 6 Distributor components for 1992–93 3.0L engine

Diagnosis and Testing

SERVICE PRECAUTIONS

- Always turn the key **OFF** and isolate both ends of a circuit whenever testing for shorts or continuity.
- Never measure voltage or resistance directly at the processor connector.
- Always disconnect solenoids and switches from the harness before measuring for continuity, resistance or energizing by way of a 12-volt source.
- When disconnecting connectors, inspect for damaged or pushed-out pins, corrosion, loose wires, etc. Service if required.

TESTING PROCEDURES

▶ See Figure 7

You will need an accurate ohmmeter, a jumper wire and a test light. Before proceeding with troubleshooting, make sure that all connections are tight and all wiring is intact.

1. Check for spark at the coil high tension lead by removing the lead from the distributor cap and holding it about 1/4 inch (6mm) from the engine block or other good ground. Use a heavy rubber glove or non-conductive clamp, such as a fuse puller or clothes pin, to hold the wire. Crank the engine and check for spark. If a good spark is noted, check the cap and rotor; if the spark is weak or nonexistent, replace the high tension lead, clean and tighten the connections and retest. If a weak spark is still noted, proceed to Step 2.

2. Check the coil primary and secondary resistance. Refer to those procedures later in this section.

3. Next, remove the distributor cap and rotor. Crank the engine until a spoke on the rotor is aligned with the pickup coil contact. Use a flat feeler gauge to check the gap. Gap should be 0.20–0.60mm. The gap is not adjustable (procedure for 1987–88 B2600 air gap adjustment is covered later in this section). On these models (except 1987–88 B2600), gap is corrected by parts replacement.

4. On B2200 carbureted models, check the pickup coil resistance. If resistance is not correct, replace the pickup coil.

5. Finally, test the ignition module. The only way to test the module is to substitute a known good module in its place.

IGNITION COIL SECONDARY VOLTAGE (SPARK) TEST

Crank Mode

▶ See Figure 8

1. Connect a spark tester between the ignition coil wire and a good engine ground.
2. Crank the engine and check for spark at the tester.
3. Turn the ignition switch **OFF**.
4. If no spark occurs, check the following:
   a. Inspect the ignition coil for damage or carbon tracking.
   b. Check that the distributor shaft is rotating when the engine is being cranked.
5. If a spark did occur, check the distributor cap and rotor for damage or carbon tracking. Go to the Ignition Coil Secondary Voltage (Run Mode) Test.

Run Mode

1. Fully apply the parking brake. Place the gear shift lever in Neutral (manual transmission) or Park (automatic transmission).
2. Disconnect the **S** terminal wire at the starter relay. Attach a remote starter switch.
3. Turn the ignition switch to the **RUN** position.
4. Using the remote starter switch, crank the engine and check for spark.
5. Turn the ignition switch **OFF**.
6. If no spark occurred, the problem lies with the wiring harness. Inspect the wiring harness for short circuits, open circuits and other defects.
7. If a spark did occur, the problem is not in the ignition system.

PICKUP COIL RESISTANCE

Unplug the primary ignition wire connector and connect an ohmmeter across the two prongs of the pickup coil connector. Resistance, at 68°F (20°C), should be 900–1,200 ohms.

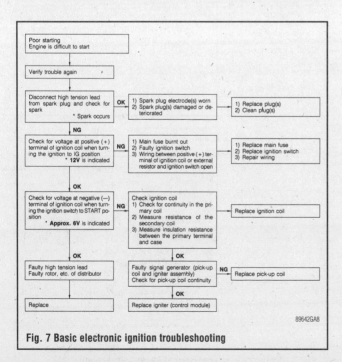

Fig. 7 Basic electronic ignition troubleshooting

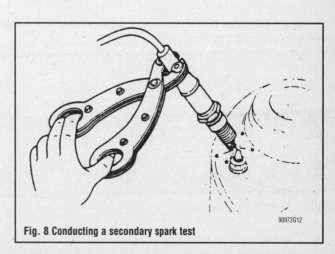

Fig. 8 Conducting a secondary spark test

ENGINE ELECTRICAL

EXTERNAL RESISTOR

▶ See Figure 9

1987–1988 B2600 models are equipped with an external resister. Measure the resistance at the two end contacts on the resistor. Resistance should be 1.0–1.5 ohms.

SPARK ADVANCE CONTROL

Distributors not equipped with a crank angle sensor are equipped with two methods of spark advance. They are, an internal centrifugal advance consisting of governor weights and springs; and an external advance controlled by engine vacuum. Models equipped with fuel injection are equipped with an electronic advance device.

To test the internal advance:
1. Connect a timing light to the engine.
2. Warm up the engine to normal operating temperature.
3. Check that the idle speed and ignition timing are correct.
4. Disconnect and plug the vacuum lines at the vacuum advance.
5. Gradually increase the engine speed while checking the timing mark with the timing light.
6. If the mark move to what seems like an excessive amount of advance, the cause could be a weak governor spring. If the mark moves slowly and seems like an insufficient amount of advance, the cause could be a governor weight or advance cam malfunction.

To test the vacuum advance:
7. Warm the engine to normal operating temperature.
8. Check that the idle speed and ignition timing are correct.
9. Disconnect the vacuum lines from the vacuum control and plug them.
10. Connect a vacuum hand pump to the vacuum control, slowly apply vacuum and check the advance with a timing light.
11. If the spark advance seems sluggish or not operating, replace the vacuum control.

To test fuel injected models equipped with an electronic advance:
12. Connect a timing light and verify that the timing advances with engine acceleration.

Adjustments

For information on timing, valve lash, idle speed and mixture adjustments, refer to Section 1. The only other adjustment possible is the air gap adjustment on the 1987–88 2.6L engine, which is outlined below:

AIR GAP ADJUSTMENT

▶ See Figure 10

Air gap adjustment is possible on 1987–1988 2.6L engines.
1. Using a wrench on the crankshaft pulley nut, turn the distributor shaft until one of the high points on the reluctor is aligned with the pickup coil face.
2. Loosen the set screws and move the pickup coil until the gap between it and the signal rotor is 0.8mm, measured with a brass or plastic feeler gauge.
3. Tighten the set screws and recheck the gap.

Ignition Coil

TESTING

▶ See Figures 11 thru 14

1. Connect an ohmmeter across the coil primary terminals and check resistance on the low scale. Resistance (@ 68°F; 20°C) should be:

Pickups:
- 1987–89 B2200 carbureted: 1.0–1.3 ohms
- 1990–93 B2200 injected: 0.81–0.99 ohms
- 1987–88 B2600: 1.0–1.3 ohms
- 1989 B2600i; Point 1: 0.77–0.95 ohms ; Point 2: 0.9–1.1k ohms
- 1990–93 B2600i: 0.81–0.99 ohms

Fig. 9 Testing resistance of the external resistor

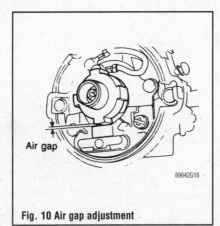

Fig. 10 Air gap adjustment

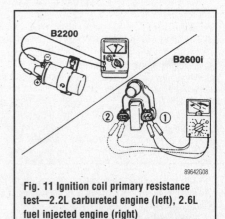

Fig. 11 Ignition coil primary resistance test—2.2L carbureted engine (left), 2.6L fuel injected engine (right)

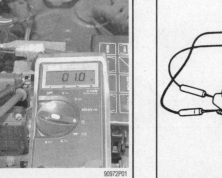

Fig. 12 Checking the primary ignition coil reesistance—3.0L engine

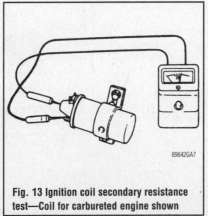

Fig. 13 Ignition coil secondary resistance test—Coil for carbureted engine shown

Fig. 14 Checking the ignition coil secondary resistance—3.0L engine (MPV)

ENGINE ELECTRICAL 2-5

MPV
- 2.6L engine; Point 1: 0.77–0.95 ohms ; Point 2: 0.9–1.1k ohms
- 1989–91 3.0L engine: 0.81–0.99 ohms
- 1992–93 3.0L engine: 0.72—0.88 ohms

2. If resistance is much higher, replace the coil.
3. Check the coil secondary resistance. Connect an ohmmeter across the distributor side of the coil and the coil center tower. Read resistance on the high scale. Resistance at 68°F; 20°C) should be:

Pickups
- 1987–88 B2200: 6–30k ohms
- 1987–88 B2600: 10–20k ohms
- 1989–93 B2200 and B2600i: 6–30k ohms

MPV
- 1989–93 2.6L engine—Point 1: 0.77–0.95 ohms ; Point 2: 0.9–1.1k ohms
- 1989–91 3.0L engine: 0.81–0.99 ohms
- 1992–93 3.0L engine: 0.72—0.88 ohms

4. If resistance is much higher, replace the coil.
5. Disconnect the high tension wire and move the connector leading to the negative terminal over to the metallic connector inside the coil tower. Set the ohmmeter to the X1000 scale. Resistance must be:

Pickups
- 1987–88 B2200: 6–30k ohms
- 1987–88 B2600: 10–20k ohms
- 1989—93 B2200 and B2600i: 6–30k ohms

MPV
- 2.6L engine; Point 1: 0.77–0.95 ohms ; Point 2: 0.9–1.1k ohms
- 1989–91 3.0L engine: 0.81–0.99 ohms
- 1992–93 3.0L engine: 0.72–0.88 ohms

6. You can also check for bad coil insulation by measuring the resistance between the coil (-) primary connection and the metal body (case) of the coil. If resistance is less then 10m ohms, replace the coil. This test may not be entirely satisfactory unless you have a megaohm tester that records 500 volts. If the tests below do not reveal the problem and, especially, if operating the engine at night may produce some bluish sparks around the coil, you may want to remove the coil and have it tested at a diagnostic center.
7. If the coil resistances are not as specified, replace the coil.

If the coil tests out ok, replace the igniter and pickup coil. However, you should make sure before doing this work that there are no basic maintenance problems in the secondary circuit of the system, since it is often impossible to return electrical parts. We suggest that before you replace the igniter and pickup coil, you carefully inspect the cap and rotor for carbon tracks or cracks and disconnect the wires and measure their resistance with an ohmmeter. Resistance should be 16k ohms per 3.28 ft. (1m). Also, check for cracks in the insulation. Replace secondary parts as inspection/testing deems necessary before replacing the igniter and pickup coil.

REMOVAL & INSTALLATION

1. Disconnect the negative battery cable and make sure the ignition switch is off. Remove any protective boots from the top of the coil, if necessary by sliding them back the coil-to-distributor wire.
2. Carefully pull the high tension wire out of the coil, twisting it gently as near as possible to the tower to get it started.
3. Note the routing and colors of the primary wires, and then remove nuts and lockwashers, retaining all parts for installation. On models equipped with plug-in connections, carefully pull the connecter from the coil assembly. Clean the primary terminals, if necessary, to ensure a clean connection. Then, loosen the through bolt or bolts which clamp the coil in place and slide the coil out of its mount. If the mounting bracket is part of the coil, remove the mounting bolts and the coil.
4. Install the new coil in exact reverse order, making sure the primary connections are tight. Ensure also that the coil-to-distributor wire is fully seated in the tower and that the protective boot is fully installed on the outside of the tower.

Ignition Module (Igniter)

REMOVAL & INSTALLATION

1. Disconnect the harness connector from the top of the igniter.
2. Remove the two attaching screws and remove the igniter.
3. Install the new igniter and secure it with the two attaching screws.
4. Connect the harness connector.

Distributor

REMOVAL & INSTALLATION

The Navajo has no distributor. Refer to Distributorless Ignition System, later in this section.

1. Disconnect the negative battery cable.
2. Remove the distributor cap from the distributor, leaving the spark plug wires attached. If spark plug wire removal is necessary to remove the distributor cap, tag the wires prior to removal so they can be reinstalled in the correct position.
3. Disconnect the electrical connectors and vacuum hose(s), if equipped, from the distributor.
4. Mark the position of the rotor in relation to the distributor housing and the position of the distributor housing on the cylinder head.
5. Remove the distributor hold-down bolt(s) and remove the distributor.
6. Check the distributor O-ring for cuts or other damage and replace, if necessary.

To install:

TIMING NOT DISTURBED

▶ See Figure 15

1. Lubricate the distributor O-ring with clean engine oil.
2. Install the distributor with the hold-down bolt(s), aligning the marks that were made during removal. Tighten the hold-down bolt(s) to 14–19 ft. lbs. (19–25 Nm).
3. Connect the electrical connectors and vacuum hose(s), if equipped.
4. Install the distributor cap on the distributor. Connect the spark plug wires, if removed.
5. Connect the negative battery cable. Start the engine and check the ignition timing.

TIMING DISTURBED

▶ See Figure 15

1. Disconnect the spark plug wire from the No. 1 cylinder spark plug and remove the spark plug. Make sure the engine is cool enough to touch, place a finger over the spark plug hole.

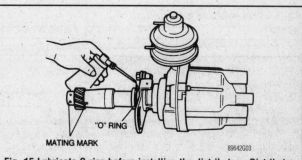

Fig. 15 Lubricate O-ring before installing the distributor—Distributor for carbureted engine shown

2-6 ENGINE ELECTRICAL

2. Turn the crankshaft in the normal direction of rotation until compression is felt at the spark plug hole.
3. Align the mark on the crankshaft pulley with the TDC mark on the timing belt cover.
4. Lubricate the distributor O-ring with clean engine oil.
5. Turn the distributor shaft until the rotor points to the No. 1 spark plug tower on the distributor cap and install the distributor. Install the distributor hold-down bolt(s) and align the distributor housing with the mark made on the cylinder head during removal. Snug the bolt(s).

6. Connect the electrical connectors and vacuum hose(s), if equipped.
7. Install the distributor cap on the distributor. Connect the spark plug wires, if removed.
8. Install the spark plug in the No. 1 cylinder and connect the spark plug wire.
9. Connect the negative battery cable. Start the engine and adjust the ignition timing. Tighten the distributor hold-down bolt(s) to 14–19 ft. lbs. (19–25 Nm) after the timing has been set.

DISTRIBUTORLESS IGNITION SYSTEM

General Information

An Electronic Distributorless Ignition System (EDIS) is used on the Navajo. The system consists of a crankshaft timing sensor (VRS), EDIS ignition module, ignition coil pack, the part of the main computer (ECU) that deals with spark angle and related wiring. The crankshaft timing sensor, used on the Navajo, is a variable reluctance-type sensor, triggered by a trigger wheel machined into the rear of the front crankshaft damper. The signal generated by this sensor is called a Variable Reluctance Sensor signal (VRS). The VRS signal provides the timing information to the EDIS module. The EDIS Module receives the VRS signal from the crankshaft timing sensor. The module processes the VRS signal into Profile Ignition Pickup (PIP) information and transmits it to the ECU (Engine Control Unit). The ECU processes the PIP signal along with signals received from other engine sensors according to a spark advance map programmed into the ECU. Using this information, the ECU produces a Spark Angle Word (SAW) signal which it sends back to the EDIS module. Using the SAW and VRS information, the EDIS switches primary current to the ignition coils like the contact points in a breaker point type ignition system, except that it sequences the spark among the three coils and provides the optimum amount of spark advance and dwell. The EDIS module also sends back to the ECU and Ignition Diagnostic Monitor signal (IDM) which the ECU uses to indicate a failure mode and which is also used to provide an rpm signal to the instrument panel tachometer. The ignition coil pack contains three separate ignition coils which are controlled by the EDIS module through three coil leads. Each ignition coil fires two spark plugs simultaneously; one plug on the compression stroke and one on the exhaust stroke. The spark plug fired on the exhaust stroke uses very little of the coil's stored energy. The majority of the energy is used by the spark plug on the compression stroke. During some EDIS faults, the failure mode effects management (FMEM) portion of the EDIS ignition module will maintain vehicle operation. If the EDIS module does not receive the SAW input, it will result in a fixed spark timing of 10° BTDC (before top dead center). If the EDIS module does not receive the VRS input, synchronization cannot be achieved, and the engine will not start.

Diagnosis and Testing

SERVICE PRECAUTIONS

1. Examine all wiring harnesses and connectors for insulation damage, burned, overheated, loose or broken connections.
2. Be certain that the battery is fully charged and that all accessories are **OFF** during the diagnosis.
   - Always turn the ignition key **OFF** and isolate both ends of a circuit whenever testing for shorts or continuity.
   - Never measure voltage or resistance directly at the processor connector.
   - Always disconnect solenoids and switches from the harness before measuring for continuity, resistance or energizing by way of a 12-volt source.
   - When disconnecting connectors, inspect for damaged or pushed-out pins, corrosion, loose wires, etc. Service if required.

PRELIMINARY CHECKS

1. Visually inspect the engine compartment to ensure that all vacuum lines and spark plug wires are properly routed and securely connected.

SECONDARY SPARK TEST

▶ See Figure 8

1. Remove the spark plug from the engine.
2. Connect the spark plug to the high tension lead.
3. Using insulated pliers, hold the high tension lead and spark plug approximately 0.20–0.39 inch (5–10mm) from a good ground.
4. Crank the engine and verify that there is a strong blue spark.

Ignition Coil Pack

TESTING

Primary and Secondary Circuit Tests

▶ See Figure 16

1. Turn the ignition switch **OFF**, disconnect the battery, then detach the wiring harness connector from the ignition coil to be tested.
2. Check for dirt, corrosion or damage on the terminals.

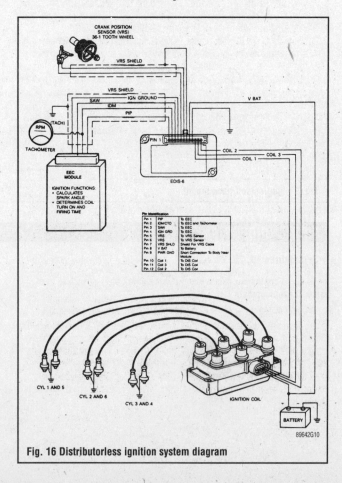

Fig. 16 Distributorless ignition system diagram

ENGINE ELECTRICAL 2-7

PRIMARY RESISTANCE

♦ See Figure 16

1. Use an ohmmeter to measure the resistance between the following terminals on the ignition coil, and note the resistance:
   - B+ to Coil 1
   - B+ to Coil 2
   - B+ to Coil 3

The resistance between all of these terminals should have been between 0.3–1.0 ohms. If the resistance was more or less than this value, the coil should be replaced with a new one.

SECONDARY RESISTANCE

♦ See Figure 16

1. Measure, using the ohmmeter, and note the resistance between each corresponding coil terminal and the two spark plug wire towers on the ignition coil. The coil terminals and plug wires towers are grouped as follows:

4.0L engine
- Terminal 3 (coil 1)—spark plugs 1 and 5
- Terminal 2 (coil 3)—spark plugs 2 and 6
- Terminal 1 (coil 2)—spark plugs 3 and 4

If the resistance for all of the readings was between 6,500–11,500 ohms, the ignition coils are OK. If any of the readings was less than 6,500 ohms or more than 11,500 ohms, replace the corresponding coil pack.

REMOVAL & INSTALLATION

♦ See Figure 17

1. Disconnect the negative battery cable.
2. Disconnect the electrical harness connector from the coil pack.
3. Remove the spark plug wires by squeezing the locking tabs to release the coil boot retainers.
4. Remove the coil pack retaining bolts and remove the coil pack.
5. Place the coil pack in position and install the retaining bolts. Tighten the bolts to 40–62 inch lbs. (4.5–7.0 Nm).

➡ Be sure to place some dielectric compound into each spark plug boot prior to installation of the spark plug wire.

Ignition Module

REMOVAL & INSTALLATION

♦ See Figure 18

1. Disconnect the battery cables and remove the battery.
2. Disconnect the electrical connector at the module.
3. Remove the module retaining bolt and remove the module.
4. Position the rivet on the module into the teardrop shaped mounting hole on the fender apron. Push the module into position and secure it with the mounting bolt. Tighten the mounting bolt to 22–31 inch lbs. (2.5–3.5 Nm).

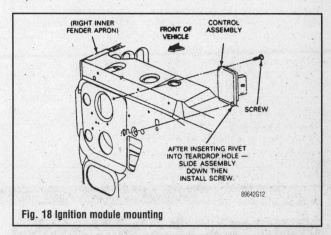

Fig. 18 Ignition module mounting

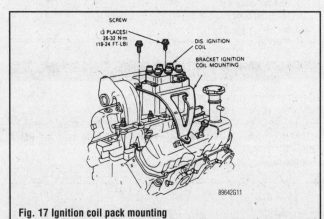

Fig. 17 Ignition coil pack mounting

Camshaft Position (CMP) and Crankshaft Position (CKP) Sensors

For procedures on these sensors, please refer to Section 4 in this manual.

FIRING ORDERS

♦ See Figures 19, 20, 21 and 22

➡ To avoid confusion, remove and tag the spark plug wires one at a time, for replacement.

If a distributor is not keyed for installation with only one orientation, it could have been removed previously and rewired. The resultant wiring would hold the correct firing order, but could change the relative placement of the plug towers in relation to the engine. For this reason it is imperative that you label all wires before disconnecting any of them. Also, before removal, compare the current wiring with the accompanying illustrations. If the current wiring does not match, make notes in your book to reflect how your engine is wired.

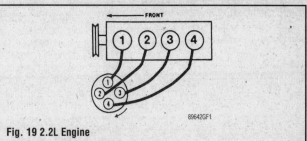

Fig. 19 2.2L Engine
Engine Firing Order: 1-3-4-2
Distributor Rotation: Clockwise

2-8 ENGINE ELECTRICAL

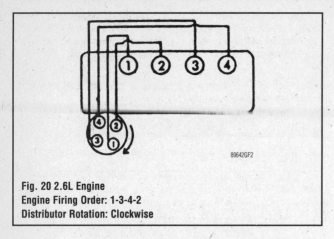

Fig. 20 2.6L Engine
Engine Firing Order: 1-3-4-2
Distributor Rotation: Clockwise

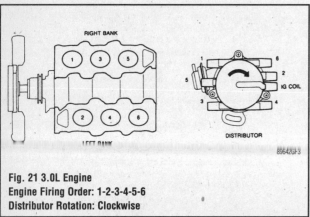

Fig. 21 3.0L Engine
Engine Firing Order: 1-2-3-4-5-6
Distributor Rotation: Clockwise

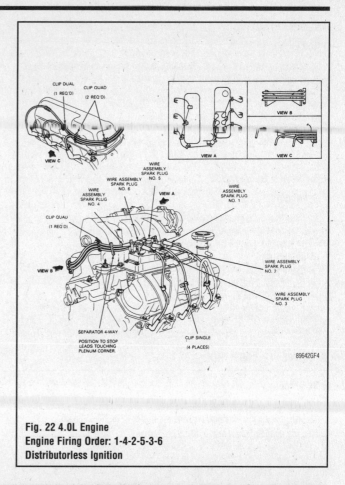

Fig. 22 4.0L Engine
Engine Firing Order: 1-4-2-5-3-6
Distributorless Ignition

CHARGING SYSTEM

General Information

♦ See Figure 23

The automobile charging system provides electrical power for operation of the vehicle's ignition and starting systems and all the electrical accessories. The battery services as an electrical surge or storage tank, storing (in chemical form) the energy originally produced by the engine driven generator. The system also provides a means of regulating generator output to protect the battery from being overcharged and to avoid excessive voltage to the accessories.

The storage battery is a chemical device incorporating parallel lead plates in a tank containing a sulfuric acid/water solution. Adjacent plates are slightly dissimilar, and the chemical reaction of the two dissimilar plates produces electrical energy when the battery is connected to a load such as the starter motor. The chemical reaction is reversible, so that when the generator is producing a voltage (electrical pressure) greater than that produced by the battery, electricity is forced into the battery, and the battery is returned to its fully charged state.

The vehicle's generator is driven mechanically, through V-belts, by the engine crankshaft. It consists of two coils of fine wire, one stationary (the stator), and one movable (the rotor). The rotor may also be known as the armature, and consists of fine wire wrapped around an iron core which is mounted on a shaft. The electricity which flows through the two coils of wire (provided initially by the battery in some cases) creates an intense magnetic field around both rotor and stator, and the interaction between the two fields creates voltage, allowing the generator to power the accessories and charge the battery.

There are two types of generators: the earlier is the direct current (DC) type. The current produced by the DC generator is generated in the armature and carried off the spinning armature by stationary brushes contacting the commutator. The commutator is a series of smooth metal contact plates on the end of the armature. The commutator is a series of smooth metal contact plates on the end of the armature. The commutator plates, which are separated from one another by a very short gap, are connected to the armature circuits so that current will flow in one directions only in the wires carrying the generator output. The generator stator consists of two stationary coils of wire which draw some of the output current of the generator to form a powerful magnetic field and create the interaction of fields which generates the voltage. The generator field is wired in series with the regulator.

Today's vehicles use alternating current generators or alternators, because

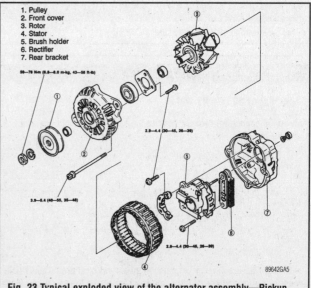

1. Pulley
2. Front cover
3. Rotor
4. Stator
5. Brush holder
6. Rectifier
7. Rear bracket

Fig. 23 Typical exploded view of the alternator assembly—Pickup alternator shown

they are more efficient, can be rotated at higher speeds, and have fewer brush problems. In an alternator, the field rotates while all the current produced passes only through the stator winding. The brushes bear against continuous slip rings rather than a commutator. This causes the current produced to periodically reverse the direction of its flow. Diodes (electrical one-way switches) block the flow of current from traveling in the wrong direction. A series of diodes is wired together to permit the alternating flow of the stator to be converted to a pulsating, but unidirectional flow at the alternator output. The alternator's field is wired in series with the voltage regulator.

The regulator consists of several circuits. Each circuit has a core, or magnetic coil of wire, which operates a switch. Each switch is connected to ground through one or more resistors. The coil of wire responds directly to system voltage. When the voltage reaches the required level, the magnetic field created by the winding of wire closes the switch and inserts a resistance into the generator field circuit, thus reducing the output. The contacts of the switch cycle open and close many times each second to precisely control voltage.

While alternators are self-limiting as far as maximum current is concerned, DC generators employ a current regulating circuit which responds directly to the total amount of current flowing through the generator circuit rather than to the output voltage. The current regulator is similar to the voltage regulator except that all system current must flow through the energizing coil on its way to the various accessories.

Alternator Precautions

To prevent damage to the alternator and regulator, the following precautionary measures must be taken when working with the electrical system.

If the battery is removed for any reason, make sure it is reconnected with the correct polarity. Reversing the battery connections may result in damage to the one-way rectifiers.

When utilizing a booster battery as a starting aid, always connect the positive to positive terminals and the negative terminal from the booster battery to a good engine ground on the vehicle being started.

Never use a fast charger as a booster to start vehicles.

Disconnect the battery cables when charging the battery with a fast charger.

Never attempt to polarize the alternator.

Do not use test lights of more than 12V when checking diode continuity.

Do not short across or ground any of the alternator terminals.

The polarity of the battery, alternator and regulator must be matched and considered before making any electrical connections within the system.

Never separate the alternator on an open circuit. Make sure all connections within the circuit are clean and tight.

Disconnect the battery ground terminal when performing any service on electrical components.

Disconnect the battery if arc welding is to be done on the vehicle.

Alternator

TESTING

General Information

There are many possible ways in which the charging system can malfunction. Often the source of a problem is difficult to diagnose, requiring special equipment and a good deal of experience. This is usually not the case, however, where the charging system fails completely and causes the dash board warning light to come on or the battery to discharge. To troubleshoot a complete system failure, only two pieces of equipment are needed: a test light, to determine that current is reaching a certain point and a current indicator (ammeter), to determine the direction of the current flow and its measurement in amps. This test works under three assumptions:

1. The battery is known to be good and fully charged.
2. The alternator belt is in good condition and adjusted to the proper tension.
3. All connections in the system are clean and tight.

➡ In order for the current indicator to give a valid reading, the vehicle must be equipped with battery cables which are of the same gauge size and quality as original equipment battery cables.

Before commencing with the following tests, turn off all electrical components on the vehicle. Make sure the doors of the vehicle are closed. If the vehicle is equipped with a clock, disconnect the clock by removing the lead wire from the rear of the clock.

Battery No-Load Test

1. Ensure that the ignition switch is turned **OFF**.
2. Connect a tachometer to the engine by following the manufacturer's instructions.
3. Using a Digital Volt Ohmmeter (DVOM) measure the voltage across the positive (+) and negative (-) battery terminals. Note the voltage reading for future reference.

Ensure that all electrical components on the vehicle are turned off. Be sure the doors of the vehicle are closed. If the vehicle is equipped with a clock, disconnect the clock by removing the lead wire from the rear of the clock.

4. Start the engine and have an assistant run it at 1500 rpm.
5. Read the voltage across the battery terminals again. The voltage should now be between 14.1–14.7 volts.
   a. If the voltage increase is less than 2.5 volts over the base voltage measured in Step 3, perform the Battery Load test.
   b. If there was no voltage increase, or the voltage increase was greater than 2.5 volts, perform the Alternator Load and No-Load tests.

Battery Load Test

1. With the engine running, turn the air conditioner ON (if equipped) or the blower motor on high speed and the headlights on high beam.
2. Have your assistant increase the engine speed to approximately 2000 rpm.
3. Read the voltage across the battery terminals again.
   a. If the voltage increase is 0.5 volts over the base voltage measured in Battery No-Load test Step 3, the charging system is working properly. If your problem continues, there may be a problem with the battery.
   b. If the voltage does not increase as indicated, perform the Alternator Load and No-Load tests.

Alternator Load Test

✹✹ WARNING

Do NOT use a normal Digital Volt Ohmmeter (DVOM) for this test; your DVOM will be destroyed by the large amounts of amperage from the car's battery. Use a tester designed for charging system analysis, such an Alternator, Regulator, Battery and Starter Motor Tester unit.

1. Switch the tester to the ammeter setting.
2. Attach the positive (+) and negative (-) leads of the tester to the battery terminals.
3. Connect the current probe to the **B+** terminal on the alternator.
4. Start the engine and have an assistant run the engine at 2000 rpm. Adjust the tester load bank to determine the output of the alternator. Alternator output should be within ten percent of the alternator's output rating; if so, continue with the Alternator No-Load test. If the output is not within ten percent of the alternator's output rating, there is a problem in the charging system. Have the system further tested by a Ford qualified automotive technician.

Alternator No-Load Test

1. Using the same tester as in the Alternator Load Test, switch the tester to the voltmeter function.
2. Connect the voltmeter positive (+) lead to the alternator **B+** terminal and the negative (-) lead to a good engine ground.
3. Turn all of the electrical accessories off and shut the doors.
4. While an assistant operates the engine at 2000 rpm, check the alternator output voltage. The voltage should be between 13.0–15.0 volts. If the alternator does not produce voltage within this range there is a problem in the charging system. Have the system further tested by a Ford qualified automotive technician.

2-10 ENGINE ELECTRICAL

REMOVAL & INSTALLATION

2.2L Engine

1. Disconnect the negative (ground) cable. On some models, it may be necessary to remove the battery.
2. Remove the nut holding the alternator wire to the terminal at the rear of the alternator.
3. Pull the multiple connector from the rear of the alternator.
4. Remove the alternator adjusting arm bolt. Swing the alternator in and disengage the fan belt.
5. Remove the alternator pivot bolt and remove the alternator from the truck.
6. Place the alternator in position and install the pivot bolt hand tight. Install the adjusting arm bolt. Place the drive belt in position and tension it properly. Secure the adjusting and pivot bolts. Be sure to adjust the drive belt tension and to connect the battery properly.

2.6L Engine

1987-88

1. Disconnect the negative battery cable.
2. Disconnect the wiring and label for easy installation.
3. Remove the adjusting strap mounting bolt.
4. Remove the drive belts.
5. Remove the support mounting bolt and nut.
6. Remove the alternator assembly.
7. Position the alternator assembly against the engine and install the support bolt.
8. Install the adjusting strap mounting bolt.
9. Install the alternator belts and adjust to specification.
10. Tighten all the support bolts and nuts.
11. Connect all alternator terminals.
12. Connect the negative battery cable.

1989-93

1. Disconnect the negative battery cable.
2. Disconnect the wiring and label for easy installation.
3. Remove the power steering pump pulley, if it interferes with alternator removal.
4. Remove the drive belts.
5. Remove the support mounting bolt and nut.
6. Remove the alternator assembly.
7. Position the alternator assembly against the engine and install the support bolt.
8. Install the adjusting strap mounting bolt.
9. Install the alternator belts and adjust to specification.
10. Install the power steering pump pulley.
11. Tighten all the support bolts and nuts.
12. Connect all alternator terminals.
13. Connect the negative battery cable.

3.0L Engine

1. Disconnect the negative battery cable.
2. Disconnect the wiring and label for easy installation.
3. Remove the mounting and adjusting bolts.
4. Remove the drive belts.
5. Remove the alternator assembly.
6. Position the alternator assembly against the engine and install the support bolt.
7. Install the adjusting strap mounting bolt.
8. Install the alternator belts and adjust to specification.
9. Tighten all the support bolts and nuts.
10. Connect all alternator terminals.
11. Connect the negative battery cable.

4.0L Engine

1. Disconnect the battery ground cable.
2. Disconnect the wire harness connector from the alternator. Remove the wiring connector bracket.
3. Remove the drive belt.
4. Remove the mounting bolts and the alternator.
5. Position the alternator and install the mounting bolts. Install the drive belt. Secure the wiring connector bracket to the alternator and connect the wiring harness. Connect the battery ground cable.

Regulator

REMOVAL & INSTALLATION

The internal voltage regulator used on alternators used with these vehicles is not removable or, in any other way, serviceable. If the voltage regulator is found to be defective, a new alternator must be installed.

STARTING SYSTEM

General Information

♦ See Figures 24 and 25

The starting system is designed to rotate the engine at a speed fast enough for the engine to start. The starting system is comprised of the following components:

- Permanent magnet gear-reduction starter motor with a solenoid-actuated drive
- Battery
- Remote control starter switch (part of the ignition switch)
- Park/Neutral Position (PNP) switch
- Clutch Pedal Position (CPP) switch (on manual transmission models)
- Starter relay (Navajo)
- Heavy circuit wiring

Heavy cables, connectors and switches are utilized by the starting system because of the large amount of amperage this system is required to handle while cranking the engine. For premium starter motor function, the resistance in the starting system must be kept to an absolute minimum.

A discharged or faulty battery, loose or corroded connections, or partially broken cables will result in slower-than-normal cranking speeds. The amount of damage evident may even prevent the starter motor from rotating the engine at all.

Vehicles equipped with a manual transmission are equipped with a Clutch

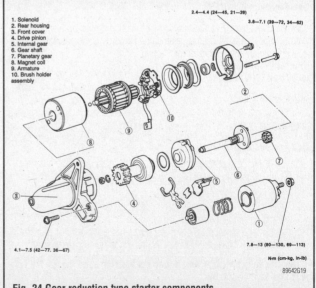

Fig. 24 Gear reduction type starter components

ENGINE ELECTRICAL 2-11

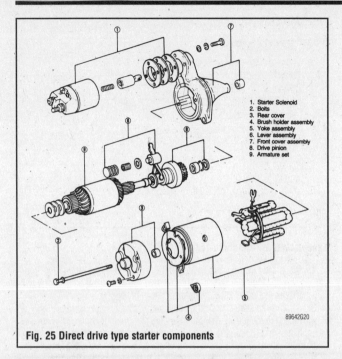

Fig. 25 Direct drive type starter components

1. Starter Solenoid
2. Bolts
3. Rear cover
4. Brush holder assembly
5. Yoke assembly
6. Lever assembly
7. Front cover assembly
8. Drive pinion
9. Armature set

Pedal Position (CPP) switch in the starter circuit, which is designed to prevent the starter motor from operating unless the clutch pedal is depressed. Vehicles equipped with automatic transmissions are equipped with a Park/Neutral Position (PNP) switch in the starter circuit. This switch prevents the starter motor from functioning unless the transmission range selector lever is in Neutral (**N**) or Park (**P**).

The starter motor is a 12 volt assembly, which has the starter solenoid mounted on the drive end-housing. The starter solenoid energizes when the relay contacts are closed. When the solenoid energizes, the starter drive engages with the flywheel ring gear, rotating the crankshaft and starting the engine. An overrunning clutch in the starter drive assembly protects the starter motor from excessive speed when the engine starts.

Starter

TESTING

Navajo Models

▶ See Figures 26 and 27

Use the charts to help locate and diagnose starting system problems. Remember that the starter uses large amounts of current during operation, so use all appropriate precautions during testing.

Pickup and MPV Models

▶ See Figure 28

ON VEHICLE TEST

▶ See Figures 28 and 29

Be sure that the battery is fully charged before starting this test
1. Turn the ignition switch to the **START** position.
2. Verify that the starter motor operates.
3. If the starter motor does not operate, measure the voltage between terminal S and ground, using a voltmeter.
4. If the voltage measures 8V or more, the starter is not working properly.
5. If the voltage measures less than 8V, the wiring harness is the fault.

➡ If the magnetic switch is hot, it may not function even though the voltage is standard or greater.

System Inspection

CAUTION: When disconnecting the plastic hardshell connector at the solenoid "S" terminal, grasp the plastic connector and pull lead off. DO NOT pull separately on lead wire.

WARNING: WHEN SERVICING STARTER OR PERFORMING OTHER UNDERHOOD WORK IN THE VICINITY OF THE STARTER, BE AWARE THAT THE HEAVY GAUGE BATTERY INPUT LEAD AT THE STARTER SOLENOID IS "ELECTRICALLY HOT" AT ALL TIMES.

A protective cap or boot is provided over this terminal on all carlines and must be replaced after servicing. Be sure to disconnect battery negative cable before servicing starter.

1. Inspect starting system for loose connections.
2. If system does not operate properly, note condition and continue diagnosis using the symptom chart.

WARNING: WHEN WORKING IN AREA OF THE STARTER, BE CAREFUL TO AVOID TOUCHING HOT EXHAUST COMPONENTS.

CONDITION	POSSIBLE SOURCE	ACTION
Starter solenoid does not pull-in and starter does not crank (Audible click may or may not be heard).	• Open fuse. • Low battery. • Inoperative fender apron relay. • Open circuit or high resistance in external feed circuit to starter solenoid. • Inoperative starter.	• Check fuse continuity. • Refer to appropriate battery section in this manual. • Go to Evaluation Procedure 2. • Go to Test A. • Replace starter. See removal and installation procedure.
Unusual starter noise during starter overrun.	• Starter not mounted flush (cocked). • Noise from other components. • Ring gear tooth damage or excessive ring gear runout. • Defective starter.	• Realign starter on transmission bell housing. Investigate other powertrain accessory noise contributors. • Refer to appropriate engine section in this manual. • Replace starter. See removal and installation procedure.
Starter cranks but engine does not start.	• Problem in fuel system. • Problem in ignition system. • Engine related concern.	• Refer to appropriate fuel system section in this manual. • Refer to appropriate ignition system section in this manual. • Refer to appropriate engine section in this manual.
Starter cranks slowly.	• Low battery. • High resistance or loose connections in starter solenoid battery feed or ground circuit. • Ring gear runout excessive. • Inoperative starter.	• Refer to appropriate battery section in this manual. • Check that all connections are secure. • Refer to appropriate engine section in this manual. • Replace Starter. See removal and installation procedure.
Starter remains engaged and runs with engine.	• Shorted ignition switch. • Battery cable touching solenoid 'S' terminal (inoperative or mispositioned cable). • Inoperative starter.	• Refer to appropriate ignition system section in this manual. • Replace or relocate cable and replace starter. • Replace starter. See removal and installation procedure.

Fig. 26 Starter system inspection chart

Evaluation Procedure 1

NOTE: Hoist vehicle (if necessary) to access starter solenoid terminals.

CAUTION: Remove plastic safety cap on starter solenoid and disconnect hardshell connector at solenoid 'S' terminal.

CHECK STARTER MOTOR — TEST A

	TEST STEP	RESULT	ACTION TO TAKE
A1	CHECK FOR VOLTAGE TO STARTER • Key OFF. Transmission in Park or Neutral. • Check for voltage between starter B+ terminal and starter drive housing. • Is voltage OK? (12-12.45V)	Yes ▶ No ▶	GO to A2. CHECK wire connections between battery and starter solenoid and the ground circuit for open or short.
A2	CHECK STARTER MOTOR • Key OFF. Transmission in Park or Neutral. • Connect one end of a jumper wire to the starter B+ terminal and momentarily touch the other end to solenoid 'S' terminal. • Does starter crank?	Yes ▶ No ▶	CHECK connections from output of fender apron relay to 'S' terminal for open or short. Defective starter. REPLACE starter.

Evaluation Procedure 2

CHECK FENDER APRON RELAY — TEST B

	TEST STEP	RESULT	ACTION TO TAKE
B1	CHECK FENDER APRON RELAY • Key in START. Transmission in Park or Neutral. • Is case ground OK?	Yes ▶ No ▶	GO to B2. SERVICE ground. GO to B2.
B2	CHECK VOLTAGE AT FENDER APRON RELAY START TERMINAL • Key in START. Transmission in Park or Neutral. • Check for voltage between fender apron relay start terminal and case ground. • Is voltage OK? (12-12.45 V)	Yes ▶ No ▶	GO to B3. Open circuit or high resistance exists in external circuit wiring or components. Check the following: • All circuit connections including plastic hardshell connector at solenoid 'S' terminal to make sure it is not broken or distorted. • Ignition switch. • Neutral switch or manual lever position sensor. • Anti-theft contact.
B3	CHECK OUTPUT TERMINAL VOLTAGE • Key in START. Transmission in Park or Neutral. • Check for voltage at output terminal of fender relay. • Is voltage OK?	Yes ▶ No ▶	REFER to Starter System Diagnosis in this section. Defective fender apron relay. REMOVE and REPLACE relay.

Fig. 27 Starter system evaluation procedure chart

2-12 ENGINE ELECTRICAL

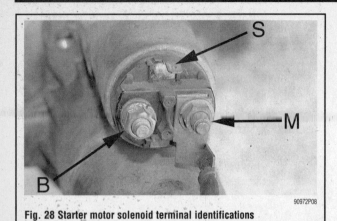

Fig. 28 Starter motor solenoid terminal identifications

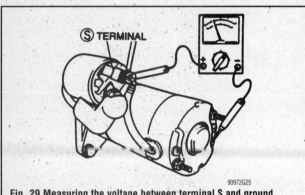

Fig. 29 Measuring the voltage between terminal S and ground, using a voltmeter

MAGNETIC SWITCH PULL OUT TEST

▶ See Figures 28, 30 and 31

1. Apply battery positive voltage to terminal S and body ground, verify that the pinion is pulled out.
2. On 3.0L engines only, measure the pinion gap while the pinion is pulled out. It should measure 0.02–0.08 inch (0.5–2.0 mm).
3. If not within specifications, adjust with an adjustment washer (drive housing front cover-magnetic switch).

RETURN TEST

▶ See Figure 28

1. Disengage the motor wire from terminal M, and then connect battery power to terminal M and ground the body.
2. Using a small prying tool, pull out the overrunning clutch. Verify that the overrunning clutch returns to its original position when released.

NO-LOAD TEST

▶ See Figures 28 and 32

1. Verify that the battery is fully charged.
2. Connect the battery, starter, ammeter and voltmeter as illustrated.
3. Operate the starter and verify that it turns smoothly.
4. Measure the voltage and current while the starter is operating.

2.6L engine:
- Voltage: 11.5V
- Current (amps): 100 max.

3.0L engine:
- Voltage: 11.0V
- Current (amps): 90 max.

5. If not as specified, replace the starter motor.

REMOVAL & INSTALLATION

Except 4WD MPV Models

1. Disconnect the negative battery cable. Raise and safely support the vehicle.
2. Disconnect the electrical connectors from the starter.
3. Remove the starter mounting bolts and remove the starter.
4. Place the starter in position, after cleaning the mounting flange surfaces. Install and tighten the mounting bolts. Tighten the starter mounting bolts to 24–33 ft. lbs. (32–46 Nm) on 1987–88 vehicles, 27–38 ft. lbs. (37–52 Nm) on 1989–92 vehicles except 4.0L engine or 15–20 ft. lbs. (21–27 Nm) on 4.0L engine. Connect all wiring connectors. Connect the negative battery cable.

4WD MPV Models

▶ See Figure 33

1. Disconnect the negative battery cable.
2. Remove the drive belts. Remove the power steering pump pulley. Remove the alternator.
3. Raise and safely support the vehicle. Remove the splash shields.
4. Remove the power steering pump mounting bolts and position the pump aside, without disconnecting the power steering hoses.
5. Remove the automatic transmission cooler line brackets.
6. Mark the position of the driveshaft on the axle flange and remove the front driveshaft.
7. Remove the wiring harness bracket and the automatic transmission cooler line bracket that is next to the starter.
8. Disconnect the electrical connectors from the starter.
9. Remove the fuel and brake line shield.
10. Remove the starter mounting bolts and remove the starter.
11. Clean the mounting surface flanges. Place the starter motor into position and install the mounting bolts. Tighten the starter mounting bolts to 27–38 ft. lbs. (37–52 Nm).
12. Install the fuel and brake line shield. Connect the starter motor wiring. Install the wiring harness bracket.

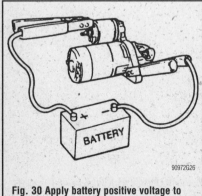

Fig. 30 Apply battery positive voltage to terminal S and body ground

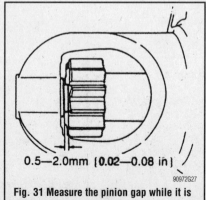

Fig. 31 Measure the pinion gap while it is pulled out—3.0L engine

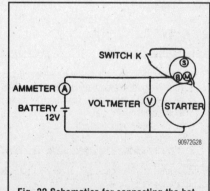

Fig. 32 Schematics for connecting the battery, starter, ammeter and voltmeter

ENGINE ELECTRICAL 2-13

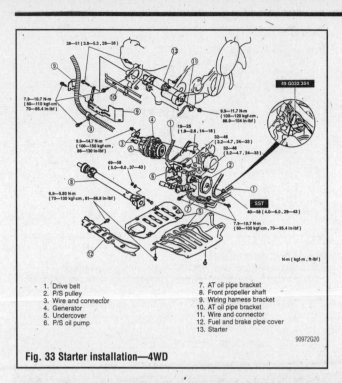

Fig. 33 Starter installation—4WD

1. Drive belt
2. P/S pulley
3. Wire and connector
4. Generator
5. Undercover
6. P/S oil pump
7. AT oil pipe bracket
8. Front propeller shaft
9. Wiring harness bracket
10. AT oil pipe bracket
11. Wire and connector
12. Fuel and brake pipe cover
13. Starter

13. Connect the driveshaft. Install the transmission cooler line brackets. Install the power steering pump and the splash shields.
14. Lower the vehicle. Install the alternator. Install the power steering pump pulley. Install and tension the drive belts. Connect the negative battery cable.

SOLENOID REPLACEMENT

▶ See Figures 24 and 25

1. Remove the starter from the truck.
2. Disconnect the field strap from the solenoid terminal.
3. Remove the two solenoid attaching screws.
4. Disengage the solenoid plunger from the shift fork and remove the solenoid.
5. Install the solenoid on the drive end housing, making sure that the solenoid plunger hook is engaged with the shift fork.
6. Apply 12v to the solenoid **S** terminal and measure the clearance between the starter drive and the stop-ring retainer. It should be 2.0–5.0mm (0.0787–0.1969 in.). If not, remove the solenoid and adjust the clearance by inserting an adjusting shim between the solenoid body and drive end housing.
7. Check the solenoid for proper operation and install the starter.
8. Check the operation of the starter.

SENDING UNITS AND SENSORS

➥This section describes the operating principles of sending units, warning lights and gauges. Sensors which provide information to the Electronic Control Module (ECM) are covered in Section 4 of this manual.

Instrument panels contain a number of indicating devices (gauges and warning lights). These devices are composed of two separate components. One is the sending unit, mounted on the engine or other remote part of the vehicle, and the other is the actual gauge or light in the instrument panel.

Several types of sending units exist, however most can be characterized as being either a pressure type or a resistance type. Pressure type sending units convert liquid pressure into an electrical signal which is sent to the gauge. Resistance type sending units are most often used to measure temperature and use variable resistance to control the current flow back to the indicating device. Both types of sending units are connected in series by a wire to the battery (through the ignition switch). When the ignition is turned **ON**, current flows from the battery through the indicating device and on to the sending unit.

Coolant Temperature Sender

The coolant temperature sender is located in the following positions:
- 2.2L engine—left side rear of the engine, below the cylinder head
- 2.6L engine—top front of the engine.
- 3.0L engine (MPV)—top front of the engine, on the right-hand side of the intake manifold
- 4.0L engine—top left front of the engine, on the intake manifold

TESTING

1. Remove the coolant temperature sender from the engine block.
2. Attach an ohmmeter to the sender unit as follows:
    a. Attach one lead to the metal body of the sender unit (near the sender unit's threads).
    b. Attach the other lead to the sender unit's wiring harness connector terminal.
3. With the leads still attached, place the sender unit in a pot of cold water so that neither of the leads is immersed in the water. The portion of the sender unit which normally makes contact with the engine coolant should be submerged.
4. Measure and note the resistance.

5. Slowly heat the pot up (on the stove) to 190–210° F (88–99° C) and observe the resistance of the sender unit. The resistance should evenly and steadily decrease as the water temperature increases. The resistance should not jump drastically or decrease erratically.
6. If the sender unit did not function as described, replace the sender unit with a new one.

REMOVAL & INSTALLATION

▶ See Figure 34

: CAUTION

Ensure that the engine is cold prior to opening the cooling system or removing the sender from the engine. The cooling system on a hot engine is under high pressures, and released hot coolant or steam can cause severe burns.

1. Disconnect the negative battery cable.
2. Remove the radiator cap to relieve any system pressure.
3. Disconnect the wiring at the sender.
4. Remove the coolant temperature sender from the engine.

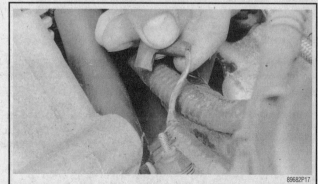

Fig. 34 The sending unit uses a push-on wire connector. To remove, simply pull straight up from the sender

2-14 ENGINE ELECTRICAL

To install:

5. Coat the threads on the sender with Teflon® tape or electrically conductive sealer, then install the sender. Tighten the sender to 107–143 inch lbs. (12–16 Nm) on Navajo/B Series Pick-up, and 57–82 inch lbs. (6–9 Nm) on MPV.
6. Attach the wiring to the sender and connect the negative battery cable.
7. If necessary, add antifreeze to replace any lost coolant, then install the radiator cap.

Oil Pressure Sender Switch

➡ Oil pressure senders are used for oil pressure gauges, whereas the oil pressure switches are used for vehicles equipped only with a low oil pressure warning lamp.

The oil pressure senders/switches are located as follows:
- 2.2L engine—Left side rear of the engine, in the cylinder head
- 2.6L engine—Right-hand side of the engine block, above the oil filter
- 3.0L engine (MPV)—Left-hand lower side of the engine, on the oil filter adapter
- 4.0L engine—Left side front of the engine, below the cylinder head in the engine block

TESTING

1. To test the oil pressure switch, open the hood and locate the switch.
2. Disconnect the wire from the switch. Attach one end of a jumper wire to the terminal on the end of the wire, then touch the other end of the jumper wire to a good engine ground (any bare metal engine surface). Have an assistant observe the instrument gauge cluster while you do this and tell you if the low oil warning lamp illuminates or not; the low oil warning lamp should illuminate.
    a. If the lamp does not illuminate, skip to Step 3.
    b. If the lamp does illuminate, replace the switch with a new one.
3. Before jumping to any bad conclusions, try a different area for grounding the jumper wire on the engine. If the lamp still does not illuminate, touch the jumper wire end to the negative (-) battery post.
    a. If the lamp illuminates, the problem lies with the engine not being properly grounded.
    b. If the lamp does not illuminate, skip to Step 4.
4. Connect the original wire to the oil pressure switch. While sitting in the vehicle, turn the ignition switch to the **ON** position without actually starting the engine. Observe the other lights on the instrument cluster.
    a. If all of the other lights illuminate when turning the ignition switch **ON**, the oil pressure switch is defective and must be replaced.
    b. If none of the other lights illuminate, there is a problem with power supply to the instrument cluster and gauges.

REMOVAL & INSTALLATION

▶ See Figure 35

1. Disconnect the negative battery cable.
2. Disconnect the wiring at the sender/switch.
3. Remove the oil pressure sender/switch from the engine.

To install:

4. Coat the threads with electrically conductive sealer and thread the unit into place. Tighten the sender/switch to 10–13 ft. lbs (13–17 Nm).
5. Attach the wiring to the sender/switch and connect the negative battery cable.

Low Oil Level Sensor

▶ See Figure 35

The low oil level sensor is located in the engine oil pan of all Navajo models.

TESTING

▶ See Figures 36 and 37

Use the accompanying diagnostic chart to help pinpoint low oil level sensor malfunctioning.

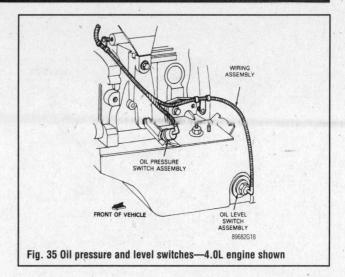

Fig. 35 Oil pressure and level switches—4.0L engine shown

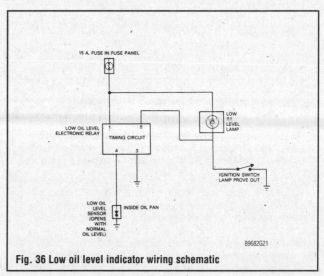

Fig. 36 Low oil level indicator wiring schematic

➡ The ignition switch should be turned OFF for a minimum of 5 minutes between checks to ensure that the electronic relay, which has a 5 minute timer, has reset.

REMOVAL & INSTALLATION

▶ See Figure 38

➡ Always install a new gasket whenever the oil level sensor is removed

1. Turn the engine **OFF**.
2. Raise the front of the vehicle and install jackstands beneath the frame. Firmly apply the parking brake and place blocks in back of the rear wheels.
3. Drain at least 2 quarts (1.9 liters) of engine oil out of the pan.
4. Disconnect the sensor wiring.
5. Remove the sensor from the oil pan using a 1 in. (26mm) socket or wrench. Discard the old gasket.

To install:

6. Install a new gasket onto the sensor.

➡ When installing the new gasket, the flange faces the sensor and the words "panside" should face the oil pan.

7. Install the sensor and gasket assembly into the oil pan. Tighten the sensor to 13–20 ft. lbs. (17–27 Nm).
8. Connect the electrical wire to the sensor.
9. Lower the front of the vehicle and remove the wheel blocks.
10. Refill the crankcase to the proper level.
11. Start the engine and check for leaks.

ENGINE ELECTRICAL 2-15

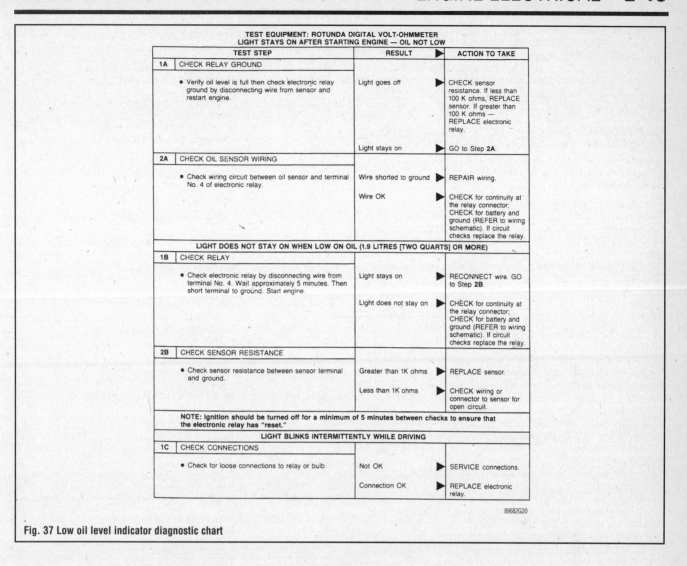

Fig. 37 Low oil level indicator diagnostic chart

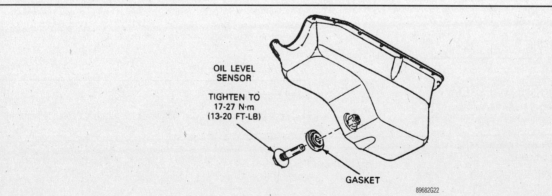

Fig. 38 Typical low oil level sensor mounting—note: the sensor is constantly submerged in oil, therefore, drain some oil out of the engine before removing it

ENGINE ELECTRICAL

Troubleshooting Basic Starting System Problems

Problem	Cause	Solution
Starter motor rotates engine slowly	• Battery charge low or battery defective	• Charge or replace battery
	• Defective circuit between battery and starter motor	• Clean and tighten, or replace cables
	• Low load current	• Bench-test starter motor. Inspect for worn brushes and weak brush springs.
	• High load current	• Bench-test starter motor. Check engine for friction, drag or coolant in cylinders. Check ring gear-to-pinion gear clearance.
Starter motor will not rotate engine	• Battery charge low or battery defective	• Charge or replace battery
	• Faulty solenoid	• Check solenoid ground. Repair or replace as necessary.
	• Damaged drive pinion gear or ring gear	• Replace damaged gear(s)
	• Starter motor engagement weak	• Bench-test starter motor
	• Starter motor rotates slowly with high load current	• Inspect drive yoke pull-down and point gap, check for worn end bushings, check ring gear clearance
	• Engine seized	• Repair engine
Starter motor drive will not engage (solenoid known to be good)	• Defective contact point assembly	• Repair or replace contact point assembly
	• Inadequate contact point assembly ground	• Repair connection at ground screw
	• Defective hold-in coil	• Replace field winding assembly
Starter motor drive will not disengage	• Starter motor loose on flywheel housing	• Tighten mounting bolts
	• Worn drive end busing	• Replace bushing
	• Damaged ring gear teeth	• Replace ring gear or driveplate
	• Drive yoke return spring broken or missing	• Replace spring
Starter motor drive disengages prematurely	• Weak drive assembly thrust spring	• Replace drive mechanism
	• Hold-in coil defective	• Replace field winding assembly
Low load current	• Worn brushes	• Replace brushes
	• Weak brush springs	• Replace springs

ENGINE MECHANICAL 3-2
ENGINE 3-2
    REMOVAL & INSTALLATION 3-2
ROCKER ARM (VALVE) COVER 3-2
    REMOVAL & INSTALLATION 3-2
ROCKER ARMS/SHAFTS 3-4
    REMOVAL & INSTALLATION 3-4
THERMOSTAT 3-6
    REMOVAL & INSTALLATION 3-6
INTAKE MANIFOLD 3-8
    REMOVAL & INSTALLATION 3-8
EXHAUST MANIFOLD 3-10
    REMOVAL & INSTALLATION 3-10
RADIATOR 3-12
    REMOVAL & INSTALLATION 3-12
ENGINE FAN 3-13
    REMOVAL & INSTALLATION 3-13
WATER PUMP 3-14
    REMOVAL & INSTALLATION 3-14
CYLINDER HEAD 3-16
    REMOVAL & INSTALLATION 3-16
OIL PAN 3-20
    REMOVAL & INSTALLATION 3-20
OIL PUMP 3-22
    REMOVAL & INSTALLATION 3-22
    INSPECTION 3-25
CRANKSHAFT PULLEY (VIBRATION
    DAMPER) 3-26
    REMOVAL & INSTALLATION 3-26
TIMING BELT COVERS 3-26
    REMOVAL & INSTALLATION 3-26
TIMING CHAIN COVER 3-27
    REMOVAL & INSTALLATION 3-27
FRONT COVER OIL SEAL 3-29
    REMOVAL & INSTALLATION 3-29
TIMING BELT AND SPROCKETS 3-29
    REMOVAL & INSTALLATION 3-29
TIMING CHAIN AND TENSIONER 3-31
    REMOVAL & INSTALLATION 3-32
CAMSHAFT 3-35
    REMOVAL & INSTALLATION 3-35
    INSPECTION 3-37
CAMSHAFT BEARINGS 3-38
    REMOVAL & INSTALLATION 3-38
VALVE LIFTERS (TAPPETS) 3-39
    REMOVAL & INSTALLATION 3-39
BALANCE (COUNTERSHAFTS)
    SHAFTS 3-39
    REMOVAL & INSTALLATION 3-39
REAR MAIN OIL SEAL 3-40
    REPLACEMENT 3-40
EXHAUST SYSTEM 3-41
SAFETY PRECAUTIONS 3-41
INSPECTION 3-42
    REPLACEMENT 3-42
ENGINE RECONDITIONING 3-44
DETERMINING ENGINE CONDITION 3-44
    COMPRESSION TEST 3-44
    OIL PRESSURE TEST 3-45

BUY OR REBUILD? 3-45
ENGINE OVERHAUL TIPS 3-45
    TOOLS 3-45
    OVERHAUL TIPS 3-46
    CLEANING 3-46
    REPAIRING DAMAGED
        THREADS 3-46
ENGINE PREPARATION 3-47
CYLINDER HEAD 3-47
    DISASSEMBLY 3-48
    INSPECTION 3-51
    REFINISHING & REPAIRING 3-52
    ASSEMBLY 3-53
ENGINE BLOCK 3-54
    GENERAL INFORMATION 3-54
    DISASSEMBLY 3-54
    INSPECTION 3-55
    REFINISHING 3-57
    ASSEMBLY 3-57
ENGINE START-UP AND BREAK-IN 3-60
    STARTING THE ENGINE 3-60
    BREAKING IT IN 3-60
    KEEP IT MAINTAINED 3-60
SPECIFICATIONS CHARTS
ENGINE MECHANICAL
    SPECIFICATIONS 3-60
TORQUE SPECIFICATIONS 3-69

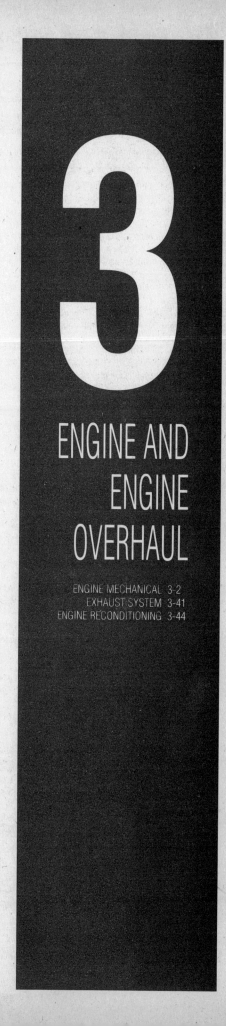

3
ENGINE AND ENGINE OVERHAUL

ENGINE MECHANICAL 3-2
EXHAUST SYSTEM 3-41
ENGINE RECONDITIONING 3-44

3-2 ENGINE AND ENGINE OVERHAUL

ENGINE MECHANICAL

Engine

REMOVAL & INSTALLATION

In the process of removing the engine, you will come across a number of steps which call for the removal of a separate component or system, such as "disconnect the exhaust system" or "remove the radiator." In most instances, a detailed removal procedure can be found elsewhere in this manual.

It is virtually impossible to list each individual wire and hose which must be disconnected, simply because so many different model and engine combinations have been manufactured. Careful observation and common sense are the best possible approaches to any repair procedure.

Removal and installation of the engine can be made easier if you follow these basic points:

- If you have to drain any of the fluids, use a suitable container.
- Always tag any wires or hoses and, if possible, the components they came from before disconnecting them.
- Because there are so many bolts and fasteners involved, store and label the retainers from components separately in muffin pans, jars or coffee cans. This will prevent confusion during installation.
- After unbolting the transmission or transaxle, always make sure it is properly supported.
- If it is necessary to disconnect the air conditioning system, have this service performed by a qualified technician using a recovery/recycling station. If the system does not have to be disconnected, unbolt the compressor and set it aside.
- When unbolting the engine mounts, always make sure the engine is properly supported. When removing the engine, make sure that any lifting devices are properly attached to the engine. It is recommended that if your engine is supplied with lifting hooks, your lifting apparatus be attached to them.
- Lift the engine from its compartment slowly, checking that no hoses, wires or other components are still connected.
- After the engine is clear of the compartment, place it on an engine stand or workbench.
- After the engine has been removed, you can perform a partial or full teardown of the engine using the procedures outlined in this manual.

1. Relieve the fuel system pressure on fuel injected models. Disconnect the negative battery cable first, the positive battery cable second, then remove the battery from the vehicle.
2. Mark the position of the hood on the hinges and remove the hood.
3. Raise and safely support the vehicle. Drain the engine oil and coolant. Remove the splash shields, as necessary.
4. Remove the starter and the transmission.
5. Disconnect the exhaust system from the exhaust manifold.
6. Remove the converter inspection cover and disconnect the converter from the flywheel.
7. Remove the converter housing-to-engine block bolts and the adapter plate-to-converter housing bolt. Lower the vehicle.
8. Remove the air cleaner assembly, if carburetor equipped. Disconnect the accelerator cable.
9. Remove the cooling fan and the radiator shroud. Disconnect the radiator hoses and transmission oil cooler lines, if equipped, and remove the radiator.
10. Disconnect the fuel lines, heater hoses and brake vacuum hose.
11. Tag and disconnect the necessary electrical connectors and vacuum hoses. Remove the ground wires from the cylinder block.
12. If carburetor equipped, disconnect the secondary air pipe assembly. On 1989-93 2.6L engine, remove the resonance chamber.
13. Remove the accessory drive belt(s). Remove the alternator and mounting bracket, then position the alternator aside.
14. If equipped, remove the power steering pump pulley and the power steering pump. Position the pump aside, leaving the hoses connected.
15. If equipped, remove the air conditioning compressor and position aside, leaving the hoses attached.
16. Remove the gusset plates, if equipped. Remove the transmission oil cooler line retainers, if equipped.
17. If necessary, remove the radiator grille. Remove the shroud upper plate and the additional condenser fan, if equipped.
18. Position a jack under the transmission and install suitable engine lifting equipment to the engine. Remove the engine mount nuts and raise the engine slightly, then carefully pull it from the transmission. Carefully lift the engine out of the engine compartment.
19. Install the engine on a workstand.

To install:

➡ Lightly oil all bolts and stud threads, except those specifying special sealant, prior to installation.

20. Using the hoist or engine crane, slowly and carefully position the engine in the vehicle. Make sure the exhaust manifolds are properly aligned with the exhaust pipes.
21. Align the engine to the transmission and install two engine-to-transmission bolts.

➡ Seat the left-hand side, front engine mount insulator locating pin prior to the right-hand side, front engine mount insulator.

22. Lower the engine onto the front engine mount insulators.
23. Detach the engine crane or hoist from the engine.
24. Remove the floor jack from beneath the transmission.
25. Tighten the two installed engine-to-transmission bolts, then raise and securely support the vehicle on jackstands.
26. Install and tighten the remaining engine-to-transmission bolts.
27. The remainder of installation is the reverse of the removal procedure. Be sure to tighten the fasteners to the values presented in the torque specification chart.

※※ WARNING

Do NOT start the engine without first filling it with the proper type and amount of clean engine oil, and installing a new oil filter. Otherwise, severe engine damage will result.

28. Fill the crankcase with the proper type and quantity of engine oil. If necessary, adjust the transmission and/or throttle linkage.
29. Check to make sure that all electrical wiring harnesses, linkages, ground wires, hoses, vacuum and fuel lines have been properly connected.
30. Check to make sure that all adjustments have been performed, as well as all fasteners tightened properly.
31. Install the air intake duct assembly.
32. Install the battery into the vehicle. Connect the positive battery cable first, then connect the negative battery cable.
33. Fill and bleed the cooling system.
34. Start the engine and allow it to reach normal operating temperature, then check for leaks and proper operation.
35. Stop the engine and check all fluid levels.
36. Install the hood, aligning the marks that were made during removal.
37. If equipped, have the A/C system properly leak-tested, evacuated and charged by a MVAC-trained, EPA-certified, automotive technician.

Rocker Arm (Valve) Cover

REMOVAL & INSTALLATION

※※ CAUTION

On fuel injected models, fuel in the system remains under high pressure, even when the engine is not running. Before disconnecting any fuel line, release the pressure from the fuel system to reduce the possibility of injury or fire.

2.2L Engine

▶ See Figure 1

1. If so equipped, disconnect the choke cable and the air bypass valve cable.

ENGINE AND ENGINE OVERHAUL

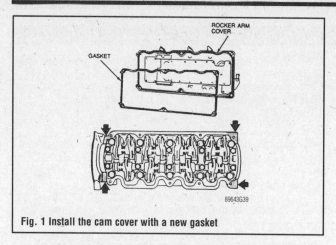

Fig. 1 Install the cam cover with a new gasket

Fig. 2 Install the cam cover with new gaskets and seal washers for the bolts

2. On carbureted models, remove the air cleaner. On fuel injected models, loosen the clamps and remove the air intake crossover.
3. Disconnect the PCV valve at the cover.
4. Remove the retaining bolts and remove the cam cover.
5. Clean all gasket mounting surfaces. To install the cover, first install new gaskets.
6. Place the cover in position and tighten the bolts in several stages, going back and forth across the cover, tighten to 26–35 inch lbs. (3–4 Nm).
7. Once all the associated components are installed, start the engine and allow it to reach normal operating temperature. Retighten the bolts. Check for leaks.

2.6L Engine

1987-88

1. Disconnect the negative battery cable.
2. Remove the air cleaner assembly. Remove or relocate any hoses or cables that will interfere with rocker cover removal.
3. Disconnect the hoses to the PCV tube.
4. Remove the cover mounting bolts and remove the rocker cover from the cylinder head. The water pump pulley belt shield is attached at rear of rocker cover.
5. Clean the cover and head mounting surfaces.
6. Apply RTV sealant to the top of the rubber cam seal and install the rocker cover.
7. With rocker cover installed, apply RTV sealant to top of semi-circular packing.
8. Tighten screws to 55 inch lbs. (6 Nm).
9. Install the vacuum hoses and spark plug wires.
10. Install the air cleaner assembly.
11. Connect the battery cable.

1989-93

1. Disconnect the battery ground cable.
2. Disconnect the accelerator cable.
3. Remove the air intake pipe and resonance chamber.
4. Tag and disconnect any wires and hoses in the way.
5. Unbolt and remove the cover.
6. Thoroughly clean all mating surfaces of old gasket material and/or sealer. The cover may use a gasket or RTV gasket sealer. If a gasket is used, coat the new gasket with sealer and position it on the head. If RTV material is used, squeeze a 1/8 inch bead on the head sealing surface.
7. Install the rocker cover. Tighten the bolts to 52–78 inch lbs. (6–9 Nm).

3.0L Engine

▶ See Figure 2

1. Disconnect the air bypass valve cable.
2. Loosen the clamps and remove the air intake crossover.
3. Disconnect the PCV valve at the cover.
4. Remove the retaining bolts and remove the cam cover.
5. To install the cover, first supply new gaskets and seal washers for the bolts.

6. Tighten the bolts in several stages, going back and forth across the cover to 30–40 inch lbs. (4–5 Nm).
7. Once all the associated components are installed, start the engine and allow it to reach normal operating temperature. Check for oil leaks.

4.0L Engine

▶ See Figure 3

➡Failure to install new rocker cover gaskets and rocker cover reinforcement pieces will result in oil leaks.

1. Disconnect the negative battery cable. Tag and remove the spark plug wires.
2. Relieve fuel system pressure. Disconnect and remove the fuel supply and return lines.
3. For left rocker cover removal, remove the upper intake manifold.
4. For right rocker cover removal, remove air inlet duct and hose to oil fill tube, drive belt from alternator, alternator. Drain cooling system remove the upper radiator hose from the engine. Remove the EDIS ignition coil and bracket assembly. Remove the A/C low pressure hose bracket if so equipped. Remove the PCV valve hose and breather.

✲✲ CAUTION

When draining the coolant, keep in mind that cats and dogs are attracted by the ethylene glycol antifreeze, and are quite likely to drink any that is left in an uncovered container or in puddles on the ground. This will prove fatal in sufficient quantity. Always drain the coolant into a sealable container. Coolant should be reused unless it is contaminated or several years old.

5. Remove the rocker cover bolts and load distribution pieces. The washers must be installed in their original positions, so keep track of them.

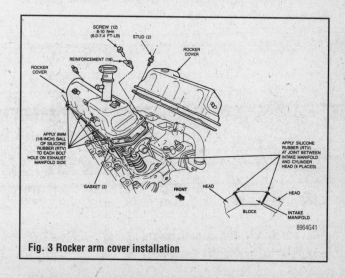

Fig. 3 Rocker arm cover installation

3-4 ENGINE AND ENGINE OVERHAUL

6. Remove the rocker cover. It will probably be necessary to tap the cover loose with a plastic or rubber mallet.
7. Remove the rocker covers.
8. Clean all gasket material from the cover and head.
9. Installation is the reverse of removal. Always use a new gasket coated with sealer. If any of the RTV silicone gasket material was removed from the mating area of the head(s) and intake manifold, replace it. Tighten the bolts to 3–5 ft. lbs. (4–7 Nm).
10. Reconnect the negative battery cable. Start the engine and run to normal operating temperature and check for oil and fuel leaks.

Rocker Arms/Shafts

REMOVAL & INSTALLATION

2.2L Engine

♦ See Figures 4 and 5

1. Disconnect the negative battery cable.
2. Remove the air cleaner assembly or air intake hose, as required.
3. Remove the rocker arm cover.
4. Loosen the rocker arm/shaft assembly mounting bolts in 2–3 steps in the proper sequence. Remove the rocker arm/shaft assembly together with the bolts.
5. If necessary, disassemble the rocker arm/shaft assembly, noting the position of each component to ease reassembly.
6. Check for wear or damage to the contact surfaces of the shafts and rocker arms; replace as necessary.
7. Measure the rocker arm inner diameter, it should be 0.6300–0.6310 in. (16.000–16.027mm). Measure the rocker arm shaft diameter, it should be 0.6286–0.6293 in. (15.966–15.984mm).
8. Subtract the shaft diameter from the rocker arm diameter to get the oil clearance. The oil clearance should be 0.0006–0.0024 in. (0.016–0.061mm) and should not exceed 0.004 in. (0.10mm). Replace parts, as necessary, if the oil clearance is not within specification.

To install:

9. Apply clean engine oil to the rocker arm shafts and rocker arms and assemble the rocker arm/shaft assembly in the reverse order of disassembly. Make sure the rocker arm shaft oil holes in the center camshaft cap face each other.

➡ Use the mounting bolts for alignment.

10. Apply silicone sealant to the cylinder head on the front and rear camshaft cap mounting surface. Apply clean engine oil to the camshaft journals and valve stem tips.
11. Install the rocker arm/shaft assembly and tighten the bolts, in sequence, in 2–3 steps to a maximum torque of 13–20 ft. lbs. (18–26 Nm).
12. Apply silicone sealant to each side of the front and rear camshaft cap and the cylinder head in the area where the caps meet the cylinder head.
13. Install the rocker arm cover and tighten the mounting bolts to 26–35 inch lbs. (3–4 Nm).
14. Install the air cleaner assembly or air intake tube. Connect the negative battery cable, start the engine and check for leaks and proper operation.

2.6L Engine

1987–88

♦ See Figure 6

1. Disconnect the negative battery cable.
2. Remove the air cleaner assembly.
3. Remove the rocker arm cover.
4. Install hydraulic lash adjuster holder tool 49 U012 001 or equivalent, over the ends of each rocker arm to keep the hydraulic lash adjuster from falling out when the rocker arm/shaft assembly is removed.
5. Loosen the rocker arm/shaft assembly mounting bolts and remove the assembly. Leave the mounting bolts in place in the shaft to keep the assembly together, if there will be no further disassembly.
6. If necessary, disassemble the rocker arm/shaft assembly, noting the position of each component to ease reassembly.
7. Check for wear or damage to the contact surfaces of the shafts and rocker arms; replace as necessary.
8. Measure the rocker arm inner diameter, it should be 0.7445–0.7452 in. (18.910–18.928mm). Measure the rocker arm shaft diameter, it should be 0.7435–0.7440 in. (18.885–18.898mm).
9. Subtract the shaft diameter from the rocker arm diameter to get the oil clearance. The oil clearance should be 0.0005–0.0017 in. (0.012–0.043mm) and should not exceed 0.004 in. (0.10mm). Replace parts, as necessary, if the oil clearance is not within specification.

To install:

10. Apply clean engine oil to the rocker arm shafts and rocker arms and assemble the rocker arm/shaft assembly in the reverse order of disassembly. Be sure to align the mating mark on the front of the shaft with the mating mark on the front camshaft cap.
11. Apply clean engine oil to the camshaft journals and valve stem tips.
12. Install the rocker arm/shaft assembly and tighten the mounting bolts, in sequence, in 2–3 steps to a maximum torque of 14–15 ft. lbs. (19–20 Nm).
13. Remove the hydraulic lash adjuster holder tools. Adjust the jet valve lash.
14. Install the rocker arm cover and tighten to 43–61 inch lbs. (5–7 Nm). Install the air cleaner assembly.
15. Connect the negative battery cable, start the engine and bring to normal operating temperature. Check for leaks and proper operation. With the engine warm, readjust the jet valve clearance.

1989–93

♦ See Figures 7 and 8

1. Disconnect the negative battery cable.
2. Remove the air intake hose.
3. Remove the rocker arm cover.
4. Loosen the rocker arm/shaft assembly mounting bolts in 2–3 steps in the proper sequence. Remove the rocker arm/shaft assembly together with the bolts.

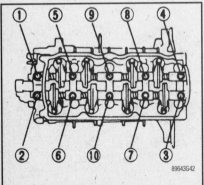

Fig. 4 Rocker arm/shaft assembly mounting bolt removal sequence

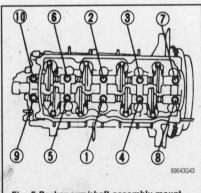

Fig. 5 Rocker arm/shaft assembly mounting bolt tightening sequence

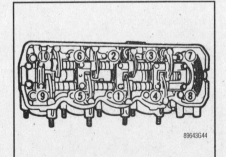

Fig. 6 Rocker arm/shaft assembly mounting bolt tightening sequence

ENGINE AND ENGINE OVERHAUL 3-5

5. If necessary, disassemble the rocker arm/shaft assembly, noting the position of each component to ease reassembly.

6. Check for wear or damage to the contact surfaces of the shafts and rocker arms; replace as necessary.

7. Measure the rocker arm inner diameter, it should be 0.8268–0.8281 in. (21.000–21.033mm). Measure the rocker arm shaft diameter, it should be 0.8252–0.8260 in. (20.959–20.980mm).

8. Subtract the shaft diameter from the rocker arm diameter to get the oil clearance. The oil clearance should be 0.0008–0.0029 in. (0.020–0.074mm) and should not exceed 0.004 in. (0.10mm). Replace parts, as necessary, if the oil clearance is not within specification.

To install:

9. Apply clean engine oil to the rocker arm shafts and rocker arms and assemble the rocker arm/shaft assembly in the reverse order of disassembly, noting the following:

   a. The intake side shaft has twice as many oil holes as the exhaust side shaft.

   b. The No. 4 camshaft cap has an oil hole from the cylinder head; make sure it is installed correctly.

10. Apply clean engine oil to the camshaft journals and valve stem tips.

11. Install the rocker arm/shaft assembly and tighten the mounting bolts, in sequence, in 2–3 steps to a maximum torque of 14–19 ft. lbs. (19–25 Nm).

12. Coat a new gasket with silicone sealer and install on the rocker arm cover. Apply sealer to the cylinder head in the area of the half circle seals and install the rocker arm cover. Install the mounting bolts and tighten to 52–78 inch lbs. (6–9 Nm).

13. Install the air intake hose. Connect the negative battery cable, start the engine and check for leaks and proper operation.

3.0L Engine

▶ See Figures 9 and 10

1. Disconnect the negative battery cable.
2. If removing the driver's side rocker arm/shaft assembly, proceed as follows:

   a. Remove the air inlet tube.

   b. Disconnect the necessary electrical connectors and vacuum hoses from the throttle body and intake air pipe.

   c. Disconnect the throttle cable.

   d. Remove the throttle body and intake air pipe.

3. Remove the rocker arm cover.

4. Loosen the rocker arm/shaft assembly mounting bolts in sequence, in 2–3 steps. Remove the assembly with the bolts.

5. If necessary, disassemble the rocker arm/shaft assembly, noting the position of each component to ease reassembly.

6. Check for wear or damage to the contact surfaces of the shafts and rocker arms; replace as necessary.

7. Measure the rocker arm inner diameter, it should be 0.7480–0.7493 in. (19.000–19.033mm). Measure the rocker arm shaft diameter, it should be 0.7464–0.7472 in. (18.959–18.980mm).

8. Subtract the shaft diameter from the rocker arm diameter to get the oil clearance. The oil clearance should be 0.0008–0.0029 in. (0.020–0.074mm) and should not exceed 0.004 in. (0.10mm). Replace parts, as necessary, if the oil clearance is not within specification.

To install:

9. Apply clean engine oil to the rocker arm shafts and rocker arms and assemble the rocker arm/shaft assembly in the reverse order of disassembly, noting the following. The intake side shaft has twice as many oil holes as the exhaust side shaft.

10. Apply clean engine oil to the camshaft journals and valve stem tips.

11. Install the rocker arm/shaft assembly and tighten the mounting bolts, in sequence, in 2–3 steps to a maximum torque of 14–19 ft. lbs. (19–25 Nm).

➡ Be careful that the rocker arm shaft spring does not get caught between the shaft and mounting boss during installation.

12. Coat a new gasket with silicone sealant and install on the rocker arm cover. Install the rocker arm cover with new seal washers and tighten the bolts to 30–39 inch lbs. (3–4 Nm).

13. Install the intake air pipe, throttle body and air intake tube, if removed. Connect the throttle cable and the necessary electrical connectors and vacuum hoses.

14. Connect the negative battery cable, start the engine and check for leaks and proper operation.

4.0L Engine

▶ See Figure 11

1. Disconnect the negative battery cable. Remove the intake shield and air intake tube.

2. If removing the right rocker arm/shaft assembly, proceed as follows:

   a. Remove the alternator and the coil pack.

   b. Remove the retaining bolt from the air conditioning pipe over the upper intake manifold.

   c. Remove the spark plug wires from the clips on the valve cover.

   d. Remove the 2 wiring harnesses from the right rocker arm cover.

   e. Disconnect the vacuum hose at the coupling over the rocker arm cover.

   f. Remove the engine wiring harness clip from the rocker arm cover-to-intake manifold stud. Do not pull on the harness but rather lift up on the clip.

   g. Remove the right rocker arm cover bolts, reinforcement plate and the cover.

3. If removing the right rocker arm/shaft assembly, proceed as follows:

   a. Remove the bolt from the air conditioning pipe over the upper intake manifold, if not already done.

   b. Disconnect the air conditioning compressor clutch connector and remove the wiring harness from the back of the compressor.

   c. Remove the air conditioning compressor bolts, pull up the tube that goes around the back of the engine and reposition the compressor and tube aside.

   d. Tag and disconnect the power brake vacuum hose and other hoses from the vacuum tee on the plenum.

   e. Remove the PCV hose and valve. Remove the wiring harness from the valve cover and position aside.

   f. Tag and disconnect the spark plug wires from the spark plugs and clips on the rocker arm cover.

   g. Remove the engine wiring harness clip from the rocker arm cover-to-intake manifold stud. Do not pull on the harness but rather lift up on the clip.

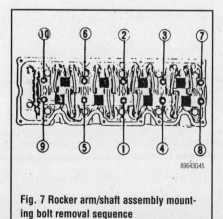

Fig. 7 Rocker arm/shaft assembly mounting bolt removal sequence

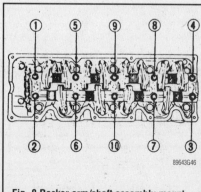

Fig. 8 Rocker arm/shaft assembly mounting bolt tightening sequence

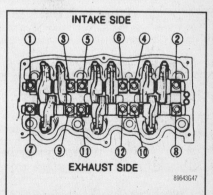

Fig. 9 Rocker arm/shaft assembly mounting bolt removal sequence

3-6 ENGINE AND ENGINE OVERHAUL

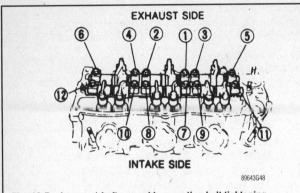

Fig. 10 Rocker arm/shaft assembly mounting bolt tightening sequence

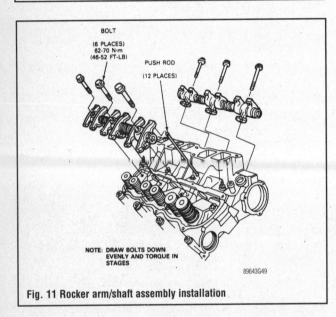

Fig. 11 Rocker arm/shaft assembly installation

h. Remove the retaining bolt from the fuel hose clip to the front of the engine to allow the fuel hoses to be moved enough to gain access to the upper front rocker arm cover bolt.

i. Remove the bolts, reinforcement plates and left rocker arm cover.

4. Remove the rocker arm shafts by loosening the support bolts 2 turns at a time until the shaft can be removed.

5. If necessary, disassemble the rocker arm/shaft assembly by removing the spring washer and pin from each end of the shaft and sliding the rocker arms, springs and rocker arm shaft supports off the shaft. Note the position of each component to ease reassembly.

6. Inspect all components for wear and replace, as necessary.

To install:

7. Coat the rocker arms and shafts with clean engine oil and reassemble. The oil holes in the shaft must point down when the shaft is installed. This position can be recognized by a notch on the front face of the shaft.

8. Lubricate the pushrod ends and valve stem tips.

9. Install the rocker arm/shaft assembly and draw the shaft support bolts down evenly, 2 turns at a time, until the rocker arm/shaft assembly is fully down. Tighten the shaft support bolts to 46–52 ft. lbs. (62–70 Nm).

10. Clean all gasket mating surfaces.

11. Apply silicone sealant to the intake manifold-to-cylinder head parting seam and an 1/8 in. ball of sealer to the rocker arm cover bolt holes on the exhaust side.

12. Install a new gasket on the rocker arm cover and install the cover, reinforcing plates and bolts. Tighten the bolts to 53–70 inch lbs. (6–8 Nm) working in a criss-cross pattern and starting at the center bolts.

13. Install the remaining components. Connect the negative battery cable, start the engine and check for leaks and proper operation.

Thermostat

REMOVAL & INSTALLATION

→If the replacement thermostat is equipped with a jiggle pin, the pin must be install facing upwards toward the top of the engine (12 o'clock), and should be on the side facing the water outlet. When installing the thermostat gasket, the seal print side should face the cylinder head.

2.2L Engine

◆ See Figure 12

1. Drain enough coolant to bring the coolant level down below the thermostat housing. the thermostat housing is located on the left front side of the cylinder block.

※※ CAUTION

When draining the coolant, keep in mind that cats and dogs are attracted by the ethelyne glycol antifreeze, and are quite likely to drink any that is left in an uncovered container or in puddles on the ground. This will prove fatal in sufficient quantity. Always drain the coolant into a sealable container. Coolant should be reused unless it is contaminated or several years old.

2. Disconnect the temperature sending unit wire.
3. Remove the coolant outlet elbow.
4. If so equipped, position the vacuum control valve out of the way. The vacuum control valve is not used on California trucks
5. Disconnect the coolant by-pass hose from the thermostat housing.
6. Remove the thermostat and housing from the engine.
7. Note the position of the jiggle pin and remove the thermostat from the housing.
8. Remove all gasket material from the parts.

To install:

9. Position the thermostat in the housing with the jiggle pin up. Coat a new gasket with sealer and install it on the thermostat housing.
10. Install the thermostat housing using a new gasket with water resistant sealer. Tighten the bolts to 14–19 ft. lbs. (19–25 Nm).
11. Install the coolant outlet elbow and vacuum control valve (if equipped).
12. Connect the by-pass and radiator hoses.
13. Connect the temperature sending unit wire.
14. Fill the cooling system with the proper coolant. Operate the engine and check the coolant lever. Check for leaks.

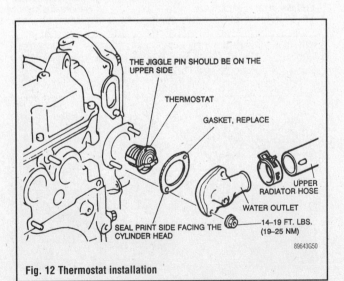

Fig. 12 Thermostat installation

ENGINE AND ENGINE OVERHAUL

2.6L Engine

1987-88

The thermostat is located in a water box at the timing belt end of the intake manifold.

1. Drain the cooling system to a level below the thermostat.

> **CAUTION**
>
> When draining the coolant, keep in mind that cats and dogs are attracted by the ethylene glycol antifreeze, and are quite likely to drink any that is left in an uncovered container or in puddles on the ground. This will prove fatal in sufficient quantity. Always drain the coolant into a sealable container. Coolant should be reused unless it is contaminated or several years old.

2. Remove the hoses from the thermostat housing.
3. Remove the thermostat housing.
4. Remove the thermostat and discard the gasket. Clean the gasket surfaces thoroughly.

To install:

5. Position the gasket on the water box. Center the thermostat in the water box and attached housing. Tighten the bolts to 15 ft. lbs. (20 Nm).
6. Connect the radiator hose to the thermostat housing. Tighten the hose clamp to 35 inch lbs. (4 Nm).
7. Fill the cooling system.

1989-93

The thermostat housing is at the end of the upper hose, on the cylinder head side, above the alternator.

1. Drain the cooling system to a point below the housing.

> **CAUTION**
>
> When draining the coolant, keep in mind that cats and dogs are attracted by the ethylene glycol antifreeze, and are quite likely to drink any that is left in an uncovered container or in puddles on the ground. This will prove fatal in sufficient quantity. Always drain the coolant into a sealable container. Coolant should be reused unless it is contaminated or several years old.

2. Remove the upper hose.
3. Remove the upper nut and lower bolt and remove the housing.
4. Remove the gasket and thermostat. Discard the gasket.
5. Thoroughly clean the mating surfaces of the head and housing.

To install:

6. Position the new thermostat in the head with the jiggle pin on the upper side.
7. Coat the gasket with an adhesive sealer and stick it in place on the head.
8. Install the housing and tighten the nut and bolt to 19 ft. lbs. (26 Nm).
9. Fill the cooling system.

3.0L Engine

▶ See Figures 13, 14, 15, 16 and 17

The thermostat housing is located at the engine end of the lower radiator hose.

1. Raise and support the front end on jackstands.
2. Drain the cooling system.

Fig. 13 After moving the hose clamp away from the thermostat housing pipe, disconnect the lower radiator hose

Fig. 14 Using a socket, remove the three thermostat housing mounting bolts

Fig. 15 After the mounting bolts are removed, remove the thermostat housing from the engine

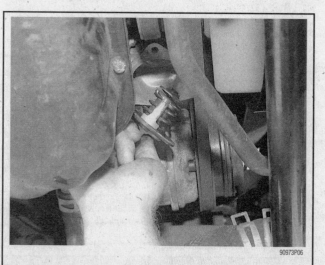

Fig. 16 Pull the thermostat down out of the opening in the engine

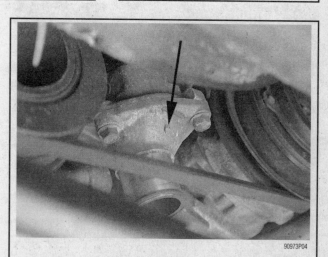

Fig. 17 During assembly, make sure that the notch sticking out of the side of the housing is facing outward

3-8 ENGINE AND ENGINE OVERHAUL

3. Remove the lower hose.
4. Unbolt and remove the housing and thermostat.

➡ Some engines have a housing which incorporates an O-ring, eliminating the need for a gasket. On these engines, use no sealer when replacing the housing. On engines which incorporate a gasket, thoroughly clean the mating surfaces and use a new gasket coated with adhesive sealer.

5. Install a new thermostat in the housing and position the housing on the engine. Some housings are equipped with a location mark on the side, the mark should face the front of the engine when the housing is installed. Tighten the bolts to 19 ft. lbs. (26 Nm).
6. Install the lower hose.
7. Fill the cooling system.

4.0L Engine

1. Drain the cooling system.

✳✳ CAUTION

When draining the coolant, keep in mind that cats and dogs are attracted by the ethylene glycol antifreeze, and are quite likely to drink any that is left in an uncovered container or in puddles on the ground. This will prove fatal in sufficient quantity. Always drain the coolant into a sealable container. Coolant should be reused unless it is contaminated or several years old.

2. Disconnect the battery ground.
3. Remove the air cleaner duct assembly.
4. Remove the upper radiator hose.
5. Remove the 3 thermostat housing attaching bolts.
6. Remove the thermostat housing. You may have to tap it loose with a plastic mallet or your hand.

To install:

7. Clean all mating surfaces thoroughly. Don't use a sharp metal tool! The housing and engine are aluminum.
8. Make sure that the sealing ring is properly installed on the thermostat rim. Position the thermostat in the housing making sure that the air release valve is in the **up** (12 o'clock) position.
9. Coat the mating surfaces of the housing and engine with an adhesive type sealer. Position the new gasket on the housing and place the housing on the engine. Tighten the bolts to 7–10 ft. lbs. (9–14 Nm).

Intake Manifold

REMOVAL & INSTALLATION

✳✳ CAUTION

On fuel injected models, fuel in the system remains under high pressure, even when the engine is not running. Before disconnecting any fuel line, release the pressure from the fuel system to reduce the possibility of injury or fire.

Never smoke when working around gasoline! Avoid all sources of sparks or ignition. Gasoline vapors are EXTREMELY volatile!

2.2L and 2.6L Engines

CARBURETED

▸ See Figure 18

1. Disconnect the negative battery cable. Drain the cooling system.
2. Remove the air cleaner assembly.
3. Disconnect the accelerator cable. Tag and disconnect the necessary electrical connectors and vacuum hoses.
4. Disconnect the coolant hoses and fuel line.
5. Remove the intake manifold mounting nuts and remove the intake manifold.

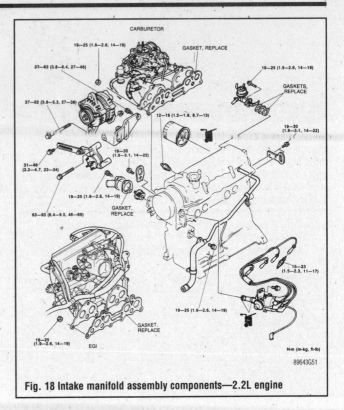

Fig. 18 Intake manifold assembly components—2.2L engine

To install:

6. Clean all gasket mating surfaces.
7. Position a new intake manifold gasket on the cylinder head and install the intake manifold.
8. Install the intake manifold mounting nuts and tighten, in 2–3 steps, to 14–19 ft. lbs. (19–25 Nm) on 2.2L engine or 11–14 ft. lbs. (15–19 Nm) on 2.6L engine. Tighten the nuts at the center of the manifold first and work towards the ends.
9. Connect the fuel line and the coolant hoses.
10. Connect the electrical connectors and vacuum hoses. Connect the accelerator cable.
11. Install the air cleaner assembly and connect the negative battery cable.
12. Fill and bleed the cooling system. Run the engine and check for leaks.

FUEL INJECTED

▸ See Figure 19

1. Relieve the fuel system pressure and disconnect the negative battery cable. Drain the cooling system.
2. Disconnect the air intake tube and ventilation hose. Remove the air pipe and resonance chamber on 2.6L engine.
3. Disconnect the accelerator cable and coolant hoses. Tag and disconnect the electrical connectors to the solenoid valve, throttle sensor and idle switch.
4. Remove the throttle body.
5. Remove the upper intake manifold brackets.
6. Tag and disconnect the vacuum hoses and PCV hose. Tag and disconnect the intake air thermosensor connector and ground wire.
7. Remove the injector harness bracket and remove the upper intake manifold.
8. Tag and disconnect the vacuum hoses from the lower intake manifold. Disconnect the fuel lines.
9. Remove the fuel supply manifold and the injectors. Remove the injector harness and bracket.
10. Remove the pulsation damper and the intake manifold bracket. Remove the attaching nuts and remove the lower intake manifold.

To install:

11. Clean all gasket mating surfaces.
12. Position a new intake manifold-to-cylinder head gasket and install the lower intake manifold. Tighten the nuts to 14–19 ft. lbs. (19–25 Nm).

ENGINE AND ENGINE OVERHAUL 3-9

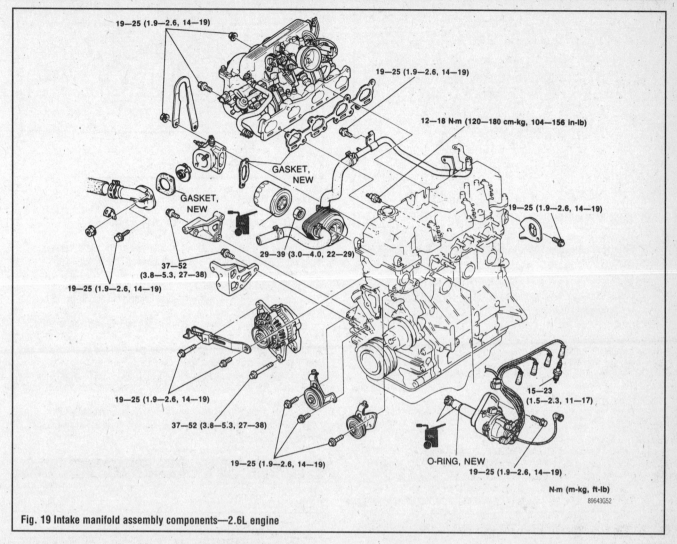

Fig. 19 Intake manifold assembly components—2.6L engine

13. Install the intake manifold bracket and pulsation damper. Install the injector harness and bracket. Tighten the pulsation damper and injector harness bracket bolts to 69–95 inch lbs. (7.8–11.0 Nm).
14. Install the injectors and the fuel supply manifold. Tighten the fuel supply manifold attaching bolts and tighten to 14–19 ft. lbs. (19–25 Nm).
15. Connect the fuel lines. Connect the vacuum hoses to the lower intake manifold.
16. Position a new gasket and install the upper intake manifold. Tighten the attaching bolts/nuts to 14–19 ft. lbs. (19–25 Nm).
17. Install the injector harness bracket. Connect the ground wire and air thermosensor electrical connector. Connect the PCV hose and the vacuum hoses to the upper intake manifold.
18. Install the upper intake manifold brackets.
19. Position a new gasket and install the throttle body. Tighten the mounting nuts to 14–19 ft. lbs. (19–25 Nm).
20. Connect the electrical connectors at the idle switch, throttle sensor and solenoid valve.
21. Connect the coolant hoses and the accelerator cable. On 2.6L engine, install the air pipe and resonance chamber.
22. Connect the ventilation hose and air intake hose. Connect the negative battery cable.
23. Fill and bleed the cooling system. Run the engine and check for leaks and proper operation.

3.0L Engine

♦ See Figures 20 and 21

1. Relieve the fuel system pressure and disconnect the negative battery cable. Drain the cooling system.
2. Remove the air intake tube from the throttle body. Disconnect the accelerator cable.
3. Disconnect the throttle sensor connector and the coolant hoses. Remove the throttle body.
4. Tag and disconnect the vacuum hoses. Remove the bypass air control valve and the intake air pipe.

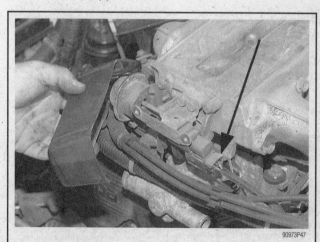

Fig. 20 Remove the front cover from the dynamic chamber and disconnect the electrical connector to the shutter valve actuator before removing the dynamic chamber

3-10 ENGINE AND ENGINE OVERHAUL

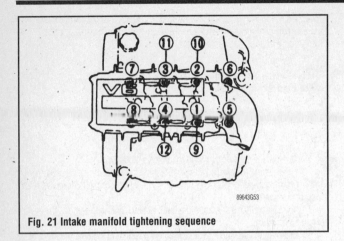

Fig. 21 Intake manifold tightening sequence

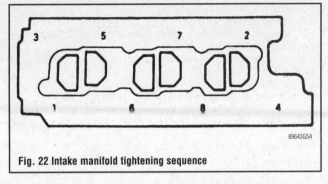

Fig. 22 Intake manifold tightening sequence

5. Remove the extension manifolds. Remove the upper intake plenum with the shutter valve actuator.
6. Remove the fuel supply manifold and the injectors. Disconnect the coolant hoses.
7. Loosen the lower intake manifold nuts, in sequence, in 2 steps and remove the lower intake manifold.

To install:
8. Clean all gasket mating surfaces.
9. Position new lower intake manifold-to-cylinder head gaskets and install the lower intake manifold.
10. Install the intake manifold washers with the white paint mark upward. Install the nuts and tighten, in sequence, in 2 steps to a maximum torque of 14–19 ft. lbs. (19–25 Nm).
11. Install the injectors and the fuel supply manifold. Tighten the attaching bolts to 14–19 ft. lbs. (19–25 Nm).
12. Connect the coolant hoses.
13. Install a new O-ring on the lower intake manifold and install the upper intake plenum. Apply clean engine oil to new O-rings and install on the extension manifolds. Position new gaskets and install the extension manifolds. Tighten the attaching nuts to 14–19 ft. lbs. (19–25 Nm).
14. Position a new gasket and install the intake air pipe. Install the bypass air control valve. Tighten the attaching bolts/nuts to 14–19 ft. lbs. (19–25 Nm).
15. Position a new gasket and install the throttle body. Tighten the attaching nuts to 14–19 ft. lbs. (19–25 Nm).
16. Connect the coolant and vacuum hoses. Connect the throttle sensor connector and accelerator cable.
17. Adjust the accelerator cable deflection to 0.039–0.118 inch (1–3mm).
18. Connect the air intake tube and the negative battery cable.
19. Fill and bleed the cooling system. Run the engine and check for leaks and proper operation.

4.0L Engine

▶ See Figure 22

1. Disconnect the negative battery cable and relieve the fuel system pressure.
2. Remove the air cleaner air intake duct from the throttle body.
3. Remove the snow/ice shield and disconnect the throttle cable and bracket assembly.
4. Tag and disconnect the vacuum hoses from the fittings on the upper intake manifold.
5. Tag and disconnect the electrical connectors at the throttle body, upper intake manifold, lower intake manifold and injectors.
6. Disconnect the fuel lines from the fuel supply manifold.
7. Remove the ignition coil and bracket assembly.
8. Remove the mounting nuts and remove the upper intake manifold.
9. Remove the rocker arm covers.
10. Remove the intake manifold attaching bolts and nuts. Tap the manifold lightly with a plastic mallet to break the gasket seal and remove the manifold.

To install:
11. Clean all gasket mating surfaces.
12. Apply silicone sealer to the block and cylinder head mating surfaces at the 4 corners of the lifter valley opening. Install the intake manifold gaskets and again apply sealer to the same locations.

13. Position the intake manifold on the 2 guide studs and install the nuts and bolts hand tight. Tighten the bolts, in sequence, in 4 steps, first to 3–6 ft. lbs. (4–8 Nm), then to 6–11 ft. lbs. (8–15 Nm), then to 11–15 ft. lbs. (15–21 Nm) and finally to 15–18 ft. lbs. (21–25 Nm).
14. Apply silicone sealer to the 4 locations where the intake manifold and the cylinder heads meet. Install the rocker arm covers with new gaskets and tighten evenly to 3–5 ft. lbs. (4–7 Nm). Wait 2 minutes and tighten the bolts again to the same specification.
15. Install the upper intake manifold and tighten the nuts to 15–18 ft. lbs. (20–25 Nm).
16. Install the ignition coil and bracket assembly. Connect the fuel lines to the fuel supply manifold.
17. Connect the electrical connectors at the throttle body, upper intake manifold, lower intake manifold and injectors.
18. Connect the vacuum hoses to the fittings on the upper intake manifold.
19. Install the throttle cable and bracket assembly and the snow/ice shield to the throttle body.
20. Connect the air cleaner air intake duct to the throttle body.
21. Connect the negative battery cable. Fill and bleed the cooling system. Run the engine and check for leaks.

Exhaust Manifold

REMOVAL & INSTALLATION

2.2L Engine

▶ See Figure 23

1. Raise and support the truck.
2. Remove the two attaching nuts from the exhaust pipe at the manifold.

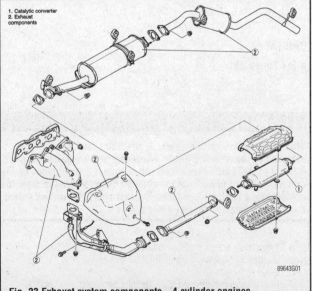

Fig. 23 Exhaust system components—4 cylinder engines

ENGINE AND ENGINE OVERHAUL 3-11

3. Remove the manifold attaching nuts.
4. Remove the manifold.
5. Apply a light film of graphite grease to the exhaust manifold mating surfaces before installation.
6. Install the manifold on the studs and install the attaching nuts. Torque the attaching nuts to specifications.
7. Install a new exhaust pipe gasket. Connect the exhaust pipe gasket. Connect the exhaust pipe and tighten the nuts to 16–21 ft. lbs. (22–28 Nm).

2.6L Engine

1987-88

▶ See Figure 23

1. Disconnect the negative battery cable.
2. Remove the air cleaner assembly.
3. Remove the belt from the power steering pump.
4. Raise the vehicle and make sure it is supported safely.
5. Remove the exhaust pipe from the manifold.
6. Disconnect the air injection tube assembly from the exhaust manifold and lower the vehicle.
7. Remove the power steering pump assembly and move to one side.
8. Remove the heat cowl from the exhaust manifold.
9. Remove the exhaust manifold retaining nuts and remove the assembly from the vehicle.
10. Remove the carburetor air heater from the manifold.
11. Separate the exhaust manifold from the catalytic converter by removing the retaining screws.
12. Clean gasket material from cylinder head and exhaust manifold gasket surfaces. Check mating surfaces for cracks or distortion.

To install:

13. Install a new gasket between the exhaust manifold and catalytic converter. Install mounting screws and tighten to 24 ft. lbs. (32 Nm).
14. Install the carburetor air heater on manifold and tighten to 80 inch lbs. (9 Nm).
15. Lightly coat the new exhaust manifold gasket with sealant (P/N 3419115) or equivalent on cylinder head side.
16. Install the exhaust manifold and mounting nuts. Tighten to 16–21 ft. lbs. (22–28 Nm).
17. Install the heat cowl to manifold and tighten screws to 80 inch lbs. (9 Nm).
18. Install the air cleaner support bracket.
19. Install the power steering pump assembly.
20. Install the air injection tube assembly to air pump.
21. Raise the vehicle and install air injection tube assembly to exhaust manifold.
22. Install the exhaust pipe to manifold.
23. Lower the vehicle and install power steering belt.
24. Fill the cooling system.
25. Install the air cleaner assembly.
26. Connect the negative battery cable.

1989–93

▶ See Figure 23

1. Disconnect the battery ground.
2. Drain the cooling system.

> **✲✲ CAUTION**
>
> When draining the coolant, keep in mind that cats and dogs are attracted by the ethylene glycol antifreeze, and are quite likely to drink any that is left in an uncovered container or in puddles on the ground. This will prove fatal in sufficient quantity. Always drain the coolant into a sealable container. Coolant should be reused unless it is contaminated or several years old.

3. Remove the oil dipstick tube.
4. It may make the job easier to remove the coolant bypass pipe.
5. Disconnect the exhaust pipe from the manifold.
6. Remove the heat shield.
7. Unbolt and remove the manifold. Discard the gasket.
8. Clean all mating surfaces. Place a new gasket and exhaust manifold in position and loosely install the mounting bolts. After all of the manifold bolts and attached components are loosely installed, secure all mounting bolts. Tighten the bolts to 18 ft. lbs. (24 Nm).

MPV

▶ See Figures 23 and 24

1. Disconnect the negative battery cable. Raise and safely support the vehicle.
2. Disconnect the exhaust pipe from the exhaust manifold. If necessary, disconnect the electrical connector from the oxygen sensor.
3. Lower the vehicle.
4. Remove the exhaust manifold insulator, if equipped.
5. If necessary, remove the engine oil dipstick and the dipstick tube.
6. On models with the 3.0L engine, disconnect or remove the exhaust manifold crossover pipe.
7. Remove the exhaust manifold attaching bolts/nuts and remove the exhaust manifold.
8. Clean all mating surfaces. Place a new gasket and exhaust manifold in position and loosely install the mounting bolts. After all of the manifold bolts and attached components are loosely installed, secure all mounting bolts. Tighten the exhaust manifold attaching bolts/nuts to 19 ft. lbs. (25 Nm).

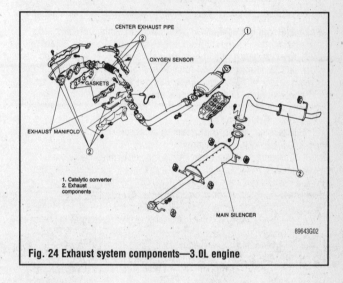

Fig. 24 Exhaust system components—3.0L engine

4.0L Engine

LEFT SIDE

1. Disconnect the negative battery cable. Remove the oil level indicator tube bracket.
2. Remove the power steering pump hoses, if necessary for working clearance.
3. Remove the exhaust pipe-to-manifold bolts.
4. Unbolt and remove the manifold.
5. Clean and lightly oil all fastener threads.
6. Clean the mating surfaces of the manifold and cylinder head. Position the manifold on the cylinder head and install the fasteners, replace all gaskets, if equipped. Install the remaining removed components. Tighten the manifold bolts to 19 ft. lbs (26 Nm).; the exhaust pipe nuts to 20 ft. lbs. (27 Nm). Tighten both exhaust pipe retaining nuts in equal amounts to correctly seat the inlet pipe flange.

RIGHT SIDE

1. Drain the cooling system.

> **✲✲ CAUTION**
>
> When draining the coolant, keep in mind that cats and dogs are attracted by the ethylene glycol antifreeze, and are quite likely to drink any that is left in an uncovered container or in puddles on the ground. This will prove fatal in sufficient quantity. Always drain the coolant into a sealable container. Coolant should be reused unless it is contaminated or several years old.

3-12 ENGINE AND ENGINE OVERHAUL

2. Remove the heater hose support bracket.
3. Disconnect the heater hoses.
4. Remove the exhaust pipe-to-manifold nuts.
5. Unbolt and remove the manifold.
6. Clean the mating surfaces of the manifold and cylinder head. Position the manifold on the cylinder head and install the fasteners, replace all gaskets, if equipped. Install the remaining removed components. Tighten the manifold bolts to 19 ft. lbs. (26 Nm).; the exhaust pipe nuts to 20 ft. lbs. (27 Nm). Tighten both exhaust pipe retaining nuts in equal amounts to correctly seat the inlet pipe flange.

Radiator

REMOVAL & INSTALLATION

Pickup and MPV

♦ See Figures 25 thru 34

1. Drain the cooling system.

> **※ CAUTION**
>
> When draining the coolant, keep in mind that cats and dogs are attracted by the ethelyne glycol antifreeze, and are quite likely to drink any that is left in an uncovered container or in puddles on the ground. This will prove fatal in sufficient quantity. Always drain the coolant into a sealable container. Coolant should be reused unless it is contaminated or several years old.

2. Remove the fresh air duct.
3. Disconnect the upper and lower radiator hoses.
4. Disconnect the coolant reservoir hose.
5. If equipped with automatic transmission, disconnect the cooler lines.

6. Remove the fan shroud.
7. Remove the fan. Don't lay the fan, if equipped with a fan clutch, on its side. Fluid will be lost and the fan clutch will have to be replaced.
8. Unbolt and remove the radiator.
9. Install the radiator against the supports and tighten the mounting bolts.
10. Install the hoses and cooler lines on the radiator. Tighten the clamps.
11. Install the fan.
12. Install the fan shroud.
10. Refill the cooling system with the specified amount and type of coolant. Run the engine and check for leaks.

Navajo

♦ See Figures 35 and 36

1. Drain the cooling system. Remove the overflow tube from the coolant recovery bottle and from the radiator.

> **※ CAUTION**
>
> When draining the coolant, keep in mind that cats and dogs are attracted by the ethylene glycol antifreeze, and are quite likely to drink any that is left in an uncovered container or in puddles on the ground. This will prove fatal in sufficient quantity. Always drain the coolant into a sealable container. Coolant should be reused unless it is contaminated or several years old.

2. Disconnect the transmission cooling lines from the bottom of the radiator, if so equipped.
3. Remove the retaining bolts at the top of the shroud, and position the shroud over the fan, clear of the radiator.
4. Disconnect the upper and lower hoses from the radiator.
5. Remove the radiator retaining bolts or the upper supports and lift the radiator from the vehicle.
6. Install the radiator in the reverse order of removal. Fill the cooling system and check for leaks.

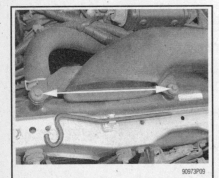

Fig. 25 If necessary, remove the two air inlet duct mounting bolts . . .

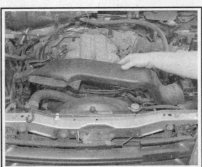

Fig. 26 . . . then remove the air inlet duct from the engine compartment

Fig. 27 Using pliers or an equivalent tool, compress the upper radiator hose clamp and move the clamp away from the radiator, then disconnect the hose from the radiator

Fig. 28 Disconnect the rubber radiator-to-overflow tank hose at the radiator

Fig. 29 When removing the radiator, if equipped with an automatic transmission, disconnect the fluid cooler lines . . .

Fig. 30 . . . then plug the cooler lines and the openings in the radiator

ENGINE AND ENGINE OVERHAUL

Fig. 31 On 4 cylinder equipped models, remove the purge solenoid valve and mounting bracket

Fig. 32 Remove the left side radiator mounting bolts . . .

Fig. 33 . . . then remove the two right side radiator mounting bolts

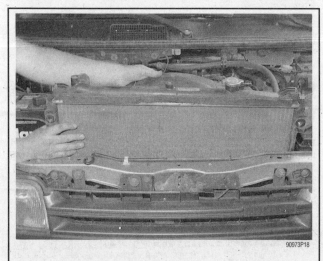

Fig. 34 Lift the radiator and shroud assembly out of the vehicle, with the engine fan in the shroud opening

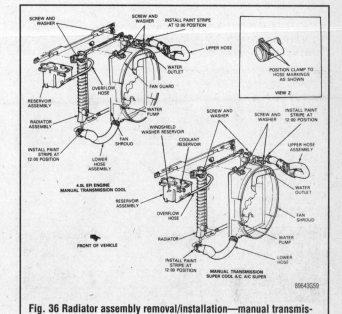

Fig. 36 Radiator assembly removal/installation—manual transmission

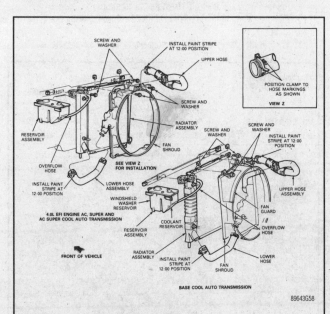

Fig. 35 Radiator assembly removal/installation—automatic transmission

Engine Fan

REMOVAL & INSTALLATION

Pickup and MPV

♦ See Figures 37 and 38

1. Loosen and remove the drive belts.
2. Remove the bolts that retain the fan shroud to the radiator support. Remove (if there is enough room between the fan blades and radiator) or position the shroud back over the water pump and fan assembly, if necessary to gain working room.
3. Loosen and remove the fan to water pump mounting bolts or nuts and remove the fan assembly. Don't lay the fan, if equipped with a fan clutch, on its side. Fluid could be lost and the fan clutch might require replacement. Inspect the condition of the fan blades, if any are cracked or damaged, replace the fan.
4. Install the fan assembly in position on the water pump and pulley. Secure the fan assembly with the mounting nuts or bolts. Tighten the nuts to 69–70 inch lbs. (8–11 Nm). Install the shroud and drive belts. Adjust the drive belts to the proper tension.

3-14　ENGINE AND ENGINE OVERHAUL

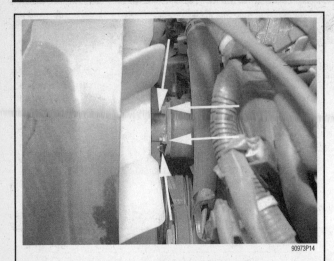

Fig. 37 Remove the four fan assembly-to-water pump pulley mounting nuts

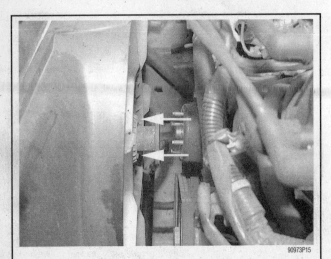

Fig. 38 After the fan mounting nuts have been removed, slide the fan assembly forward and place inside the radiator shroud

Navajo

♦ See Figure 39

1. Remove the fan shroud.
2. Using Fan Clutch Pulley Holder T84T-6312-C and Fan Clutch Nut Wrench T84T-6312-D, or their equivalents, loosen the large nut attaching the clutch to the water pump hub.

➡ The nut is loosened clockwise.

3. Remove the fan/clutch assembly.
4. Remove the fan-to-clutch bolts.

To install:

5. Installation is the reverse of removal. Torque the fan-to-clutch bolts to 55-70 inch lbs.; the hub nut to 50-100 ft. lbs. Don't forget, the hub is tightened counterclockwise.

Water Pump

REMOVAL & INSTALLATION

2.2L Engine

♦ See Figure 40

➡ Use special tool No. 49E301060 or equivalent on the engine flywheel gear to stop the engine from rotating during removal and installation of the crankshaft pulley.

1. Drain the cooling system.

✴✴ CAUTION

When draining the coolant, keep in mind that cats and dogs are attracted by the ethylene glycol antifreeze, and are quite likely to drink any that is left in an uncovered container or in puddles on the ground. This will prove fatal in sufficient quantity. Always drain the coolant into a sealable container. Coolant should be reused unless it is contaminated or several years old.

2. Disconnect the negative battery cable.
3. Remove the alternator belt.
4. Remove the power steering pump belt.
5. Remove the upper and the lower timing belt covers.
6. Turn the crankshaft to position the **A** mark on the camshaft pulley with the mark on the housing.
7. Remove the crankshaft pulley mounting bolts and the pulley.
8. Remove the tensioner pulley lock bolt, the pulley and the spring.
9. Remove the water inlet pipe from the water pump.
10. Remove the water pump retaining bolts. Remove the water pump from its mounting.
11. Thoroughly clean the mounting surfaces of the pump and engine.

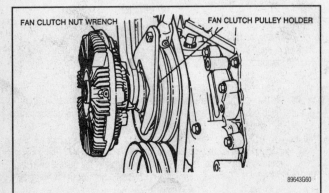

Fig. 39 Fan clutch removal/installation—remember that the nut is left-handed

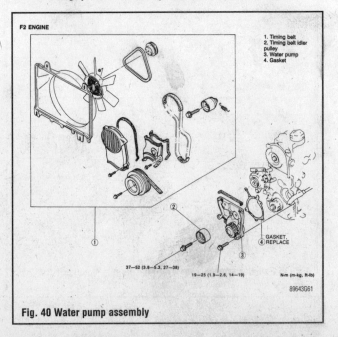

Fig. 40 Water pump assembly

ENGINE AND ENGINE OVERHAUL 3-15

12. Install the components in the reverse order.
13. Use new O-rings and gaskets when installing the pump. Use a new O-ring and 3 rubber seals.
14. Tighten the water pump mounting bolts to 14–19 ft. lbs. (19–25 Nm).
15. Install the timing belt, tensioner and spring.
16. Fill the cooling system and check the timing.

2.6L Engine

1987-88

♦ See Figure 41

1. Drain the cooling system.

> **※ CAUTION**
>
> When draining the coolant, keep in mind that cats and dogs are attracted by the ethylene glycol antifreeze, and are quite likely to drink any that is left in an uncovered container or in puddles on the ground. This will prove fatal in sufficient quantity. Always drain the coolant into a sealable container. Coolant should be reused unless it is contaminated or several years old.

2. Remove the radiator hose, by-pass hose and heater hose from the water pump.
3. Remove the drive pulley shield.
4. Remove the locking screw and pivot screws.
5. Remove the drive belt and water pump from the engine.

6. Install a new O-ring gasket in O-ring groove of pump body assembly to cylinder block.
7. Position the water pump assembly against the engine and install the mounting fasteners.
8. Tighten the water pump mounting fasteners to 14–19 ft. lbs. (19–25 Nm).
9. Install the water pump drive belt and adjust to specification.
10. Install drive belt pulley cover.
11. Install the radiator hose, by-pass hose and heater hose.
12. Fill the cooling system.

1989-93

♦ See Figure 41

1. Disconnect the battery ground.
2. Drain the cooling system.

> **※ CAUTION**
>
> When draining the coolant, keep in mind that cats and dogs are attracted by the ethylene glycol antifreeze, and are quite likely to drink any that is left in an uncovered container or in puddles on the ground. This will prove fatal in sufficient quantity. Always drain the coolant into a sealable container. Coolant should be reused unless it is contaminated or several years old.

3. Remove the accessory drive belts.
4. Remove the fan and shroud.
5. Remove the water pump pulley.
6. Unbolt and remove the water pump.
7. Thoroughly clean the gasket mounting surfaces.
8. Using a new gasket coated with sealer, position the water pump on the engine. Tighten the bolts to 14–19 ft. lbs. (19–25 Nm).
9. Install all the remaining parts.
10. Fill the cooling system.
11. Run the engine and check for leaks.

3.0L Engine

♦ See Figures 42, 43 and 44

1. Position the engine at TDC on the compression stroke.
2. Properly relieve the fuel system pressure.
3. Disconnect the negative battery cable.
4. Remove the air cleaner assembly.
5. Drain the cooling system.
6. Remove the spark plug wires.
7. Remove the fresh air duct assembly.
8. Remove the cooling fan and radiator cowling.
9. Remove the drive belts.
10. Remove the air conditioning compressor idler pulley. If necessary, remove the compressor and position it to the side.
11. Remove the crankshaft pulley and baffle plate.
12. Remove the coolant bypass hose.
13. Remove the upper radiator hose.

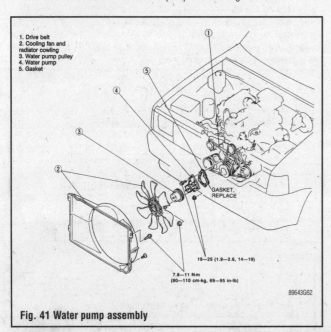

Fig. 41 Water pump assembly

Fig. 42 Remove the water pump pulley

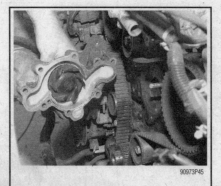

Fig. 43 Remove the water pump assembly from the engine

Fig. 44 Clean the water pump-to-engine mating surface of all gasket material

3-16 ENGINE AND ENGINE OVERHAUL

14. Remove the timing belt cover assembly retaining bolts.
15. Remove the timing belt cover assembly and gasket.
16. Turn the crankshaft to align the mating marks of the pulleys.
17. Remove the upper idler pulley.
18. Remove the timing belt. If reusing the belt be sure to mark the direction of rotation.
19. Remove the timing belt auto tensioner.
20. Unbolt and remove the water pump. Discard the gasket.
21. Thoroughly clean the mating surfaces of the pump and engine.

To install:

22. Position the pump and a new gasket, coated with sealer, onto the engine. Tighten the bolts to 14–19 ft. lbs. (19–25 Nm).
23. Install the timing belt.
24. Install the timing belt cover assembly.
25. Install the upper radiator hose.
26. Install the coolant bypass hose.
27. Install the crankshaft pulley and baffle plate.
28. Install the compressor.
29. Install the air conditioning compressor idler pulley.
30. Install and adjust the drive belts.
31. Install the cooling fan and radiator cowling.
32. Install the fresh air duct assembly.
33. Install the spark plug wires.
34. Fill the cooling system.
35. Install the air cleaner assembly.
36. Connect the negative battery cable.

4.0L Engine

1. Drain the cooling system.

> ✱✱ **CAUTION**
>
> When draining the coolant, keep in mind that cats and dogs are attracted by the ethylene glycol antifreeze, and are quite likely to drink any that is left in an uncovered container or in puddles on the ground. This will prove fatal in sufficient quantity. Always drain the coolant into a sealable container. Coolant should be reused unless it is contaminated or several years old.

2. Remove the lower radiator hose.
3. Disconnect the heater hose at the pump.
4. Remove the fan and fan clutch assembly.
5. Loosen the alternator mounting bolts and remove the belt. On vehicles with air conditioning, remove the alternator and bracket.
6. Remove the water pump pulley.
7. Remove the attaching bolts and remove the water pump.

To install:

8. Clean the mounting surfaces of the pump and front cover thoroughly. Remove all traces of gasket material.
9. Apply adhesive gasket sealer to both sides of a new gasket and place the gasket on the pump.
10. Position the pump on the cover and install the bolts finger-tight. When all bolts are in place, torque them to 72–108 inch lbs. (8–12 Nm).
11. Install the pulley.
12. On vehicles with air conditioning, install the alternator and bracket.
13. Install and adjust the drive belt.
14. Connect the hoses and tighten the clamps.
15. Install the fan and clutch assembly.
16. Fill and bleed the cooling system. Start the engine and check for leaks.

Cylinder Head

REMOVAL & INSTALLATION

2.2L Engine

▶ See Figures 45 and 46

1. Relieve the fuel system pressure if an injected model. Disconnect the negative battery cable and drain the cooling system.

> ✱✱ **CAUTION**
>
> When draining the coolant, keep in mind that cats and dogs are attracted by the ethylene glycol antifreeze, and are quite likely to drink any that is left in an uncovered container or in puddles on the ground. This will prove fatal in sufficient quantity. Always drain the coolant into a sealable container. Coolant should be reused unless it is contaminated or several years old.

2. Disconnect the spark plug wires and remove the spark plugs.
3. Disconnect the accelerator cable. If equipped with automatic transmission, disconnect the throttle cable.
4. Remove the air intake pipe.
5. Remove the air cleaner and fuel hose. Cover the fuel hose to prevent leakage, if equipped with a carburetor. If injected, remove the air intake hose. If equipped with a carburetor, remove the fuel pump.

> ✱✱ **CAUTION**
>
> Never smoke when working around gasoline! Avoid all sources of sparks or ignition. Gasoline vapors are EXTREMELY volatile!

6. Remove the upper radiator hose, water by-pass hose, heater hose, and brake vacuum hose.
7. Remove the 3-way and EGR solenoid valve assemblies.
8. Disconnect the engine harness connector and ground wire.
9. Remove the vacuum chamber and exhaust manifold insulator.
10. Remove the EGR pipe and exhaust pipe.
11. Remove the exhaust manifold.
12. Remove the intake manifold bracket and the intake manifold.
13. Remove the distributor.
14. Loosen the air conditioning compressor and bracket, position it off to the side and tie it out of the way. Do not disconnect the refrigerant lines!
15. Remove the upper timing belt cover and the timing belt tensioner spring.
16. To remove the timing belt, perform the following:

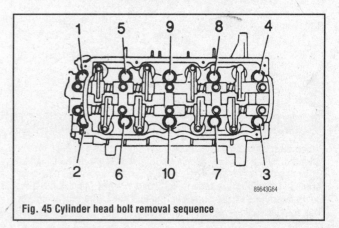

Fig. 45 Cylinder head bolt removal sequence

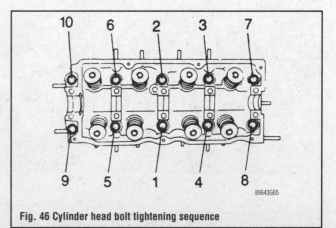

Fig. 46 Cylinder head bolt tightening sequence

ENGINE AND ENGINE OVERHAUL

    a. Rotate the crankshaft so that the **1** on the camshaft pulley is aligned with the timing mark on the front housing.
    b. When timing marks are aligned, loosen the timing belt tensioner lock bolt. Pull the tensioner as far out as will go and then temporarily tighten the lock bolt to hold it there.
    c. Lift the timing belt from the camshaft pulley and position it out of the way.
17. Remove the cylinder head cover and cover gasket.
18. Loosen the cylinder head bolts in the proper sequence and remove the cylinder head and head gasket.
19. Once the cylinder head has been removed, it should be cleaned and inspected before installing it onto the engine or being replaced. Refer to the "Engine Reconditioning" later in this section.

To install:
20. Position a new head gasket on the cylinder block and install the cylinder head. Apply oil to the threads and seat faces of the cylinder head bolts and install.
21. Tighten the cylinder head bolts in 2–3 steps in the proper sequence. The final torque specification is 59–64 ft. lbs. (80–86 Nm).
22. Apply silicone sealer to each side of the front and rear camshaft bearing caps where the cap meets the cylinder head. Install the rocker arm cover.
23. Install the timing belt and tensioner. Install the timing belt front cover.
24. Install the intake and exhaust manifolds.
25. Install the upper radiator and water by-pass hoses. If carburetor equipped, install the secondary air pipe.
26. Install the distributor. Install the spark plugs and connect the spark plug wires.
27. Connect all vacuum hoses and electrical connectors. Connect the heater hoses and the brake vacuum hose.
28. If equipped, install the fuel pump. Connect the fuel lines.
29. Install the cooling fan and shroud.
30. Connect the accelerator cable. Install the air cleaner or air intake hose.
31. Install the splash shield. Connect the negative battery cable.
32. Fill and bleed the cooling system. Run the engine and check for leaks.
33. Check and adjust the ignition timing.

2.6L Engine

1987-88

♦ See Figures 47 and 48

1. Relieve fuel system pressure, if injected. Disconnect the negative battery cable. Drain the cooling system.

> ※※ **CAUTION**
>
> When draining the coolant, keep in mind that cats and dogs are attracted by the ethylene glycol antifreeze, and are quite likely to drink any that is left in an uncovered container or in puddles on the ground. This will prove fatal in sufficient quantity. Always drain the coolant into a sealable container. Coolant should be reused unless it is contaminated or several years old.

2. Remove the air cleaner assembly.
3. Remove the exhaust manifold.
4. Remove the water by-pass pipe. Disconnect all necessary vacuum hoses and electrical connectors. Disconnect the fuel lines.

> ※※ **CAUTION**
>
> Never smoke when working around gasoline! Avoid all sources of sparks or ignition. Gasoline vapors are EXTREMELY volatile!

5. Remove the carburetor and intake manifold. Remove the distributor.
6. Remove the rocker arm cover. Remove the rocker arm and shaft assembly, leaving the bolts in place in the shaft to keep the assembly together. Install hydraulic lash adjuster holder tools 49 U012 001 or equivalent to keep the adjusters from falling.
7. Hold the crankshaft pulley bolt with a wrench to keep it from turning, then remove the camshaft sprocket bolt. Remove the camshaft. With the camshaft sprocket and timing chain meshed together, place the sprocket on the sprocket holder.

➡ Do not rotate the engine with the camshaft removed. Once the cylinder head is removed, do not allow the sprocket to fall from the sprocket holder.

8. Remove the cylinder head-to-timing chain case bolts.
9. Loosen the cylinder head bolts in 2–3 steps in the proper sequence. Remove the bolts and remove the cylinder head.
Clean all gasket mating surfaces.
10. Once the cylinder head has been removed, it should be cleaned and inspected before installing it onto the engine or being replaced. Refer to the "Engine Reconditioning" later in this section.

To install:
11. Apply a thin coat of sealer to the cylinder block and timing chain cover contact surfaces, install a new head gasket and the cylinder head. Do not apply sealer to the cylinder head gasket.
12. Apply oil to the threads and seat faces of the cylinder head bolts and install. With the engine cold, tighten the cylinder head bolts, in sequence in 2–3 steps. Final torque on bolts 1–10 should be 65–72 ft. lbs. (88–98 Nm) and on bolts 11 and 12, 11–16 ft. lbs. (15–22 Nm).
13. Apply engine oil to the camshaft journals and install the camshaft, aligning the dowel pin with the camshaft sprocket. Install the distributor drive gear and the lock bolt. Tighten the lock bolt by hand only at this time.
14. Coat the circular packing with sealant and install in the rear of the cylinder head. Install the rocker arms/shafts assembly and tighten the mounting bolts, in sequence, to 14–15 ft. lbs. (19–20 Nm) except for the No. 5 bearing cap rear bolt, which should be tightened to 15–20 ft. lbs. (20–26 Nm).
15. Tighten the distributor drive gear/camshaft sprocket lock bolt to 36–43 ft. lbs. (49–58 Nm).
16. Remove the hydraulic lash adjuster holders. Adjust the jet valve clearance.
17. Install the half-circle packing in the front of the cylinder head and apply sealer across the top of the packing and on both sides of the packing on the cylinder head. Install the rocker arm cover and tighten the bolts to 43–61 inch lbs. (5–7 Nm).

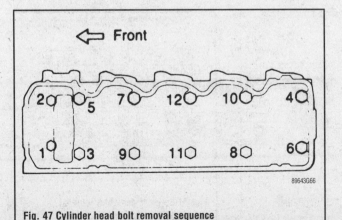

Fig. 47 Cylinder head bolt removal sequence

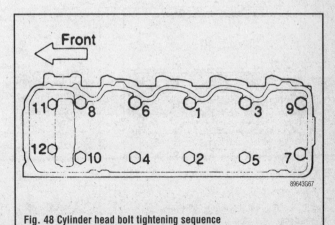

Fig. 48 Cylinder head bolt tightening sequence

3-18 ENGINE AND ENGINE OVERHAUL

18. Install the remaining components. Fill and bleed the cooling system. Run the engine and check for leaks. Check and adjust the ignition timing.

1989-93

♦ See Figures 49 and 50

1. Properly relieve the fuel system pressure.

> **CAUTION**
>
> Never smoke when working around gasoline! Avoid all sources of sparks or ignition. Gasoline vapors are EXTREMELY volatile!

2. Disconnect the negative battery cable.
3. Remove the air cleaner assembly.
4. Drain the coolant.

> **CAUTION**
>
> When draining the coolant, keep in mind that cats and dogs are attracted by the ethylene glycol antifreeze, and are quite likely to drink any that is left in an uncovered container or in puddles on the ground. This will prove fatal in sufficient quantity. Always drain the coolant into a sealable container. Coolant should be reused unless it is contaminated or several years old.

5. Position the engine at TDC on the compression stroke so that all the pulley matchmarks are aligned.
6. Remove the accelerator cable. Remove the air intake pipe and resonance chamber.
7. Remove the accessory drive belts and AC belt idler.
8. Remove the upper radiator hose.
9. Remove the brake vacuum hose.
10. Remove the spark plug wires.
11. Remove the spark plugs.
12. Remove the oil cooler coolant hose.
13. Remove the canister hose.
14. Remove the fuel lines.
15. Disconnect the oxygen sensor.
16. Remove the solenoid valves.
17. Disconnect the emissions harness.
18. Remove the rocker cover. Check to ensure the engine is set on TDC. The timing mark on the camshaft sprocket should be 90 degrees to the right, parallel to the top of the cylinder head. Make sure the yellow crankshaft pulley timing mark is aligned with the indicator pin.
19. Remove the distributor. Do not rotate the engine after distributor removal.
20. Hold the crankshaft pulley with a suitable tool and remove the distributor drive gear/camshaft pulley retaining bolt and the drive gear. Remove the upper timing cover assembly.
21. Push the timing chain adjuster sleeve in towards the left, and insert a pin (2mm [0.0787 in.] diameter by 45mm [1.77 in.] long) into the lever hole to hold it in place.
22. Wire the chain to the pulley and remove the pulley from the camshaft.

Do not allow the sprocket and chain to fall down into the engine and cause the chain to become disengaged from the crankshaft sprocket.

23. Remove the intake manifold bracket.
24. Disconnect the exhaust pipe.
25. Remove the 2 front head bolts.
26. Remove the remaining head bolts starting from the middle and working outward toward the ends of the head.
27. Lift off the head.
28. Discard the head gasket.
29. Once the cylinder head has been removed, it should be cleaned and inspected before installing it onto the engine or being replaced. Refer to the "Engine Reconditioning" later in this section.

To install:

30. Apply RTV sealer to the top front of the block as shown.
31. Place a new head gasket on the block.
32. Position the head on the block.
33. Clean the head bolts and apply oil to the threads.
34. Tighten the head bolts, in 2 even steps, to 64 ft. lbs. (87 Nm).
35. Torque the two front bolts, to 17 ft. lbs. (23 Nm).
36. Place the camshaft pulley on the camshaft and tighten the bolt to to 95 inch lbs. (11 Nm).; the nut to 87 inch lbs. (10 Nm).
37. Connect the exhaust pipe.
38. Install the intake manifold bracket.
39. Install the upper timing cover assembly.
40. Install the distributor.
41. Install the rocker cover.
42. Connect the emissions harness.
43. Install the solenoid valves.
44. Connect the oxygen sensor.
45. Install the fuel lines.
46. Install the canister hose.
47. Install the oil cooler coolant hose.
48. Install the spark plugs.
49. Install the spark plug wires.
50. Install the brake vacuum hose.
51. Install the upper radiator hose.
52. Install the accessory drive belts.
53. Install the accelerator cable.
54. Fill the cooling system.
55. Install the air cleaner assembly.
56. Connect the negative battery cable.

3.0L Engine

♦ See Figures 51 and 52

1. Position the engine at TDC on the compression stroke.
2. Properly relieve the fuel system pressure.
3. Disconnect the negative battery cable.
4. Remove the air cleaner assembly.
5. Drain the cooling system.
6. Remove the spark plug wires.
7. Remove the fresh air duct assembly.
8. Remove the cooling fan and radiator cowling.

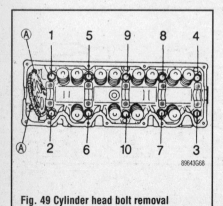

Fig. 49 Cylinder head bolt removal sequence

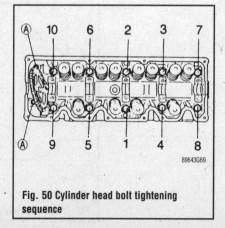

Fig. 50 Cylinder head bolt tightening sequence

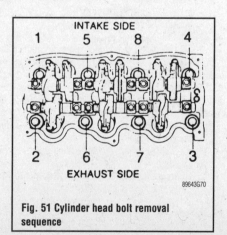

Fig. 51 Cylinder head bolt removal sequence

ENGINE AND ENGINE OVERHAUL 3-19

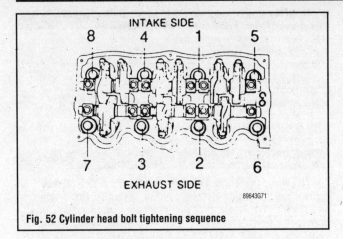

Fig. 52 Cylinder head bolt tightening sequence

9. Remove the drive belts.
10. Remove the air conditioning compresser idler pulley. If necessary, remove the compressor and position it to the side.
11. Remove the crankshaft pulley and baffle plate.
12. Remove the coolant bypass hose.
13. Remove the upper radiator hose.
14. Remove the timing belt cover assembly retaining bolts. Remove the timing belt cover assembly and gasket.
15. Turn the crankshaft to align the mating marks of the pulleys.
16. Remove the upper idler pulley.
17. Remove the timing belt. If reusing the belt be sure to mark the direction of rotation.
18. Disconnect and plug canister, brake vacuum and fuel hoses. If equipped with automatic transmission, disconnect the automatic transmission vacuum hose.
19. Remove the 3-way soleniod valve assembly and disconnect all engine harness connector and grounds.
20. If equipped with automatic transmission, remove the dipstick. Disconnect the required vacuum hoses. Disconnect the accelerator linkage.
21. Remove the distributor and the EGR pipe.
22. Remove the six extension manifolds. Remove the O-rings from the extension manifolds and purchase new ones. Remove the intake manifold by loosening the retaining bolts in the proper sequence.
23. Remove the cylinder head cover, gasket and seal washers.
24. Remove the center exhaust pipe insulator and pipe. Disconnect the exhaust manifold retaining bolts. Remove the exhaust manifold with insulator.
25. Remove the seal plate.
26. Remove the cylinder head retaining bolts in the proper sequence in 2 or 3 stages. Remove the cylinder head from the vehicle.
27. Once the cylinder head has been removed, it should be cleaned and inspected before installing it onto the engine or being replaced. Refer to the "Engine Reconditioning" later in this section.
28. Thoroughly clean the cylinder head bolts. After the bolts are cleaned, measure the length of each bolt and replace out of specifications bolts as required.
- Intake: 108mm
- Exhaust: 138mm**Maximum**
- Intake: 109mm
- Exhaust: 139mm
29. Check the oil control plug projection at the cylinder block. Projection should be 0.53–0.57mm. If correct, apply clean engine oil to a new O-ring and position it on the control plug.

To install:
30. Place the new cylinder head gasket on the left bank with the **L** mark facing up. Place the new cylinder head gasket on the right bank with the **R** mark facing up. Install the cylinder onto the block. Tighten the head bolts in the following manner:
 a. Coat the threads and the seating faces of the head bolts with clean engine oil.
 b. Torque the bolts in the proper sequence to 14 ft. lbs. (19 Nm).
 c. Paint a mark on the head of each bolt.
 d. Using this mark as a reference, tighten the bolts in the proper sequence an additional 90°.
 e. Repeat Step d.

31. Install the seal plate.
32. Install the exhaust manifold with insulator.
33. Connect the exhaust manifold retaining bolts.
34. Install the center exhaust pipe insulator and pipe.
35. Install the cylinder head cover, gasket and seal washers.
36. Install the intake manifold by loosening the retaining bolts in the proper sequence.
37. Install the O-rings from the extension manifolds.
38. Install the six extension manifolds.
39. Install the distributor and the EGR pipe.
40. If equipped with automatic transmission, install the dipstick. Connect the required vacuum hoses. Connect the accelerator linkage.
41. Install the 3-way soleniod valve assembly and connect all engine harness connector and grounds.
42. Connect the canister, brake vacuum and fuel hoses. If equipped with automatic transmission, connect the automatic transmission vacuum hose.
43. Install the timing belt.
44. Install the timing belt cover assembly and new gasket.
45. Install the upper radiator hose.
46. Install the coolant bypass hose.
47. Install the crankshaft pulley and baffle plate.
48. Install the compressor.
49. Install the air conditioning compresser idler pulley.
50. Install the accessory drive belts.
51. Install the cooling fan and radiator cowling.
52. Install the fresh air duct assembly.
53. Install the spark plug wires.
54. Fill the cooling system.
55. Install the air cleaner assembly.
56. Connect the negative battery cable.

4.0L Engine

▶ See Figure 53

1. Relieve fuel system pressure. Disconnect the negative battery cable. Drain the cooling system (engine cold) into a clean container and save the coolant for reuse.

✳✳ CAUTION

When draining the coolant, keep in mind that cats and dogs are attracted by the ethylene glycol antifreeze, and are quite likely to drink any that is left in an uncovered container or in puddles on the ground. This will prove fatal in sufficient quantity. Always drain the coolant into a sealable container. Coolant should be reused unless it is contaminated or several years old.

2. Disconnect the battery ground cable.
3. Remove the air cleaner.
4. Remove the upper and lower intake manifolds as described earlier.
5. If the left cylinder head is being removed:
 a. Remove the accessory drive belt.
 b. Remove the air conditioning compressor.

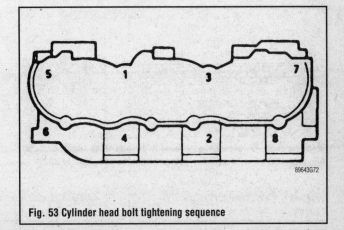

Fig. 53 Cylinder head bolt tightening sequence

3-20 ENGINE AND ENGINE OVERHAUL

    c. Remove the power steering pump and bracket assembly. DO NOT disconnect the hoses. Tie the assembly out of the way.
    d. Remove the spark plugs.
  6. If the right head is being removed:
    a. Remove the accessory drive belt.
    b. Remove the alternator and bracket.
    c. Remove the EDIS ignition coil and bracket.
    d. Remove the spark plugs.
  7. Remove the exhaust manifold(s).
  8. Remove the rocker arm covers as previously described.
  9. Remove the rocker shaft assembly.
  10. Remove the pushrods, keeping them in order so they may be installed in their original locations.
  11. Loosen the cylinder head attaching bolts in reverse of the torque sequence, then remove the bolts and discard them. They cannot be re-used.
  12. Lift off the cylinder head(s).
  13. Remove and discard the old cylinder head gasket(s).
  14. Once the cylinder head has been removed, it should be cleaned and inspected before installing it onto the engine or being replaced. Refer to the "Engine Reconditioning" later in this section.

To install:
  15. Lightly oil all bolt and stud bolt threads except those specifying special sealant. Position the new head gasket(s) on the cylinder block, using the dowels for alignment. The dowels should be replaced if damaged.

➡ **The cylinder head(s) and intake manifold are torqued alternately and in sequence, to assure a correct fit and gasket crush.**

  16. Position the cylinder head(s) on the block.
  17. Apply a bead of RTV silicone gasket material to the mating joints of the head and block at the 4 corners. Install the intake manifold gasket and again apply the sealer.

➡ **This sealer sets within 15 minutes, so work quickly!**

  18. Install the lower intake manifold and install the bolts and nuts for the manifold and head(s). Tighten all fasteners finger-tight.
  19. Tighten the intake manifold fasteners, in sequence, to 36–72 inch lbs. (4–8 Nm).

✱✱ WARNING

Do not re-use the old head bolts. ALWAYS use new head bolts!

  20. Torque the head bolts, in sequence, to 59 ft. lbs. (80 Nm).
  21. Tighten the intake manifold fasteners, in sequence, to 6–11 ft. lbs. (8–15 Nm).
  22. Tighten the head bolts, in sequence, an additional 80–85 DEGREES tighter. 85 degrees is a little less than ¼ turn. ¼ turn would equal 90 degrees.
  23. Torque the intake manifold fasteners, in sequence, to 11–15 ft. lbs. (15–20 Nm).; then, in sequence, to 15–18 ft. lbs. (20–24 Nm).
  24. Dip each pushrod in heavy engine oil then install the pushrods in their original locations.
  25. Install the rockerarm/shaft assembly(ies).
  26. Apply another bead of RTV sealer at the 4 corners where the intake manifold and heads meet.
  27. Install the rocker covers, using new gaskets coated with sealer.
  28. Install the upper intake manifold.
  29. Install the exhaust manifold(s).
  30. Install the spark plugs and wires.
  31. If the left head was removed, install the power steering pump, compressor and drive belt.
  32. If the right head was removed, install the EDIS coil and bracket, alternator and bracket, and the drive belt.
  33. Install the air cleaner.
  34. Fill the cooling system.

➡ **At this point, it's a good idea to change the engine oil. Coolant contamination of the engine oil often occurs during cylinder head removal.**

  35. Connect the battery ground cable.
  36. Start the engine and check for leaks.

Oil Pan

REMOVAL & INSTALLATION

2.2L Engine

◆ See Figure 54

  1. Disconnect the battery ground cable.
  2. Raise and support the truck on jackstands. Drain the oil.

✱✱ CAUTION

The EPA warns that prolonged contact with used engine oil may cause a number of skin disorders, including cancer! You should make every effort to minimize your exposure to used engine oil. Protective gloves should be worn when changing the oil. Wash your hands and any other exposed skin areas as soon as possible after exposure to used engine oil. Soap and water, or waterless hand cleaner should be used.

  3. Remove the skid plate.
  4. Place a floor jack under the front of the engine at the crankshaft pulley and take up the weight of the engine. Or use a shop crane to support the engine.
  5. Remove the crossmember.
  6. Remove the cotter pin and nut and, with a puller, disconnect the idler arm from the center link.
  7. Remove the engine mount gusset plates from the sides of the engine.
  8. Remove the bell housing front cover.
  9. Unbolt and remove the oil pan. A flat tipped screwdriver may be used to break the seal between the pan and block.
  10. Clean all the gasket surfaces. Straighten and portion of the pan rim that is bent.
  11. Clean the oil pan, oil pump pickup tube and oil pump screen.

To install:
  12. If you are using a gasket, install a new oil pan gasket coated with oil resistant sealer. Place RTV silicone sealer at the points shown in the accompanying illustration. If you are using RTV silicone gasket material in place of a

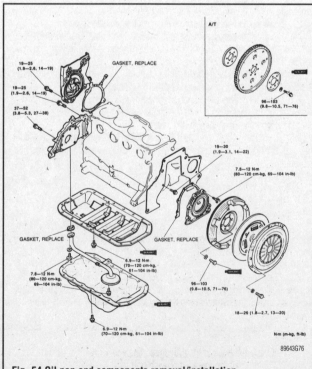

Fig. 54 Oil pan and components removal/installation

ENGINE AND ENGINE OVERHAUL 3-21

conventional gasket, run a ⅛ in. bead around the rim of the pan, going inboard of each bolt hole. Tighten the pan bolts within 30 minutes of application. Tighten the pan bolts to 61–104 inch lbs. (7–12 Nm).

13. Install the bell housing front cover. Tighten the bell housing cover to 15–20 ft. lbs. (20–27 Nm).
14. Install the engine mount gusset plates from the sides of the engine.
15. Install the idler arm on the center link.
16. Install the cotter pin and nut. Tighten the idler arm nut to 25–30 ft. lbs. (34–41 Nm).
17. Install the crossmember.
18. Remove the shop crane.
19. Install the skid plate.
20. Fill the engine with the proper amount of oil.
21. Install the battery ground cable.

2.6L Engine

1987-88

♦ See Figure 55

1. Raise and safely support the vehicle on jackstands. Drain the oil pan.

❉❉ CAUTION

The EPA warns that prolonged contact with used engine oil may cause a number of skin disorders, including cancer! You should make every effort to minimize your exposure to used engine oil. Protective gloves should be worn when changing the oil. Wash your hands and any other exposed skin areas as soon as possible after exposure to used engine oil. Soap and water, or waterless hand cleaner should be used.

2. Remove the oil pan attaching bolts and remove oil pan.
3. Clean oil pan and engine block gasket surfaces thoroughly.
4. Install a new pan gasket.
5. Install oil pan and tighten the bolts to 52–61 inch lbs. (6–7 Nm).
6. Install all other parts in reverse order of removal.

1989-93

♦ See Figure 55

1. Raise and support the front end on jackstands.
2. Drain the oil.

❉❉ CAUTION

The EPA warns that prolonged contact with used engine oil may cause a number of skin disorders, including cancer! You should make every effort to minimize your exposure to used engine oil. Protective gloves should be worn when changing the oil. Wash your hands and any other exposed skin areas as soon as possible after exposure to used engine oil. Soap and water, or waterless hand cleaner should be used.

3. Remove the splash pan.
4. Remove the engine braces.
5. Remove the stabilizer bracket.
6. Unbolt and remove the oil pan.
7. Clean the oil pan and engine block gasket surfaces thoroughly.
8. Install a new pan gasket.
9. Install oil pan and tighten the bolts to 69–95 inch lbs. (8–11 Nm).
10. Install all other parts in reverse order of removal.

3.0L Engine

♦ See Figure 56

1. Disconnect the negative battery cable. Raise and support the vehicle safely.
2. Drain the engine oil.
3. Remove the engine under cover.
4. Remove the oil pan retaining bolts.
5. With a scraper or suitable prying tool, separate the oil pan from the block and remove it. Some oil pans may or may not have a gasket.

➡ Be careful not to bend the oil pan when separating it from the block.

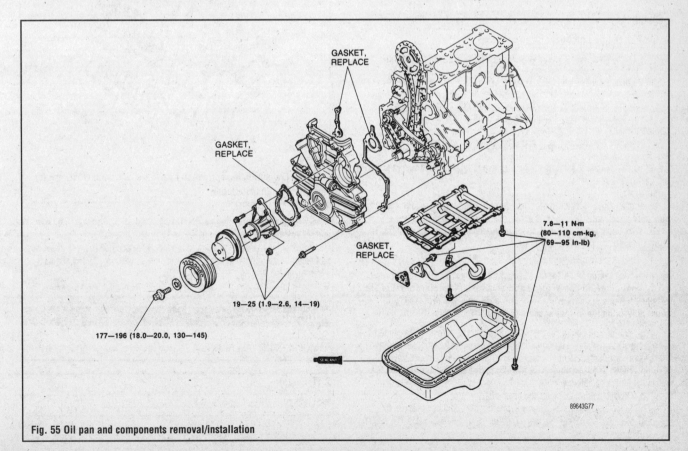

Fig. 55 Oil pan and components removal/installation

3-22 ENGINE AND ENGINE OVERHAUL

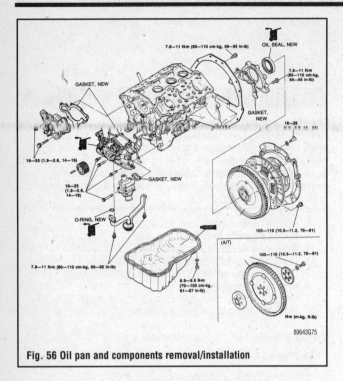

Fig. 56 Oil pan and components removal/installation

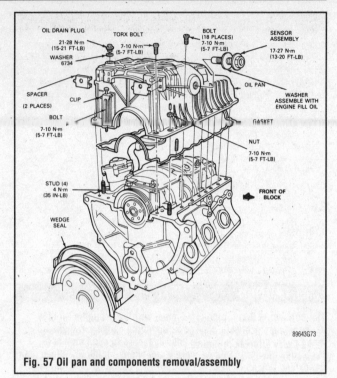

Fig. 57 Oil pan and components removal/assembly

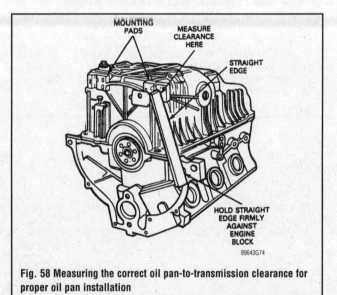

Fig. 58 Measuring the correct oil pan-to-transmission clearance for proper oil pan installation

6. On oil pans with a gasket, apply sealer to the joints between the front cover and the block and the rear main seal housing and the block. On gasketless oil pans, apply a continuous bead of sealant to the oil pan flange around the inside of the bolt holes and overlap the ends.

7. Raise the oil pan onto the block and install the retaining bolts. Torque the bolts to 61–87 inch lbs. (7–10 Nm).

8. Install the engine undercover and lower the vehicle. Fill the crankcase to the proper level. Start the engine and check for leaks.

4.0L Engine

◆ See Figures 57 and 58

➡Review the complete service procedure before starting this repair.

1. Disconnect the negative battery cable. Remove the complete engine assembly from the vehicle. Refer to the necessary service procedures in this Chapter.
2. Mount the engine on a suitable engine stan with oil pan facing up.
3. Remove the oil pan attaching bolts (note location of 2 spacers) and remove the pan from the engine block.
4. Remove the oil pan gasket and crankshaft rear main bearing cap wedge seal.
5. Clean all gasket surfaces on the engine and oil pan. Remove all traces of old gasket and/or sealer.

To install:

6. Install a new crankshaft rear main bearing cap wedge seal. The seal should fit snugly into the sides of the rear main bearing cap.
7. Position the oil pan gasket to the engine block and place the oil pan in correct position on the 4 locating studs.
8. Torque the oil pan retaining bolts EVENLY to 5–7 ft. lbs. (7–10 Nm).

➡The transmission bolts to the engine and oil pan. There are 2 spacers on the rear of the oil pan to allow proper mating of the transmission and oil pan. If these spacers were lost, or the oil pan was replaced, you must determine the proper spacers to install. To do this:

 a. With the oil pan installed, place a straightedge across the machined mating surface of the rear of the block, extending over the oil pan-to-transmission mounting surface.
 b. Using a feeler gauge, measure the gap bewteen the oil pan mounting pad and the straightedge.
 c. Repeat the procedure for the other side.
 d. Select the spacers as follows: Gap = 0.011–0.020 in.; spacer = 0.010 in. Gap = 0.021–0.029 in.; spacer = 0.020 in. Gap = 0.030–0.039 in.; spacer = 0.030 in.

9. Failure to use the correct spacers will result in damage to the oil pan and oil leakage.
10. Install the selected spacers to the mounting pads on the rear of the oil pan before bolting the engine and transmission together. Install the engine assembly in the vehicle.
11. Connect the negative battery cable. Start the engine and check for leaks.

Oil Pump

REMOVAL & INSTALLATION

2.2L Engine

◆ See Figure 59

1. Disconnect the battery ground.
2. Drain the cooling system.

ENGINE AND ENGINE OVERHAUL 3-23

** CAUTION

When draining the coolant, keep in mind that cats and dogs are attracted by the ethelyne glycol antifreeze, and are quite likely to drink any that is left in an uncovered container or in puddles on the ground. This will prove fatal in sufficient quantity. Always drain the coolant into a sealable container. Coolant should be reused unless it is contaminated or several years old.

3. Remove the distributor.
4. Remove the fan shroud and fan.
5. Remove the alternator.
6. Disconnect the air injection pipes.
7. Remove the fan pulley, hub and bracket.
8. If so equipped, remove the air conditioning compressor drive belt.
9. If so equipped, remove the power steering pump drive belt.
10. Remove the crankshaft pulley and baffle plate.
11. Remove the upper, then the lower, timing belt covers.
12. Remove the timing belt.
13. Unbolt and remove the crankshaft sprocket.
14. Drain the oil.

** CAUTION

The EPA warns that prolonged contact with used engine oil may cause a number of skin disorders, including cancer! You should make every effort to minimize your exposure to used engine oil. Protective gloves should be worn when changing the oil. Wash your hands and any other exposed skin areas as soon as possible after exposure to used engine oil. Soap and water, or waterless hand cleaner should be used.

15. Remove the skid plate.
16. Place a floor jack under the front of the engine at the crankshaft pulley and take up the weight of the engine. Or use a shop crane to support the engine.
17. Remove the crossmember.
18. Remove the cotter pin and nut and, with a puller, disconnect the idler arm from the center link.
19. Remove the engine mount gusset plates from the sides of the engine.
20. Remove the bell housing front cover.
21. Unbolt and remove the oil pan. A flat tipped screwdriver may be used to break the seal between the pan and block.
22. Remove the oil pick-up tube.
23. Unbolt and remove the oil pump.

To install:

24. Apply a thin coating of grease to the O-ring and install it in its recess in the pump body.
25. Apply a thin bead of RTV silicone sealer to the pump mounting surface.
26. Coat the oil seal lip with clean engine oil and install the pump. Tighten the oil pump mounting bolts.
27. Clean all the gasket surfaces. Straighten and portion of the pan rim that is bent.
28. Clean the oil pan, oil pump pickup tube and oil pump screen.
29. Install the oil pan.
30. Install the bell housing front cover. Tighten the bolts to 20 ft. lbs. (27 Nm).
31. Install the engine mount gusset plates on the sides of the engine. Tighten the bolts to 35 ft. lbs. (47 Nm).
32. Install the idler arm on the center link. Tighten the nut to 30 ft. lbs. (41 Nm). Install a new cotter pin.
33. Install the crossmember.
34. Remove the floor jack or shop crane used to support the engine.
35. Install the skid plate.
36. Install the crankshaft sprocket.
37. Install the timing belt.
38. Install the upper, then the lower, belt covers.
39. Install the crankshaft pulley and baffle plate.
40. Install the power steering pump drive belt.
41. Install the air conditioning compressor drive belt.
42. Install the fan pulley, hub and bracket.
43. Install the air injection pipes.
44. Install the alternator.
45. Install the fan shroud and fan.
46. Install the distributor.
47. Fill the engine with the proper amount of oil.
48. Fill the cooling system.
49. Install the battery ground cable.

Install the drive belts.

2.6L Engine

1987-88

◆ See Figure 60

1. Disconnect the battery ground (negative) cable. Drain the coolant. Disconnect and remove the radiator hoses. Remove the radiator.

** CAUTION

When draining the coolant, keep in mind that cats and dogs are attracted by the ethylene glycol antifreeze, and are quite likely to drink any that is left in an uncovered container or in puddles on the ground. This will prove fatal in sufficient quantity. Always drain the coolant into a sealable container. Coolant should be reused unless it is contaminated or several years old.

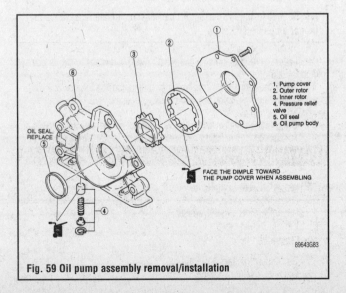

Fig. 59 Oil pump assembly removal/installation

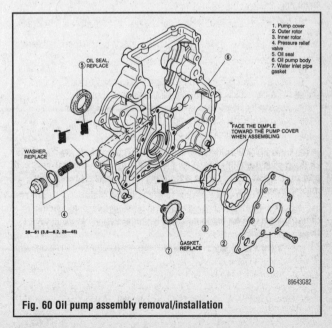

Fig. 60 Oil pump assembly removal/installation

3-24 ENGINE AND ENGINE OVERHAUL

2. Remove the alternator and accessory belts.
3. Remove the distributor.
4. Remove the crankshaft pulley.
5. Remove the water pump assembly.
6. Remove the cylinder head.

➡ It may be possible to replace the timing chain without removing the cylinder head, however removing the head will make the job easier.

7. Raise the front of the truck and support it safely on jackstands.
8. Drain the engine oil and remove the oil pan and screen.

✱✱ CAUTION

The EPA warns that prolonged contact with used engine oil may cause a number of skin disorders, including cancer! You should make every effort to minimize your exposure to used engine oil. Protective gloves should be worn when changing the oil. Wash your hands and any other exposed skin areas as soon as possible after exposure to used engine oil. Soap and water, or waterless hand cleaner should be used.

9. Remove the timing case cover.
10. Remove the chain guides, Side (A), Top (B), Bottom (C), from the B chain (outer).
11. Remove the locking bolts from the B chain sprockets.
12. Remove the crankshaft sprocket, counterbalance shaft sprocket and the outer chain.
13. Remove the crankshaft and camshaft sprockets and the A (inner) chain.
14. Remove the camshaft sprocket holder and the chain guides, both left and right. Remove the tensioner spring and sleeve from the oil pump.
15. Remove the oil pump by first removing the bolt locking the oil pump driven gear and the right counterbalance shaft, and then remove the oil pump mounting bolts. Remove the counterbalance shaft from the engine block.

➡ If the bolt locking the oil pump driven gear and the counterbalance shaft is hard to loosen, remove the oil pump and the shaft as a unit.

16. Remove the gaskets and the oil pump backing plate.
17. Clean the gasket mounting surfaces.

To install:

18. Install new gaskets/seals, align the mating marks of the oil pump gears, refill the pump with oil, and install the pump assembly.
19. Install the oil pump silent shaft sprocket and sprocket bolt. Tighten the sprocket bolt to 34 Nm (25 ft. lbs.).
20. Install a new O-ring on the thrust plate and install the unit into the engine block, using a pair of bolts without heads, as alignment guides.

➡ If the thrust plate is turned to align the bolts holes, the O-ring may be damaged.

21. Remove the guide bolts and install the regular bolts into the thrust plate and tighten securely.
22. Rotate the crankshaft to bring No. 1 piston to TDC.
23. Using a new head gasket install the cylinder head.
24. Install the sprocket holder and the right and left chain guides.
25. Install the tensioner spring and sleeve on the oil pump body.
26. Install the camshaft and crankshaft sprockets on the timing chain, aligning the sprocket punch marks to the plate chain links.
27. While holding the sprocket and chain as a unit, install the crankshaft sprocket over the crankshaft and align it with the keyway.
28. Keeping the dowel pin hole on the camshaft in a vertical position, install the camshaft sprocket and chain on the camshaft.

➡ The sprocket timing mark and the plated chain link should be at the 2 to 3 o'clock position when correctly installed. The chain must be aligned in the right and left chain guides with the tensioner pushing against the chain. The tension for the inner chain is predetermined by spring tension.

29. Install the crankshaft sprocket for the outer or **B** chain.
30. Install the two counterbalance shaft sprockets and align the punched mating marks with the plated links of the chain.
31. Holding the two shaft sprockets and chain, install the outer chain in alignment with the mark on the crankshaft sprocket. Install the shaft sprockets on the counterbalance shaft and the oil pump driver gear. Install the lock bolts and recheck the alignment of the punch marks and the plated links.
32. Temporarily install the chain guides: Side (A), Top (B), and Bottom (C).
33. Tighten Side (A), chain guide securely.
34. Tighten Bottom (B) chain guide securely.
35. Adjust the position of the Top (B) chain guide, after shaking the right and left sprockets to collect any chain slack, so that when the chain is moved toward the center, the clearance between the chain guide and the chain links will be approximately 3.5mm (0.13779 in.). Tighten the Top (B) chain guide bolts.
36. Install the timing chain cover using a new gasket, being careful not to damage the front seal.
37. Install the oil screen and the oil pan, using a new gasket.
38. Install the crankshaft pulley, alternator and accessory belts, and the distributor.
39. Install the oil pressure switch, if removed, and install the battery ground cable.
40. Install the fan blades, radiator, fill the system with coolant and start the engine.

1989-93

▶ See Figure 60

1. Disconnect the battery ground.
2. Drain the cooling system.

✱✱ CAUTION

When draining the coolant, keep in mind that cats and dogs are attracted by the ethylene glycol antifreeze, and are quite likely to drink any that is left in an uncovered container or in puddles on the ground. This will prove fatal in sufficient quantity. Always drain the coolant into a sealable container. Coolant should be reused unless it is contaminated or several years old.

3. Remove the accessory drive belts.
4. Remove the fan and shroud.
5. Remove the water pump pulley.
6. Unbolt and remove the water pump.
7. Remove the crankshaft pulley.
8. Remove the oil pan.
9. Remove the timing chain cover.

➡ The pump is built into the cover

10. Remove the oil pick-up tube.
11. Remove the pump cover from the case.
12. Remove the inner and outer rotors.
13. Remove the pressure relief valve.
14. Remove and discard the water inlet pipe gasket.

To install:

15. Install a new water inlet pipe gasket using adhesive sealer.
16. Install the oil pick-up tube using a new gasket. Tighten the bolts to 95 inch lbs. (11 Nm).
17. Install the pressure relief valve. Tighten the plug to 45 ft. lbs. (61 Nm).
18. Install the inner and outer rotors.
19. Install the pump cover.
20. Using new gaskets coated with sealer, install the timing chain cover.
21. Tighten the oil pick-up brace bolt to 95 inch lbs. (11 Nm).
22. Install the oil pan.
23. Install the crankshaft pulley.
24. Install the water pump.
25. Install the water pump pulley.
26. Install the fan and shroud.
27. Install the accessory drive belts.
28. Fill the cooling system.
29. Connect the battery ground.

ENGINE AND ENGINE OVERHAUL 3-25

3.0L Engine

♦ See Figure 61

1. Disconnect the negative battery cable. Raise and support the vehicle safely.
2. Drain the engine oil and the cooling system.
3. Remove the timing belt and the timing belt pulley and key as decribed in this chapter. Remove the thermostat and gasket.
4. Remove the oil pan, oil strainer and O-ring.
5. Unbolt and remove the oil pump and gasket.
6. To install, press in a new oil seal and coat the seal lip with clean engine oil. Use a new gasket, O-ring and sealant as required. Tighten the oil pump retaining bolts to 14–19 ft. lbs. (19–25 Nm).

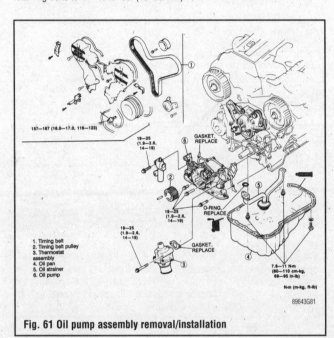

Fig. 61 Oil pump assembly removal/installation

4.0L Engine

♦ See Figure 62

➡ The oil pumps are not serviceable. If defective, they must be replaced.

1. Follow the service procedures under Oil Pan Removal and remove the oil pan assembly.

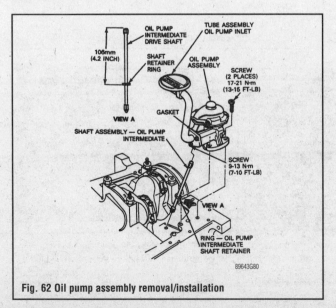

Fig. 62 Oil pump assembly removal/installation

2. Remove the oil pick-up and tube assembly from the pump.
3. Remove the oil pump retainer bolts and remove the oil pump.

To install:

4. Prime the oil pump with clean engine oil by filling either the inlet or outlet port with clean engine oil. Rotate the pump shaft to distribute the oil within the pump body.
5. Install the pump and tighten the mounting bolts to 13–15 ft. lbs. (17–21 Nm).

※※ WARNING

Do not force the oil pump if it does not seat readily. The oil pump driveshaft may be misaligned with the distributor or shaft assembly. If the pump is tightened down with the driveshaft misaligned, damage to the pump could occur. To align, rotate the intermediate driveshaft into a new position.

6. Install the oil pick-up and tube assembly to the pump. If there is a gasket between the pump and the pick-up, use a new gasket when installing.
7. Install the oil pan as previously described.

INSPECTION

2.2L Engine

➡ The crescent referred to in these procedures is the crescent-shaped slinger between the inner and outer pump gears.

OUTER GEAR TOOTH TIP TO CRESCENT CLEARANCE

♦ See Figure 63

Check this clearance with a feeler gauge. If the clearance exceeds 0.33mm (0.01299 in.), replace the gear.

INNER GEAR TOOTH TIP TO CRESCENT CLEARANCE

♦ See Figure 63

Check this clearance with a feeler gauge. If the clearance exceeds 0.40mm (0.01575 in.), replace the gear.

SIDE CLEARANCE

♦ See Figure 63

Lay a straightedge across the pump body and, using a feeler gauge, measure between the gear faces and straightedge. If the clearance exceeds 0.10mm (0.003937 in.), replace the pump.

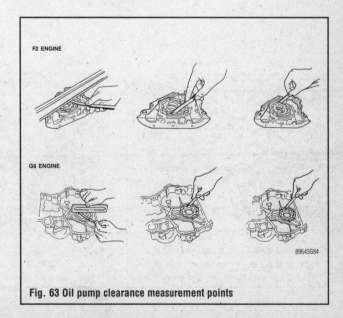

Fig. 63 Oil pump clearance measurement points

3-26 ENGINE AND ENGINE OVERHAUL

OUTER GEAR TO PUMP BODY CLEARANCE

▶ See Figure 63

Insert a feeler gauge between the outer gear and the pump body. If the clearance exceeds 0.20mm (0.00787 in.), replace the gear or pump body.

2.6L Engine (1987-88)

▶ See Figure 63

1. Thoroughly clean all parts in a safe solvent and check for wear and damage.
2. Clean all orifices and passages.
3. Place the gear back in the pump body and check clearances.
- Gear teeth-to-body: 0.10–0.15mm (0.0039–0.0059 in.)
- Driven gear end play: 0.06–0.12mm (0.0024–0.0047 in.)
- Drive gear-to-bearing (front end): 0.020–0.045mm (0.00079–0.00177 in.)
- Drive gear-to-bearing (rear end): 0.043–0.066mm (0.00169–0.00259 in.)

➡ If gear replacement is necessary, the entire pump body must be replaced.

4. Check the relief valve spring for wear or damage.

1987–88
- Free length: 47mm (1.85 in.)
- Load/length: 9.5 lb @ 40mm (1.5748 in.)

1989–93
- Relief valve spring free length: 46.4mm (1.8267 in.)
- Rotor side clearance: 0.01mm (0.00039 in.)
- Tooth tip clearance: 0.18mm (0.00708 in.)
- Outer rotor-to-pump body: 0.20mm (0.00787 in.)

Crankshaft Pulley (Vibration Damper)

REMOVAL & INSTALLATION

▶ See Figures 64, 65 and 66

1. Remove the fan shroud, as required.
2. On those engines with a separate pulley, remove the retaining bolts and separate the pulley from the vibration damper.
3. Remove the vibration damper/pulley retaining bolt from the crankshaft end.
4. Using a puller, remove the damper/pulley from the crankshaft.
5. Upon installation, align the key slot of the pulley hub to the crankshaft key. Complete the assembly in the reverse order of removal. Tighten the crankshaft pulley/damper retaining bolt(s) to the following specifications:
- 4.0L engine: 30–37 ft. lbs. (30–50 Nm), plus an additional 80–90°
- 3.0L engine: 116–123 ft. lbs. (158–167 Nm)
- 2.6L engine (1987–88): 80–94 ft. lbs. (109–128 Nm)
- 2.6L engine (1989–93): 145 ft. lbs. (197 Nm)
- 2.2L engine: 9–13 ft. lbs. (12–18 Nm)

Timing Belt Covers

REMOVAL & INSTALLATION

2.2L Engine

▶ See Figure 67

1. Disconnect the negative battery.
2. Remove the cooling fan.
3. Remove the cooling fan shroud.
4. Remove the drive belts.
5. Remove the cooling fan pulley and bracket.
6. If carbureted, remove the secondary air pipe assembly.
7. Unbolt the crankshaft pulley and remove the upper and lower timing belt covers.
8. Installation is the reverse of removal.

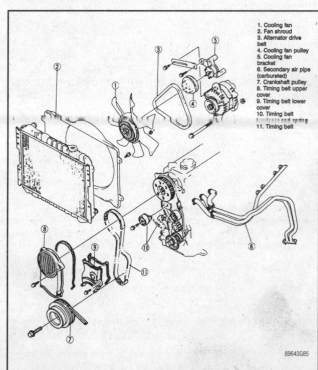

Fig. 67 Timing belt covers and timing belt removal/installation

1. Cooling fan
2. Fan shroud
3. Alternator drive belt
4. Cooling fan pulley
5. Cooling fan bracket
6. Secondary air pipe (carbureted)
7. Crankshaft pulley
8. Timing belt upper cover
9. Timing belt lower cover
10. Timing belt tensioner and spring
11. Timing belt

Fig. 64 Loosen the crankshaft pulley center nut by installing a small bolt on the outside of the pulley, using a prybar for holding force and a socket and breaker bar

Fig. 65 Remove the crankshaft pulley retaining bolt

Fig. 66 Remove the crankshaft pulley, keep the small bolt in for installation purposes

ENGINE AND ENGINE OVERHAUL

3.0L Engine

♦ See Figures 68 thru 73

1. Disconnect the negative battery cable, and drain the cooling system.
2. Drain the engine coolant.
3. Remove the air cleaner duct assembly.
4. Remove the cooling fan and radiator shroud.
5. Remove the accessory drive belt.
6. Remove the A/C compressor idler pulley.
7. Remove the crankshaft pulley.
8. Remove the coolant bypass and upper radiator hoses.
9. Remove the timing belt covers.

To install:

10. Install the timing belt covers and cover gaskets. Tighten the timing belt cover fasteners to 70–95 inch lbs. (8–11 Nm).
11. Install the coolant bypass and upper radiator hoses. Tighten the hose clamps.
12. Install the crankshaft pulley.
13. Install the A/C compressor idler pulley.
14. Install the accessory drive belt.
15. Install the cooling fan and the radiator shroud.
16. Install the air cleaner duct assembly.
17. Fill and bleed the cooling system.
18. Connect the negative battery cable.
19. Run the engine and check for leaks and proper operation. Check the idle speed and the ignition timing.

Timing Chain Cover

REMOVAL & INSTALLATION

2.6L Engine

1987-88

1. Disconnect the negative battery cable.
2. Remove the air cleaner assembly.
3. Remove the accessory drive belts.
4. Remove the alternator mounting bolts and remove alternator.
5. Remove the power steering mounting bolts and set power steering pump aside.
6. Remove the air condition compressor mounting bolts and set compressor aside.
7. Support the vehicle on jackstands and remove right inner splash shield.
8. Drain the engine oil.

※※ CAUTION

The EPA warns that prolonged contact with used engine oil may cause a number of skin disorders, including cancer! You should make every effort to minimize your exposure to used engine oil. Protective gloves should be worn when changing the oil. Wash your hands and any other exposed skin areas as soon as possible after exposure to used engine oil. Soap and water, or waterless hand cleaner should be used.

9. Remove the crankshaft pulley.
10. Lower the vehicle and place a jack under the engine with a piece of wood between jack and lifting point.
11. Raise the jack until contact is made with the engine. Relieve pressure by jacking slightly and remove the center bolt from the right engine mount. Remove right engine mount.
12. Remove the engine oil dipstick.
13. Remove the engine valve cover.
14. Remove the front (2) cylinder head to timing chain cover bolts. DO NOT LOOSEN ANY OTHER CYLINDER HEAD BOLTS.
15. Remove the oil pan retaining bolts and lower the oil pan.
16. Remove the screws holding the timing indicator and engine mounting plate.
17. Remove the bolts holding the timing chain case cover and remove cover.
18. Clean and inspect chain case cover for crack or other damage.

To install:

19. Position a new timing chain case cover gasket on case cover. Trim as required to assure fit at top and bottom.
20. Coat the cover gasket with sealant (P/N 3419115) or equivalent. Install chain case cover and tighten mounting bolts to 13 ft. lbs. (18 Nm).
21. Install the (2) front cylinder head to timing chain case cover mounting bolts.
22. Install the engine oil pan.
23. Install the engine mounting plate and timing indicator.
24. Install the crankshaft pulley.
25. Install the right engine mount, lower engine and install right engine mount center bolt.
26. Install the engine valve cover.
27. Install the engine oil dipstick.
28. Install the air conditioner compressor.
29. Install the power steering pump.
30. Install the alternator.
31. Install the accessory drive belts.
32. Fill the engine crankcase with recommended engine oil.
33. Install the air cleaner assembly.
34. Connect the negative battery cable.

1989-93

1. Disconnect the battery ground.
2. Drain the cooling system.

※※ CAUTION

When draining the coolant, keep in mind that cats and dogs are attracted by the ethylene glycol antifreeze, and are quite likely to drink any that is left in an uncovered container or in puddles on the ground. This will prove fatal in sufficient quantity. Always drain the coolant into a sealable container. Coolant should be reused unless it is contaminated or several years old.

Fig. 68 Remove the A/C compressor belt idler pulley assembly

Fig. 69 Remove the cooling system bypass hose between the upper and lower hoses at the engine

Fig. 70 Disconnect the upper radiator hose at the engine

3-28 ENGINE AND ENGINE OVERHAUL

Fig. 71 Remove the mounting fasteners, then remove the left side timing belt cover

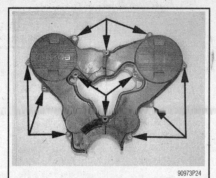

Fig. 72 Note the locations of the timing belt cover mounting fasteners

Fig. 73 After removing the mounting fasteners, remove the right side timing belt cover

3. Remove the accessory drive belts.
4. Remove the fan and shroud.
5. Remove the water pump pulley.
6. Unbolt and remove the water pump.
7. Remove the crankshaft pulley.
8. Remove the oil pan.
9. Loosen the mounting bolts and remove the timing chain cover.

To install:

10. Install a new water inlet pipe gasket using adhesive sealer.
11. Install the oil pick-up tube using a new gasket. Tighten the bolts to 95 inch lbs. (11 Nm).
12. Using new gaskets coated with sealer, install the timing chain cover. Tighten the bolts to 19 ft. lbs. (26 Nm).
13. Tighten the oil pick-up brace bolt to 95 inch lbs. (11 Nm).
14. Install the oil pan.
15. Install the crankshaft pulley.
16. Install the water pump.
17. Install the water pump pulley.
18. Install the fan and shroud.
19. Install the accessory drive belts.
20. Fill the cooling system. Fill the engine with the correct type and amount of engine oil.
21. Connect the battery ground.

4.0L Engine

▶ See Figure 74

➜ Review the complete service procedure before starting this repair. Refer to the necessary service procedures in this Chapter.

1. Disconnect the negative battery cable. Remove the oil pan.
2. Drain the cooling system.
3. Remove the radiator.

⁂ CAUTION

When draining the coolant, keep in mind that cats and dogs are attracted by the ethylene glycol antifreeze, and are quite likely to drink any that is left in an uncovered container or in puddles on the ground. This will prove fatal in sufficient quantity. Always drain the coolant into a sealable container. Coolant should be reused unless it is contaminated or several years old.

4. Remove the air conditioning compressor and position it out of the way. DO NOT disconnect the refrigerant lines!
5. Remove the power steering pump and position it out of the way. DO NOT disconnect the hoses!
6. Remove the alternator.
7. Remove the fan.
8. Remove the water pump.
9. Remove the drive pulley/damper from the crankshaft.
10. Remove the crankshaft timing sensor.
11. Remove the front cover attaching bolts. It may be necessary to tap the cover loose with a plastic mallet.

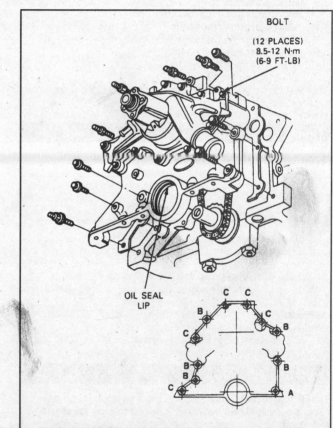

FRONT COVER BOLTS				
FASTENER	LOCATION	SIZE	QUANTITY	TORQUE N·M (FT-LB)
BOLT	A	M8 × 47	(1)	17-21 (13-15)
BOLT	B	M8 × 25	(5)	17-21 (13-15)
STUD	C	M8 × 25	(5)	17-21 (13-15)

Fig. 74 Timing chain cover installation and bolt tightening specifications

To install:

12. Install the front cover and attaching bolts. Refer to the necessary illustration.
13. Install the radiator.
14. Install the crankshaft position sensor.
15. Install the drive pulley/damper.
16. Install the water pump.
17. Install the fan.

ENGINE AND ENGINE OVERHAUL

18. Install the alternator.
19. Install the power steering pump.
20. Install the air conditioning compressor.
21. Fill the cooling system.
22. Install the oil pan.
23. Fill the crankcase to the proper level with engine oil. Connect the negative battery cable. Start engine check for leaks and roadtest the vehicle for proper operation.

Front Cover Oil Seal

REMOVAL & INSTALLATION

All Engines

WITH FRONT COVER REMOVED FROM THE ENGINE

If the specialized tools aren't available, the front seal can be replaced using normal hand tools. Before installing the cover, use a punch to carefully knock the old seal out of the cover, and a seal driver to seat the new seal.

WITH FRONT COVER ON THE ENGINE

The front cover oil seal can be removed and a new one installed without removing the front cover.
1. Drain the cooling system.

✱✱ CAUTION

When draining the coolant, keep in mind that cats and dogs are attracted by the ethylene glycol antifreeze, and are quite likely to drink any that is left in an uncovered container or in puddles on the ground. This will prove fatal in sufficient quantity. Always drain the coolant into a sealable container. Coolant should be reused unless it is contaminated or several years old.

2. Disconnect the upper and lower radiator hoses and remove the radiator.
3. Remove the drive belt(s).
4. Remove the crankshaft pulley.
5. Pry the front oil seal from the front cover.
6. Clean the pulley and seal area.

To install:
7. Press a new front seal into position (flush).
8. Install the crankshaft pulley and torque the bolt to specifications.
9. Install the drive belt(s) and adjust the tension.
10. Install the radiator and connect the upper and lower hoses. Fill the cooling system.
11. Start the engine and check for leaks.

Timing Belt and Sprockets

The correct installation and adjustment of the camshaft drive belt is mandatory if the engine is to run properly. The camshaft controls the opening of the engine valves through coordination of the movement of the crankshaft and camshaft. When any given piston is on the intake stroke the corresponding intake valve must open to admit air/fuel mixture into the cylinder. When the same piston is on the compression and power strokes, both valves in that cylinder must be closed. When the piston is on the exhaust stroke, the exhaust valve for that cylinder must be open. If the opening and closing of the valves is not coordinated with the movements of the pistons, the engine will run vey poorly, if at all.

REMOVAL & INSTALLATION

2.2L Engine

♦ See Figures 67, 75 thru 81

1. Disconnect the negative battery cable. Remove the accessory drive belts.
2. Remove the upper and the lower timing belt cover.
3. Turn the crankshaft to position the **A** mark on the camshaft pulley with

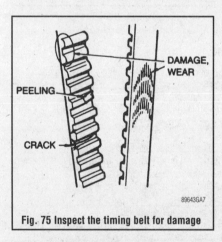

Fig. 75 Inspect the timing belt for damage

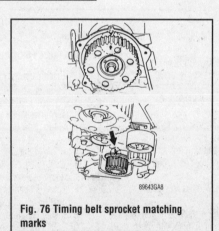

Fig. 76 Timing belt sprocket matching marks

Fig. 77 Removing the timing belt sprocket lock bolt

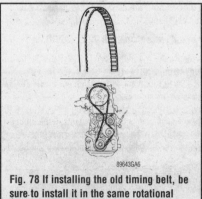

Fig. 78 If installing the old timing belt, be sure to install it in the same rotational direction as before removal

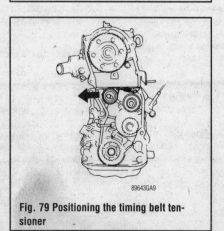

Fig. 79 Positioning the timing belt tensioner

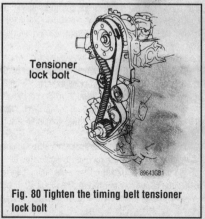

Fig. 80 Tighten the timing belt tensioner lock bolt

3-30 ENGINE AND ENGINE OVERHAUL

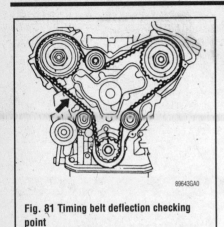

Fig. 81 Timing belt deflection checking point

Fig. 82 Remove the upper idler pulley

Fig. 83 Location of the timing marks on the timing belt cover

Fig. 84 Before removing the timing belt, line up the timing marks on the crankshaft sprocket and the engine block . . .

Fig. 85 . . . the right side camshaft sprocket and the mark on the cylinder head . . .

Fig. 86 . . . and the left side camshaft sprocket and the cylinder head

the mark on the housing. Remove the crankshaft pulley mounting bolts and the pulley.

4. With brake pliers or another suitable tool, remove the timing belt tensioner spring. Now, remove the bolt from the tensioner and remove the tensioner.

5. Inspect the tensioner for bearing wear. It should rotate freely and smoothly. If not, replace it.

6. Remove the timing belt and mark an arrow in the direction of rotation on the timing belt.

7. To remove the camshaft pulley, insert a T-wrench through the camshaft pulley onto a housing bolt, place another wrench on the pulley center bolt, hold the T-wrench securely and remove the pulley center bolt.

8. Pull the camshaft pulley from the camshaft. To remove the crankshaft pulley, remove the center bolt and the pulley.

9. Installation is the reverse of the removal procedure. Install the timing belt and be sure the timing mark on the timing belt pulley is aligned with the matching mark. Make sure that the mark (A) of the camshaft pulley is aligned with the timing mark. If it is not, turn the camshaft to align it.

10. If using the old timing belt, be sure it is reinstalled in the same direction of previous rotation. Also make sure there is no oil, grease or dirt on the timing belt.

11. Install the timing belt tensioner and spring.

12. Make sure the belt remains engaged at all three sprockets and that crankshaft and camshaft sprockets are properly in time. Locate the tensioner on its swivel pin with the slot through which the mounting bolt passes centered over the bolt hole. Install the bolt, but do not tighten it.

13. Install the tensioner spring. Temporarily secure it as the spring is fully extended.

14. Loosen the tensioner lock bolt. Turn the crankshaft twice in the diretion of rotation. Align the timing marks. Tighten the timing belt tensioner lock bolt to 28–38 ft. lbs. (38–52 Nm). Check the timing belt tension. The timing belt deflection should be 11–13mm (0.433–0.510 in.) at 22 lbs.

3.0L Engine

▶ See Figures 82 thru 96

1. Disconnect the negative battery cable, and drain the cooling system.
2. Remove the timing belt covers.
3. Remove the upper idler pulley.
4. Turn the crankshaft to align the matching marks on the sprockets. If the timing belt is to be reused, make an arrow on the belt to indicate rotation direction.
5. Remove the timing belt and automatic tensioner.
6. Using SST 49 H012 010 tool or its equivalent, loosen and remove the camshaft sprocket lock bolts. Remove the camshaft sprockets.
7. Using a suitable puller, remove the crankshaft sprocket.

To install:

8. Install the crankshaft sprocket to the crankshaft.
9. Install the camshaft sprockets with the lock bolts and torque the bolts to 52–59 ft. lbs. (71–80 Nm).
10. Set a plane washer at the bottom of the tensioner body to prevent damage to the body plug. Press in the tensioner rod slowly, using a press or a vise.

➡ Do not press the tensioner rod more than 2200 lbs. (9800N).

11. Insert a pin to hold the tensioner rod in the body. Install the automatic tensioner and tighten the mounting bolts to 14–19 ft. lbs. (19–25 Nm).
12. Install the crankshaft pulley lock bolt and loosely tighten. Check the alignment of the matching marks on the sprockets.
13. With the upper idler pulley removed, install the timing belt, making sure there is no slack between the crankshaft and camshaft sprockets. If the timing belt is being reused, it must be installed in the same direction of rotation.
14. Install the upper idler pulley and tighten the attaching bolt to 27–38 ft. lbs. (37–52 Nm).
15. Turn the crankshaft twice in the direction of rotation and align the matching marks. If the marks do not align, repeat the previous three steps.

ENGINE AND ENGINE OVERHAUL   3-31

Fig. 87 Mark the direction of rotation on the timing belt if installing the same belt

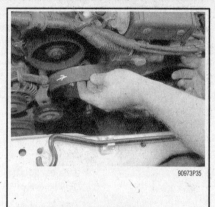

Fig. 88 Remove the timing belt

Fig. 89 Remove the timing belt tensioner from the front of the engine

Fig. 90 Remove the cranshaft sprocket by pulling straight off of the crankshaft

Fig. 91 After removing the crankshaft sprocket, remove the woodruff ke. Be careful not to lose this

Fig. 92 Place the washer on the tensioner allowing the plug on the tensioner to fit through the hole in the washer

Fig. 93 Mount the tensioner assembly and washer in the bench vise as shown

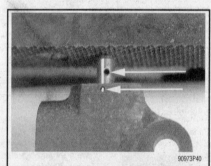

Fig. 94 When compressing the tensioner and plunger, be sure to line up the holes of the plunger to the hole in the tensioner body

Fig. 95 Once the plunger and tensioner holes are aligned . . .

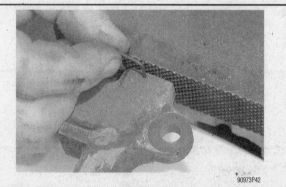

Fig. 96 . . . place a small hex key into the holes to keep the plunger compressed

16. Remove the pin from the automatic tensioner. Turn the crankshaft twice and align the matching marks. Make sure the marks are aligned.
17. Check the timing belt deflection. The deflection should be 0.20–0.28 in. (5–7mm). Do not apply tension other than that of the automatic tensioner.
18. If the deflection is not correct, repeat the previous three steps.
19. Remove the crankshaft pulley lock bolt.
20. Install the timing belt cover, along with the related components.
21. Reinstall the crankshaft pulley and torque the lock bolt to 116–123 ft. lbs. (37–52 Nm).
22. Fill and bleed the cooling system.
23. Run the engine and check for leaks and proper operation. Check the idle speed and the ignition timing.

Timing Chain and Tensioner

The correct installation and adjustment of the camshaft drive chain is mandatory if the engine is to run properly. The camshaft controls the opening of the

3-32 ENGINE AND ENGINE OVERHAUL

engine valves through coordination of the movement of the crankshaft and camshaft. When any given piston is on the intake stroke the corresponding intake valve must open to admit air/fuel mixture into the cylinder. When the same piston is on the compression and power strokes, both valves in that cylinder must be closed. When the piston is on the exhaust stroke, the exhaust valve for that cylinder must be open. If the opening and closing of the valves is not coordinated with the movements of the pistons, the engine will run vey poorly, if at all.

REMOVAL & INSTALLATION

2.6L Engine

1987-88

♦ See Figures 97 and 98

The following procedures are the recommended removal and installion procedures for the timing chain. Some modifications to the procedures may be necessary due to added accessories, sheetmetal parts, or emission control units and connecting hoses.

➡The timing chain case is cast aluminum, so exercise caution when handling this part.

1. Disconnect the battery ground (negative) cable. Drain the coolant. Disconnect and remove the radiator hoses. Remove the radiator.

※※ CAUTION

When draining the coolant, keep in mind that cats and dogs are attracted by the ethylene glycol antifreeze, and are quite likely to drink any that is left in an uncovered container or in puddles on the ground. This will prove fatal in sufficient quantity. Always drain the coolant into a sealable container. Coolant should be reused unless it is contaminated or several years old.

2. Remove the alternator and accessory belts.
3. Rotate the crankshaft to bring No.1 piston to TDC, on the compression stroke.
4. Mark and remove the distributor.
5. Remove the crankshaft pulley.
6. Remove the water pump assembly.
7. Remove the cylinder head.

➡It may be possible to replace the timing chain without removing the cylinder head, however removing the head will make the job easier.

8. Raise the front of the vehicle and support it safely on jackstands.
9. Drain the engine oil and remove the oil pan and screen.

※※ CAUTION

The EPA warns that prolonged contact with used engine oil may cause a number of skin disorders, including cancer! You should make every effort to minimize your exposure to used engine oil. Protective gloves should be worn when changing the oil. Wash your hands and any other exposed skin areas as soon as possible after exposure to used engine oil. Soap and water, or waterless hand cleaner should be used.

10. Remove the timing case cover.
11. Remove the balancer chain and sprockets.
12. Remove the crankshaft and camshaft sprockets and the timing chain.
13. Remove the camshaft sprocket holder and the chain guides, both left and right. Remove the tensioner spring and sleeve from the oil pump.
14. Remove the oil pump assembly. Remove the balance shafts from the engine block.

➡If the bolt locking the oil pump driven gear and the balance shaft is hard to loosen, remove the oil pump and the shaft as a unit.

➡If the tensioner rubber nose, or chain guide show wear, they should be replaced.

To install:

15. Install the right counterbalance shaft into the engine block.
16. Install the oil pump assembly if removed. Do not lose the woodruff key from the end of the counterbalance shaft.
17. If they have been removed, tighten the counterbalance shaft and the oil pump driven gear mounting bolts.

➡The counterbalance shaft and the oil pump can be installed as a unit, if necessary.

18. Install the left counterbalance shaft into the engine block if removed.
19. Install a new O-ring on the thrust plate and install the unit into the engine block, using a pair of bolts without heads, as alignment guides.

➡If the thrust plate is turned to align the bolts holes, the O-ring may be damaged.

20. Remove the guide bolts and install the regular bolts into the thrust plate and tighten securely.
21. Rotate the crankshaft to bring No. 1 piston to TDC.
22. Using a new head gasket install the cylinder head.
23. Install the sprocket holder and the right and left chain guides.
24. Install the tensioner spring and sleeve on the oil pump body.
25. Install the camshaft and crankshaft sprockets on the timing chain, aligning the sprocket punch marks to the plate chain links.
26. While holding the sprocket and chain as a unit, install the crankshaft sprocket over the crankshaft and align it with the keyway.
27. Keeping the dowel pin hole on the camshaft in a vertical position, install the camshaft sprocket and chain on the camshaft.

➡The sprocket timing mark and the plated chain link should be at the 2–3 o'clock position when correctly installed. The chain must be aligned in the right and left chain guides with the tensioner pushing against the chain. The tension for the inner chain is predetermined by spring tension.

28. Install the crankshaft sprocket for the outer or **B** chain.
29. Install the two counterbalance shaft sprockets and align the punched mating marks with the plated links of the chain.

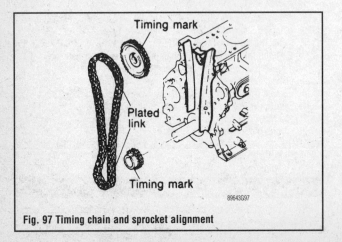

Fig. 97 Timing chain and sprocket alignment

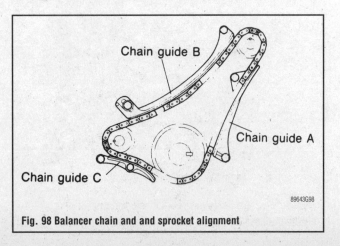

Fig. 98 Balancer chain and and sprocket alignment

ENGINE AND ENGINE OVERHAUL   3-33

30. Holding the two shaft sprockets and chain, install the outer chain in alignment with the mark on the crankshaft sprocket. Install the shaft sprockets on the counterbalance shaft and the oil pump driver gear. Install the lock bolts and recheck the alignment of the punch marks and the plated links.
31. Temporarily install the chain guides: Side (A), Top (B), and Bottom (C).
32. Tighten Side (A), chain guide securely.
33. Tighten Bottom (B) chain guide securely.
34. Adjust the position of the Top (B) chain guide, after shaking the right and left sprockets to collect any chain slack, so that when the chain is moved toward the center, the clearance between the chain guide anmd the chain links will be approximately 3.5mm (0.13779 inch). Tighten the Top (B) chain guide bolts.
35. Install the timing chain cover using a new gasket. being careful not to damage the front seal.
36. Install the oil screen and the oil pan.
37. Install the crankshaft pulley, alternator and accessory belts, and the distributor.
38. Install the oil pressure switch, if removed, and install the battery ground cable.
39. Install the fan blades, radiator, fill the system with coolant and start the engine.

1989-93

▶ See Figures 99, 100, 101, 102 and 103

1. Disconnect the battery ground.
2. Drain the cooling system.

✱✱ CAUTION

When draining the coolant, keep in mind that cats and dogs are attracted by the ethylene glycol antifreeze, and are quite likely to drink any that is left in an uncovered container or in puddles on the ground. This will prove fatal in sufficient quantity. Always drain the coolant into a sealable container. Coolant should be reused unless it is contaminated or several years old.

3. Remove the accessory drive belts.
4. Remove the fan and shroud.
5. Remove the water pump pulley.
6. Unbolt and remove the water pump.
7. Remove the crankshaft pulley.
8. Remove the oil pan.
9. Remove the timing chain cover.
10. Remove the oil pick-up tube.
11. Remove and discard the water inlet pipe gasket.
12. Remove the crankshaft sprocket spacer.
13. Remove the idler sprocket lockbolt.
14. Remove chain guide **A**.
15. Remove chain guide **B**.
16. Remove chain guide **C**.
17. Remove the idler sprocket and crankshaft sprocket bolts and pull off the sprockets along with the balancer chain.
18. Remove the left balance shaft.

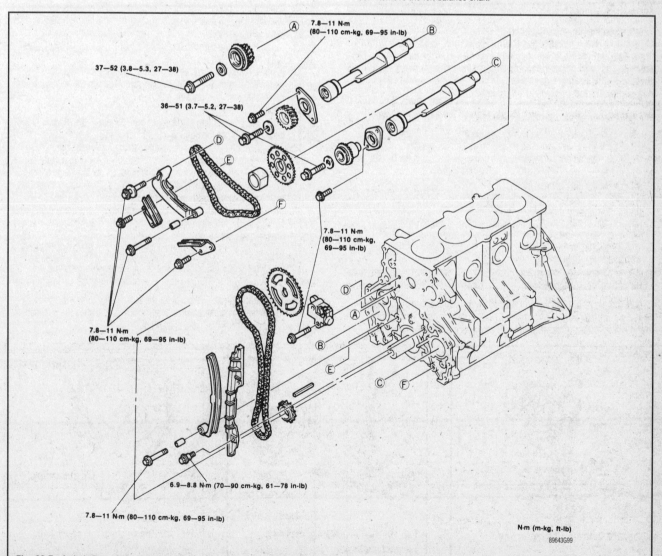

Fig. 99 Exploded view of the balancer chain and timing chain assemblies

ENGINE AND ENGINE OVERHAUL

19. Remove the right balance shaft.
20. Remove the crankshaft timing gear bolt.
21. Remove the timing chain tensioner.
22. Pull the timing chain and sprockets off.
23. Remove the chain guide and lever.

To install:

24. Install the chain guide and lever. Tighten the chain guide and lever bolts to 69–95 inch lbs. (8–11 Nm). Make sure that the lever moves smoothly.
25. Install the chain adjuster. Tighten the bolts to 69–95 inch lbs. (8–11 Nm). Move the adjuster sleeve towards the left and install a pin to hold it in place.
26. Install the timing gear sprocket and chain. With the timing mark on the sprocket at the 6 o'clock position, the white links on the chain should be on either side of the mark.
27. Install the camshaft sprocket on the camshaft with the timing mark at the 3 o'clock position. The white link should align with the mark. Temporarily secure the chain to the sprocket with a piece of wire.
28. Install the balance shafts.
29. Install the crankshaft sprocket.
30. Install the idler shaft and sprocket.
31. Set the balance chain on the balance shaft sprocket aligns with the brown link on the chain.
32. Install the balancer chain, aligning all the marks as shown.
33. Install chain guides **A** and **B**. Tighten the bolts to 69–95 inch lbs. (8–11 Nm).
34. Install chain guide **C**. Tighten the bolt to 69–95 inch lbs. (8–11 Nm).
35. Torque the idler sprocket lockbolt to 38 ft. lbs. (52 Nm).
36. Install the spacer.
37. Loosen chain guide **C** adjusting bolt.
38. Push chain guide **C** with a force of about 10 lb. downward, against the chain, then, pull it back 3.5mm (0.13779 in.) and tighten the bolt to 95 in. lbs. When properly adjusted, there should be about 3mm (0.1181 in.) of slack in the chain at the mid-point of the guide.

→If, when applying the downward force on the chain guide, it bottoms against the adjusting bolt, you should replace the balancer chain.

39. Remove the tensioner adjuster retaining pin.
40. Install a new water inlet pipe gasket using adhesive sealer.
41. Install the oil pick-up tube using a new gasket. Tighten the bolts to 95 in. lbs. (11 Nm).
42. Install the timing chain cover.
43. Install the camshaft sprocket service cover.
44. Remove the wire from the camshaft sprocket.
45. Tighten the oil pick-up brace bolt to 95 inch lbs. (11 Nm).
46. Install the oil pan.
47. Install the crankshaft pulley.
48. Install the water pump.
49. Install the water pump pulley.
50. Install the fan and shroud.
51. Install the accessory drive belts.
52. Fill the cooling system.
53. Connect the battery ground.

4.0L Engine

♦ See Figure 104

→Review the complete service procedure before starting this repair. Refer to the necessary service procedures in this Chapter.

1. Disconnect the negative battery cable. Remove the oil pan.
2. Drain the cooling system.

CAUTION

When draining the coolant, keep in mind that cats and dogs are attracted by the ethylene glycol antifreeze, and are quite likely to drink any that is left in an uncovered container or in puddles on the ground. This will prove fatal in sufficient quantity. Always drain the coolant into a sealable container. Coolant should be reused unless it is contaminated or several years old.

3. Remove the air conditioning compressor and position it out of the way. DO NOT disconnect the refrigerant lines!
4. Remove the power steering pump and position it out of the way. DO NOT disconnect the hoses!
5. Remove the alternator.
6. Remove the fan.
7. Remove the water pump.
8. Remove the drive pulley/damper from the crankshaft.
9. Remove the crankshaft position sensor.
10. Remove the front cover attaching bolts. It may be necessary to tap the cover loose with a plastic mallet.
11. Remove the radiator.
12. Rotate the engine by hand until the No.1 cylinder is at TDC compression, and the timing marks are aligned.
13. Remove the camshaft sprocket bolt and sprocket retaining key.
14. Remove the camshaft and crankshaft sprockets with the timing chain.
15. If necessary, remove the tensioner and guide.

To install:

16. Install the timing chain guide. Make sure the pin of the guide is in the hole in the block. Tighten the bolts to 84–96 inch lbs (9–11 Nm).
17. Align the timing marks on the crankshaft and camshaft sprockets and install the sprockets and chain.
18. Install the camshaft sprocket bolt and sprocket retaining key. Make sure that the timing marks are still aligned.
19. Install the tensioner with the clip in place to keep it retracted.
20. Install the crankshaft key. Make sure the timing marks are still aligned.
21. Make sure the tensioner side of the chain is held inward and the other side is straight and tight.
22. Install the camshaft sprocket bolt and tighten it to 50 ft. lbs. (68 Nm).
23. Remove the tensioner clip.
24. Check camshaft endplay.
25. Install the front cover and attaching bolts.
26. Install the radiator.
27. Install the crankshaft position sensor.

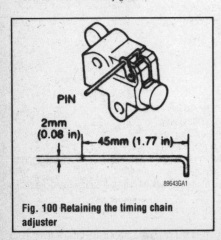

Fig. 100 Retaining the timing chain adjuster

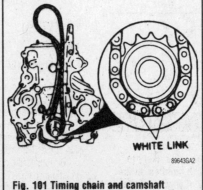

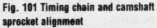

Fig. 101 Timing chain and camshaft sprocket alignment

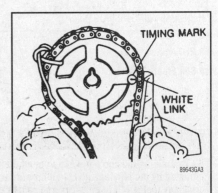

Fig. 102 Timing chain and crankshaft sprocket alignment

ENGINE AND ENGINE OVERHAUL 3-35

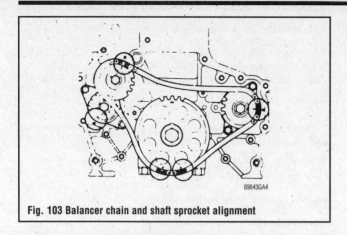

Fig. 103 Balancer chain and shaft sprocket alignment

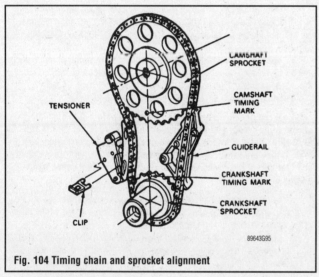

Fig. 104 Timing chain and sprocket alignment

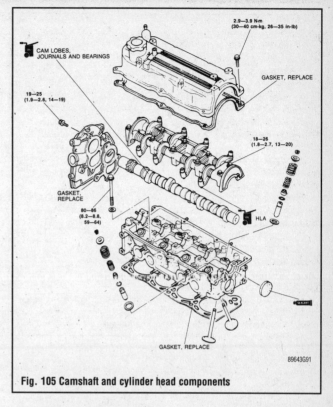

Fig. 105 Camshaft and cylinder head components

28. Install the drive pulley/damper.
29. Install the water pump.
30. Install the fan.
31. Install the alternator.
32. Install the power steering pump.
33. Install the air conditioning compressor.
34. Fill the cooling system.
35. Install the oil pan.
36. Fill the crankcase to the proper level. Connect the negative battery cable. Start engine check for leaks and roadtest the vehicle for proper operation.

Camshaft

REMOVAL & INSTALLATION

2.2L Engine

▶ See Figure 105

1. Disconnect the negative battery cable.
2. Remove the air cleaner assembly.
3. Drain the cooling system.

✳✳ CAUTION

When draining the coolant, keep in mind that cats and dogs are attracted by the ethylene glycol antifreeze, and are quite likely to drink any that is left in an uncovered container or in puddles on the ground. This will prove fatal in sufficient quantity. Always drain the coolant into a sealable container. Coolant should be reused unless it is contaminated or several years old.

4. Remove the front cover assembly.
5. Remove the cam gear.
6. Remove the thermostat housing.
7. Remove the distributor assembly.
8. Remove the rocker cover.
9. Remove the rear housing.
10. Remove the rocker arm assembly.
11. If equipped, remove the thrust plate.
12. Remove the camshaft from the cylinder head.

To install:

13. Coat the camshaft with clean engine oil and position it on the cylinder head.
14. Install the thrust plate and new gasket.
15. Install the rocker arm assembly.
16. Install the rear housing and new gasket.
17. Install the rocker cover and new gasket.
18. Install the distributor assembly.
19. Install the thermostat housing and new gasket.
20. Install the cam gear.
21. Install the front cover assembly.
22. Fill the cooling system.
23. Install the air cleaner assembly.
24. Connect the negative battery cable.

2.6L Engine

1987-88

▶ See Figure 106

1. Disconnect the negative battery cable.
2. Remove the air cleaner assembly.
3. Remove the rocker arm/valve cover.
4. Remove the water pump belt and pulley.
5. Rotate the crankshaft until number 1 piston is at the top of its compression stroke (both valves closed).
6. Record the position of mating mark on camshaft sprocket and plated link on timing chain.
7. Remove the camshaft sprocket bolt, washer and distributor drive gear.
8. Remove the timing chain and camshaft sprocket assembly and lay aside.

3-36 ENGINE AND ENGINE OVERHAUL

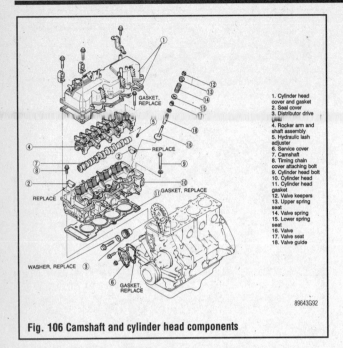

Fig. 106 Camshaft and cylinder head components

16. With the rocker arms/shafts assembly and bearing caps torqued down, rotate the camshaft so that dowel pin is on the top center of cylinder head.
17. Install the timing chain and camshaft sprocket assembly. Make sure that the timing mark and the white link align. Removing the wire.
18. Install the distributor drive gear, a new washer, and the lockbolt. Hold the crankshaft to keep it from turning and tighten the lockbolt to 45 ft. lbs. (61 Nm).
19. Release the chain adjuster.
20. Install the service cover. Tighten the bolt to 95 inch lbs.(11 Nm); the nut to 87 inch lbs. (10 Nm).
21. Install the valve cover.
22. Install the air cleaner assembly.
23. Connect the negative battery cable.

3.0L Engine

♦ See Figure 107

1. Remove the cylinder heads from the vehicle as described in this chapter.
2. Engage the camshaft sprocket with a suitable spanner wrench type holding tool and loosen the retaining bolt. Pull the sprocket from the end of the camshaft.
3. Insert the blade of a flat tipped screwdriver between the camshaft oil seal and gently pry the seal from the cylinder head bore. Be careful not to damage the seal bore. Discard the seal and purchase a new one.
4. Remove the rocker arm and shaft assemblies.

➥Do not remove the hydraulic lash adjusters unless it is absolutely necessary to do so. The lash adjusters are sealed in the rocker arms by an O-ring. If this O-ring is disturbed or damaged, the lash adjusters may leak. If they are removed, make sure that a new O-ring(s) is installed and that the oil reservoirs in the rocker arms are filled with clean engine oil.

5. Before removing the camshaft thrust plate, proceed to the inspection section to measure the camshaft endplay. This will determine if the thrust plate or the camshaft have to be replaced.
6. Remove the thrust plate and slowly and carefully withdraw the camshaft from the cylinder head. Clean off the lobe and journal surfaces and proceed to the insection section.

To install:

7. Apply a liberal coating of clean engine oil to the surfaces of the cam lobes, bearing and journal surfaces. Slowly and carefully insert the camshaft into the cylinder head. Install the camsahft thrust plate and torque the retaining bolt to 6–8 ft. lbs. (8–11 Nm).
8. Wipe down the surface of the seal bore with a clean rag and coat the lip of the new oil seal with clean engine oil. Install the seal into the cylinder head using a socket or a length of pipe that closely approximates the diameter of the seal as an installation tool. Tap the seal evenly into the seal bore.

9. Remove the rocker arms/shafts assembly.
10. Carefully remove camshaft without cocking to prevent damage to cam.

To install:
11. Lubricate the camshaft and set in place.
12. Install the rocker arms/shafts assembly.
13. With the rocker arms/shafts assembly and bearing caps torqued down, rotate the camshaft so that dowel hole is on vertical centerline of cylinder head.
14. Install the timing chain and camshaft sprocket assembly. Make sure mating mark on camshaft sprocket and plated link on timing chain are lined up.
15. Install the distributor drive gear, washer and camshaft sprocket bolt. Tighten the bolt to 40 ft. lbs. (54 Nm).
16. Install the water pump pulley and belt.

➥After servicing rocker shaft assembly, Jet Valve Clearance (if used) adjustment must be performed. See Valve Adjusting Procedure.

17. Install the rocker arm/valve cover.
18. Install the air cleaner assembly.
19. Connect the negative battery cable.

1989–93

♦ See Figure 106

1. Disconnect the negative battery cable.
2. Remove the air cleaner assembly.
3. Remove the rocker arm/valve cover.
4. Rotate the crankshaft until number 1 piston is at the top of its compression stroke (both valves closed).
5. Remove the distributor drive gear.
6. Record the position of mating mark on camshaft sprocket and plated link on timing chain.
7. Remove the camshaft sprocket service cover.
8. Push the chain adjuster sleeve towards the left and insert a 2mm (0.0787 inch) diameter by 47mm (1.85 inch) long pin in the lever hole to hold the adjuster.
9. Wire the camshaft sprocket and timing chain together and remove the timing chain and camshaft sprocket assembly from the camshaft, and lay aside.
10. Remove the rocker arms/shafts assembly.
11. Remove the cover end seals.
12. Carefully remove camshaft without cocking to prevent damage to cam.

To install:
13. Lubricate the camshaft and set in place.
14. Coat the end seal with RTV silicone gasket mater and install them in their recesses.
15. Install the rocker arms/shafts assembly.

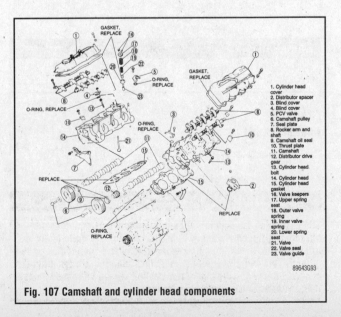

Fig. 107 Camshaft and cylinder head components

ENGINE AND ENGINE OVERHAUL 3-37

9. Install the rockerarm/shaft assembly.
10. Complete the installation of the cylinder head by reversing the removal procedure.

4.0L Engine

▶ See Figure 108

➥Review the complete service procedure before starting this repair. Refer to the necessary service procedures in this Chapter.

1. Disconnect the negative battery cable.
2. Drain the engine oil into a suitable container and dispose of it properly.

✳✳ CAUTION

The EPA warns that prolonged contact with used engine oil may cause a number of skin disorders, including cancer! You should make every effort to minimize your exposure to used engine oil. Protective gloves should be worn when changing the oil. Wash your hands and any other exposed skin areas as soon as possible after exposure to used engine oil. Soap and water, or waterless hand cleaner should be used.

3. Drain the cooling system.

✳✳ CAUTION

When draining the coolant, keep in mind that cats and dogs are attracted by the ethylene glycol antifreeze, and are quite likely to drink any that is left in an uncovered container or in puddles on the ground. This will prove fatal in sufficient quantity. Always drain the coolant into a sealable container. Coolant should be reused unless it is contaminated or several years old.

4. Remove the radiator.
5. Remove the condenser.
6. Remove the cooling fan and shroud.
7. Remove the air cleaner hoses.
8. Tag and remove the spark plug wires.
9. Remove the EDIS ignition coil and bracket.
10. Remove the crankshaft pulley/damper.
11. Remove the clamp, bolt and oil pump drive from the rear of the block.
12. Remove the alternator.
13. Relieve the fuel system pressure.
14. Remove the fuel lines at the fuel supply manifold.

15. Remove the intake manifold assembly.
16. Remove the rocker arm covers.
17. Remove the rocker arm/shaft assemblies.
18. Remove the pushrods. Identify them for installation. They must be installed in their original positions!
19. Remove the tappets. Identify them for installation.
20. Remove the oil pan.
21. Remove the engine front cover and water pump.
22. Place the timing chain tensioner in the retracted position and install the retaining clip.
23. Turn the engine by hand until the timing marks align at TDC of the power stroke on No.1 piston.
24. Check the camshaft endplay. If excessive, you'll have to replace the thrust plate.
25. Remove the camshaft gear attaching bolt and washer, then slide the gear off the camshaft.
26. Remove the camshaft thrust plate.
27. Carefully slide the camshaft out of the engine block, using caution to avoid any damage to the camshaft bearings.

To install:

28. Oil the camshaft journals and cam lobes with heavy SG engine oil (50W).
29. Install the camshaft in the block, using caution to avoid any damage to the camshaft bearings.
30. Install the thrust plate. Make sure that it covers the main oil gallery. Tighten the attaching screws to 7–10 ft. lbs. (9–13 Nm).
31. Rotate the camshaft and crankshaft as necessary to align the timing marks. Install the camshaft gear and chain. Tighten the attaching bolt to 50 ft. lbs. (68 Nm).
32. Remove the clip from the chain tensioner.
33. Install the engine front cover and water pump assembly. Refer to the necessary service procedures in this Chapter.
34. Install the crankshaft damper/pulley.
35. Install the oil pan.
36. Coat the tappets with 50W engine oil and place them in their original locations.
37. Apply 50W engine oil to both ends of the pushrods. Install the pushrods in their original locations.
38. Install the intake manifold assembly.
39. Install the rocker arm/shaft assemblies.
40. Install the rocker covers.
41. Install the cooling fan.
42. Install the fuel lines.
43. Install the oil pump drive.
44. Install the alternator.
45. Install the EDIS coil and plug wires. Coat the inside of each wire boot with silicone lubricant.
46. Install the radiator and condenser.
47. Refill the cooling system.
48. Replace the oil filter and refill the crankcase with the specified amount of engine oil.
49. Reconnect the battery ground cable.
50. Start the engine and check the ignition timing and idle speed. Adjust if necessary. Run the engine at fast idle and check for coolant, fuel, vacuum or oil leaks.

INSPECTION

Camshaft Lobe Lift

2.2L, 2.6L AND 3.0L ENGINES

▶ See Figure 109

Check the lift of each lobe in consecutive order and make a note of the readings. Camshaft assembly specifications are sometimes modified by Mazda after production. Refer to a local reputable machine shop as necessary.

1. Remove the valve cover.
2. Measure the distance between the major (A–A) and minor (B–B) diameters of each cam lobe with a Vernier caliper and record the readings. The difference in the readings on each cam diameter is the lobe lift.
3. If the readings do not meet specifications, replace the camshaft and all rocker arms/cam followers.
4. Install the valve cover.

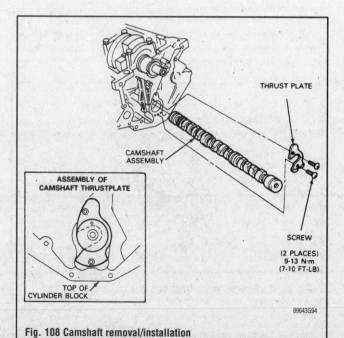

Fig. 108 Camshaft removal/installation

3-38 ENGINE AND ENGINE OVERHAUL

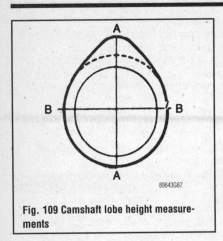

Fig. 109 Camshaft lobe height measurements

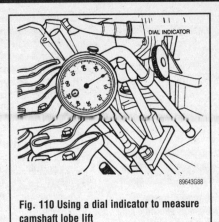

Fig. 110 Using a dial indicator to measure camshaft lobe lift

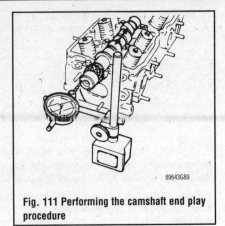

Fig. 111 Performing the camshaft end play procedure

4.0L ENGINE

♦ See Figures 109 and 110

Check the lift of each lobe in consecutive order and make a note of the reading. Camshaft assembly specifications are sometimes modify by Ford after production. Refer to a local reputable machine shop as necessary.

1. Remove the fresh air inlet tube and the air cleaner. Remove the heater hose and crankcase ventilation hoses. Remove valve rocker arm cover(s).
2. Remove the rocker arm stud nut or fulcrum bolts, fulcrum seat and rocker arm.
3. Make sure the push rod is in the valve tappet socket. Install a dial indicator D78P–4201–B (or equivalent) so that the actuating point of the indicator is in the push rod socket (or the indicator ball socket adapter Tool 6565–AB is on the end of the push rod) and in the same plane as the push rod movement.
4. Disconnect the I terminal and the S terminal at the starter relay. Install an auxiliary starter switch between the battery and S terminals of the starter relay. Crank the engine with the ignition switch off. Turn the crankshaft over until the tappet is on the base circle of the camshaft lobe. At this position, the push rod will be in its lowest position.
5. Zero the dial indicator. Continue to rotate the crankshaft slowly until the push rod is in the fully raised position.
6. Compare the total lift recorded on the dial indicator with the specification. To check the accuracy of the original indicator reading, continue to rotate the crankshaft until the indicator reads zero. If the lift on any lobe is below specified wear limits, the camshaft and the valve tappet operating on the worn lobe(s) must be replaced.
7. Remove the dial indicator and auxiliary starter switch.
8. Install the rocker arm, fulcrum seat and stud nut or fulcrum bolts. Check the valve clearance. Adjust if required (refer to procedure in Chapter 2).
9. Install the valve rocker arm covers and the air cleaner.

Camshaft End Play

2.2L, 2.6L AND 3.0L ENGINES

♦ See Figure 111

➥On engines with an aluminum or nylon camshaft sprocket, prying against the sprocket, with the valve train load on the camshaft, can break or damage the sprocket. Therefore, the rocker arm adjusting nuts must be backed off, or the rocker arm and shaft assembly must be loosened sufficiently to free the camshaft. After checking the camshaft end play, check the valve clearance. Adjust if required (refer to procedure in this chapter).

1. Push the camshaft toward the rear of the engine. Install a dial indicator so that the indicator point is on the camshaft sprocket attaching screw.
2. Zero the dial indicator. Position a prybar between the camshaft gear and the block. Pull the camshaft forward and release it. Compare the dial indicator reading with the specifications.

3. If the end play is excessive, check the spacer for correct installation before it is removed. If the spacer is correctly installed, replace the thrust plate.
4. Remove the dial indicator.

4.0L ENGINES

♦ See Figure 112

1. Push the camshaft toward the rear of the engine. Install a dial indicator (Tool D78P–4201–C or equivalent so that the indicator point is on the camshaft sprocket attaching screw.
2. Zero the dial indicator. Position a prybar between the camshaft gear and the block. Pull the camshaft forward and release it. Compare the dial indicator reading with the specification. The camshaft endplay specification is 0.0008–0.004 inch and the service limit is 0.009 inch (0.007 inch on 3.0L engine). Camshaft specifications are sometimes modified by the manufacturer after production.
3. If the end play is excessive, check the spacer for correct installation before it is removed. If the spacer is correctly installed, replace the thrust plate.

➥The spacer ring and thrust plate are available in two thicknesses to permit adjusting the end play.

4. Remove the dial indicator.

Camshaft Bearings

REMOVAL & INSTALLATION

On all engines, except the 4.0L, excessive oil clearance between the cylinder head/bearing caps-to-camshaft bearing surfaces, indicates that cylinder head and/or camshaft replacement is required.

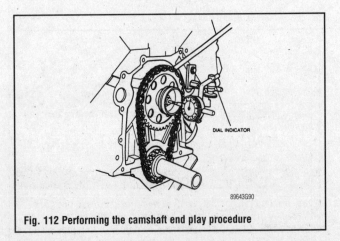

Fig. 112 Performing the camshaft end play procedure

ENGINE AND ENGINE OVERHAUL 3-39

4.0L Engines

1. Remove the engine and place it on a work stand.
2. Remove the flywheel.
3. Remove the camshaft.
4. Using a sharp punch and hammer, drive a hole in the rear bearing bore plug and pry it out.
5. Using the special tools and instructions in Cam Bearing Replacer Kit T65L–6250–A, or their equivalents, remove the bearings.
6. To remove the front bearing, install the tool from the rear of the block.

To install:

7. Following the instructions in the tool kit, install the bearings. Make sure that you follow the instructions carefully. Failure to use the correct expanding collets can cause severe bearing damage!

➡ Make sure that the oil holes in the bearings and block are aligned! Make sure that the front bearing is installed 0.51–0.89mm below the face of the block.

8. Install a new bearing bore plug coated with sealer.
9. Install the camshaft.
10. Install the flywheel.
11. Install the engine.

Valve Lifters (Tappets)

All engines, except the 4.0L, are equipped with hydraulic lash adjusters located in the rocker arms. The hydraulic lash adjusters can be removed bt hand or with suitable pliers after the rocker arm/shaft assembly has been removed.

REMOVAL & INSTALLATION

4.0L Engine

1. Remove the upper and lower intake manifolds.
2. Remove the rocker covers.
3. Remove the rocker shaft assembly.
4. Remove and mark the pushrods for installation.
5. Remove the tappets with a magnet. If they are to be re-used, identify them.

➡ If the tappets are stuck in their bores, you'll need a claw-type removal tool.

6. Coat the new tappets with clean engine oil and insert them in their bores.
7. Coat the pushrods with heavy engine oil and insert them into the bores from which they came.
8. Install the rocker shaft assembly.
9. Install the rocker covers.
10. Install the upper and lower manifold.

Balance (Countershafts) Shafts

REMOVAL & INSTALLATION

2.6L Engine

1987-88

◆ See Figure 98

➡ The following procedures are to be performed with engine removed from vehicle.

1. Disconnect the battery ground (negative) cable. Drain the coolant. Disconnect and remove the radiator hoses. Remove the radiator.

✱✱ CAUTION

When draining the coolant, keep in mind that cats and dogs are attracted by the ethylene glycol antifreeze, and are quite likely to drink any that is left in an uncovered container or in puddles on the ground. This will prove fatal in sufficient quantity. Always drain the coolant into a sealable container. Coolant should be reused unless it is contaminated or several years old.

2. Remove the alternator and accessory belts.
3. Rotate the crankshaft to bring No.1 piston to TDC, on the compression stroke.
4. Mark and remove the distributor.
5. Remove the crankshaft pulley.
6. Remove the water pump assembly.
7. Remove the cylinder head (See cylinder head removal section).

➡ It may be possible to replace the timing chain without removing the cylinder head, however removing the head will make the job easier.

8. Raise the front of the truck and support it safely on jackstands.
9. Drain the engine oil and remove the oil pan and screen.

✱✱ CAUTION

The EPA warns that prolonged contact with used engine oil may cause a number of skin disorders, including cancer! You should make every effort to minimize your exposure to used engine oil. Protective gloves should be worn when changing the oil. Wash your hands and any other exposed skin areas as soon as possible after exposure to used engine oil. Soap and water, or waterless hand cleaner should be used.

10. Remove the timing case cover.
11. Remove the chain guides, Side (A), Top (B), Bottom (C), from the B chain (outer). See illustration.
12. Remove the locking bolts from the B chain sprockets.
13. Remove the crankshaft sprocket, counterbalance shaft sprocket and the outer chain.
14. Remove the crankshaft and camshaft sprockets and the A (inner) chain.
15. Remove the camshaft sprocket holder and the chain guides, both left and right. Remove the tensioner spring and sleeve from the oil pump.

➡ For further service to shafts and oil pump follow step 16 and 17.

16. Remove the oil pump by first removing the bolt locking the oil pump driven gear and the right counterbalance shaft, and then remove the oil pump mounting bolts. Remove the counterbalance shaft from the engine block.

➡ If the bolt locking the oil pump driven gear and the counterbalance shaft is hard to loosen, remove the oil pump and the shaft as a unit.

17. Remove the left silent shaft thrust plate by screwing two 8mm screws into tapped holes in thrust plate. Remove left silent shaft.

➡ If the tensioner rubber nose, or chain guide show wear, they should be replaced.

Silent shaft clearance specifications are the following:
- Outer diameter to outer bearing clearance: 0.02-0.06mm (0.000787-0.00236 in.)
- Inner diameter to inner bearing clearance: 0.05-0.09mm (0.00197-0.00354 in.)

To install:

18. Install the right counterbalance shaft into the engine block if removed.
19. Install the oil pump assembly if removed. Do not lose the woodruff key from the end of the counterbalance shaft. Torque the oil pump mounting bolts to 6 to 7 ft. lbs.
20. If they have been removed tighten the counterbalance shaft and the oil pump driven gear mounting bolts.

➡ The counterbalance shaft and the oil pump can be installed as a unit, if necessary.

21. Install the left counterbalance shaft into the engine block if removed.
22. Install a new O-ring on the thrust plate and install the unit into the engine block, using a pair of bolts without heads, as alignment guides.

➡ If the thrust plate is turned to align the bolts holes, the O-ring may be damaged.

23. Remove the guide bolts and install the regular bolts into the thrust plate and tighten securely.
24. Rotate the crankshaft to bring No. 1 piston to TDC.
25. Using a new head gasket install the cylinder head.
26. Install the sprocket holder and the right and left chain guides.
27. Install the tensioner spring and sleeve on the oil pump body.

3-40 ENGINE AND ENGINE OVERHAUL

28. Install the camshaft and crankshaft sprockets on the timing chain, aligning the sprocket punch marks to the plate chain links.
29. While holding the sprocket and chain as a unit, install the crankshaft sprocket over the crankshaft and align it with the keyway.
30. Keeping the dowel pin hole on the camshaft in a vertical position, install the camshaft sprocket and chain on the camshaft.

➡ The sprocket timing mark and the plated chain link should be at the 2 to 3 o'clock position when correctly installed. The chain must be aligned in the right and left chain guides with the tensioner pushing against the chain. The tension for the inner chain is predetermined by spring tension.

31. Install the crankshaft sprocket for the outer or **B** chain.
32. Install the two counterbalance shaft sprockets and align the punched mating marks with the plated links of the chain.
33. Holding the two shaft sprockets and chain, install the outer chain in alignment with the mark on the crankshaft sprocket. Install the shaft sprockets on the counterbalance shaft and the oil pump driver gear. Install the lock bolts and recheck the alignment of the punch marks and the plated links.
34. Temporarily install the chain guides: Side (A), Top (B), and Bottom (C).
35. Tighten Side (A), chain guide securely.
36. Tighten Bottom (B) chain guide securely.
37. Adjust the position of the Top (B) chain guide, after shaking the right and left sprockets to collect any chain slack, so that when the chain is moved toward the center, the clearance between the chain guide anmd the chain links will be approximately 3.5mm (0.13779 in.). Tighten the Top (B) chain guide bolts.
38. Install the timing chain cover using a new gasket. being careful not to damage the front seal.
39. Install the oil screen and the oil pan, using a new gasket. Torque the bolts to 54–66 in. lbs. (6–7 Nm).
40. Install the crankshaft pulley, alternator and accessory belts, and the distributor.
41. Install the oil pressure switch, if removed, and install the battery ground cable.
42. Install the fan blades, radiator, fill the system with coolant and start the engine.

1989-93

♦ See Figures 99 and 103

1. Disconnect the battery ground.
2. Drain the cooling system.

✱✱ CAUTION

When draining the coolant, keep in mind that cats and dogs are attracted by the ethylene glycol antifreeze, and are quite likely to drink any that is left in an uncovered container or in puddles on the ground. This will prove fatal in sufficient quantity. Always drain the coolant into a sealable container. Coolant should be reused unless it is contaminated or several years old.

3. Remove the accessory drive belts.
4. Remove the fan and shroud.
5. Remove the water pump pulley.
6. Unbolt and remove the water pump.
7. Remove the crankshaft pulley.
8. Remove the oil pan.
9. Remove the timing chain cover.
10. Remove the oil pick-up tube.
11. Remove and discard the water inlet pipe gasket.
12. Remove the crankshaft sprocket spacer.
13. Remove the idler sprocket lockbolt.
14. Remove chain guide **A**.
15. Remove chain guide **B**.
16. Remove chain guide **C**.
17. Remove the idler sprocket and crankshaft sprocket bolts and pull off the sprockets along with the balancer chain.
18. Remove the left balance shaft.
19. Remove the right balance shaft.
20. Remove the crankshaft timing gear bolt.
21. Remove the timing chain tensioner.
22. Pull the timing chain and sprockets off.
23. Remove the chain guide and lever.

To install:

24. Install the chain guide and lever. Torque the chain guide bolts to 78 in. lbs. (9 Nm); the lever bolts to 95 in. lbs. (11 Nm). Make sure that the lever moves smoothly.
25. Install the chain adjustor. Torque the bolts to 95 in. lbs. (11 Nm). Move the adjuster sleeve towards the left and install a pin to hold it in place.
26. Install the timing gear sprocket and chain. With the timing mark on the sprocket at the 6 o'clock position, the white links on the chain should be on either side of the mark.
27. Install the camshaft sprocket on the camshaft with the timing mark at the 3 o'clock position. The white link should align with the mark. Temporarily secure the chain to the sprocket with a piece of wire.
28. Install the balance shafts. Torque the thust plate bolts to 95 in. lbs. (11 Nm).
29. Install the crankshaft sprocket.
30. Install the idler shaft and sprocket.
31. Set the balance chain on the balance shaft sprocket aligns with the brown link on the chain.
32. Install the balancer chain, aligning all the marks as shown.
33. Install chain guides **A** and **B**. Torque the bolts to 95 in. lbs. (11 Nm).
34. Install chain guide **C**. Tighten the bolt to 95 in. lbs. (11 Nm).
35. Torque the idler sprocket lockbolt to 38 ft. lbs. (52 Nm).
36. Install the spacer.
37. Loosen chain guide **C** adjusting bolt.
38. Push chain guide **C** with a force of about 10 lb., downward, against the chain, then, pull it back 3.5mm (0.13779 in.) and tighten the bolt to 95 in. lbs. (11 Nm). When properly adjusted, there should be about 3mm (0.1181 in.) of slack in the chain at the mid-point of the guide.

➡ If, when applying the downward force on the chain guide, it bottoms against the adjusting bolt, you should replace the balancer chain.

39. Remove the tensioner adjuster retaining pin.
40. Install a new water inlet pipe gasket using adhesive sealer.
41. Install the oil pick-up tube using a new gasket. Torque the bolts to 95 in. lbs. (11 Nm).
42. Using new gaskets coated with sealer, install the timing chain cover. Torque the bolts to 19 ft. lbs. (26 Nm).
43. Install the camshaft sprocket service cover.
44. Remove the wire from the camshaft sprocket.
45. Tighten the oil pick-up brace bolt to 95 in. lbs. (11 Nm).
46. Install the oil pan. Torque the bolts to 95 in. lbs. (11 Nm).
47. Install the crankshaft pulley. Torque the bolt to 145 ft. lbs. (197 Nm).
48. Install the water pump.
49. Install the water pump pulley.
50. Install the fan and shroud.
51. Install the accessory drive belts.
52. Fill the cooling system.
53. Connect the battery ground.

Rear Main Oil Seal

REPLACEMENT

2.2L and 2.6L (1989–93) Engines

1. Disconnect the negative battery cable. Raise and support the vehicle safely. Remove the transmission.
2. If equipped with a manual transmission, remove the pressure plate, the clutch disc and the flywheel. If equipped with an automatic transmission, remove the drive plate from the crankshaft.
3. Remove the rear oil pan-to-seal housing bolts.
4. Remove the rear main seal housing bolts and the housing from the engine.
5. Remove the oil seal from the rear main housing.
6. Clean the gasket mounting surfaces.
7. To install, use a new seal, coat the seal and the housing with oil. Drive the seal into the housing, using a seal driver.

ENGINE AND ENGINE OVERHAUL 3-41

8. To complete the installation, use new gaskets, apply sealant to the oil pan mounting surface and reverse the removal procedure. Torque the rear seal housing to 6–8 ft. lbs. (8–11 Nm) on the 2.2L engine and 95 in. lbs. (11 Nm) on the 2.6L engine.

1987-88 2.6L Engine

The rear main seal is located in a housing on the rear of the block. To replace the seal, it is necessary to remove the transmission and perform the work from underneath the truck or remove the engine and perform the work on an engine stand or work bench.

1. Unscrew the retaining bolts and remove the housing from the block.
2. Remove the separator from the housing.
3. Using a small pry bar, pry out the old seal.
4. Clean the housing and the separator.
5. Lightly oil the replacement seal. Tap the seal into housing using a canister top or other circular piece of metal. The oil seal should be installed so that the seal plate fits into the inner contact surface of the seal case.
6. Install the separator into the housing so that the oil hole faces down.
7. Oil the lips of the seal and install the housing on the rear of the engine block.

3.0L Engine

1. Raise and support the vehicle safely. Remove the transmission from the vehicle.
2. If equipped with manual transmission, remove the clutch pressure plate and flywheel.
3. If equipped with automatic transmission, remove the flywheel assembly.
4. Drain the engine oil. Remove the engine oil pan.
5. Remove the rear main seal cover retaining bolts. Remove the rear main seal cover. Remove the seal from the rear cover.
6. Installation is the reverse of the removal procedure. Apply clean engine oil to the seal before pressing it into the cover.
7. After installing the rear cover cut away the portion of the gasket that projects out toward the oil pan side.

4.0L Engine

▶ See Figures 113 and 114

If the crankshaft rear oil seal replacement is the only operation being performed, it can be done in the vehicle as detailed in the following procedure. If the oil seal is being replaced in conjunction with a rear main bearing replacement, the engine must be removed from the vehicle and installed on a work stand.

1. Remove the starter.
2. Remove the transmission from the vehicle, following the procedures in Chapter 7.
3. On a manual shift transmission, remove the pressure plate and cover assembly and the clutch disc following the procedure in Chapter 7.
4. Remove the flywheel attaching bolts and remove the flywheel and engine rear cover plate.
5. Use an awl to punch two holes in the crankshaft rear oil seal. Punch the holes on opposite sides of the crankshaft and just above the bearing cap to cylinder block split line. Install a sheet metal screw in each hole. Use two large screwdrivers or small pry bars and pry against both screws at the same time to remove the crankshaft rear oil seal. It may be necessary to place small blocks of wood against the cylinder block to provide a fulcrum point for the pry bars. Use caution throughout this procedure to avoid scratching or otherwise damaging the crankshaft oil seal surface.

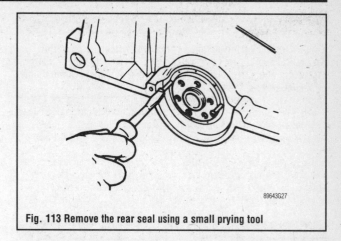

Fig. 113 Remove the rear seal using a small prying tool

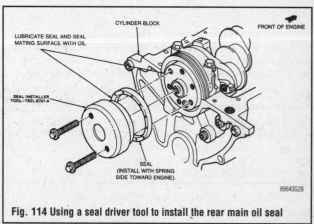

Fig. 114 Using a seal driver tool to install the rear main oil seal

To install:

6. Clean the oil seal recess in the cylinder block and main bearing cap.
7. Clean, inspect and polish the rear oil seal rubbing surface on the crankshaft. Coat a new oil seal and the crankshaft with a light film of engine oil. Start the seal in the recess with the seal lip facing forward and install it with a seal driver. Keep the tool, T72C–6165, straight with the centerline of the crankshaft and install the seal until the tool contacts the cylinder block surface. Remove the tool and inspect the seal to be sure it was not damaged during installation.
8. Install the engine rear cover plate. Position the flywheel on the crankshaft flange. Coat the threads of the flywheel attaching bolts with oil-resistant sealer and install the bolts. Tighten the bolts in sequence across from each other to the specifications listed in the Torque chart at the beginning of this Chapter.
9. On a manual shift transmission, install the clutch disc and the pressure plate assembly following the procedure in Chapter 7.
10. Install the transmission, following the procedure in Chapter 7.

EXHAUST SYSTEM

Safety Precautions

For a number of reasons, exhaust system work can be the most dangerous type of work you can do on your vehicle. Always observe the following precautions:

• Support the vehicle extra securely. Not only will you often be working directly under it, but you'll frequently be using a lot of force, say, heavy hammer blows, to dislodge rusted parts. This can cause a vehicle that's improperly supported to shift and possibly fall.
• Wear goggles. Exhaust system parts are always rusty. Metal chips can be dislodged, even when you're only turning rusted bolts. Attempting to pry pipes apart with a chisel makes the chips fly even more frequently.
• If you're using a cutting torch, keep it a great distance from either the fuel tank or lines. Stop what you're doing and feel the temperature of the fuel bearing pipes on the tank frequently. Even slight heat can expand and/or vaporize fuel, resulting in accumulated vapor, or even a liquid leak, near your torch.
• Watch where your hammer blows fall and make sure you hit squarely. You could easily tap a brake or fuel line when you hit an exhaust system part with a glancing blow. Inspect all lines and hoses in the area where you've been working.

3-42 ENGINE AND ENGINE OVERHAUL

** CAUTION

Be very careful when working on or near the catalytic converter. External temperatures can reach 1,500°F (816°C) and more, causing severe burns. Removal or installation should be performed only on a cold exhaust system.

A number of special exhaust system tools can be rented from auto supply houses or local stores that rent special equipment. A common one is a tail pipe expander, designed to enable you to join pipes of identical diameter.

It may also be quite helpful to use solvents designed to loosen rusted bolts or flanges. Soaking rusted parts the night before you do the job can speed the work of freeing rusted parts considerably. Remember that these solvents are often flammable. Apply only to parts after they are cool!

Inspection

▶ See Figures 115 thru 121

→Safety glasses should be worn at all times when working on or near the exhaust system. Older exhaust systems will almost always be covered with loose rust particles which will shower you when disturbed. These particles are more than a nuisance and could injure your eye.

** CAUTION

DO NOT perform exhaust repairs or inspection with the engine or exhaust hot. Allow the system to cool completely before attempting any work. Exhaust systems are noted for sharp edges, flaking metal and rusted bolts. Gloves and eye protection are required. A healthy supply of penetrating oil and rags is highly recommended.

Your vehicle must be raised and supported safely to inspect the exhaust system properly. By placing 4 safety stands under the vehicle for support should provide enough room for you to slide under the vehicle and inspect the system completely. Start the inspection at the exhaust manifold or turbocharger pipe where the header pipe is attached and work your way to the back of the vehicle. On dual exhaust systems, remember to inspect both sides of the vehicle. Check the complete exhaust system for open seams, holes loose connections, or other deterioration which could permit exhaust fumes to seep into the passenger compartment. Inspect all mounting brackets and hangers for deterioration, some models may have rubber O-rings that can be overstretched and non-supportive. These components will need to be replaced if found. It has always been a practice to use a pointed tool to poke up into the exhaust system where the deterioration spots are to see whether or not they crumble. Some models may have heat shield covering certain parts of the exhaust system, it

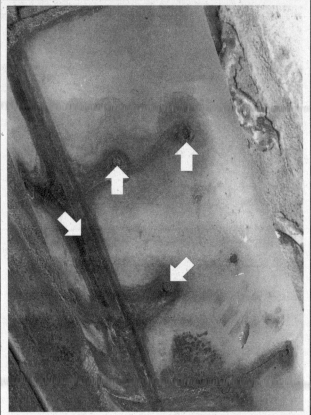
Fig. 116 Check the muffler for rotted spot welds and seams

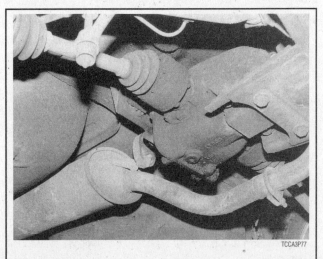
Fig. 117 Make sure the exhaust components are not contacting the body or suspension

will be necessary to remove these shields to have the exhaust visible for inspection also.

REPLACEMENT

▶ See Figures 122 and 123

There are basically two types of exhaust systems. One is the flange type where the component ends are attached with bolts and a gasket in-between. The other exhaust system is the slip joint type. These components slip into one another using clamps to retain them together.

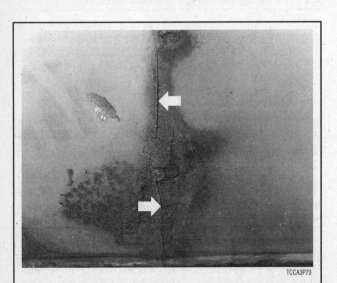
Fig. 115 Cracks in the muffler are a guaranteed leak

ENGINE AND ENGINE OVERHAUL 3-43

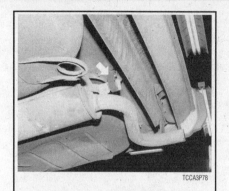

Fig. 118 Check for overstretched or torn exhaust hangers

Fig. 119 Example of a badly deteriorated exhaust pipe

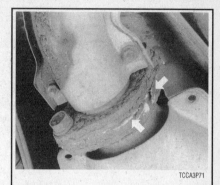

Fig. 120 Inspect flanges for gaskets that have deteriorated and need replacement

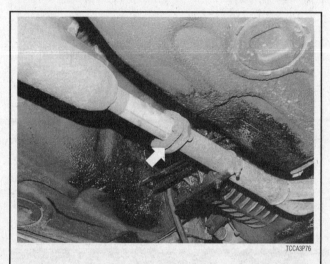

Fig. 121 Some systems, like this one, use large O-rings (donuts) in between the flanges

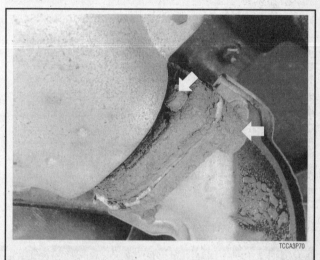

Fig. 123 Nuts and bolts will be extremely difficult to remove when deteriorated with rust

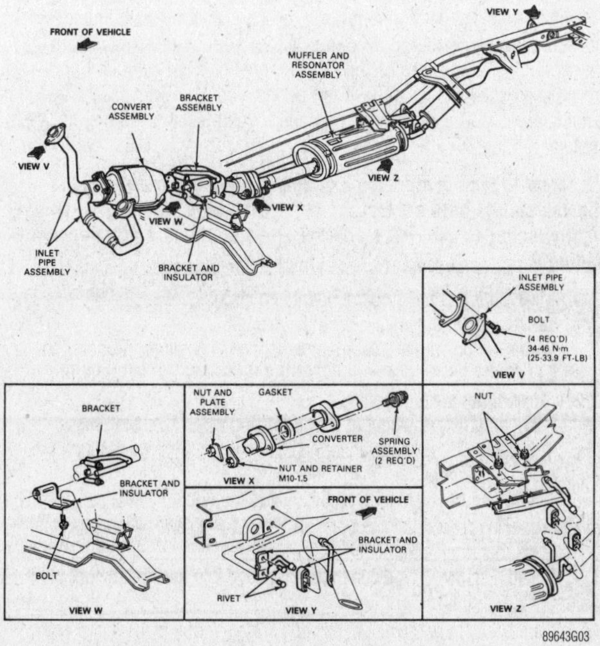

Fig. 122 Exhaust system components

✱✱ CAUTION

Allow the exhaust system to cool sufficiently before spraying a solvent exhaust fasteners. Some solvents are highly flammable and could ignite when sprayed on hot exhaust components.

Before removing any component of the exhaust system, ALWAYS squirt a liquid rust dissolving agent onto the fasteners for ease of removal. A lot of knuckle skin will be saved by following this rule. It may even be wise to spray the fasteners and allow them to sit overnight.

Flange Type

♦ See Figure 124

✱✱ CAUTION

Do NOT perform exhaust repairs or inspection with the engine or exhaust hot. Allow the system to cool completely before attempting any work. Exhaust systems are noted for sharp edges, flaking metal and rusted bolts. Gloves and eye protection are required. A healthy supply of penetrating oil and rags is highly recommended. Never spray liquid rust dissolving agent onto a hot exhaust component.

Before removing any component on a flange type system, ALWAYS squirt a liquid rust dissolving agent onto the fasteners for ease of removal. Start by unbolting the exhaust piece at both ends (if required). When unbolting the headpipe from the manifold, make sure that the bolts are free before trying to remove them. if you snap a stud in the exhaust manifold, the stud will have to be removed with a bolt extractor, which often means removal of the manifold itself. Next, disconnect the

3-44 ENGINE AND ENGINE OVERHAUL

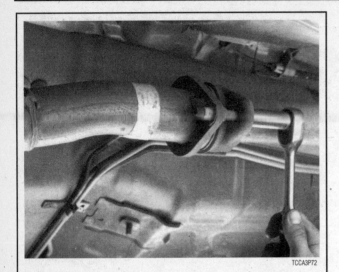

Fig. 124 Example of a flange type exhaust system joint

removal. Start by unbolting the exhaust piece at both ends (if required). When unbolting the headpipe from the manifold, make sure that the bolts are free before trying to remove them. if you snap a stud in the exhaust manifold, the stud will have to be removed with a bolt extractor, which often means removal of the manifold itself. Next, remove the mounting U-bolts from around the exhaust pipe you are extracting from the vehicle. Don't be surprised if the U-bolts break while removing the nuts. Loosen the exhaust pipe from any mounting brackets retaining it to the floor pan and separate the components.

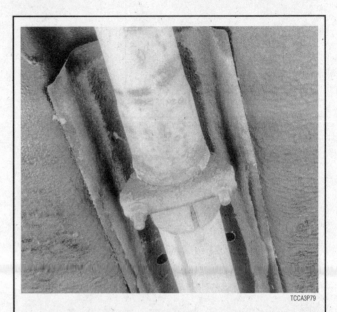

Fig. 125 Example of a common slip joint type system

component from the mounting; slight twisting and turning may be required to remove the component completely from the vehicle. You may need to tap on the component with a rubber mallet to loosen the component. If all else fails, use a hacksaw to separate the parts. An oxy-acetylene cutting torch may be faster but the sparks are DANGEROUS near the fuel tank, and at the very least, accidents could happen, resulting in damage to the under-car parts, not to mention yourself.

Slip Joint Type

▶ See Figure 125

Before removing any component on the slip joint type exhaust system, ALWAYS squirt a liquid rust dissolving agent onto the fasteners for ease of

ENGINE RECONDITIONING

Determining Engine Condition

Anything that generates heat and/or friction will eventually burn or wear out (for example, a light bulb generates heat, therefore its life span is limited). With this in mind, a running engine generates tremendous amounts of both; friction is encountered by the moving and rotating parts inside the engine and heat is created by friction and combustion of the fuel. However, the engine has systems designed to help reduce the effects of heat and friction and provide added longevity. The oiling system reduces the amount of friction encountered by the moving parts inside the engine, while the cooling system reduces heat created by friction and combustion. If either system is not maintained, a break-down will be inevitable. Therefore, you can see how regular maintenance can affect the service life of your vehicle. If you do not drain, flush and refill your cooling system at the proper intervals, deposits will begin to accumulate in the radiator, thereby reducing the amount of heat it can extract from the coolant. The same applies to your oil and filter; if it is not changed often enough it becomes laden with contaminates and is unable to properly lubricate the engine. This increases friction and wear.

There are a number of methods for evaluating the condition of your engine. A compression test can reveal the condition of your pistons, piston rings, cylinder bores, head gasket(s), valves and valve seats. An oil pressure test can warn you of possible engine bearing, or oil pump failures. Excessive oil consumption, evidence of oil in the engine air intake area and/or bluish smoke from the tailpipe may indicate worn piston rings, worn valve guides and/or valve seals. As a general rule, an engine that uses no more than one quart of oil every 1000 miles is in good condition. Engines that use one quart of oil or more in less than 1000 miles should first be checked for oil leaks. If any oil leaks are present, have them fixed before determining how much oil is consumed by the engine, especially if blue smoke is not visible at the tailpipe.

COMPRESSION TEST

▶ See Figure 126

A noticeable lack of engine power, excessive oil consumption and/or poor fuel mileage measured over an extended period are all indicators of internal engine wear. Worn piston rings, scored or worn cylinder bores, blown head gaskets, sticking or burnt valves, and worn valve seats are all possible culprits. A check of each cylinder's compression will help locate the problem.

➥**A screw-in type compression gauge is more accurate than the type you simply hold against the spark plug hole. Although it takes slightly longer to use, it's worth the effort to obtain a more accurate reading.**

1. Make sure that the proper amount and viscosity of engine oil is in the crankcase, then ensure the battery is fully charged.
2. Warm-up the engine to normal operating temperature, then shut the engine **OFF**.
3. Disable the ignition system.
4. Label and disconnect all of the spark plug wires from the plugs.
5. Thoroughly clean the cylinder head area around the spark plug ports, then remove the spark plugs.
6. Set the throttle plate to the fully open (wide-open throttle) position. You can block the accelerator linkage open for this, or you can have an assistant fully depress the accelerator pedal.
7. Install a screw-in type compression gauge into the No. 1 spark plug hole until the fitting is snug.

ENGINE AND ENGINE OVERHAUL

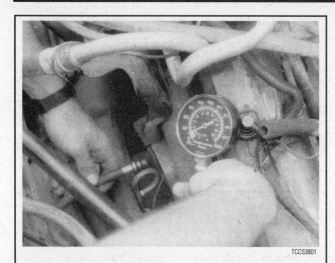

Fig. 126 A screw-in type compression gauge is more accurate and easier to use without an assistant

⚠️ WARNING

Be careful not to crossthread the spark plug hole.

8. According to the tool manufacturer's instructions, connect a remote starting switch to the starting circuit.

9. With the ignition switch in the **OFF** position, use the remote starting switch to crank the engine through at least five compression strokes (approximately 5 seconds of cranking) and record the highest reading on the gauge.

10. Repeat the test on each cylinder, cranking the engine approximately the same number of compression strokes and/or time as the first.

11. Compare the highest readings from each cylinder to that of the others. The indicated compression pressures are considered within specifications if the lowest reading cylinder is within 75 percent of the pressure recorded for the highest reading cylinder. For example, if your highest reading cylinder pressure was 150 psi (1034 kPa), then 75 percent of that would be 113 psi (779 kPa). So the lowest reading cylinder should be no less than 113 psi (779 kPa).

12. If a cylinder exhibits an unusually low compression reading, pour a tablespoon of clean engine oil into the cylinder through the spark plug hole and repeat the compression test. If the compression rises after adding oil, it means that the cylinder's piston rings and/or cylinder bore are damaged or worn. If the pressure remains low, the valves may not be seating properly (a valve job is needed), or the head gasket may be blown near that cylinder. If compression in any two adjacent cylinders is low, and if the addition of oil doesn't help raise compression, there is leakage past the head gasket. Oil and coolant in the combustion chamber, combined with blue or constant white smoke from the tailpipe, are symptoms of this problem. However, don't be alarmed by the normal white smoke emitted from the tailpipe during engine warm-up or from cold weather driving. There may be evidence of water droplets on the engine dipstick and/or oil droplets in the cooling system if a head gasket is blown.

OIL PRESSURE TEST

Check for proper oil pressure at the sending unit passage with an externally mounted mechanical oil pressure gauge (as opposed to relying on a factory installed dash-mounted gauge). A tachometer may also be needed, as some specifications may require running the engine at a specific rpm.

1. With the engine cold, locate and remove the oil pressure sending unit.
2. Following the manufacturer's instructions, connect a mechanical oil pressure gauge and, if necessary, a tachometer to the engine.
3. Start the engine and allow it to idle.
4. Check the oil pressure reading when cold and record the number. You may need to run the engine at a specified rpm, so check the specifications chart located earlier in this section.
5. Run the engine until normal operating temperature is reached (upper radiator hose will feel warm).
6. Check the oil pressure reading again with the engine hot and record the number. Turn the engine **OFF**.
7. Compare your hot oil pressure reading to that given in the chart. If the reading is low, check the cold pressure reading against the chart. If the cold pressure is well above the specification, and the hot reading was lower than the specification, you may have the wrong viscosity oil in the engine. Change the oil, making sure to use the proper grade and quantity, then repeat the test.

Low oil pressure readings could be attributed to internal component wear, pump related problems, a low oil level, or oil viscosity that is too low. High oil pressure readings could be caused by an overfilled crankcase, too high of an oil viscosity or a faulty pressure relief valve.

Buy or Rebuild?

Now that you have determined that your engine is worn out, you must make some decisions. The question of whether or not an engine is worth rebuilding is largely a subjective matter and one of personal worth. Is the engine a popular one, or is it an obsolete model? Are parts available? Will it get acceptable gas mileage once it is rebuilt? Is the car it's being put into worth keeping? Would it be less expensive to buy a new engine, have your engine rebuilt by a pro, rebuild it yourself or buy a used engine from a salvage yard? Or would it be simpler and less expensive to buy another car? If you have considered all these matters and more, and have still decided to rebuild the engine, then it is time to decide how you will rebuild it.

➡ **The editors at Chilton feel that most engine machining should be performed by a professional machine shop. Don't think of it as wasting money, rather, as an assurance that the job has been done right the first time. There are many expensive and specialized tools required to perform such tasks as boring and honing an engine block or having a valve job done on a cylinder head. Even inspecting the parts requires expensive micrometers and gauges to properly measure wear and clearances. Also, a machine shop can deliver to you clean, and ready to assemble parts, saving you time and aggravation. Your maximum savings will come from performing the removal, disassembly, assembly and installation of the engine and purchasing or renting only the tools required to perform the above tasks. Depending on the particular circumstances, you may save 40 to 60 percent of the cost doing these yourself.**

A complete rebuild or overhaul of an engine involves replacing all of the moving parts (pistons, rods, crankshaft, camshaft, etc.) with new ones and machining the non-moving wearing surfaces of the block and heads. Unfortunately, this may not be cost effective. For instance, your crankshaft may have been damaged or worn, but it can be machined undersize for a minimal fee.

So, as you can see, you can replace everything inside the engine, but, it is wiser to replace only those parts which are really needed, and, if possible, repair the more expensive ones. Later in this section, we will break the engine down into its two main components: the cylinder head and the engine block. We will discuss each component, and the recommended parts to replace during a rebuild on each.

Engine Overhaul Tips

Most engine overhaul procedures are fairly standard. In addition to specific parts replacement procedures and specifications for your individual engine, this section is also a guide to acceptable rebuilding procedures. Examples of standard rebuilding practice are given and should be used along with specific details concerning your particular engine.

Competent and accurate machine shop services will ensure maximum performance, reliability and engine life. In most instances it is more profitable for the do-it-yourself mechanic to remove, clean and inspect the component, buy the necessary parts and deliver these to a shop for actual machine work.

Much of the assembly work (crankshaft, bearings, piston rods, and other components) is well within the scope of the do-it-yourself mechanic's tools and abilities. You will have to decide for yourself the depth of involvement you desire in an engine repair or rebuild.

TOOLS

The tools required for an engine overhaul or parts replacement will depend on the depth of your involvement. With a few exceptions, they will be the tools

3-46 ENGINE AND ENGINE OVERHAUL

found in a mechanic's tool kit (see Section 1 of this manual). More in-depth work will require some or all of the following:
- A dial indicator (reading in thousandths) mounted on a universal base
- Micrometers and telescope gauges
- Jaw and screw-type pullers
- Scraper
- Valve spring compressor
- Ring groove cleaner
- Piston ring expander and compressor
- Ridge reamer
- Cylinder hone or glaze breaker
- Plastigage®
- Engine stand

The use of most of these tools is illustrated in this section. Many can be rented for a one-time use from a local parts jobber or tool supply house specializing in automotive work.

Occasionally, the use of special tools is called for. See the information on Special Tools and the Safety Notice in the front of this book before substituting another tool.

OVERHAUL TIPS

Aluminum has become extremely popular for use in engines, due to its low weight. Observe the following precautions when handling aluminum parts:
- Never hot tank aluminum parts (the caustic hot tank solution will eat the aluminum.
- Remove all aluminum parts (identification tag, etc.) from engine parts prior to the tanking.
- Always coat threads lightly with engine oil or anti-seize compounds before installation, to prevent seizure.
- Never overtighten bolts or spark plugs especially in aluminum threads.

When assembling the engine, any parts that will be exposed to frictional contact must be prelubed to provide lubrication at initial start-up. Any product specifically formulated for this purpose can be used, but engine oil is not recommended as a prelube in most cases.

When semi-permanent (locked, but removable) installation of bolts or nuts is desired, threads should be cleaned and coated with Loctite® or another similar, commercial non-hardening sealant.

CLEANING

▶ See Figures 127, 128, 129 and 130

Before the engine and its components are inspected, they must be thoroughly cleaned. You will need to remove any engine varnish, oil sludge and/or carbon deposits from all of the components to insure an accurate inspection. A crack in the engine block or cylinder head can easily become overlooked if hidden by a layer of sludge or carbon.

Most of the cleaning process can be carried out with common hand tools and readily available solvents or solutions. Carbon deposits can be chipped away using a hammer and a hard wooden chisel. Old gasket material and varnish or sludge can usually be removed using a scraper and/or cleaning solvent. Extremely stubborn deposits may require the use of a power drill with a wire brush. If using a wire brush, use extreme care around any critical machined surfaces (such as the gasket surfaces, bearing saddles, cylinder bores, etc.). USE OF A WIRE BRUSH IS NOT RECOMMENDED ON ANY ALUMINUM COMPONENTS. Always follow any safety recommendations given by the manufacturer of the tool and/or solvent. You should always wear eye protection during any cleaning process involving scraping, chipping or spraying of solvents.

An alternative to the mess and hassle of cleaning the parts yourself is to drop them off at a local garage or machine shop. They will, more than likely, have the necessary equipment to properly clean all of the parts for a nominal fee.

✲✲ CAUTION

Always wear eye protection during any cleaning process involving scraping, chipping or spraying of solvents.

Remove any oil galley plugs, freeze plugs and/or pressed-in bearings and carefully wash and degrease all of the engine components including the fasteners and bolts. Small parts such as the valves, springs, etc., should be placed in a metal basket and allowed to soak. Use pipe cleaner type brushes, and clean all passageways in the components. Use a ring expander and remove the rings from the pistons. Clean the piston ring grooves with a special tool or a piece of broken ring. Scrape the carbon off of the top of the piston. You should never use a wire brush on the pistons. After preparing all of the piston assemblies in this manner, wash and degrease them again.

✲✲ WARNING

Use extreme care when cleaning around the cylinder head valve seats. A mistake or slip may cost you a new seat.

When cleaning the cylinder head, remove carbon from the combustion chamber with the valves installed. This will avoid damaging the valve seats.

REPAIRING DAMAGED THREADS

▶ See Figures 131, 132, 133, 134 and 135

Several methods of repairing damaged threads are available. Heli-Coil® (shown here), Keenserts® and Microdot® are among the most widely used. All involve basically the same principle—drilling out stripped threads, tapping the hole and installing a prewound insert—making welding, plugging and oversize fasteners unnecessary.

Two types of thread repair inserts are usually supplied: a standard type for most inch coarse, inch fine, metric course and metric fine thread sizes and a spark lug type to fit most spark plug port sizes. Consult the individual tool manufacturer's catalog to determine exact applications. Typical thread repair kits will contain a selection of prewound threaded inserts, a tap (corresponding to the outside diameter threads of the insert) and an installation tool. Spark plug inserts usually differ because they require a tap equipped with pilot threads and a combined reamer/tap section. Most manufacturers also supply blister-packed thread repair inserts separately in addition to a master kit containing a variety of taps and inserts plus installation tools.

Before attempting to repair a threaded hole, remove any snapped, broken or damaged bolts or studs. Penetrating oil can be used to free frozen threads. The

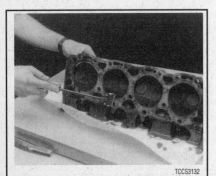

Fig. 127 Use a gasket scraper to remove the old gasket material from the mating surfaces

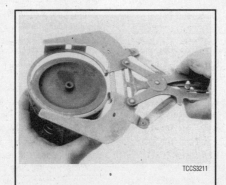

Fig. 128 Use a ring expander tool to remove the piston rings

Fig. 129 Clean the piston ring grooves using a ring groove cleaner tool, or . . .

ENGINE AND ENGINE OVERHAUL

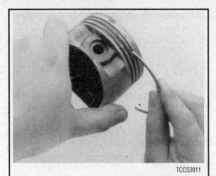

Fig. 130 . . . use a piece of an old ring to clean the grooves. Be careful, the ring can be quite sharp

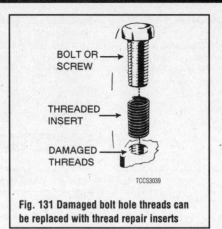

Fig. 131 Damaged bolt hole threads can be replaced with thread repair inserts

Fig. 132 Standard thread repair insert (left), and spark plug thread insert

Fig. 133 Drill out the damaged threads with the specified size bit. Be sure to drill completely through the hole or to the bottom of a blind hole

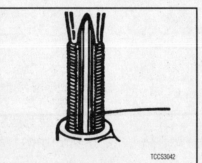

Fig. 134 Using the kit, tap the hole in order to receive the thread insert. Keep the tap well oiled and back it out frequently to avoid clogging the threads

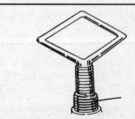

Fig. 135 Screw the insert onto the installer tool until the tang engages the slot. Thread the insert into the hole until it is ¼–½ turn below the top surface, then remove the tool and break off the tang using a punch

offending item can usually be removed with locking pliers or using a screw/stud extractor. After the hole is clear, the thread can be repaired, as shown in the series of accompanying illustrations and in the kit manufacturer's instructions.

Engine Preparation

To properly rebuild an engine, you must first remove it from the vehicle, then disassemble and diagnose it. Ideally you should place your engine on an engine stand. This affords you the best access to the engine components. Follow the manufacturer's directions for using the stand with your particular engine. Remove the flywheel or flexplate before installing the engine to the stand.

Now that you have the engine on a stand, and assuming that you have drained the oil and coolant from the engine, it's time to strip it of all but the necessary components. Before you start disassembling the engine, you may want to take a moment to draw some pictures, or fabricate some labels or containers to mark the locations of various components and the bolts and/or studs which fasten them. Modern day engines use a lot of little brackets and clips which hold wiring harnesses and such, and these holders are often mounted on studs and/or bolts that can be easily mixed up. The manufacturer spent a lot of time and money designing your vehicle, and they wouldn't have wasted any of it by haphazardly placing brackets, clips or fasteners on the vehicle. If it's present when you disassemble it, put it back when you assemble, you will regret not remembering that little bracket which holds a wire harness out of the path of a rotating part.

You should begin by unbolting any accessories still attached to the engine, such as the water pump, power steering pump, alternator, etc. Then, unfasten any manifolds (intake or exhaust) which were not removed during the engine removal procedure. Finally, remove any covers remaining on the engine such as the rocker arm, front or timing cover and oil pan. Some front covers may require the vibration damper and/or crank pulley to be removed beforehand. The idea is to reduce the engine to the bare necessities (cylinder head(s), valve train, engine block, crankshaft, pistons and connecting rods), plus any other `in block' components such as oil pumps, balance shafts and auxiliary shafts.

Finally, remove the cylinder head(s) from the engine block and carefully place on a bench. Disassembly instructions for each component follow later in this section.

Cylinder Head

There are two basic types of cylinder heads used on today's automobiles: the Overhead Valve (OHV) and the Overhead Camshaft (OHC). The latter can also be broken down into two subgroups: the Single Overhead Camshaft (SOHC) and the Dual Overhead Camshaft (DOHC). Generally, if there is only a single camshaft on a head, it is just referred to as an OHC head. Also, an engine with an OHV cylinder head is also known as a pushrod engine.

Most cylinder heads these days are made of an aluminum alloy due to its light weight, durability and heat transfer qualities. However, cast iron was the material of choice in the past, and is still used on many vehicles today. Whether made from aluminum or iron, all cylinder heads have valves and seats. Some use two valves per cylinder, while the more hi-tech engines will utilize a multi-valve configuration using 3, 4 and even 5 valves per cylinder. When the valve contacts the seat, it does so on precision machined surfaces, which seals the combustion chamber. All cylinder heads have a valve guide for each valve. The guide centers the valve to the seat and allows it to move up and down within it. The clearance between the valve and guide can be critical. Too much clearance and the engine may consume oil, lose vacuum and/or damage the seat. Too little, and the valve can stick in the guide causing the engine to run poorly if at all, and possibly causing severe damage. The last component all cylinder heads have are valve springs. The spring holds the valve against its seat. It also returns the valve to this position when the valve has been opened by the valve train or camshaft. The spring is fastened to the valve by a retainer and valve locks (sometimes called keepers). Aluminum heads will also have a valve spring shim to keep the spring from wearing away the aluminum.

An ideal method of rebuilding the cylinder head would involve replacing all of the valves, guides, seats, springs, etc. with new ones. However, depending on how the engine was maintained, often this is not necessary. A major cause of valve, guide and seat wear is an improperly tuned engine. An engine that is running too rich, will often wash the lubricating oil out of the guide with gasoline, causing it to wear rapidly. Conversely, an engine which is running too lean will place higher combustion temperatures on the valves and seats allowing them to wear or even burn. Springs fall victim to the driving habits of the individual. A driver who often runs the engine rpm to the redline will wear out or break the

3-48 ENGINE AND ENGINE OVERHAUL

springs faster then one that stays well below it. Unfortunately, mileage takes it toll on all of the parts. Generally, the valves, guides, springs and seats in a cylinder head can be machined and re-used, saving you money. However, if a valve is burnt, it may be wise to replace all of the valves, since they were all operating in the same environment. The same goes for any other component on the cylinder head. Think of it as an insurance policy against future problems related to that component.

Unfortunately, the only way to find out which components need replacing, is to disassemble and carefully check each piece. After the cylinder head(s) are disassembled, thoroughly clean all of the components.

DISASSEMBLY

OHV Heads

▶ See Figures 136 thru 141

Before disassembling the cylinder head, you may want to fabricate some containers to hold the various parts, as some of them can be quite small (such as keepers) and easily lost. Also keeping yourself and the components organized will aid in assembly and reduce confusion. Where possible, try to maintain a components original location; this is especially important if there is not going to be any machine work performed on the components.

1. If you haven't already removed the rocker arms and/or shafts, do so now.
2. Position the head so that the springs are easily accessed.
3. Use a valve spring compressor tool, and relieve spring tension from the retainer.

→Due to engine varnish, the retainer may stick to the valve locks. A gentle tap with a hammer may help to break it loose.

4. Remove the valve locks from the valve tip and/or retainer. A small magnet may help in removing the locks.
5. Lift the valve spring, tool and all, off of the valve stem.
6. If equipped, remove the valve seal. If the seal is difficult to remove with the valve in place, try removing the valve first, then the seal. Follow the steps below for valve removal.
7. Position the head to allow access for withdrawing the valve.

→Cylinder heads that have seen a lot of miles and/or abuse may have mushroomed the valve lock grove and/or tip, causing difficulty in removal of the valve. If this has happened, use a metal file to carefully remove the high spots around the lock grooves and/or tip. Only file it enough to allow removal.

8. Remove the valve from the cylinder head.
9. If equipped, remove the valve spring shim. A small magnetic tool or screwdriver will aid in removal.
10. Repeat Steps 3 though 9 until all of the valves have been removed.

OHC Heads

▶ See Figures 142 and 143

Whether it is a single or dual overhead camshaft cylinder head, the disassembly procedure is relatively unchanged. One aspect to pay attention to is careful labeling of the parts on the dual camshaft cylinder head. There will be an intake camshaft and followers as well as an exhaust camshaft and followers and they must be labeled as such. In some cases, the components are identical and could easily be installed incorrectly. DO NOT MIX THEM UP! Determining which is which is very simple; the intake camshaft and components are on the same side of the head as was the intake manifold. Conversely, the exhaust camshaft and components are on the same side of the head as was the exhaust manifold.

CUP TYPE CAMSHAFT FOLLOWERS

▶ See Figures 144, 145 and 146

Most cylinder heads with cup type camshaft followers will have the valve spring, retainer and locks recessed within the follower's bore. You will need a C-clamp style valve spring compressor tool, an OHC spring removal tool (or equivalent) and a small magnet to disassemble the head.

Fig. 136 When removing an OHV valve spring, use a compressor tool to relieve the tension from the retainer

Fig. 137 A small magnet will help in removal of the valve locks

Fig. 138 Be careful not to lose the small valve locks (keepers)

Fig. 139 Remove the valve seal from the valve stem—O-ring type seal shown

Fig. 140 Removing an umbrella/positive type seal

Fig. 141 Invert the cylinder head and withdraw the valve from the valve guide bore

ENGINE AND ENGINE OVERHAUL

Fig. 142 Exploded view of a valve, seal, spring, retainer and locks from an OHC cylinder head

Fig. 143 Example of a multi-valve cylinder head. Note how it has 2 intake and 2 exhaust valve ports

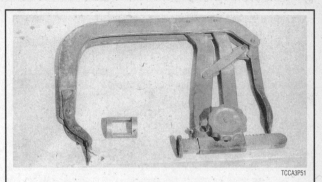

Fig. 144 C-clamp type spring compressor and an OHC spring removal tool (center) for cup type followers

Fig. 145 Most cup type follower cylinder heads retain the camshaft using bolt-on bearing caps

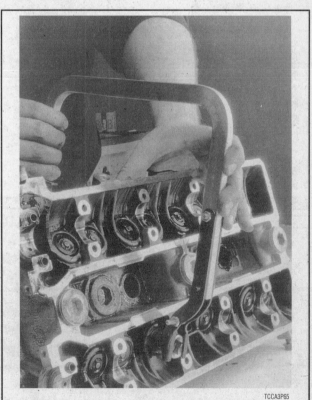

Fig. 146 Position the OHC spring tool in the follower bore, then compress the spring with a C-clamp type tool

ENGINE AND ENGINE OVERHAUL

1. If not already removed, remove the camshaft(s) and/or followers. Mark their positions for assembly.
2. Position the cylinder head to allow use of a C-clamp style valve spring compressor tool.

➡ It is preferred to position the cylinder head gasket surface facing you with the valve springs facing the opposite direction and the head laying horizontal.

3. With the OHC spring removal adapter tool positioned inside of the follower bore, compress the valve spring using the C-clamp style valve spring compressor.
4. Remove the valve locks. A small magnetic tool or screwdriver will aid in removal.
5. Release the compressor tool and remove the spring assembly.
6. Withdraw the valve from the cylinder head.
7. If equipped, remove the valve seal.

➡ Special valve seal removal tools are available. Regular or needlenose type pliers, if used with care, will work just as well. If using ordinary pliers, be sure not to damage the follower bore. The follower and its bore are machined to close tolerances and any damage to the bore will effect this relationship.

8. If equipped, remove the valve spring shim. A small magnetic tool or screwdriver will aid in removal.
9. Repeat Steps 3 through 8 until all of the valves have been removed.

ROCKER ARM TYPE CAMSHAFT FOLLOWERS

▶ See Figures 147 thru 155

Most cylinder heads with rocker arm-type camshaft followers are easily disassembled using a standard valve spring compressor. However, certain models may not have enough open space around the spring for the standard tool and may require you to use a C-clamp style compressor tool instead.

1. If not already removed, remove the rocker arms and/or shafts and the camshaft. If applicable, also remove the hydraulic lash adjusters. Mark their positions for assembly.

Fig. 147 Example of the shaft mounted rocker arms on some OHC heads

Fig. 148 Another example of the rocker arm type OHC head. This model uses a follower under the camshaft

Fig. 149 Before the camshaft can be removed, all of the followers must first be removed . . .

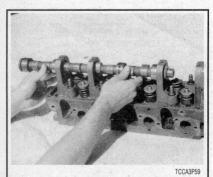
Fig. 150 . . . then the camshaft can be removed by sliding it out (shown), or unbolting a bearing cap (not shown)

Fig. 151 Compress the valve spring . . .

Fig. 152 . . . then remove the valve locks from the valve stem and spring retainer

Fig. 153 Remove the valve spring and retainer from the cylinder head

Fig. 154 Remove the valve seal from the guide. Some gentle prying or pliers may help to remove stubborn ones

Fig. 155 All aluminum and some cast iron heads will have these valve spring shims. Remove all of them as well

ENGINE AND ENGINE OVERHAUL 3-51

2. Position the cylinder head to allow access to the valve spring.
3. Use a valve spring compressor tool to relieve the spring tension from the retainer.

➡ Due to engine varnish, the retainer may stick to the valve locks. A gentle tap with a hammer may help to break it loose.

4. Remove the valve locks from the valve tip and/or retainer. A small magnet may help in removing the small locks.
5. Lift the valve spring, tool and all, off of the valve stem.
6. If equipped, remove the valve seal. If the seal is difficult to remove with the valve in place, try removing the valve first, then the seal. Follow the steps below for valve removal.
7. Position the head to allow access for withdrawing the valve.

➡ Cylinder heads that have seen a lot of miles and/or abuse may have mushroomed the valve lock grove and/or tip, causing difficulty in removal of the valve. If this has happened, use a metal file to carefully remove the high spots around the lock grooves and/or tip. Only file it enough to allow removal.

8. Remove the valve from the cylinder head.
9. If equipped, remove the valve spring shim. A small magnetic tool or screwdriver will aid in removal.
10. Repeat Steps 3 though 9 until all of the valves have been removed.

INSPECTION

Now that all of the cylinder head components are clean, it's time to inspect them for wear and/or damage. To accurately inspect them, you will need some specialized tools:

- A 0–1 in. micrometer for the valves
- A dial indicator or inside diameter gauge for the valve guides
- A spring pressure test gauge

If you do not have access to the proper tools, you may want to bring the components to a shop that does.

Valves

▶ See Figures 156 and 157

The first thing to inspect are the valve heads. Look closely at the head, margin and face for any cracks, excessive wear or burning. The margin is the best place to look for burning. It should have a squared edge with an even width all around the diameter. When a valve burns, the margin will look melted and the edges rounded. Also inspect the valve head for any signs of tulipping. This will show as a lifting of the edges or dishing in the center of the head and will usually not occur to all of the valves. All of the heads should look the same, any that seem dished more than others are probably bad. Next, inspect the valve lock grooves and valve tips. Check for any burrs around the lock grooves, especially if you had to file them to remove the valve. Valve tips should appear flat, although slight rounding with high mileage engines is normal. Slightly worn valve tips will need to be machined flat. Last, measure the valve stem diameter with the micrometer. Measure the area that rides within the guide, especially towards the tip where most of the wear occurs. Take several measurements along its length and compare them to each other. Wear should be even along the length with little to no taper. If no minimum diameter is given in the specifications, then the stem should not read more than 0.001 in. (0.025mm) below the specification. Any valves that fail these inspections should be replaced.

Springs, Retainers and Valve Locks

▶ See Figures 158 and 159

The first thing to check is the most obvious, broken springs. Next check the free length and squareness of each spring. If applicable, insure to distinguish between intake and exhaust springs. Use a ruler and/or carpenter's square to measure the length. A carpenter's square should be used to check the springs for squareness. If a spring pressure test gauge is available, check each springs rating and compare to the specifications chart. Check the readings against the specifications given. Any springs that fail these inspections should be replaced.

The spring retainers rarely need replacing, however they should still be checked as a precaution. Inspect the spring mating surface and the valve lock retention area for any signs of excessive wear. Also check for any signs of cracking. Replace any retainers that are questionable.

Valve locks should be inspected for excessive wear on the outside contact area as well as on the inner notched surface. Any locks which appear worn or broken and its respective valve should be replaced.

Cylinder Head

There are several things to check on the cylinder head: valve guides, seats, cylinder head surface flatness, cracks and physical damage.

VALVE GUIDES

▶ See Figure 160

Now that you know the valves are good, you can use them to check the guides, although a new valve, if available, is preferred. Before you measure anything, look at the guides carefully and inspect them for any cracks, chips or breakage. Also if the guide is a removable style (as in most aluminum heads), check them for any looseness or evidence of movement. All of the guides should appear to be at the same height from the spring seat. If any seem lower (or higher) from another, the guide has moved. Mount a dial indicator onto the spring side of the cylinder head. Lightly oil the valve stem and insert it into the cylinder head. Position the dial indicator against the valve stem near the tip and zero the gauge. Grasp the valve stem and wiggle towards and away from the dial indicator and observe the readings. Mount the dial indicator 90 degrees from the initial point and zero the gauge and again take a reading. Compare the two readings for a out of round condition. Check the readings against the specifications given. An Inside Diameter (I.D.) gauge designed for valve guides will give you an accurate valve guide bore measurement. If the I.D. gauge is used, compare the readings with the specifications given. Any guides that fail these inspections should be replaced or machined.

VALVE SEATS

A visual inspection of the valve seats should show a slightly worn and pitted surface where the valve face contacts the seat. Inspect the seat carefully for severe pitting or cracks. Also, a seat that is badly worn will be recessed into the

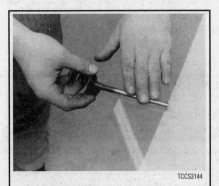

Fig. 156 Valve stems may be rolled on a flat surface to check for bends

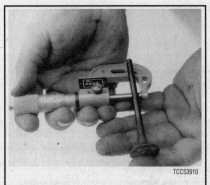

Fig. 157 Use a micrometer to check the valve stem diameter

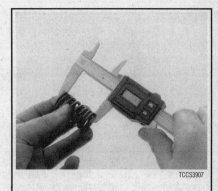

Fig. 158 Use a caliper to check the valve spring free-length

3-52 ENGINE AND ENGINE OVERHAUL

Fig. 159 Check the valve spring for squareness on a flat surface; a carpenter's square can be used

Fig. 160 A dial gauge may be used to check valve stem-to-guide clearance; read the gauge while moving the valve stem

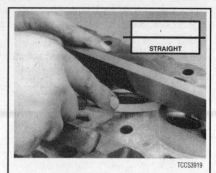

Fig. 161 Check the head for flatness across the center of the head surface using a straightedge and feeler gauge

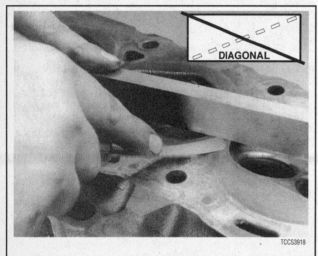

Fig. 162 Checks should also be made along both diagonals of the head surface

cylinder head. A severely worn or recessed seat may need to be replaced. All cracked seats must be replaced. A seat concentricity gauge, if available, should be used to check the seat run-out. If run-out exceeds specifications the seat must be machined (if no specification is given use 0.002 in. or 0.051mm).

CYLINDER HEAD SURFACE FLATNESS

▶ See Figures 161 and 162

After you have cleaned the gasket surface of the cylinder head of any old gasket material, check the head for flatness.

Place a straightedge across the gasket surface. Using feeler gauges, determine the clearance at the center of the straightedge and across the cylinder head at several points. Check along the centerline and diagonally on the head surface. If the warpage exceeds 0.003 in. (0.076mm) within a 6.0 in. (15.2cm) span, or 0.006 in. (0.152mm) over the total length of the head, the cylinder head must be resurfaced. After resurfacing the heads of a V-type engine, the intake manifold flange surface should be checked, and if necessary, milled proportionally to allow for the change in its mounting position.

CRACKS AND PHYSICAL DAMAGE

Generally, cracks are limited to the combustion chamber, however, it is not uncommon for the head to crack in a spark plug hole, port, outside of the head or in the valve spring/rocker arm area. The first area to inspect is always the hottest: the exhaust seat/port area.

A visual inspection should be performed, but just because you don't see a crack does not mean it is not there. Some more reliable methods for inspecting for cracks include Magnaflux®, a magnetic process or Zyglo®, a dye penetrant. Magnaflux® is used only on ferrous metal (cast iron) heads. Zyglo® uses a spray on fluorescent mixture along with a black light to reveal the cracks. It is strongly recommended to have your cylinder head checked professionally for cracks, especially if the engine was known to have overheated and/or leaked or consumed coolant. Contact a local shop for availability and pricing of these services.

Physical damage is usually very evident. For example, a broken mounting ear from dropping the head or a bent or broken stud and/or bolt. All of these defects should be fixed or, if unrepairable, the head should be replaced.

Camshaft and Followers

Inspect the camshaft(s) and followers as described earlier in this section.

REFINISHING & REPAIRING

Many of the procedures given for refinishing and repairing the cylinder head components must be performed by a machine shop. Certain steps, if the inspected part is not worn, can be performed yourself inexpensively. However, you spent a lot of time and effort so far, why risk trying to save a couple bucks if you might have to do it all over again?

Valves

Any valves that were not replaced should be refaced and the tips ground flat. Unless you have access to a valve grinding machine, this should be done by a machine shop. If the valves are in extremely good condition, as well as the valve seats and guides, they may be lapped in without performing machine work.

It is a recommended practice to lap the valves even after machine work has been performed and/or new valves have been purchased. This insures a positive seal between the valve and seat.

LAPPING THE VALVES

➡Before lapping the valves to the seats, read the rest of the cylinder head section to insure that any related parts are in acceptable enough condition to continue.

➡Before any valve seat machining and/or lapping can be performed, the guides must be within factory recommended specifications.

1. Invert the cylinder head.
2. Lightly lubricate the valve stems and insert them into the cylinder head in their numbered order.
3. Raise the valve from the seat and apply a small amount of fine lapping compound to the seat.
4. Moisten the suction head of a hand-lapping tool and attach it to the head of the valve.
5. Rotate the tool between the palms of both hands, changing the position of the valve on the valve seat and lifting the tool often to prevent grooving.
6. Lap the valve until a smooth, polished circle is evident on the valve and seat.
7. Remove the tool and the valve. Wipe away all traces of the grinding compound and store the valve to maintain its lapped location.

ENGINE AND ENGINE OVERHAUL 3-53

⚠ WARNING

Do not get the valves out of order after they have been lapped. They must be put back with the same valve seat with which they were lapped.

Springs, Retainers and Valve Locks

There is no repair or refinishing possible with the springs, retainers and valve locks. If they are found to be worn or defective, they must be replaced with new (or known good) parts.

Cylinder Head

Most refinishing procedures dealing with the cylinder head must be performed by a machine shop. Read the sections below and review your inspection data to determine whether or not machining is necessary.

VALVE GUIDE

➡ If any machining or replacements are made to the valve guides, the seats must be machined.

Unless the valve guides need machining or replacing, the only service to perform is to thoroughly clean them of any dirt or oil residue.

There are only two types of valve guides used on automobile engines: the replaceable-type (all aluminum heads) and the cast-in integral-type (most cast iron heads). There are four recommended methods for repairing worn guides.
- Knurling
- Inserts
- Reaming oversize
- Replacing

Knurling is a process in which metal is displaced and raised, thereby reducing clearance, giving a true center, and providing oil control. It is the least expensive way of repairing the valve guides. However, it is not necessarily the best, and in some cases, a knurled valve guide will not stand up for more than a short time. It requires a special knurlizer and precision reaming tools to obtain proper clearances. It would not be cost effective to purchase these tools, unless you plan on rebuilding several of the same cylinder head.

Installing a guide insert involves machining the guide to accept a bronze insert. One style is the coil-type which is installed into a threaded guide. Another is the thin-walled insert where the guide is reamed oversize to accept a split-sleeve insert. After the insert is installed, a special tool is then run through the guide to expand the insert, locking it to the guide. The insert is then reamed to the standard size for proper valve clearance.

Reaming for oversize valves restores normal clearances and provides a true valve seat. Most cast-in type guides can be reamed to accept an valve with an oversize stem. The cost factor for this can become quite high as you will need to purchase the reamer and new, oversize stem valves for all guides which were reamed. Oversizes are generally 0.003 to 0.030 in. (0.076 to 0.762mm), with 0.015 in. (0.381mm) being the most common.

To replace cast-in type valve guides, they must be drilled out, then reamed to accept replacement guides. This must be done on a fixture which will allow centering and leveling off of the original valve seat or guide, otherwise a serious guide-to-seat misalignment may occur making it impossible to properly machine the seat.

Replaceable-type guides are pressed into the cylinder head. A hammer and a stepped drift or punch may be used to install and remove the guides. Before removing the guides, measure the protrusion on the spring side of the head and record it for installation. Use the stepped drift to hammer out the old guide from the combustion chamber side of the head. When installing, determine whether or not the guide also seals a water jacket in the head, and if it does, use the recommended sealing agent. If there is no water jacket, grease the valve guide and its bore. Use the stepped drift, and hammer the new guide into the cylinder head from the spring side of the cylinder head. A stack of washers the same thickness as the measured protrusion may help the installation process.

VALVE SEATS

➡ Before any valve seat machining can be performed, the guides must be within factory recommended specifications.

➡ If any machining or replacements were made to the valve guides, the seats must be machined.

If the seats are in good condition, the valves can be lapped to the seats, and the cylinder head assembled. See the valves section for instructions on lapping.

If the valve seats are worn, cracked or damaged, they must be serviced by a machine shop. The valve seat must be perfectly centered to the valve guide, which requires very accurate machining.

CYLINDER HEAD SURFACE

If the cylinder head is warped, it must be machined flat. If the warpage is extremely severe, the head may need to be replaced. In some instances, it may be possible to straighten a warped head enough to allow machining. In either case, contact a professional machine shop for service.

➡ Any OHC cylinder head that shows excessive warpage should have the camshaft bearing journals align bored after the cylinder head has been resurfaced.

⚠ WARNING

Failure to align bore the camshaft bearing journals could result in severe engine damage including but not limited to: valve and piston damage, connecting rod damage, camshaft and/or crankshaft breakage.

CRACKS AND PHYSICAL DAMAGE

Certain cracks can be repaired in both cast iron and aluminum heads. For cast iron, a tapered threaded insert is installed along the length of the crack. Aluminum can also use the tapered inserts, however welding is the preferred method. Some physical damage can be repaired through brazing or welding. Contact a machine shop to get expert advice for your particular dilemma.

ASSEMBLY

The first step for any assembly job is to have a clean area in which to work. Next, thoroughly clean all of the parts and components that are to be assembled. Finally, place all of the components onto a suitable work space and, if necessary, arrange the parts to their respective positions.

OHV Engines

1. Lightly lubricate the valve stems and insert all of the valves into the cylinder head. If possible, maintain their original locations.
2. If equipped, install any valve spring shims which were removed.
3. If equipped, install the new valve seals, keeping the following in mind:
   - If the valve seal presses over the guide, lightly lubricate the outer guide surfaces.
   - If the seal is an O-ring type, it is installed just after compressing the spring but before the valve locks.
4. Place the valve spring and retainer over the stem.
5. Position the spring compressor tool and compress the spring.
6. Assemble the valve locks to the stem.
7. Relieve the spring pressure slowly and insure that neither valve lock becomes dislodged by the retainer.
8. Remove the spring compressor tool.
9. Repeat Steps 2 through 8 until all of the springs have been installed.

OHC Engines

▶ See Figure 163

CUP TYPE CAMSHAFT FOLLOWERS

To install the springs, retainers and valve locks on heads which have these components recessed into the camshaft follower's bore, you will need a small screwdriver-type tool, some clean white grease and a lot of patience. You will also need the C-clamp style spring compressor and the OHC tool used to disassemble the head.

1. Lightly lubricate the valve stems and insert all of the valves into the cylinder head. If possible, maintain their original locations.
2. If equipped, install any valve spring shims which were removed.
3. If equipped, install the new valve seals, keeping the following in mind:
   - If the valve seal presses over the guide, lightly lubricate the outer guide surfaces.

3-54 ENGINE AND ENGINE OVERHAUL

Fig. 163 Once assembled, check the valve clearance and correct as needed

• If the seal is an O-ring type, it is installed just after compressing the spring but before the valve locks.
4. Place the valve spring and retainer over the stem.
5. Position the spring compressor and the OHC tool, then compress the spring.
6. Using a small screwdriver as a spatula, fill the valve stem side of the lock with white grease. Use the excess grease on the screwdriver to fasten the lock to the driver.
7. Carefully install the valve lock, which is stuck to the end of the screwdriver, to the valve stem then press on it with the screwdriver until the grease squeezes out. The valve lock should now be stuck to the stem.
8. Repeat Steps 6 and 7 for the remaining valve lock.
9. Relieve the spring pressure slowly and insure that neither valve lock becomes dislodged by the retainer.
10. Remove the spring compressor tool.
11. Repeat Steps 2 through 10 until all of the springs have been installed.
12. Install the followers, camshaft(s) and any other components that were removed for disassembly.

ROCKER ARM TYPE CAMSHAFT FOLLOWERS

1. Lightly lubricate the valve stems and insert all of the valves into the cylinder head. If possible, maintain their original locations.
2. If equipped, install any valve spring shims which were removed.
3. If equipped, install the new valve seals, keeping the following in mind:
• If the valve seal presses over the guide, lightly lubricate the outer guide surfaces.
• If the seal is an O-ring type, it is installed just after compressing the spring but before the valve locks.
4. Place the valve spring and retainer over the stem.
5. Position the spring compressor tool and compress the spring.
6. Assemble the valve locks to the stem.
7. Relieve the spring pressure slowly and insure that neither valve lock becomes dislodged by the retainer.
8. Remove the spring compressor tool.
9. Repeat Steps 2 through 8 until all of the springs have been installed.
10. Install the camshaft(s), rockers, shafts and any other components that were removed for disassembly.

Engine Block

GENERAL INFORMATION

A thorough overhaul or rebuild of an engine block would include replacing the pistons, rings, bearings, timing belt/chain assembly and oil pump. For OHV engines also include a new camshaft and lifters. The block would then have the cylinders bored and honed oversize (or if using removable cylinder sleeves, new sleeves installed) and the crankshaft would be cut undersize to provide new wearing surfaces and perfect clearances. However, your particular engine may not have everything worn out. What if only the piston rings have worn out and the clearances on everything else are still within factory specifications? Well, you could just replace the rings and put it back together, but this would be a very rare example. Chances are, if one component in your engine is worn, other components are sure to follow, and soon. At the very least, you should always replace the rings, bearings and oil pump. This is what is commonly called a "freshen up".

Cylinder Ridge Removal

Because the top piston ring does not travel to the very top of the cylinder, a ridge is built up between the end of the travel and the top of the cylinder bore.

Pushing the piston and connecting rod assembly past the ridge can be difficult, and damage to the piston ring lands could occur. If the ridge is not removed before installing a new piston or not removed at all, piston ring breakage and piston damage may occur.

➡ It is always recommended that you remove any cylinder ridges before removing the piston and connecting rod assemblies. If you know that new pistons are going to be installed and the engine block will be bored oversize, you may be able to forego this step. However, some ridges may actually prevent the assemblies from being removed, necessitating its removal.

There are several different types of ridge reamers on the market, none of which are inexpensive. Unless a great deal of engine rebuilding is anticipated, borrow or rent a reamer.
1. Turn the crankshaft until the piston is at the bottom of its travel.
2. Cover the head of the piston with a rag.
3. Follow the tool manufacturers instructions and cut away the ridge, exercising extreme care to avoid cutting too deeply.
4. Remove the ridge reamer, the rag and as many of the cuttings as possible. Continue until all of the cylinder ridges have been removed.

DISASSEMBLY

♦ See Figures 164 and 165

The engine disassembly instructions following assume that you have the engine mounted on an engine stand. If not, it is easiest to disassemble the engine on a bench or floor with it resting on the bell housing or transmission mounting surface. You must be able to access the connecting rod fasteners and turn the crankshaft during disassembly. Also, all engine covers (timing, front, side, oil pan, whatever) should have already been removed. Engines which are seized or locked up may not be able to be completely disassembled, and a core (salvage yard) engine should be purchased.

Fig. 164 Place rubber hose over the connecting rod studs to protect the crankshaft and cylinder bores from damage

ENGINE AND ENGINE OVERHAUL

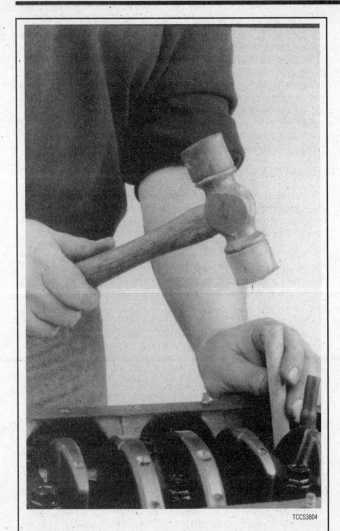

Fig. 165 Carefully tap the piston out of the bore using a wooden dowel

Pushrod Engines

If not done during the cylinder head removal, remove the pushrods and lifters, keeping them in order for assembly. Remove the timing gears and/or timing chain assembly, then remove the oil pump drive assembly and withdraw the camshaft from the engine block. Remove the oil pick-up and pump assembly. If equipped, remove any balance or auxiliary shafts. If necessary, remove the cylinder ridge from the top of the bore. See the cylinder ridge removal procedure earlier in this section.

OHC Engines

If not done during the cylinder head removal, remove the timing chain/belt and/or gear/sprocket assembly. Remove the oil pick-up and pump assembly and, if necessary, the pump drive. If equipped, remove any balance or auxiliary shafts. If necessary, remove the cylinder ridge from the top of the bore. See the cylinder ridge removal procedure earlier in this section.

All Engines

Rotate the engine over so that the crankshaft is exposed. Use a number punch or scribe and mark each connecting rod with its respective cylinder number. The cylinder closest to the front of the engine is always number 1. However, depending on the engine placement, the front of the engine could either be the flywheel or damper/pulley end. Generally the front of the engine faces the front of the vehicle. Use a number punch or scribe and also mark the main bearing caps from front to rear with the front most cap being number 1 (if there are five caps, mark them 1 through 5, front to rear).

✱✱ WARNING

Take special care when pushing the connecting rod up from the crankshaft because the sharp threads of the rod bolts/studs will score the crankshaft journal. Insure that special plastic caps are installed over them, or cut two pieces of rubber hose to do the same.

Again, rotate the engine, this time to position the number one cylinder bore (head surface) up. Turn the crankshaft until the number one piston is at the bottom of its travel, this should allow the maximum access to its connecting rod. Remove the number one connecting rods fasteners and cap and place two lengths of rubber hose over the rod bolts/studs to protect the crankshaft from damage. Using a sturdy wooden dowel and a hammer, push the connecting rod up about 1 in. (25mm) from the crankshaft and remove the upper bearing insert. Continue pushing or tapping the connecting rod up until the piston rings are out of the cylinder bore. Remove the piston and rod by hand, put the upper half of the bearing insert back into the rod, install the cap with its bearing insert installed, and hand-tighten the cap fasteners. If the parts are kept in order in this manner, they will not get lost and you will be able to tell which bearings came form what cylinder if any problems are discovered and diagnosis is necessary. Remove all the other piston assemblies in the same manner. On V-style engines, remove all of the pistons from one bank, then reposition the engine with the other cylinder bank head surface up, and remove that banks piston assemblies.

The only remaining component in the engine block should now be the crankshaft. Loosen the main bearing caps evenly until the fasteners can be turned by hand, then remove them and the caps. Remove the crankshaft from the engine block. Thoroughly clean all of the components.

INSPECTION

Now that the engine block and all of its components are clean, it's time to inspect them for wear and/or damage. To accurately inspect them, you will need some specialized tools:

- Two or three separate micrometers to measure the pistons and crankshaft journals
- A dial indicator
- Telescoping gauges for the cylinder bores
- A rod alignment fixture to check for bent connecting rods

If you do not have access to the proper tools, you may want to bring the components to a shop that does.

Generally, you shouldn't expect cracks in the engine block or its components unless it was known to leak, consume or mix engine fluids, it was severely overheated, or there was evidence of bad bearings and/or crankshaft damage. A visual inspection should be performed on all of the components, but just because you don't see a crack does not mean it is not there. Some more reliable methods for inspecting for cracks include Magnaflux®, a magnetic process or Zyglo®, a dye penetrant. Magnaflux® is used only on ferrous metal (cast iron). Zyglo® uses a spray on fluorescent mixture along with a black light to reveal the cracks. It is strongly recommended to have your engine block checked professionally for cracks, especially if the engine was known to have overheated and/or leaked or consumed coolant. Contact a local shop for availability and pricing of these services.

Engine Block

ENGINE BLOCK BEARING ALIGNMENT

Remove the main bearing caps and, if still installed, the main bearing inserts. Inspect all of the main bearing saddles and caps for damage, burrs or high spots. If damage is found, and it is caused from a spun main bearing, the block will need to be align-bored or, if severe enough, replacement. Any burrs or high spots should be carefully removed with a metal file.

Place a straightedge on the bearing saddles, in the engine block, along the centerline of the crankshaft. If any clearance exists between the straightedge and the saddles, the block must be align-bored.

Align-boring consists of machining the main bearing saddles and caps by means of a flycutter that runs through the bearing saddles.

DECK FLATNESS

The top of the engine block where the cylinder head mounts is called the deck. Insure that the deck surface is clean of dirt, carbon deposits and old gasket material. Place a straightedge across the surface of the deck along its centerline and, using feeler gauges, check the clearance along several points. Repeat the checking procedure with the straightedge placed along both diagonals of the deck surface. If the reading exceeds 0.003 in. (0.076mm) within a 6.0 in. (15.2cm) span, or 0.006 in. (0.152mm) over the total length of the deck, it must be machined.

CYLINDER BORES

♦ See Figure 166

The cylinder bores house the pistons and are slightly larger than the pistons themselves. A common piston-to-bore clearance is 0.0015–0.0025 in. (0.0381mm–0.0635mm). Inspect and measure the cylinder bores. The bore should be checked for out-of-roundness, taper and size. The results of this inspection will determine whether the cylinder can be used in its existing size and condition, or a rebore to the next oversize is required (or in the case of removable sleeves, have replacements installed).

The amount of cylinder wall wear is always greater at the top of the cylinder than at the bottom. This wear is known as taper. Any cylinder that has a taper of 0.0012 in. (0.305mm) or more, must be rebored. Measurements are taken at a number of positions in each cylinder: at the top, middle and bottom and at two points at each position; that is, at a point 90 degrees from the crankshaft centerline, as well as a point parallel to the crankshaft centerline. The measurements are made with either a special dial indicator or a telescopic gauge and micrometer. If the necessary precision tools to check the bore are not available, take the block to a machine shop and have them mike it. Also if you don't have the tools to check the cylinder bores, chances are you will not have the necessary devices to check the pistons, connecting rods and crankshaft. Take these components with you and save yourself an extra trip.

For our procedures, we will use a telescopic gauge and a micrometer. You will need one of each, with a measuring range which covers your cylinder bore size.

1. Position the telescopic gauge in the cylinder bore, loosen the gauges lock and allow it to expand.

➡ **Your first two readings will be at the top of the cylinder bore, then proceed to the middle and finally the bottom, making a total of six measurements.**

2. Hold the gauge square in the bore, 90 degrees from the crankshaft centerline, and gently tighten the lock. Tilt the gauge back to remove it from the bore.
3. Measure the gauge with the micrometer and record the reading.
4. Again, hold the gauge square in the bore, this time parallel to the crankshaft centerline, and gently tighten the lock. Again, you will tilt the gauge back to remove it from the bore.
5. Measure the gauge with the micrometer and record this reading. The difference between these two readings is the out-of-round measurement of the cylinder.

6. Repeat steps 1 through 5, each time going to the next lower position, until you reach the bottom of the cylinder. Then go to the next cylinder, and continue until all of the cylinders have been measured.

The difference between these measurements will tell you all about the wear in your cylinders. The measurements which were taken 90 degrees from the crankshaft centerline will always reflect the most wear. That is because at this position is where the engine power presses the piston against the cylinder bore the hardest. This is known as thrust wear. Take your top, 90 degree measurement and compare it to your bottom, 90 degree measurement. The difference between them is the taper. When you measure your pistons, you will compare these readings to your piston sizes and determine piston-to-wall clearance.

Crankshaft

Inspect the crankshaft for visible signs of wear or damage. All of the journals should be perfectly round and smooth. Slight scores are normal for a used crankshaft, but you should hardly feel them with your fingernail. When measuring the crankshaft with a micrometer, you will take readings at the front and rear of each journal, then turn the micrometer 90 degrees and take two more readings, front and rear. The difference between the front-to-rear readings is the journal taper and the first-to-90 degree reading is the out-of-round measurement. Generally, there should be no taper or out-of-roundness found, however, up to 0.0005 in. (0.0127mm) for either can be overlooked. Also, the readings should fall within the factory specifications for journal diameters.

If the crankshaft journals fall within specifications, it is recommended that it be polished before being returned to service. Polishing the crankshaft insures that any minor burrs or high spots are smoothed, thereby reducing the chance of scoring the new bearings.

Pistons and Connecting Rods

PISTONS

♦ See Figure 167

The piston should be visually inspected for any signs of cracking or burning (caused by hot spots or detonation), and scuffing or excessive wear on the skirts. The wrist pin attaches the piston to the connecting rod. The piston should move freely on the wrist pin, both sliding and pivoting. Grasp the connecting rod securely, or mount it in a vise, and try to rock the piston back and forth along the centerline of the wrist pin. There should not be any excessive play evident between the piston and the pin. If there are C-clips retaining the pin in the piston then you have wrist pin bushings in the rods. There should not be any excessive play between the wrist pin and the rod bushing. Normal clearance for the wrist pin is approx. 0.001–0.002 in. (0.025mm–0.051mm).

Use a micrometer and measure the diameter of the piston, perpendicular to the wrist pin, on the skirt. Compare the reading to its original cylinder measurement obtained earlier. The difference between the two readings is the piston-to-wall clearance. If the clearance is within specifications, the piston may be used as is. If

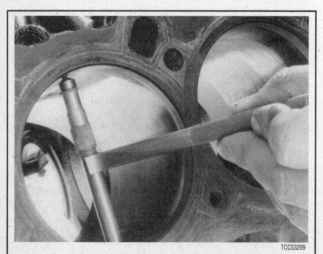

Fig. 166 Use a telescoping gauge to measure the cylinder bore diameter—take several readings within the same bore

Fig. 167 Measure the piston's outer diameter, perpendicular to the wrist pin, with a micrometer

ENGINE AND ENGINE OVERHAUL

the piston is out of specification, but the bore is not, you will need a new piston. If both are out of specification, you will need the cylinder rebored and oversize pistons installed. Generally if two or more pistons/bores are out of specification, it is best to rebore the entire block and purchase a complete set of oversize pistons.

CONNECTING ROD

You should have the connecting rod checked for straightness at a machine shop. If the connecting rod is bent, it will unevenly wear the bearing and piston, as well as place greater stress on these components. Any bent or twisted connecting rods must be replaced. If the rods are straight and the wrist pin clearance is within specifications, then only the bearing end of the rod need be checked. Place the connecting rod into a vice, with the bearing inserts in place, install the cap to the rod and torque the fasteners to specifications. Use a telescoping gauge and carefully measure the inside diameter of the bearings. Compare this reading to the rods original crankshaft journal diameter measurement. The difference is the oil clearance. If the oil clearance is not within specifications, install new bearings in the rod and take another measurement. If the clearance is still out of specifications, and the crankshaft is not, the rod will need to be reconditioned by a machine shop.

➡ You can also use Plastigage® to check the bearing clearances. The assembling section has complete instructions on its use.

Camshaft

Inspect the camshaft and lifters/followers as described earlier in this section.

Bearings

All of the engine bearings should be visually inspected for wear and/or damage. The bearing should look evenly worn all around with no deep scores or pits. If the bearing is severely worn, scored, pitted or heat blued, then the bearing, and the components that use it, should be brought to a machine shop for inspection. Full-circle bearings (used on most camshafts, auxiliary shafts, balance shafts, etc.) require specialized tools for removal and installation, and should be brought to a machine shop for service.

Oil Pump

➡ The oil pump is responsible for providing constant lubrication to the whole engine and so it is recommended that a new oil pump be installed when rebuilding the engine.

Completely disassemble the oil pump and thoroughly clean all of the components. Inspect the oil pump gears and housing for wear and/or damage. Insure that the pressure relief valve operates properly and there is no binding or sticking due to varnish or debris. If all of the parts are in proper working condition, lubricate the gears and relief valve, and assemble the pump.

REFINISHING

♦ See Figure 168

Almost all engine block refinishing must be performed by a machine shop. If the cylinders are not to be rebored, then the cylinder glaze can be removed with a ball hone. When removing cylinder glaze with a ball hone, use a light or penetrating type oil to lubricate the hone. Do not allow the hone to run dry as this may cause excessive scoring of the cylinder bores and wear on the hone. If new pistons are required, they will need to be installed to the connecting rods. This should be performed by a machine shop as the pistons must be installed in the correct relationship to the rod or engine damage can occur.

Pistons and Connecting Rods

♦ See Figure 169

Only pistons with the wrist pin retained by C-clips are serviceable by the home-mechanic. Press fit pistons require special presses and/or heaters to remove/install the connecting rod and should only be performed by a machine shop.

All pistons will have a mark indicating the direction to the front of the engine and the must be installed into the engine in that manner. Usually it is a notch or arrow on the top of the piston, or it may be the letter F cast or stamped into the piston.

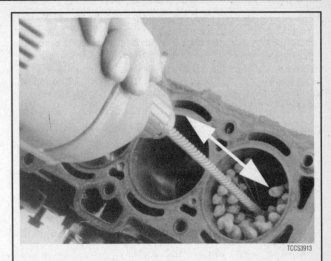

Fig. 168 Use a ball type cylinder hone to remove any glaze and provide a new surface for seating the piston rings

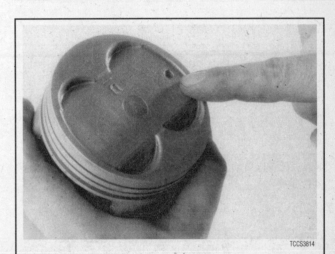

Fig. 169 Most pistons are marked to indicate positioning in the engine (usually a mark means the side facing the front)

C-CLIP TYPE PISTONS

1. Note the location of the forward mark on the piston and mark the connecting rod in relation.
2. Remove the C-clips from the piston and withdraw the wrist pin.

➡ Varnish build-up or C-clip groove burrs may increase the difficulty of removing the wrist pin. If necessary, use a punch or drift to carefully tap the wrist pin out.

3. Insure that the wrist pin bushing in the connecting rod is usable, and lubricate it with assembly lube.
4. Remove the wrist pin from the new piston and lubricate the pin bores on the piston.
5. Align the forward marks on the piston and the connecting rod and install the wrist pin.
6. The new C-clips will have a flat and a rounded side to them. Install both C-clips with the flat side facing out.
7. Repeat all of the steps for each piston being replaced.

ASSEMBLY

Before you begin assembling the engine, first give yourself a clean, dirt free work area. Next, clean every engine component again. The key to a good assembly is cleanliness.

3-58 ENGINE AND ENGINE OVERHAUL

Mount the engine block into the engine stand and wash it one last time using water and detergent (dishwashing detergent works well). While washing it, scrub the cylinder bores with a soft bristle brush and thoroughly clean all of the oil passages. Completely dry the engine and spray the entire assembly down with an anti-rust solution such as WD-40® or similar product. Take a clean lint-free rag and wipe up any excess anti-rust solution from the bores, bearing saddles, etc. Repeat the final cleaning process on the crankshaft. Replace any freeze or oil galley plugs which were removed during disassembly.

Crankshaft

▶ See Figures 170, 171, 172 and 173

1. Remove the main bearing inserts from the block and bearing caps.
2. If the crankshaft main bearing journals have been refinished to a definite undersize, install the correct undersize bearing. Be sure that the bearing inserts and bearing bores are clean. Foreign material under inserts will distort bearing and cause failure.
3. Place the upper main bearing inserts in bores with tang in slot.

➡ The oil holes in the bearing inserts must be aligned with the oil holes in the cylinder block.

4. Install the lower main bearing inserts in bearing caps.
5. Clean the mating surfaces of block and rear main bearing cap.
6. Carefully lower the crankshaft into place. Be careful not to damage bearing surfaces.

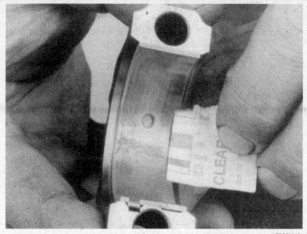
Fig. 171 After the cap is removed again, use the scale supplied with the gauging material to check the clearance

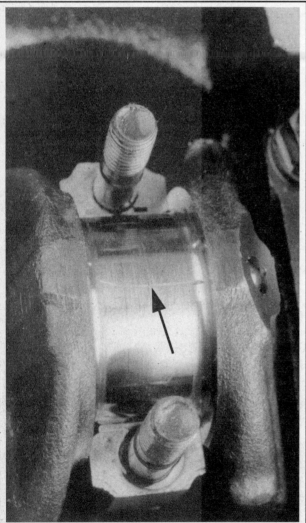
Fig. 170 Apply a strip of gauging material to the bearing journal, then install and torque the cap

Fig. 172 A dial gauge may be used to check crankshaft end-play

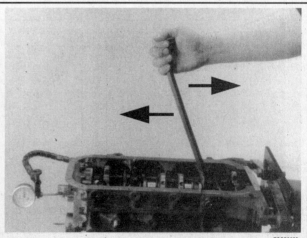
Fig. 173 Carefully pry the crankshaft back and forth while reading the dial gauge for end-play

ENGINE AND ENGINE OVERHAUL

7. Check the clearance of each main bearing by using the following procedure:

   a. Place a piece of Plastigage® or its equivalent, on bearing surface across full width of bearing cap and about ¼ in. off center.

   b. Install cap and tighten bolts to specifications. Do not turn crankshaft while Plastigage® is in place.

   c. Remove the cap. Using the supplied Plastigage® scale, check width of Plastigage® at widest point to get maximum clearance. Difference between readings is taper of journal.

   d. If clearance exceeds specified limits, try a 0.001 in. or 0.002 in. undersize bearing in combination with the standard bearing. Bearing clearance must be within specified limits. If standard and 0.002 in. undersize bearing does not bring clearance within desired limits, refinish crankshaft journal, then install undersize bearings.

8. After the bearings have been fitted, apply a light coat of engine oil to the journals and bearings. Install the rear main bearing cap. Install all bearing caps except the thrust bearing cap. Be sure that main bearing caps are installed in original locations. Tighten the bearing cap bolts to specifications.

9. Install the thrust bearing cap with bolts finger-tight.

10. Pry the crankshaft forward against the thrust surface of upper half of bearing.

11. Hold the crankshaft forward and pry the thrust bearing cap to the rear. This aligns the thrust surfaces of both halves of the bearing.

12. Retain the forward pressure on the crankshaft. Tighten the cap bolts to specifications.

13. Install the rear main seal.

14. Measure the crankshaft end-play as follows:

    a. Mount a dial gauge to the engine block and position the tip of the gauge to read from the crankshaft end.

    b. Carefully pry the crankshaft toward the rear of the engine and hold it there while you zero the gauge.

    c. Carefully pry the crankshaft toward the front of the engine and read the gauge.

    d. Confirm that the reading is within specifications. If not, install a new thrust bearing and repeat the procedure. If the reading is still out of specifications with a new bearing, have a machine shop inspect the thrust surfaces of the crankshaft, and if possible, repair it.

15. Rotate the crankshaft so as to position the first rod journal to the bottom of its stroke.

Pistons and Connecting Rods

♦ See Figures 174, 175, 176 and 177

1. Before installing the piston/connecting rod assembly, oil the pistons, piston rings and the cylinder walls with light engine oil. Install connecting rod bolt protectors or rubber hose onto the connecting rod bolts/studs. Also perform the following:

   a. Select the proper ring set for the size cylinder bore.

   b. Position the ring in the bore in which it is going to be used.

   c. Push the ring down into the bore area where normal ring wear is not encountered.

   d. Use the head of the piston to position the ring in the bore so that the ring is square with the cylinder wall. Use caution to avoid damage to the ring or cylinder bore.

   e. Measure the gap between the ends of the ring with a feeler gauge. Ring gap in a worn cylinder is normally greater than specification. If the ring gap is greater than the specified limits, try an oversize ring set.

   f. Check the ring side clearance of the compression rings with a feeler gauge inserted between the ring and its lower land according to specification. The gauge should slide freely around the entire ring circumference without binding. Any wear that occurs will form a step at the inner portion of the lower land. If the lower lands have high steps, the piston should be replaced.

2. Unless new pistons are installed, be sure to install the pistons in the cylinders from which they were removed. The numbers on the connecting rod and bearing cap must be on the same side when installed in the cylinder bore. If a connecting rod is ever transposed from one engine or cylinder to another, new bearings should be fitted and the connecting rod should be numbered to correspond with the new cylinder number. The notch on the piston head goes toward the front of the engine.

3. Install all of the rod bearing inserts into the rods and caps.

4. Install the rings to the pistons. Install the oil control ring first, then the second compression ring and finally the top compression ring. Use a piston ring expander tool to aid in installation and to help reduce the chance of breakage.

5. Make sure the ring gaps are properly spaced around the circumference of the piston. Fit a piston ring compressor around the piston and slide the piston and connecting rod assembly down into the cylinder bore, pushing it in with the wooden hammer handle. Push the piston down until it is only slightly below the top of the cylinder bore. Guide the connecting rod onto the crankshaft bearing journal carefully, to avoid damaging the crankshaft.

6. Check the bearing clearance of all the rod bearings, fitting them to the crankshaft bearing journals. Follow the procedure in the crankshaft installation above.

7. After the bearings have been fitted, apply a light coating of assembly oil to the journals and bearings.

8. Turn the crankshaft until the appropriate bearing journal is at the bottom of its stroke, then push the piston assembly all the way down until the connecting rod bearing seats on the crankshaft journal. Be careful not to allow the bearing cap screws to strike the crankshaft bearing journals and damage them.

9. After the piston and connecting rod assemblies have been installed, check the connecting rod side clearance on each crankshaft journal.

10. Prime and install the oil pump and the oil pump intake tube.

OHV Engines

CAMSHAFT, LIFTERS AND TIMING ASSEMBLY

1. Install the camshaft.
2. Install the lifters/followers into their bores.
3. Install the timing gears/chain assembly.

Fig. 174 Checking the piston ring-to-ring groove side clearance using the ring and a feeler gauge

Fig. 175 The notch on the side of the bearing cap matches the tang on the bearing insert

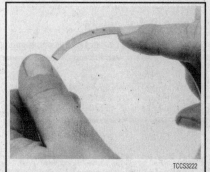

Fig. 176 Most rings are marked to show which side of the ring should face up when installed to the piston

3-60 ENGINE AND ENGINE OVERHAUL

Fig. 177 Install the piston and rod assembly into the block using a ring compressor and the handle of a hammer

CYLINDER HEAD(S)

1. Install the cylinder head(s) using new gaskets.
2. Assemble the rest of the valve train (pushrods and rocker arms and/or shafts).

OHC Engines

CYLINDER HEAD(S)

1. Install the cylinder head(s) using new gaskets.
2. Install the timing sprockets/gears and the belt/chain assemblies. If equipped, install the auxiliary/balance shaft(s).

Engine Covers and Components

Install the timing cover(s) and oil pan. Refer to your notes and drawings made prior to disassembly and install all of the components that were removed. Install the engine into the vehicle.

Engine Start-up and Break-in

STARTING THE ENGINE

Now that the engine is installed and every wire and hose is properly connected, go back and double check that all coolant and vacuum hoses are connected. Check that you oil drain plug is installed and properly tightened. If not already done, install a new oil filter onto the engine. Fill the crankcase with the proper amount and grade of engine oil. Fill the cooling system with a 50/50 mixture of coolant/water.
1. Connect the vehicle battery.
2. Start the engine. Keep your eye on your oil pressure indicator; if it does not indicate oil pressure within 10 seconds of starting, turn the vehicle off.

✽✽ WARNING

Damage to the engine can result if it is allowed to run with no oil pressure. Check the engine oil level to make sure that it is full. Check for any leaks and if found, repair the leaks before continuing. If there is still no indication of oil pressure, you may need to prime the system.

3. Confirm that there are no fluid leaks (oil or other).
4. Allow the engine to reach normal operating temperature (the upper radiator hose will be hot to the touch).
5. If necessary, set the ignition timing.
6. Install any remaining components such as the air cleaner (if removed for ignition timing) or body panels which were removed.

BREAKING IT IN

Make the first miles on the new engine, easy ones. Vary the speed but do not accelerate hard. Most importantly, do not lug the engine, and avoid sustained high speeds until at least 100 miles. Check the engine oil and coolant levels frequently. Expect the engine to use a little oil until the rings seat. Change the oil and filter at 500 miles, 1500 miles, then every 3000 miles past that.

KEEP IT MAINTAINED

Now that you have just gone through all of that hard work, keep yourself from doing it all over again by thoroughly maintaining it. Not that you may not have maintained it before, heck you could have had one to two hundred thousand miles on it before doing this. However, you may have bought the vehicle used, and the previous owner did not keep up on maintenance. Which is why you just went through all of that hard work. See?

2.2L ENGINE MECHANICAL SPECIFICATIONS

Component	U.S.	Metric
Bore x stroke	3.39 x 3.70 in.	86.0 x 94.0 mm
Total piston displacement	133.2 cu in.	2,184 cc
Compression pressure		
Standard	173 psi	1,197 KPa
Minimum	121 psi	838 KPa
Maximum difference between cylinders	28 psi	196 KPa
Valve timing		
Intake	Opens 13° BTDC	Closes 57° ABDC
Exhaust	Opens 58° BBDC	Closes 12° ATDC
Valve clearance		
Intake	0	
Exhaust	0	
Cylinder Head		
Height	3.620-3.624 in.	91.95-92.05 mm
Distortion	0.006 in. max.	0.15 mm max.
Grinding	0.008 in. max.	0.20 mm max.
Valve head diameter		
Intake	1.728-1.736 in.	43.9-44.1 mm
Exhaust	1.413-1.421 in.	35.9-36.1 mm
Valve head margin thickness		
Intake	0.031-0.047 in.	0.8-1.2 mm
Exhaust	0.051-0.067 in.	1.3-1.7 mm
Valve face angle		
Intake	45°	
Exhaust	45°	

ENGINE AND ENGINE OVERHAUL

2.2L ENGINE MECHANICAL SPECIFICATIONS

Component	U.S.	Metric
Valve length		
Intake		
Standard	4.4051 in.	111.89 mm
Minimum	4.3894 in.	111.49 mm
Exhaust		
Standard	4.3972 in.	111.69 mm
Minimum	4.3815 in.	111.29 mm
Valve stem diameter		
Intake	0.3161-0.3167 in.	8.030-8.045 mm
Exhaust	0.3159-0.3165 in.	8.025-8.040 mm
Valve guide inner diameter		
Intake	0.3177-0.3185 in.	8.07-8.09 mm
Exhaust	0.3177-0.3185 in.	8.07-8.09 mm
Valve stem-to-guide clearance		
Intake	0.0010-0.0024 in.	0.025-0.060 mm
Exhaust	0.0012-0.0026 in.	0.030-0.065 mm
Maximum	0.008 in.	0.20 mm
Valve guide projection (Height "A")	0.752-0.772 in.	19.1-19.6 mm
Valve Seat		
Intake	45°	
Exhaust	45°	
Valve seat contact width		
Intake	0.047-0.063 in.	1.2-1.6 mm
Exhaust	0.047-0.063 in.	1.2-1.6 mm
Valve seat sinking (measure valve protruding length)		
Intake		
Standard	1.831 in.	46.5 mm
Maximum	1.890 in.	48.0 mm
Exhaust		
Standard	1.831 in.	46.5 mm
Maximum	1.890 in.	48.0 mm
Valve spring		
Free length		
Outer	2.047 in.	52.0 mm
Inner	1.984 in.	50.4 mm
Standard		
Outer	1.732 in.	44.0 mm
Inner	1.681 in.	42.7 mm
Minimum		
Outer	2.047 in.	52.0 mm
Inner	1.984 in.	50.4 mm
Outer	1.732 in.	44.0 mm
Inner	1.681 in.	42.7 mm
Out-of-square		
Intake		
Outer	0.07 in. max.	1.8 mm max.
Inner	0.06 in. max.	1.5 mm max.
Exhaust		
Outer	0.07 in. max.	1.8 mm max.
Inner	0.06 in. max.	1.5 mm max.
Setting load height		
Intake		
Outer	94.6 lbs./1.22 in.	421.8 N/31.0mm
Inner	66.0 lbs./1.04 in.	294.3 N/26.5mm
Exhaust		
Outer	94.6 lbs./1.22 in.	421.8 N/31.0mm
Inner	66.0 lbs./1.04 in.	294.3 N/26.5mm

2.2L ENGINE MECHANICAL SPECIFICATIONS

Component	U.S.	Metric
Camshaft		
Cam lobe height		
Intake		
Standard	1.4984 in.	38.059 mm
Minimum	1.4905 in.	37.859 mm
Exhaust		
Standard	1.4984 in.	38.059 mm
Minimum	1.4905 in.	37.859 mm
Journal diameter		
Front and rear (No.1,5)	1.2575-1.2584 in.	31.940-31.965 mm
Center (No.2,3,4)	1.2563-1.2573 in.	31.910-31.935 mm
Out-of-round max.	0.0020 in.	0.05 mm
Camshaft bearing oil clearance		
Front and rear (No.1,5)	0.0014-0.0033 in.	0.035-0.085 mm
Center (No.2,3,4)	0.0026-0.0045 in.	0.065-0.115 mm
Maximum	0.006 in.	0.15 mm
Camshaft runout	0.0012 in. max.	0.03 mm max.
Camshaft end play		
Standard	0.0031-0.0063 in.	0.08-0.16 mm
Maximum		0.20 mm
Rocker Arms		
Rocker arm inner diameter	0.6300-0.6310 in.	16.000-16.027 mm
Rocker arm shaft diameter	0.6286-0.6293 in.	15.966-15.984 mm
Rocker arm-to-clearance	0.0006-0.0024 in.	0.016-0.061 mm
Maximum	0.004 in.	0.10 mm
Cylinder block		
Height	11.87 in.	301.5 mm
Distortion	0.006 in. max.	0.15 mm max.
Grinding	0.008 in. max.	0.20 mm max.
Cylinder bore diameter		
Standard size	3.3858-3.3866 in.	86.000-86.019 mm
0.010 in. oversize (0.25 mm)	3.3957-3.3964 in.	86.250-86.269 mm
0.020 in. oversize (0.50 mm)	3.4055-3.4063 in.	86.500-86.519 mm
Cylinder bore taper	0.0007 in. max.	0.019 mm max.
Cylinder bore out-of-round	0.0004 in. max.	0.010 mm max.
Piston and Rings		
Piston diameter (Measured at 90° to pin bore axis and 0.709 in. (18.0mm) below oil ring groove)		
Standard size	3.3836-3.3844 in.	85.944-85.964 mm
Oversize 0.010 in. (0.25 mm)	3.3935-3.3942 in.	86.194-86.214 mm
Oversize 0.020 in. (0.50 mm)	3.4033-3.4041 in.	86.444-86.464 mm
Piston-to-cylinder clearance		
Standard	0.0017-0.0024 in.	0.043-0.062 mm
Maximum	0.006 in.	0.15 mm
Piston ring		
Thickness	0.058-0.059 in.	1.47-1.49 mm
End gap measured in cylinder		
Top	0.008-0.014 in.	0.20-0.35 mm
Second	0.006-0.012 in.	0.15-0.30 mm
Oil (rail)	0.008-0.028 in.	0.20-0.70 mm
Maximum	0.039 in.	1.0 mm
Ring groove width in piston		
Top	0.0598-0.0606 in.	1.52-1.54 mm
Second	0.0598-0.0606 in.	1.52-1.54 mm
Oil	0.1583-0.1591 in.	4.02-4.04 mm
Piston ring-to-ring land clearance		
Top	0.0012-0.0028 in.	0.03-0.07 mm
Second	0.0012-0.0028 in.	0.03-0.07 mm
Maximum	0.006 in.	0.15 mm

ENGINE AND ENGINE OVERHAUL

2.2L ENGINE MECHANICAL SPECIFICATIONS

Component	U.S.	Metric
Piston pin		
Diameter	0.8651–0.8654 in.	21.974–21.980 mm
Interference in connecting rod	0.0005–0.0015 in.	0.013–0.037 mm
Piston-to-piston pin clearance	0.0003–0.0009 in.	0.008–0.024 mm
Pressure force	1,100–3,300 lbs.	4,905–14,715 N
Connecting rod		
Length (Center to center)	6.2382–6.2421 in.	158.45–158.55 mm
Bend	0.0094 in. max.	0.24 mm max.
Small end bore	0.8640–0.8646 in.	21.943–21.961 mm
Big end bore	2.1261–2.1266 in.	54.002–54.017 mm
Big end width	1.0566–1.0587 in.	26.838–26.890 mm
Connecting rod side clearance		
Standard	0.0043–0.0103 in.	0.110–0.262 mm
Maximum	0.012 in.	0.30 mm
Crankshaft		
Crankshaft runout	0.0012 in. max.	0.03 mm max.
Main journal diameter		
Standard	2.3597–2.3604 in.	59.937–59.955 mm
0.010 in. undersize (0.25 mm)		
No.1,2,4,5	2.3501–2.3508 in.	59.693–59.711 mm
No.3	2.3499–2.3506 in.	59.687–59.705 mm
0.020 in. undersize (0.50 mm)		
No.1,2,4,5	2.3403–2.3410 in.	59.443–59.461 mm
No.3	2.3400–2.3407 in.	59.437–59.455 mm
0.030 in. undersize (0.75 mm)		
No.1,2,4,5	2.3304–2.3311 in.	59.193–59.211 mm
No.3	2.3302–2.3309 in.	59.187–59.205 mm
Main journal taper	0.002 in. max.	0.05 mm max.
Main journal out-of-round	0.00012 in.	0.003 mm
Crankpin journal diameter		
Standard	2.0055–2.0061 in.	50.940–50.955 mm
0.010 in. undersize (0.25 mm)	1.9957–1.9963 in.	50.690–50.705 mm
0.020 in. undersize (0.50 mm)	1.9858–1.9864 in.	50.440–50.455 mm
0.030 in. undersize (0.75 mm)	1.9760–1.9766 in.	50.190–50.205 mm
Crankpin taper	0.0020 in. max.	0.05 mm max.
Crankpin out-of-round	0.00012 in.	0.003 mm
Main bearing		
Main journal bearing oil clearance		
Standard		
No.1,2,4,5	0.0010–0.0017 in.	0.025–0.043 mm
No.3	0.0012–0.0019 in.	0.031–0.049 mm
Maximum	0.0031 in.	0.08 mm
Available undersize bearing	0.010, 0.020, 0.030 in.	0.25, 0.50, 0.75 mm
Crankpin bearing		
Crankpin bearing oil clearance		
Standard	0.0011–0.0026 in.	0.027–0.067 mm
Maximum	0.004 in.	0.10 mm
Available undersize bearing	0.010, 0.020, 0.030 in.	0.25, 0.50, 0.75 mm
Crankshaft end play		
Standard	0.0031–0.0071 in.	0.08–0.18 mm
Maximum	0.0118 in.	0.30 mm
Thrust Bearing width		
Standard	1.100–1.102 in.	27.94–27.99 mm
0.010 undersize (0.25 mm)	1.104–1.106 in.	28.04–28.09 mm
0.020 undersize (0.50 mm)	1.107–1.109 in.	28.12–28.17 mm
0.030 undersize (0.75 mm)	1.110–1.112 in.	28.20–28.25 mm
Timing belt		
Belt deflection		
New	0.31–0.35 in.	8.0–9.0 mm
Used	0.35–0.39 in.	9.0–10.0 mm

1987-88 2.6L ENGINE MECHANICAL SPECIFICATIONS

Component	U.S.	Metric
Bore x stroke	3.59 x 3.86 in.	91.1 x 98.0 mm
Valve timing		
Intake valve		
Opens	25° BTDC	
Closes	59° ABDC	
Exhaust valve		
Opens	64° BBDC	
Closes	20° ATDC	
Compression pressure @ 250 rpm		
Standard	171 psi	1179 KPa
Minimum	119 psi	820 KPa
Limit of difference between cylinders	28 psi	196 KPa
Valve clearance (Under hot engine condition)		
Jet valve	0.010 in.	0.25 mm
Cylinder head		
Height	3.539–3.547 in.	89.9–90.1 mm
Distortion limit	0.008 in.	0.20 mm
Grinding limit	0.008 in.	0.20 mm
Valve seat		
Seat sinking Dimension "L"		
Standard		
Intake	1.691 in.	42.96 mm
Exhaust	1.691 in.	42.96 mm
Seat angle		
Intake	45°	
Exhaust	45°	
Seat width		
Intake	0.035–0.051 in.	0.9–1.3 mm
Exhaust	0.047–0.063 in.	1.2–1.6 mm
Valve guide, valve and valve spring		
Valve stem to guide clearance		
Standard		
Intake	0.0010–0.0023 in.	0.025–0.058 mm
Exhaust	0.0020–0.0035 in.	0.050–0.088 mm
Maximum	0.0079 in.	0.20 mm
Guide inner diameter	0.3150–0.3157 in.	8.000–8.018 mm
Valve stem diameter		
Intake	0.313–0.314 in.	7.960–7.975 mm
Exhaust	0.312–0.313 in.	7.930–7.950 mm
Valve head diameter		
Intake	1.811 in.	46 mm
Exhaust	1.496 in.	38 mm
Valve face angle		
Intake	45°	
Exhaust	45°	
Valve head thickness (margin)		
Intake	0.047 in.	1.2 mm
Exhaust	0.079 in.	2.0 mm
Valve spring angle limit	0.067 in.	1.7 mm
Free length of valve spring		
Standard	1.961 in.	49.8 mm
Rocker arm and rocker arm shaft		
Rocker arm inner diameter	0.7445–0.7452 in.	18.910–19.928 mm
Rocker arm shaft diameter	0.7435–0.7440 in.	18.885–18.899 mm
Clearance in rocker arm		
Standard	0.0005–0.0017 in.	0.012–0.043 mm

ENGINE AND ENGINE OVERHAUL 3-63

1987-88 2.6L ENGINE MECHANICAL SPECIFICATIONS

Component	U.S.	Metric
Camshaft		
Camshaft run-out	0.0008 in. max.	0.02 mm max.
Camshaft end play		
Standard	0.0008-0.0070 in.	0.02-0.18 mm
Wear limit	0.008 in.	0.20 mm
Journal diameter	1.3360-1.3366 in.	33.935-33.950 mm
Wear limit of journal	0.002 in.	0.05 mm
Camshaft bearing oil clearance		
Standard	0.002-0.004 in.	0.05-0.09 mm
Maximum	0.0059 in.	0.15 mm
Cam height		
Standard	1.669 in.	42.4 mm
Minimum	1.650 in.	41.9 mm
Connecting rod and connecting rod bearing		
Length (center to center)	6.536 in.	166 mm
Maximum allowable twist and bend		
Less than	0.006 in.	0.16 mm
Small end bore	0.8651-0.8655 in.	21.974-21.985 mm
Connecting rod side clearance		
Standard	0.004-0.010 in.	0.10-0.25 mm
Maximum	0.0157 in.	0.40 mm
Crankpin bearing oil clearance	0.0008-0.0024 in.	0.02-0.06 mm
Available undersize bearing	0.010, 0.020, 0.030 in.	0.25, 0.50, 0.75 mm
Crankshaft and main bearing		
Crankshaft run-out	0.0012 in. max.	0.03 mm max.
Crankpin diameter	2.0855-2.0866 in.	52.973-53.000 mm
Wear limit	0.0012 in.	0.03 mm
Grinding limit	0.030 in.	0.75 mm
Main journal diameter	2.3614-2.3622 in.	59.980-60.000 mm
Wear limit	0.0012 in.	0.03 mm
Grinding limit	0.030 in.	0.75 mm
Main journal bearing clearance	0.0008-0.0020 in.	0.02-0.05 mm
Available undersize bearing	0.010, 0.020, 0.030 in.	0.25, 0.50, 0.75 mm
Crankshaft end play		
Standard	0.0020-0.0071 in.	0.05-0.18 mm
Maximum	0.010 in.	0.25 mm
Available undersize thrust bearing	0.010, 0.020, 0.030 in.	0.25, 0.50, 0.75 mm
Cylinder block		
Limit of distortion of block	0.0039 in.	0.10 mm
Cylinder bore diameter	3.5874-3.5882 in.	91.12-91.14 mm
Boring size	0.010, 0.020, 0.030, 0.040 in.	0.25, 0.50, 0.75, 1.00mm
Piston and Ring		
Piston diameter (diameter measured at 90° to pin bore axis and 1.65 in. (42 mm) from bottom piston	3.586-3.587 in.	91.08-91.10 mm
Piston and cylinder clearance	0.0008-0.016 in.	0.02-0.04 mm
Available oversize piston	0.010, 0.020, 0.030, 0.040 in.	0.25, 0.50, 0.75, 1.00mm
Clearance between piston ring and ring groove		
Top	0.0020-0.0035 in.	0.05-0.09 mm
Second	0.0018-0.0024 in.	0.02-0.06 mm
Maximum	0.010 in.	0.25 mm
Piston ring end gap		
Top	0.012-0.018 in.	0.30-0.45 mm
Second	0.010-0.016 in.	0.25-0.40 mm
Oil	0.012-0.024 in.	0.30-0.60 mm
Maximum	0.039 in.	1.0 mm

89643G09

1987-88 2.6L ENGINE MECHANICAL SPECIFICATIONS

Component	U.S.	Metric
Piston and Ring		
Available oversize piston ring	0.010, 0.020, 0.030, 0.40 in.	0.25, 0.50, 0.75, 1.00mm
Piston pin		
Diameter	0.8662-0.8664 in.	22.001-22.007 mm
Interference in piston	0.0006-0.0013 in.	0.016-0.033 mm
Press-in load	1,650-3,850 lbs.	750-1,750 kg
Balance shaft		
Right		
Rear journal diameter	1.693 in.	43 mm
Oil clearance	0.0024-0.0039 in.	0.06-0.10 mm
Left		
Front journal diameter	0.906 in.	23 mm
Rear journal diameter	1.693 in.	43 mm
Oil clearance		
Front	0.0008-0.0024 in.	0.020-0.06 mm
Rear	0.0024-0.0039 in.	0.06-0.10 mm

89643G10

1989-92 2.6L ENGINE MECHANICAL SPECIFICATIONS

Component	U.S.	Metric
Bore x Stroke	3.62 x 3.86 in.	92.0 x 98.0 mm
Total piston displacement	158.97 cu. in.	2,606 cc
Compression pressure @ 270 rpm		
Standard	182 psi	1,255 KPa
Minimum	142 psi	981 KPa
Maximum difference between cylinders	28 psi	196 KPa
Valve timing		
Intake	Opens 10° BTDC	Closes 50° ABDC
Exhaust	Opens 55° BBDC	Closes 15° ATDC
Valve clearance		
Intake	0	0
Exhaust	0	0
Cylinder head		
Height	3.541-3.545 in.	89.95-90.05 mm
Distortion	0.006 in. max.	0.15 mm max.
Grinding	0.008 in. max.	0.20 mm max.
Valve and valve guide		
Valve head diameter		
Intake	1.307-1.315 in.	89.95-90.05 mm
Exhaust	1.413-1.421 in.	35.9-36.1 mm
Valve head margin thickness		
Intake	0.039 in.	1.0 mm
Exhaust	0.059 in.	1.5 mm
Valve face angle		
Intake	45°	
Exhaust	45°	
Valve length		
Intake		
Standard	4.4367 in.	112.69 mm
Minimum	4.4209 in.	112.29 mm
Exhaust		
Standard	4.4812 in.	113.82 mm
Minimum	4.4654 in.	113.42 mm

89643G11

3-64 ENGINE AND ENGINE OVERHAUL

1989-92 2.6L ENGINE MECHANICAL SPECIFICATIONS

Component	U.S.	Metric
Valve and valve guide		
Valve stem diameter		
Intake	0.2744-0.2750 in.	6.970-6.985 mm
Exhaust	0.2742-0.2748 in.	6.965-6.980 mm
Guide inner diameter		
Intake	0.2760-0.2768 in.	7.01-7.03 mm
Exhaust	0.2760-0.2768 in.	7.01-7.03 mm
Valve stem-to-guide clearance		
Intake	0.0010-0.0024 in.	0.025-0.060 mm
Exhaust	0.0012-0.0026 in.	0.030-0.065 mm
Maximum	0.008 in.	0.20 mm
Guide projection (Height "A")	0.925-0.953 in.	23.5-24.2 mm
Valve seat		
Seat angle		
Intake	45°	
Exhaust	45°	
Seat contact width		
Intake	0.047-0.063 in.	1.2-1.6 mm
Exhaust	0.047-0.063 in.	1.2-1.6 mm
Seat sinking (Measure valve protruding length)		
Standard		
Intake	1.929 in.	49.0 mm
Exhaust	1.949 in.	49.5 mm
Maximum		
Intake	1.929 in.	49.0 mm
Exhaust	1.949 in.	49.5 mm
Valve spring		
Free length		
Standard		
Intake	1.970 in.	50.05 mm
Exhaust	1.963 in.	49.85 mm
Minimum		
Intake	1.970 in.	50.05 mm
Exhaust	1.963 in.	49.85 mm
Out-of-square	0.069 in. max.	1.75 mm max.
Setting load/height		
Intake	43.9-49.7 lbs./1.693 in.	195-222 N/43mm
Exhaust	43.8-49.7 lbs./1.693 in.	195-222 N/43mm
Camshaft		
Cam lobe height		
Intake		
Standard	1.6423 in.	41.714 mm
Minimum	1.6344 in.	41.514 mm
Exhaust		
Standard	1.6531 in.	41.988 mm
Minimum	1.6452 in.	41.788 mm
Journal diameter		
Front and rear (No.1,5)	1.1788-1.1797 in.	29.940-29.965 mm
Center (No.2,3,4)	1.1776-1.1786 in.	29.910-29.935 mm
Out-of-round	0.002 in. max.	0.05 mm max.
Camshaft bearing oil clearance		
Front and rear (No.1,5)	0.0014-0.0033 in.	0.035-0.085 mm
Center (No.2,3,4)	0.0026-0.0045 in.	0.065-0.115 mm
Maximum	0.006 in.	0.15 mm
Camshaft runout		
Maximum	0.0012 in.	0.03 mm
Camshaft end play		
Standard	0.0008-0.0059 in.	0.02-0.15 mm
Maximum	0.008 in.	0.20 mm

1989-92 2.6L ENGINE MECHANICAL SPECIFICATIONS

Component	U.S.	Metric
Rocker arm and rocker arm shaft		
Rocker arm inner diameter	0.8268-0.8281 in.	21.000-21.033 mm
Rocker arm shaft diameter	0.8252-0.8260 in.	20.959-20.980 mm
Rocker arm to shaft clearance	0.0008-0.0029 in.	0.020-0.074 mm
Standard		
Maximum	0.004 in.	0.10 mm
Cylinder block		
Height	12.46 in.	316.5 mm
Distortion	0.006 in. max.	0.15 mm max.
Grinding	0.008 in. max.	0.20 mm max.
Cylinder bore diameter		
Standard	3.6220-3.6230 in.	92.000-92.022 mm
0.010 in. oversize (0.25 mm)	3.6320-3.6330 in.	92.250-92.272 mm
0.020 in. oversize (0.50 mm)	3.6420-3.6430 in.	92.500-92.522 mm
Cylinder bore taper and out-of-round	0.0007 in. max.	0.019 mm max.
Piston		
Piston diameter measured at 90° to pin bore axis and 0.709 in. (18.0 mm) below oil ring groove		
Standard	3.6194-3.6202 in.	91.935-91.955 mm
0.010 in. oversize (0.25 mm)	3.6293-3.6301 in.	92.185-92.205 mm
0.020 in. oversize (0.50 mm)	3.6391-3.6400 in.	92.435-92.455 mm
Piston-to-cylinder clearance	0.0023-0.0029 in.	0.058-0.074 mm
Maximum	0.006 in.	0.15 mm
Piston ring		
Thickness		
Top	0.058-0.059 in.	1.47-1.49 mm
Second	0.058-0.059 in.	1.47-1.49 mm
End gap measured in cylinder		
Top	0.008-0.014 in.	0.20-0.35 mm
Second	0.010-0.016 in.	0.25-0.40 mm
Oil (rail)	0.008-0.028 in.	0.20-0.70 mm
Maximum	0.039 in.	1.0 mm
Ring groove width in piston		
Top	0.0598-0.0606 in.	1.52-1.54 mm
Second	0.0598-0.0606 in.	1.52-1.54 mm
Oil	0.1583-0.1591 in.	4.02-4.04 mm
Piston ring-to-ring land clearance		
Top	0.0012-0.0028 in.	0.03-0.07 mm
Second	0.0012-0.0028 in.	0.03-0.07 mm
Maximum	0.006 in.	0.15 mm
Piston pin		
Diameter	0.9045-0.9047 in.	22.974-22.980 mm
Interference in connecting rod	0.0003-0.0015 in.	0.013-0.037 mm
Piston to piston pin clearance	0.0003-0.0010 in.	0.008-0.026 mm
Installation force	1,100-3,300 lbs.	500-1,360 kg
Connecting rod and connecting rod bearing		
Length (center to center)	6.553-6.557 in.	166.45-166.55 mm
Bend	0.0098 in. max.	0.249 mm max.
Small end bore	0.9033-0.9040 in.	22.943-22.961 mm
Big end bore	2.1261-2.1266 in.	54.002-54.017 mm
Big end width	1.0094-1.0114 in.	25.638-25.690 mm
Connecting rod side clearance		
Standard	0.0043-0.0103 in.	0.110-0.262 mm
Maximum	0.012 in.	0.30 mm

ENGINE AND ENGINE OVERHAUL 3-65

1989-92 2.6L ENGINE MECHANICAL SPECIFICATIONS

Component	U.S.	Metric
Crankshaft		
Crankshaft runout	0.0012 in. max.	0.03 mm max.
Main journal diameter		
Standard size	2.3597-2.3604 in.	59.937-59.955 mm
0.010 in. undersize (0.25mm)	2.3499-2.3506 in.	59.687-59.705 mm
0.020 in. undersize (0.50mm)	2.3400-2.3407 in.	59.437-59.455 mm
0.030 in. undersize (0.75mm)	2.3302-2.3309 in.	59.187-59.205 mm
Main journal taper and out-of-round	0.0020 in. max.	0.05 mm. max.
Crankpin journal diameter		
Standard	2.0055-2.0061 in.	50.940-50.955 mm
0.010 in. undersize (0.25mm)	1.9957-1.9963 in.	50.690-50.705 mm
0.020 in. undersize (0.50mm)	1.9858-1.9864 in.	50.440-50.455 mm
0.030 in. undersize (0.75mm)	1.9760-1.9766 in.	50.190-50.205 mm
Crankpin taper and out-of-round	0.0020 in. max.	0.05 mm max.
Main bearing		
Main journal bearing oil clearance		
Standard	0.0010-0.0017 in.	0.025-0.044 mm
Maximum	0.0031 in.	0.08 mm
Available undersize bearing	0.010,0.020,0.030 in.	0.25,0.50,0.75 mm
Crankpin bearing		
Crankpin bearing oil clearance		
Standard	0.0011-0.0026 in.	0.027-0.067 mm
Maximum	0.0039 in.	0.10 mm
Available undersize bearing	0.010,0.020,0.030 in.	0.25,0.50,0.75 mm
Thrust bearing (center main bearing)		
Crankshaft end play		
Standard	0.0031-0.0071 in.	0.08-0.18 mm
Maximum	0.0118 in.	0.30 mm
Bearing width		
Standard	1.021-1.023 in.	25.94-25.99 mm
0.010 in. oversize (0.25 mm)	1.025-1.027 in.	26.04-26.09 mm
0.020 in. oversize (0.50 mm)	1.028-1.030 in.	26.12-26.17 mm
0.030 in. oversize (0.75 mm)	1.031-1.033 in.	26.20-26.25 mm
Balance shaft		
Front journal diameter	1.6514-1.6520 in.	41.945-41.960 mm
Center journal diameter	1.5727-1.5732 in.	39.945-39.960 mm
Rear journal diameter	0.8247-0.8251 in.	20.945-20.960 mm
Oil clearance		
Front	0.0020-0.0045 in.	0.050-0.115 mm
Center	0.0031-0.0057 in.	0.080-0.145 mm
Rear	0.0031-0.0057 in.	0.080-0.145 mm

3.0L ENGINE MECHANICAL SPECIFICATIONS

Component	U.S.	Metric
Bore x stroke	3.54 x 3.05 in.	90.0 x 77.4 mm
Total piston displacement	180.2 cu in.	2,954 cc
Compression pressure @ 300 rpm		
Standard	164 psi	1,128 KPa
Minimum	121 psi	834 KPa
Maximum difference between cylinders	28 psi	196 KPa
Valve timing		
Intake	Opens 9° BTDC	Closes 53° ABDC
Exhaust	Opens 51° BBDC	Closes 11° ATDC
Valve clearance		
Intake	0	0
Exhaust	0	0
Cylinder head		
Height	4.931-4.935 in.	125.25-125.35 mm
Distortion	0.004 in. max.	0.10 mm max.
Grinding	0.006 in. max.	0.15 mm max.
Valve and valve guide		
Valve head diameter		
Intake	1.295-1.303 in.	32.9-33.1 mm
Exhaust	1.492-1.500 in.	37.9-38.1 mm
Valve head margin thickness		
Intake	0.030-0.049 in.	0.75-1.25 mm
Exhaust	0.047-0.071 in.	1.2-1.8 mm
Valve face angle		
Intake	45°	
Exhaust	45°	
Valve length		
Intake	4.7760 in.	121.31 mm
Exhaust	4.7602 in.	120.91 mm
Standard	4.8279 in.	122.63 mm
Minimum	4.8122 in.	122.23 mm
Valve stem diameter		
Intake	0.2744-0.2750 in.	6.970-6.985 mm
Exhaust	0.3159-0.3165 in.	8.025-8.040 mm
Guide inner diameter		
Intake	0.2760-0.2768 in.	7.01-7.03 mm
Exhaust	0.3177-0.3185 in.	8.07-8.09 mm
Valve stem-to-guide clearance		
Intake	0.0010-0.0024 in.	0.025-0.060 mm
Exhaust	0.0012-0.0026 in.	0.030-0.065 mm
Maximum	0.008 in.	0.20 mm
Guide projection (Height "A")		
Intake	0.520-0.543 in.	13.2-13.8 mm
Exhaust	0.772-0.795 in.	19.6-20.2 mm
Valve seat		
Seat angle		
Intake	45°	
Exhaust	45°	
Seat contact width		
Intake	0.047-0.063 in.	1.2-1.6 mm
Exhaust	0.047-0.063 in.	1.2-1.6 mm
Seat sinking (measure valve protruding length)		
Intake		
Standard	1.988 in.	50.5 mm
Maximum	2.047 in.	52.0 mm
Exhaust		
Standard	1.949 in.	49.5 mm
Maximum	2.008 in.	51.0 mm

3.0L ENGINE MECHANICAL SPECIFICATIONS

Component	U.S.	Metric
Valve spring		
Free length		
Intake		
Outer	2.005 in.	50.9 mm
Inner	1.73 in.	44.0 mm
Exhaust		
Outer	1.840 in.	46.73 mm
Inner	1.56 in.	39.5 mm
Standard		
Outer	2.296 in.	58.33 mm
Inner	1.77 in.	45.0 mm
Minimum		
Outer	2.092 in.	53.14 mm
Inner	1.59 in.	40.5 mm
Out-of-square		
Intake		
Outer	0.071 in. max.	1.8 mm max.
Inner	0.063 in. max.	1.6 mm max.
Exhaust		
Outer	0.080 in. max.	2.04 mm max.
Inner	0.073 in. max.	1.86 mm max.
Setting load/height		
Intake		
Outer	31-33 lbs./1.733 in.	137-148 N/44.0mm
Inner	21-22 lbs./1.555 in.	92-99 N/39.5mm
Exhaust		
Outer	51.9-58.7 lbs./1.772 in.	232-262 N/45.0mm
Inner	33.0-37.4 lbs./1.594 in.	148-167 N/40.5mm
Camshaft		
Cam lobe height		
Intake		
Standard	1.6163 in.	41.054 mm
Minimum	1.6084 in.	40.854 mm
Exhaust		
Standard	1.6257 in.	41.293 mm
Minimum	1.6178 in.	41.093 mm
Journal diameter		
Front and rear (No.1,4)	1.9261-1.9267 in.	48.923-48.938 mm
Center (No.2,3)	1.9258-1.9266 in.	48.915-48.935 mm
Out-of-round max.	0.0012 in.	0.03 mm
Camshaft bearing bore diameter		
Front and rear (No.1,4)	1.9297-1.9305 in.	49.015-49.035 mm
Center (No.2,3)	1.9297-1.9305 in.	49.015-49.035 mm
Camshaft bearing oil clearance		
Front and rear (No.1,4)	0.0031-0.0044 in.	0.077-0.112 mm
Center (No.2,3)	0.0031-0.0047 in.	0.080-0.120 mm
Maximum	0.006 in.	0.15 mm
Camshaft runout	0.0012 in. max.	0.03 mm max.
Camshaft end play		
Standard	0.0020-0.0071 in.	0.05-0.18 mm
Maximum	0.008 in.	0.20 mm
Rocker arm and rocker arm shaft		
Rocker arm inner diameter	0.7480-0.7493 in.	19.000-19.033 mm
Rocker arm shaft diameter	0.7464-0.7472 in.	18.959-19.980 mm
Rocker arm-to-shaft clearance		
Standard	0.0008-0.0029 in.	0.020-0.074 mm
Maximum	0.004 in.	0.10 mm

3.0L ENGINE MECHANICAL SPECIFICATIONS

Component	U.S.	Metric
Cylinder block		
Height	8.66 in.	220.0 mm
Distortion	0.006 in. max.	0.15 mm max.
Grinding	0.008 in. max.	0.20 mm max.
Cylinder bore diameter		
Standard size	3.5433-3.5442 in.	90.000-90.022 mm
0.010 in. oversize (0.25mm)	3.5531-3.5540 in.	90.250-90.272 mm
0.020 in. oversize (0.50mm)	3.5630-3.5639 in.	90.500-90.522 mm
Cylinder bore taper and out-of-round	0.0007 in. max.	0.019 mm max.
Piston		
Piston diameter (measured at 90° to pin bore axis and 0.866 in. (22.0 mm) below oil ring groove)		
Standard size	3.5417-3.5429 in.	89.958-89.990 mm
0.010 in. oversize (0.25mm)	3.5515-3.5527 in.	90.208-90.240 mm
0.020 in. oversize (0.50mm)	3.5614-3.5626 in.	90.458-90.490 mm
Piston-to-cylinder clearance	0.0009-0.0022 in.	0.023-0.051 mm
Maximum	0.006 in.	0.15 mm
Piston ring		
Thickness	0.058-0.059 in.	1.47-1.49 mm
End gap measured in cylinder		
Top	0.008-0.014 in.	0.20-0.35 mm
Second	0.006-0.012 in.	0.15-0.35 mm
Oil (rail)	0.008-0.028 in.	0.20-0.70 mm
Maximum	0.039 in.	1.0 mm
Ring groove width piston		
Top	0.0598-0.0606 in.	1.52-1.54 mm
Second	0.0598-0.0606 in.	1.52-1.54 mm
Oil	0.1583-0.1591 in.	4.02-4.04 mm
Piston ring-to-ring land clearance		
Top	0.0012-0.0028 in.	0.03-0.07 mm
Second	0.0012-0.0028 in.	0.03-0.07 mm
Maximum	0.006 in.	0.15 mm
Piston pin		
Diameter	0.9045-0.9047 in.	22.974-22.980 mm
Interference in connecting rod	0.0005-0.0015 in.	0.013-0.037 mm
Piston-to-piston pin clearance	0.0003-0.0010 in.	0.008-0.026 mm
Pressure force	1,100-3,300 lb.	4,905-14,715 N
Connection rod		
Length (center to center)	5.758-5.762 in.	146.25-146.35 mm
Bend	0.0092 in. max.	0.234 mm max.
Small end bore	0.9033-0.9040 in.	22.943-22.961 mm
Big end bore	2.2047-2.2053 in.	56.000-56.015 mm
Big end width	0.8374-0.8394 in.	21.270-21.322 mm
Connecting rod side clearance		
Standard	0.0070-0.0130 in.	0.178-0.330 mm
Maximum	0.016 in.	0.40 mm
Crankshaft		
Crankshaft runout	0.0012 in. max.	0.03 mm max.
Main journal diameter		
Standard size	2.4385-2.5492 in.	61.937-61.955 mm
0.010 in. undersize (0.25mm)	2.4286-2.4293 in.	61.687-61.705 mm
Crankpin journal diameter	0.0020 in. max.	0.05 mm max.
Standard size	2.0842-2.0848 in.	52.940-52.955 mm
0.010 in. undersize (0.25mm)	2.0744-2.0750 in.	52.690-52.705 mm
Crankpin taper and out-of-round	0.0020 in. max.	0.05 mm max.

ENGINE AND ENGINE OVERHAUL

3.0L ENGINE MECHANICAL SPECIFICATIONS

Component	U.S.	Metric
Main bearing		
Main journal bearing oil clearance		
Standard	0.0010-0.0015 in.	0.025-0.037 mm
Maximum	0.0031 in.	0.08 mm
Available undersize bearing	0.010 in.	0.25 mm
Crankpin bearing		
Crankpin bearing oil clearance		
Standard	0.0009-0.0025 in.	0.023-0.064 mm]
Maximum	0.004 in.	0.10 mm
Thrust bearing		
Crankshaft end play		
Standard	0.0031-0.0111 in.	0.080-0.282 mm
Maximum	0.0118 in.	0.30 mm
Bearing width		
Standard	0.0787-0.0807 in.	2.000-2.050 mm
0.010 in. oversize (0.25mm)	0.0837-0.0856 in.	2.125-2.175 mm
0.020 in. oversize (0.25mm)	0.0886-0.0906 in.	2.250-2.300 mm
Timing belt		
Belt deflection	0.20-0.28 in.	5-7 mm

4.0L ENGINE MECHANICAL SPECIFICATIONS

Component	U.S.	Metric
Displacement	242.1 cu. in.	3,958.4 cc
Bore x stroke	3.94 x 3.31 in.	100 x 84 mm
Oil pressure (hot @ 2000 RPM)	40-60 psi	276-414 KPa
Comustion chamber volume	3.49-3.61 cu. in.	57.2-59.2 cc
Valves		
Valve guide bore diameter	0.3174-0.3184 in.	8.062-8.087mm
Valve seats		
Width		
Intake	0.060-0.079 in.	1.50-2.00mm
Exhaust	0.060-0.079 in.	1.50-2.00mm
Angle	45 degrees	
Runout limit (TIR Max.)	0.0015 in.	0.038mm
Valve stem-to-guide clearance		
Intake	0.0008-0.0025 in.	0.020-0.064mm
Exhaust	0.0018-0.0035 in.	0.046-0.089mm
Service clearance	0.0055 in.	0.14mm
Valve head diameter		
Standard		
Intake	0.3159-0.3167 in.	8.024-8.044mm
Exhaust	0.3149-0.3156 in.	8.000-8.016mm
0.008 in. (0.20mm) oversize		
Intake	0.3239-0.3245 in.	8.227-8.242mm
Exhaust	0.3228-0.3235 in.	8.200-8.217mm
0.016 in. (0.40mm) oversize		
Intake	0.3318-0.3324 in.	8.428-8.443mm
Exhaust	0.3307-0.3314 in.	8.400-8.418mm
Valve springs		
Compression pressure	60-68 lbs./1.585 in.	138-149 N/1.222mm
Free length (approx.)	1.91 in.	48.5mm
Assembled height	1 37/64-1 39/64 in.	40.0-41.0mm

4.0L ENGINE MECHANICAL SPECIFICATIONS

Component	U.S.	Metric
Rocker arm		
Shaft diameter	0.7799-0.7811 in.	20.81-20.84mm
Bore diameter	0.7830-0.7842 in.	20.89-20.92mm
Valve roller tappet		
Diameter (Std.)	0.8742-0.8755 in.	22.20-22.24mm
Clearance to bore	0.0005-0.0022 in.	0.013-0.056mm
Service limit	0.005 in.	0.13mm
Camshaft		
Lobe lift		
Allowable lobe lift loss	0.005 in.	0.13mm
Theoretical valve lift @ zero lash		
Intake	0.4024 in.	10.22mm
Exhaust	0.4024 in.	10.22mm
End play	0.0008-0.004 in.	0.02-0.10mm
Service limit	0.009 in.	0.23mm
Thrust plate thickness	0.158-0.159 in.	4.00-4.04mm
Journal-to-bearing clearance	0.001-0.0026 in.	0.025-0.066mm
Service limit	0.006 in.	0.15mm
Bearing outside diameter	0.158-0.159 in.	4.00-4.04mm
#1	1.951-1.952 in.	49.57-49.59 mm
#2	1.937-1.938 in.	49.21-49.23 mm
#3	1.922-1.923 in.	48.83-48.85 mm
#4	1.907-1.908 in.	48.44-48.46 mm
Runout	0.005 in.	0.127 mm
Out-of-round	0.0003 in.	0.0076 mm
Bearing inside diameter		
#1	1.954-1.955 in.	49.635-49.655 mm
#2	1.939-1.940 in.	49.255-49.275 mm
#3	1.919-1.920 in.	48.750-48.768 mm
#4	1.924-1.925 in.	48.875-48.895 mm
Front bearing location	0.040-0.060 in.	1.00-1.50mm
Cylinder block		
Head gasket surface flatness		
Finish	60-150 RMS	
Crankshaft to rear face of block runout		
T.I.R. Max.	0.006 in.	0.15mm
Taper bore diameter	0.005 in.	0.13mm
Main bearing bore diameter	0.8750-0.8760 in.	22.22-22.25mm
Cylinder bore diameter	2.3866-2.3874 in.	60.62-60.64mm
Diameter	3.9527-3.9543 in.	100.40-100.45mm
Surface finish	18.38 RMS	
Out-of-round	0.0015 in.	0.038mm
Out-of-round service limit	0.005 in.	0.13mm
Taper service limit	0.010 in.	0.25mm
Piston		
Height		
Diameter	0.0015-0.0205 in.	0.038-0.52mm
Standard	3.9524-3.9531 in.	100.40-100.41mm
Piston-to-bore clearance	0.0008-0.0019 in.	0.020-0.050mm
Pin bore diameter	0.9450-0.9452 in.	24.00-24.24mm
Ring groove width		
Compression (Top)	0.0803-0.0811 in.	2.40-2.06mm
Compression (Bottom)	0.1197-0.1205 in.	3.04-3.06mm
Oil	0.1579-0.1587 in.	4.00-4.03mm
Piston pin		
Length	2.835-2.866 in.	72.00-72.80mm
Diameter		
Red	0.9446-0.9448 in.	23.993-23.998mm
Blue	0.9448-0.9449 in.	23.998-24.000mm
Pin-to-piston clearance	0.0003-0.0006 in.	0.0076-0.0152mm
Pin-to-rod clearance	interference fit	

3-68 ENGINE AND ENGINE OVERHAUL

4.0L ENGINE MECHANICAL SPECIFICATIONS

Component	U.S.	Metric
Piston rings		
Ring width		
Compression (top)	0.0778-0.0783 in.	1.976-1.988mm
Compression (bottom)	0.1172-0.1177 in.	2.977-2.989mm
Side clearance		
Compression (top)	0.0020-0.0033 in.	0.05-0.84mm
Compression (bottom)	0.0020-0.0033 in.	0.05-0.84mm
Oil ring	Snug fit	
Service limit	0.006 in.	0.15mm
Ring gap		
Compression (top)	0.015-0.023 in.	0.38-0.58mm
Compression (bottom)	0.015-0.023 in.	0.38-0.58mm
Oil ring (steel rail)	0.015-0.055 in.	0.38-1.40mm
Crankshaft and flywheel		
Main bearing journal diameter	2.2433-2.2441 in.	56.98-57.00mm
Out-of-round	0.0006 in.	0.015mm
Taper limit	0.0006 per in.	0.015mm
Journal runout	0.002 in. max.	0.05mm
Surface finish	12 RMS	
Runout service limit	0.005 in.	0.13mm
Thrust bearing journal		
Length	1.039-1.041 in.	26.39-26.44mm
Connecting rod journal		
Diameter	2.1252-2.1260 in.	53.98-54.00mm
Out-of-round	0.0006 in.	0.015mm
Taper limit	0.0006 per in.	0.015mm
Surface finish	12 RMS max.	
Main bearing thrust face		
Surface finish	20 RMS max.	
Runout (T.I.R)	0.001 max. 0.025mm	
Flywheel clutch face runout	0.005 in.	0.13mm
Flywheel ring gear lateral runout (T.I.R.)		
Standard transmission	0.025 in.	0.635mm
Automatic transmission	0.060 in.	1.524mm
Crankshaft free end play	0.0016-0.0126 in.	0.04-0.32mm
Service limit	0.012 in.	0.30mm
Connecting rod bearings		
Clearance to crankshaft	0.0003-0.0024 in.	0.008-0.060mm
Allowable	0.0005-0.0022 in.	0.013-0.056mm
Bearing wall thickness		
Red	0.0548-0.0552 in.	1.392-1.400mm
Blue	0.0552-0.0556 in.	1.400-1.412mm
Main bearing		
Clearance to crankshaft	0.0008-0.0015 in.	0.020-0.038mm
Allowable	0.0005-0.0019 in.	0.013-0.048mm
Bearing wall thickness		
Red	0.0707-0.0711 in.	1.795-1.800mm
Blue	0.0711-0.0714 in.	1.800-1.814mm
Connecting rod		
Connecting rod		
Piston pin bore diameter	0.9432-0.9439 in.	23.957-23.975mm
Crankshaft bearing bore diameter	2.2370-2.2378 in.	56.82-56.84mm
Out-of-round	0.0004 in.	0.010mm
Taper	0.0004 in.	0.010mm
Length (center-to-center)	5.1386-5.1413 in.	130.5-130.6mm
Alignment (Bore-to-bore max. diff.)		
Twist	0.006 in.	0.15mm
Bend	0.002 in.	0.05mm
Side clearance (assembled to crank)		
Standard	0.0002-0.0025 in.	0.005-0.064mm
Service limit	0.014 in.	0.35mm

89643G21

4.0L ENGINE MECHANICAL SPECIFICATIONS

Component	U.S.	Metric
Lubricating system		
Oil pump		
Relief valve spring tension	13.60-14.7 lbs./1.39 in.	6 kg/35.3mm
Drive shaft-to-housing bearing clearance	0.0015-0.0030 in.	0.038-0.076mm
Relief valve-to-bore clearance	0.0015-0.0030 in.	0.038-0.076mm
Rotor assembly end clearance	0.004 in. max.	0.10mm max.
Outer race-to-housing clearance	0.001-0.013 in.	0.025-0.33mm

89643G22

ENGINE AND ENGINE OVERHAUL

TORQUE SPECIFICATIONS

Component	U.S.	Metric
Air conditioning compressor bolts		
B Series	29-40 ft. lbs.	39-54 Nm
MPV	13-20 ft. lbs.	18-27 Nm
Air conditioning pipe bracket nuts		
MPV	61-87 inch lbs.	7-10 Nm
Automatic tensioner bolts		
2.2L	27-38 ft. lbs.	37-52 Nm
3.0L	14-19 ft. lbs.	19-26 Nm
Balancer shaft sprocket bolt		
2.6L (1987-88)	43-50 ft. lbs.	58-68 Nm
Bypass air control valve bolts/nuts		
fuel injected 3.0L	14-19 ft. lbs.	19-26 Nm
Cam cover		
2.2L	26-35 inch lbs.	3-4 Nm
3.0L	30-40 inch lbs.	3.3-4.5 Nm
Camshaft seal plate bolts		
3.0L	69-95 inch lbs.	8-11 Nm
Camshaft pulley		
2.2L & 2.6L (1989-92)		
bolt	95 inch lbs.	11 Nm
nut	87 inch lbs.	10 Nm
Camshaft sprocket bolt		
2.2L	35-48 ft. lbs.	48-65 Nm
2.6L (1987-88)	40 ft. lbs.	54 Nm
3.0L	45 ft. lbs.	61 Nm
4.0L	52-59 ft. lbs.	71-80 Nm
	44-50 ft. lbs.	60-68 Nm
Camshaft thrust plate		
2.6L (1989-92)	95 inch lbs.	11 Nm
3.0L	6-8 ft. lbs.	8-11 Nm
4.0L	96 inch lbs.	11 Nm
Catalytic converter connections		
pickups	23 ft. lbs.	31 Nm
Coil Pack		
Navajo	40-62 inch lbs.	4.5-20 Nm
Converter housing upper bolts		
Navajo	33-45 ft. lbs.	45-61 Nm
Coolant bypass pipe bolts		
2.6L	27-38 ft. lbs.	27-52 Nm
Crankshaft position sensor		
Navajo	75-106 inch lbs.	8.5-12 Nm
Crankshaft pulley/damper assembly bolts		
4.0L	30-37 ft. lbs.	30-50 Nm
Main bearing cap bolts		
2.2L	61-65 ft. lbs.	82-88 Nm
2.6L	61 ft. lbs.	82 Nm
3.0L	59 ft. lbs.	80 Nm
4.0L		
Step 1	25 ft. lbs.	33 Nm
Step 2	72 ft. lbs.	97 Nm
Connecting rod cap nuts		
2.2L	37-41 ft. lbs.	50-55 Nm
2.6L	54-61 ft. lbs.	73-82 Nm
3.0L	26 ft. lbs.	35 Nm
4.0L	18-24 ft. lbs.	24-32 Nm

TORQUE SPECIFICATIONS

Component	U.S.	Metric
Crankshaft pulley bolt		
2.2L	9-13 ft. lbs.	12-18 Nm
2.6L (1987-88)	80-94 ft. lbs.	109-128 Nm
2.6L (1989-92)	145 ft. lbs.	197 Nm
3.0L	116-123 ft. lbs.	158-167 Nm
Crankshaft sprocket bolt		
2.2L	116-123 ft. lbs.	158-167 Nm
Cylinder head bolts		
2.2L	59-64 ft. lbs.	80-87 Nm
2.6L (1987-88)	65-72 ft. lbs.	88-98 Nm
bolts 1-10	11-16 ft. lbs.	15-22 Nm
2.6L (1989-92) bolts 11, 12		
Large bolts	64 ft. lbs.	87 Nm
Remaining two head bolts	17 ft. lbs.	23 Nm
3.0L	14 ft. lbs.	19 Nm
4.0L	59 ft. lbs.	80 Nm
Distributor holddown bolt	14-19 ft. lbs.	19-25 Nm
Distributor drive rear/camshaft sprocket lock bolt		
2.6L (1987-88)	36-43 ft. lbs.	49-58 Nm
Distributor drive gear lock bolt		
2.6L (1989-92)	45 ft. lbs.	61 Nm
Engine mount nuts		
B Series	30-36 ft. lbs.	41-49 Nm
MPV	25-36 ft. lbs.	34-49 Nm
Exhaust pipe nuts		
2.2L	16-21 ft. lbs.	22-29 Nm
Exhaust manifold bolts/nuts		
2.6L (1987-88)	24 ft. lbs.	33 Nm
2.6L (1989-92)	18 ft. lbs.	24 Nm
3.0L	19 ft. lbs.	26 Nm
4.0L	19 ft. lbs.	26 Nm
Exhaust pipe to exhaust manifold nuts		
B Series	30-36 ft. lbs.	41-49 Nm
MPV	25-36 ft. lbs.	34-49 Nm
Fan assembly nuts/bolts		
except Navajo	70 inch lbs.	8 Nm
Navajo	55-70 inch lbs.	6-8 Nm
Fan and clutch assembly nut		
4.0L	50-100 ft. lbs.	68-136 Nm
Front housing bolts		
2.2L	14-19 ft. lbs.	19-25 Nm
Fuel supply manifold bolts		
fuel injected 2.2L, 2.6L & 3.0L	14-19 ft. lbs.	19-25 Nm
Heat cowl screws		
2.6L (1987-88)	80 inch lbs.	9 Nm
Idler sprocket assembly lock bolt		
2.6L (1989-92)	27-38 ft. lbs.	37-52 Nm
Ignition Module		
Navajo	22-31 inch lbs.	2.5-3.5Nm
Intake manifold mounting nuts		
carbureted 2.2L	14-19 ft. lbs.	19-25 Nm
carbureted 2.6L	11-14 ft. lbs.	11-19 Nm
fuel injected 3.0L	14-19 ft. lbs.	19-25 Nm
fuel injected 4.0L	see text	
Jet valves	13.5-15.5 ft. lbs.	18-21 Nm

3-70 ENGINE AND ENGINE OVERHAUL

TORQUE SPECIFICATIONS

Component	U.S.	Metric
Oil pan bolts		
2.2L	61-104 inch lbs.	7-12 Nm
1987-88 2.6L	52-61 inch lbs.	6-7 Nm
1989-92 2.6L	69-95 inch lbs.	8-11 Nm
3.0L	61-87 inch lbs.	7-10 Nm
4.0L	5-7 ft. lbs.	7-10 Nm
Oil pick-up brace bolt		
2.2L, 2.6L & 3.0L	69-95 inch lbs.	8-11 Nm
4.0L	7-10 ft. lbs.	10-14 Nm
Oil pump		
A bolts	14-19 ft. lbs.	19-25 Nm
B bolts	27-38 ft. lbs.	37-49 Nm
2.6L (1987-88)	6-7 ft. lbs.	8-10 Nm
3.0L	14-19 ft. lbs.	19-25 Nm
4.0L	13-15 ft. lbs.	18-20 Nm
Oil pump sprocket bolt		
2.6L (1987-88)	22-29 ft. lbs.	30-39 Nm
Power steering pump mounting bolts	23-34 ft. lbs.	31-46 Nm
Power steering pump pulley nut		
B Series	36-43 ft. lbs.	49-58 Nm
MPV	29-43 ft. lbs.	39-58 Nm
Pulsation damper & injector harness bracket bolts		
fuel injected 2.2L & 2.6L	69-95 inch lbs.	8-11 Nm
Radiator hose clamp	35 inch lbs.	4 Nm
Rear seal housing bolt		
2.2L	6-8 ft. lbs.	8-11 Nm
2.6L (1989-92)	95 inch lbs.	11 Nm
Rocker arm/shaft assembly bolts		
2.2L	13-20 ft. lbs.	18-27 Nm
2.6L (1987-88)	14-15 ft. lbs.	19-20 Nm
2.6L (1989-92)	14-19 ft. lbs.	19-25 Nm
3.0L	14-19 ft. lbs.	19-25 Nm
4.0L	46-52 ft. lbs.	63-71 Nm
Rocker cover bolts		
2.2L	26-35 inch lbs.	3-4 Nm
2.6L (1987-88)	55 inch lbs.	6 Nm
2.6L (1989-92)	52-78 inch lbs.	6-9 Nm
3.0L	30-39 inch lbs.	3-4 Nm
4.0L	53-70 inch lbs.	6-8 Nm
Solenoid mounting bolts	45-85 inch lbs.	5.1-9.6 Nm
Starter Mounting Bolts		
1987-88	24-33 ft. lbs.	33-45 Nm
1989-92 except 4.0L	27-38 ft. lbs.	37-52 Nm
4.0L	15-20 ft. lbs.	21-27 Nm
Starter through bolts	45-84 inch lbs.	5.0-9.5 Nm
Tensioner spring bolt		
2.2L	28-38 ft. lbs.	38-52 Nm
Tensioner body		
3.0L	14-19 ft. lbs.	19-25 Nm
Thermostat housing bolts		
2.2L	20 ft. lbs.	27 Nm
2.6L (1987-88)	15 ft. lbs.	18 Nm
2.6L (1989-92)	19 ft. lbs.	25 Nm
3.0L	19 ft. lbs.	25 Nm
4.0L	7-10 ft. lbs.	10-14 Nm
Throttle body nuts	14-19 ft. lbs.	19-25 Nm

TORQUE SPECIFICATIONS

Component	U.S.	Metric
Timing belt cover bolts		
2.2L	61-87 inch lbs.	7-10 Nm
3.0L		
6mm bolts	69-95 inch lbs.	8-11 Nm
10mm bolts	27-38 ft. lbs.	37-52 Nm
Timing belt tensioner lock bolt		
2.2L	28-38 ft. lbs.	38-52 Nm
Timing chain cover bolts		
1988 2.6L	104-122 inch lbs.	12-14 Nm
1989-92	130-145 inch lbs.	15-16 Nm
Timing chain adjuster bolts		
2.6L (1989-92)	95 inch lbs.	11 Nm
Timing chain cover bolts		
2.6L (1989-92)	19 ft. lbs.	25 Nm
Timing chain guide bolts		
2.6L (1987-88) A & C	43-69 inch lbs.	5-8 Nm
upper bolt	69-78 inch lbs.	8-9 Nm
lower bolt	11-15 ft. lbs.	15-18 Nm
2.6L (1989-92)	78 inch lbs.	9 Nm
4.0L	7-9 ft. lbs.	10-13 Nm
Timing chain lever bolts		
2.6L (1989-92)	69-95 inch lbs.	8-11 Nm
Upper idler pulley bolt		
3.0L	27-38 ft. lbs.	38-52 Nm
Upper intake manifold nuts/bolts		
fuel injected 2.2L & 2.6L	14-19 ft. lbs.	19-25 Nm
4.0L	15-18 ft. lbs.	18-24 Nm
Valve cover bolt		
2.6L (1989-92)	78 inch lbs.	9 Nm
Water pump bolts		
2.6L (1989-92)	19 ft. lbs.	25 Nm
3.0L	19 ft. lbs.	25 Nm
4.0L	72-108 inch lbs.	8-12 Nm

4

DRIVEABILITY AND EMISSION CONTROLS

EMISSION CONTROLS 4-2
ELECTRONIC ENGINE CONTROLS 4-9
TROUBLE CODES 4-20
COMPONENT LOCATIONS 4-23
VACUUM DIAGRAMS 4-30

EMISSION CONTROLS 4-2
CRANKCASE VENTILATION SYSTEM 4-2
 OPERATION 4-2
 COMPONENT TESTING 4-2
 REMOVAL & INSTALLATION 4-2
EVAPORATIVE EMISSION CONTROL
 SYSTEM 4-2
 OPERATION 4-2
 COMPONENT TESTING 4-2
 REMOVAL & INSTALLATION 4-5
EXHAUST GAS RECIRCULATION
 SYSTEM 4-5
 OPERATION 4-5
 SERVICE 4-5
 REMOVAL & INSTALLATION 4-6
AIR INJECTION SYSTEM 4-6
 OPERATION 4-6
 SERVICE 4-6
DECELERATION CONTROL SYSTEM 4-7
 OPERATION 4-7
 SERVICE 4-7
EMISSION MAINTENANCE WARNING
 LIGHT 4-8
 RESETTING 4-8
**ELECTRONIC ENGINE
CONTROLS 4-9**
POWERTRAIN CONTROL MODULE
 (PCM)/ELECTRONIC CONTROL
 MODULE (ECM) 4-9
 OPERATION 4-9
 REMOVAL & INSTALLATION 4-9
OXYGEN SENSOR 4-10
 OPERATION 4-10
 TESTING 4-10
 REMOVAL & INSTALLATION 4-10
IDLE AIR CONTROL (IAC) VALVE 4-11
 OPERATION 4-11
 TESTING 4-11
 REMOVAL & INSTALLATION 4-11
ENGINE COOLANT TEMPERATURE (ECT)
 SENSOR 4-12
 OPERATION 4-12
 TESTING 4-12
 REMOVAL & INSTALLATION 4-12
INTAKE AIR TEMPERATURE (IAT)
 SENSOR 4-13
 OPERATION 4-13
 TESTING 4-13
 REMOVAL & INSTALLATION 4-13
MASS AIR FLOW (MAF) SENSOR 4-13
 OPERATION 4-13
 TESTING 4-14
 REMOVAL & INSTALLATION 4-15
THROTTLE POSITION (TP) SENSOR 4-15
 OPERATION 4-15
 TESTING 4-16
 REMOVAL & INSTALLATION 4-16
CAMSHAFT POSITION (CMP)
 SENSOR 4-16
 OPERATION 4-16
 TESTING 4-17
 REMOVAL & INSTALLATION 4-17
CRANKSHAFT POSITION (CKP)
 SENSOR 4-18
 OPERATION 4-18
 TESTING 4-18
 REMOVAL & INSTALLATION 4-19
VEHICLE SPEED SENSOR (VSS) 4-19
 OPERATION 4-19
 TESTING 4-19
 REMOVAL & INSTALLATION 4-20
TROUBLE CODES 4-20
GENERAL INFORMATION 4-20
 MALFUNCTION INDICATOR LAMP
 (MIL) 4-20
READING CODES 4-20
 PICKUP AND MPV 4-20
 NAVAJO 4-21
CLEARING CODES 4-22
CODE DESCRIPTION 4-22
 PICKUP AND MPV 4-22
 NAVAJO 4-22
COMPONENT LOCATIONS 4-23
VACUUM DIAGRAMS 4-30
SPECIFICATIONS CHART
 IAT/ECT SENSOR VOLTAGE AND
 RESISTANCE SPECIFICATIONS 4-12

4-2 DRIVEABILITY AND EMISSION CONTROLS

EMISSION CONTROLS

Crankcase Ventilation System

OPERATION

♦ See Figure 1

The function of the PCV valve is to divert blow-by gases from the crankcase to the intake manifold to be burned in the cylinders. The system consists of a PCV valve, an oil separator and the hoses necessary to connect the components.

Ventilating air is routed into the rocker cover from the air cleaner. The air is then moved to the oil separator and from the separator to the PCV valve. The PCV valve is operated by differences in air pressure between the intake manifold and the rocker cover.

The most critical component of the system is the PCV valve. This vacuum-controlled valve regulates the amount of gases which are recycled into combustion chamber. At low engine speeds the valve is partially closed, limiting the flow of gases into the intake manifold. As engine speed increases, the valve opens to admit greater quantities of the gasses into the intake manifold. If the valve should become blocked or plugged, the gases will bee prevented from escaping the crankcase by the normal route. Since these gases are under pressure, they will find their own way out of the crankcase. This alternate route is usually a weak oil seal or gasket in the engine. As the gas escapes by the gasket, it also creates an oil leak. Besides causing oil leaks, a clogged PCV valve also allows these gases to remain in the crankcase for an extended period of time, promoting the formation of sludge in the engine.

COMPONENT TESTING

♦ See Figure 2

1. Remove the hose from the PCV valve.
2. Start the engine and run it at approximately 700–1,000 rpm.
3. Cover the end of the PCV valve with a finger. A distinct vacuum should be felt. If no vacuum is felt, replace the valve.

REMOVAL & INSTALLATION

1. Disconnect the hose from the PCV valve.
2. Remove the valve from the mounting fitting.
3. Install the valve in the fitting.
4. Connect the hose to the valve.

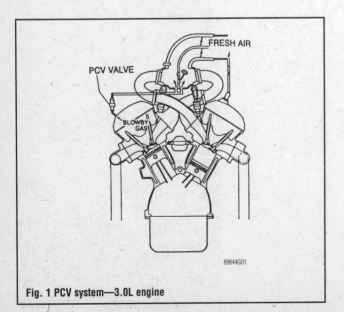

Fig. 1 PCV system—3.0L engine

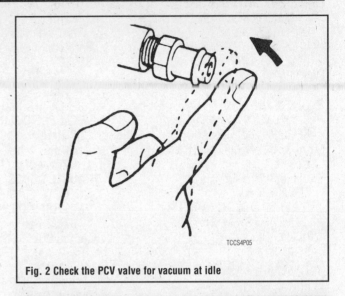

Fig. 2 Check the PCV valve for vacuum at idle

Evaporative Emission Control System

OPERATION

The evaporative emission control system is designed to control the emission of gasoline vapors into the atmosphere. On carbureted engines, the vapors rising from the gasoline in the fuel tank are vented into a separate condensing tank which is located above the fuel tank. There they condense and return to the fuel tank in liquid form when the engine is not running.

When the engine is running, the fuel vapors are sucked directly into the engine through the PCV valve and are burned along with the air/fuel mixture.

Any additional fuel vapors which are not handled by the condensing tank are stored in a charcoal canister. When the engine is running, the charcoal is purged of its stored fuel vapor. On some models, a check valve vents the fuel vapor into the atmosphere if pressure becomes too excessive in the fuel tank.

The system on the carbureted models consists of a charcoal canister, check and cut valve, purge control valves, evaporator shutter valve in air cleaner.

On the pickup and MPV with fuel injection, the system takes fuel vapor that is generated in the fuel tank and stores it in the charcoal canister when the engine is not running. This fuel vapor remains in the canister until the engine is started at which time the fuel vapor is drawn into the dynamic chamber and burned. The system on these models is made up of the charcoal canister, purge control solenoid valves, a three-way check valve, vacuum switch control valve and an electronic control unit.

On the Navajo, the Evaporative Emission Control System provides a sealed fuel system with the capability to store and condense fuel vapors. The system has three parts: a fill control vent system; a vapor vent and storage system; and a pressure and vacuum relief system (special fill cap). The fill control vent system is a modification to the fuel tank. It uses an air space within the tank which is 10–12% of the tank's volume. The air space is sufficient to provide for the thermal expansion of the fuel. The space also serves as part of the in-tank vapor vent system. The in-tank vent system consists of the air space previously described and a vapor separator assembly. The separator assembly is mounted to the top of the fuel tank and is secured by a cam-lockring, similar to the one which secures the fuel sending unit. Foam material fills the vapor separator assembly. The foam material separates raw fuel and vapors, thus retarding the entrance of fuel into the vapor line.

COMPONENT TESTING

There are several things to check if a malfunction of the evaporative emission control system is suspected.

Leaks may be traced by using an infrared hydrocarbon tester. Run the test probe along the lines an connections. The meter will indicate the presence of a leak by a high hydrocarbon (HC) reading. This method is much more accurate

DRIVEABILITY AND EMISSION CONTROLS 4-3

than a visual inspection which would indicate only the presence of a leak large enough to pass liquid.

Leaks may be caused by any of the following, so always check these areas when looking for them:
- Defective or worn lines
- Disconnected or pinched lines
- Improperly routed lines
- A defective check valve

➡ **If it becomes necessary to replace any of the lines used in the evaporative emission control system, use only those hoses which are fuel resistant or are marked EVAP.**

If the fuel tank has collapsed, it may be the fault of clogged or pinched vent lines, a defective vapor separator, or a plugged or incorrect check valve.

Carbureted Engines

♦ See Figure 3

The EEC system consists of a canister No. 3 purge control valve, water thermo valve, check-and-cut valve, purge solenoid valve and an air vent solenoid valve.

The water thermo valve opens the vacuum passage to the No. 1 and No. 3 purge control valves. The canister incorporates the No. 2 purge control valve (2 way check valve), and the No. 1 purge control valve, which opens the fuel vapor passage between the canister and the intake manifold. The No. 3 purge control valve opens the fuel vapor passage between the canister and the intake manifold when the purge solenoid valve is ON.

Port vacuum is applied to the No. 1 purge control valve while the engine is running and to the No. 3 purge control valve during running or heavy load driving.

The check-and-cut valve vents the vapors to the atmosphere if the evaporative hoses become clogged. It also prevents fuel leakage if the vehicle overturns.

1. Check the vacuum hose routing. Repair or replace if necessary.
2. Start the engine and allow it to reach normal operating temperature.
3. Disconnect the vacuum hose from the No. 1 purge control valve. Connect a suitable vacuum gauge to the hose.
4. Increase the engine speed to 2500 rpm and check that the vacuum gauge reads more than 5.9 in. Hg. If the required vacuum is not reached, check the thermo valve. Reconnect the vacuum hose to the control valve.
5. Disconnect the vacuum line from the canister. Connect a vacuum gauge to the hose. Check that vacuum is present when engine speed exceeds 1400 rpm.
6. If no vacuum exists, check the purge solenoid valve, No. 3 purge control valve and the 1V terminal on the emission control unit. Reconnect the vacuum hose to the canister.
7. Disconnect the evaporation hose from the evaporation pipe, then connect a suitable hand operated vacuum pump to the evaporation pipe. Operate the hand pump and check that no vacuum is held in the system. If so, examine the check-and-cut valve and the evaporation pipe for clogging.

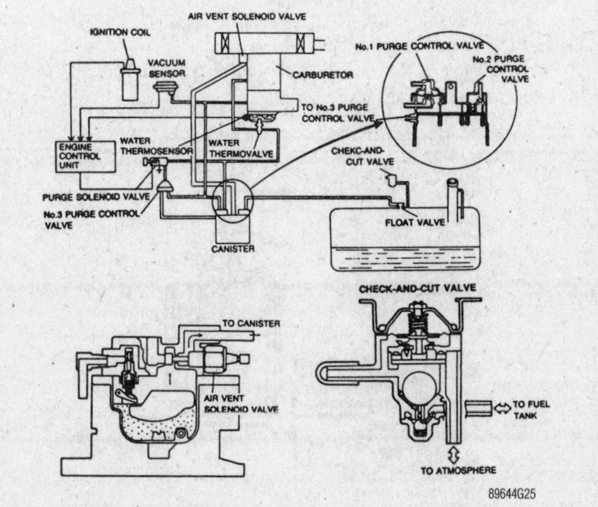

AIR VENT SOLENOID VALVE

1. Remove the air cleaner assembly.
2. Touch the air vent solenoid on the carburetor.
3. Verify that when the ignition switch is turned ON and OFF, the solenoid clicking is felt and heard.
4. If not click is heard or felt, check for voltage at the solenoid or replace the solenoid.

CHECK AND CUT VALVE

♦ See Figure 4

1. Remove the valve from its location just before the gas tank. Hold the valve in the horizontal position, otherwise the weight of the valve will cause it to move out of position and close the passage.
2. Connect a vacuum gauge in line to the passage which normally connects to the fuel tank — port **A**.
3. Blow air into port **A** and verify that the valve opens at a pressure of 0.78–1.00 psi (5.38–6.89kpa).
4. Remove the vacuum gauge and connect it to the passage to atmosphere — port **B**.
5. Blow air into port **B** and verify that the valve opens at a pressure of 0.14–0.71 psi (0.97–4.89kpa).
6. If not as specified, replace the valve.

WATER THERMO VALVE

1. Remove the water thermo valve and immerse the valve in a container of water.
2. Heat the water gradually and observe the temperature. Remember, you are working with hot metal and hot water, take steps to prevent burning yourself. Blow through the valve. If air passes through the valve at 130°F (54°C) or higher, the valve is operating correctly.

SLOW FUEL CUT SOLENOID VALVE

1. Run the engine at idle speed and disconnect the slow fuel cut solenoid valve connector.
2. If the engine stalls, the solenoid valve is operating properly. If the engine does not stall, replace the solenoid valve.

NO. 1 PURGE CONTROL VALVE

♦ See Figure 5

1. Disconnect the vacuum lines from the canister.
2. Blow through port **A** and verify that the air does not flow.
3. Connect a vacuum pump to the purge control valve.

Fig. 3 Evaporative emission control system

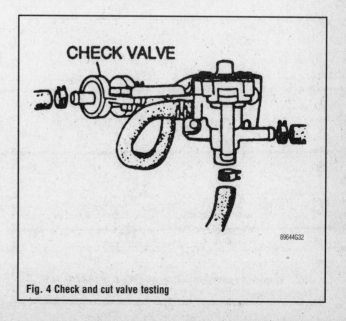

Fig. 4 Check and cut valve testing

4-4 DRIVEABILITY AND EMISSION CONTROLS

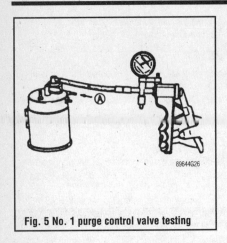

Fig. 5 No. 1 purge control valve testing

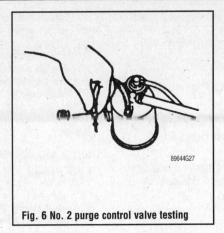

Fig. 6 No. 2 purge control valve testing

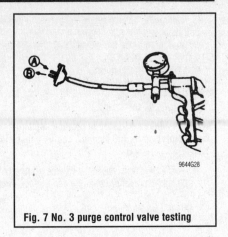

Fig. 7 No. 3 purge control valve testing

4. Apply 4.3 in. Hg of vacuum to the purge control valve.
5. Blow air through port **B** and verify that air does not flow.
6. If not as specified, replace the purge control valve.

NO. 2 PURGE CONTROL VALVE

♦ See Figure 6

1. Disconnect the vacuum hose from the fuel tank to the canister at the canister.
2. Blow air into the canister.
3. Verify that the air flows freely and there are no restrictions.
4. If not as specified, replace the canister.

NO. 3 PURGE CONTROL VALVE

♦ See Figure 7

1. Remove the No. 3 purge control valve.
2. Connect a hand vacuum pump to the thermosensor side of the valve.
3. Blow air through the valve from port **A** while applying 3-4 in. Hg of vacuum to the valve.
4. Verify that air flows through the valve from port **A** to port **B**.
5. If not as specified, replace the No. 3 purge control valve.

Pickup and MPV with Fuel Injection

♦ See Figure 8

The EEC system consists of a fuel separator, a fuel vapor valve, a check-and-cut valve, a 2 way check valve, a purge control solenoid valve, the engine control unit and input devices.

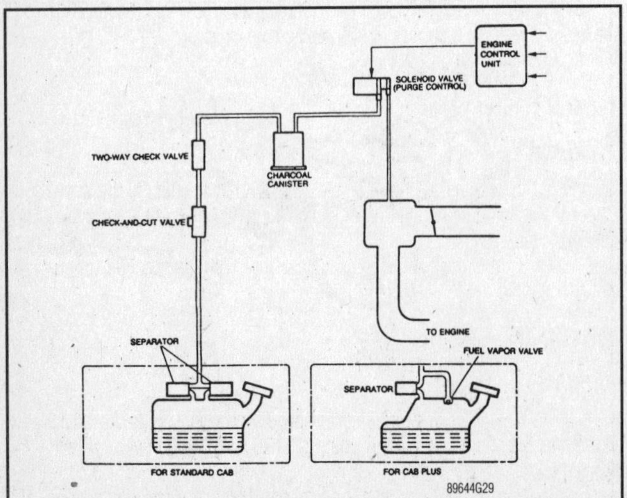

The amount of evaporative fumes introduced into and burned by the engine is controlled by the solenoid valve and control unit to correspond to the engine's operating conditions.

PURGE CONTROL SOLENOID VALVE

♦ See Figure 9

1. Start the vehicle and warm the engine.
2. Disconnect the vacuum hose (usually white) closest to the wire connector from the solenoid valve.
3. Verify there is no vacuum at the solenoid valve at idle speed.
4. If there is vacuum at the solenoid valve, turn the engine OFF.
5. Disconnect the other vacuum hose and apply air pressure. Verify that no air flows through the valve.
6. Next, supply 12 volts to the terminals of the solenoid. Apply air pressure and verify that air does flow through the valve.
7. If not as specified, replace the solenoid valve.

FUEL SEPARATOR

The fuel separator is not serviceable and should be checked periodically for damage or leaking. Replace the separator if damage or leaking is evident. The separator is located on the right rear wheelhouse on MPVs, and above the fuel tank on pickup models.

CHECK AND CUT VALVE

♦ See Figure 4

1. Remove the valve from its location just before the gas tank. Hold the valve in the horizontal position, otherwise the weight of the valve will cause it to move out of position and close the passage.
2. Connect a vacuum gauge in line to the passage which normally connects to the fuel tank — port **A**.
3. Blow air into port **A** and verify that the valve opens at a pressure of 0.78–1.00 psi (5.38–6.89kpa).

Fig. 8 Evaporative emission control system—2.2L and 2.6L fuel injected engines

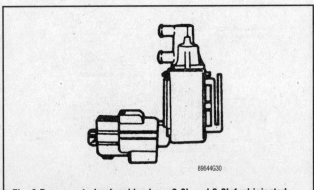

Fig. 9 Purge control solenoid valve—2.6L and 3.0L fuel injected engines

DRIVEABILITY AND EMISSION CONTROLS

4. Remove the vacuum gauge and connect it to the passage to atmosphere — port **B**.
5. Blow air into port **B** and verify that the valve opens at a pressure of 0.14–0.71 psi (0.96–4.89kpa).
6. If not as specified, replace the valve.

2 WAY CHECK VALVE

▶ See Figure 10

The 2 way check valve is not serviceable and should be replaced if damage or leaking is evident.
1. Remove the valve and blow through from port **A** and check that air flows.
2. Blow through from the opposite side port and check that air does not flow through. Replace the valve as required.

Navajo

▶ See Figure 11

Other than a visual check to determine that none of the vapor lines are broken, there is no test for this equipment.

The only maintenance on the evaporative system is to periodically check all hoses and connections for leaks and deterioration. Replace any hoses which are found to be damaged in any way. Under normal circumstances, the charcoal canister is expected to last the life of the vehicle, but it should be periodically inspected for any damage or contamination by raw gasoline. Replace any gasoline soaked canister found. Refer to the illustrations for canister mounting and evaporative hose routing on the various engines. Filler cap damage or contamination that clogs the pressure/vacuum valve may result in deformation of the fuel tank.

REMOVAL & INSTALLATION

1. Disconnect the negative battery cable.
2. Mark and disconnect the vapor hoses from the canister assembly.

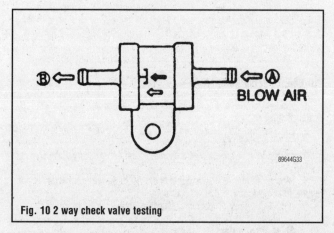

Fig. 10 2 way check valve testing

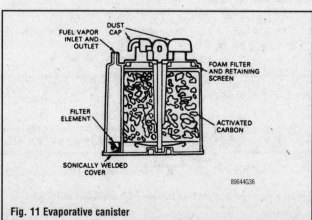

Fig. 11 Evaporative canister

3. Remove the screw securing the canister to the bracket or fender apron.
4. Lift up on the canister assembly to disengage the tab on the back side and remove the canister.
5. Installation is the reverse of the service removal procedure. Always install vapor hose in correct location.

To disconnect a vapor hose from any component securely grip component with one hand and vapor hose with the other hand as close as possible to connection. Sharply twist hose along its axis to break the connection. No adhesive is used to make hose connections during vehicle assembly, but aging of the connections causes a temporary bond to exist.

If the connection is stubborn and the above method does not work, grip the hose with a pair of small pliers directly over the joint and twist again. Remove the vapor hose from the component.

Exhaust Gas Recirculation System

OPERATION

The function of the EGR system is to reduce Oxides of Nitrogen (NOx) in the exhaust gas, which are the result of high combustion chamber temperatures. The combustion chamber temperature is reduced by directing a portion of the spent exhaust gases into the combustion chamber along with the incoming air/fuel mixture. The lower temperature in the combustion chamber results in reduced amounts of NOx.

The EGR system consists of the EGR valve and various sensors and solenoids which regulate the induction of exhaust gas.

SERVICE

1. Check all vacuum hose to make sure they are tight and not leaking.
2. Run the engine to normal operating temperature.
3. Disconnect the vacuum hose from the 1st EGR control valve and check that the engine continues to run smoothly. If not, replace the valve.
4. Connect a vacuum gauge to the hose and increase the engine speed to 2,500 rpm. The gauge should read at least 5.9 in.Hg. If not, check the water thermovalve, water temperature switch, No.1 EGR solenoid valve and, on trucks with manual transmission, the No.2 EGR vacuum switch.
5. Disconnect the vacuum hose from the 2nd EGR control valve and see that the engine continues to run smoothly. If not, replace the valve.
6. Connect a vacuum gauge to the hose and increase the engine speed to 2,500 rpm. The gauge should read at least 5.9 in.Hg. If not, check the No.2 EGR solenoid valve and the No.1 EGR vacuum switch.
7. Disconnect the hose from the EGR modulator valve and plug it.
8. Connect a length of hose to the port of the EGR modulator valve.
9. Increase the engine speed to 2,500 rpm and blow into the hose. See if the vacuum reading increases. If not, replace the valve.
10. Reconnect all hoses.

1st EGR Control Valve

PICKUPS

1. Start the engine and run it at idle speed. Apply vacuum to the valve. Check that the engine runs rough or stalls when 1.58 in.Hg of vacuum is applied to the valve.
2. If the engine does not behave as described with the specified amount of vacuum applied, replace the control valve.

2nd EGR Control Valve

PICKUPS

1. Start the engine and run it at idle speed. Apply vacuum to the valve. Check that the engine runs rough or stalls when 2.76 in.Hg of vacuum is applied to the valve.
2. If the engine does not behave as described with the specified amount of vacuum applied, replace the control valve.

4-6 DRIVEABILITY AND EMISSION CONTROLS

EGR Position Sensor

PICKUPS

1. The sensor is located on top of the EGR valve. Remove the rubber boot from the connector at the EGR valve.
2. Disconnect the vacuum hose from the EGR valve and connect a hand vacuum pump.
3. Turn the ignition switch ON. Using a voltmeter, check the voltage at each terminal in the harness side of the connector.
4. With no vacuum applied:
- Terminal **A** (B/L)—0.7 volts
- Terminal **B** (B/LG)—less than 1.5 volts
- Terminal **C** (G/Y)—4.5-5.5 volts
5. With vacuum applied:
- Terminal **A** (B/L)—4.7 volts
- Terminal **B** (B/LG)—less than 1.5 volts
- Terminal **C** (G/Y)—4.5-5.5 volts
6. If the voltage is incorrect for terminals **B** and **C**, check the wiring harness and ECU terminals **1D**, **1F** and **1G**.
7. If incorrect at terminal **A**, check the resistance of the sensor, then the wiring harness and then the ECU.
8. Check the resistance between the terminals while applying 0-5.9 in. Hg of vacuum to the EGR valve:
- Terminals **B** and **C**—5 k
- Terminals **A** and **C**—0.0-5.5 k
- Terminals **A** and **B**—0.7-6.0 k

Checking Duty Solenoid Valve

PICKUPS

1. Disconnect the vacuum hoses from the solenoid valve. Label the hoses for reinstallation.
2. Blow air through the vent hose and air should flow.
3. Disconnect the solenoid valve connector.
4. Connect 12 volts to one terminal of the solenoid valve and ground the other.
5. Blow air through the solenoid valve and verify that air does not flow.
6. Replace the duty solenoid valve if it does not perform as specified.
7. Blow air through the vacuum hose and verify that the air does not flow.
8. Connect 12 volts to one terminal of the solenoid and ground the other.
9. Blow air through the vacuum hose and verify that the air flows through. If not as specified, replace the valve.

EGR Modulator Valve Testing

1. Note the routing of all hoses leading to the modulator valve, especially the hose which is connected to the exhaust side of the EGR valve. Remove the EGR Modulator valve. Plug the No. 1 port and then attach a source of vacuum to the No. 3 port.
2. Attach a clean hose to the exhaust gas port. Blow into the end of the hose and maintain pressure. Apply vacuum to the No. 3 port and then seal off the source of vacuum. Vacuum should be maintained as long as air pressure is applied.
3. Stop applying air pressure. The vacuum should be released. If the valve fails to respond properly in either Step 2 or 3, replace it.

REMOVAL & INSTALLATION

EGR Control Valve

PICKUPS

1. Tag and disconnect the hoses at the valve.
2. Unbolt and remove the valve.
3. Installation is the reverse of removal.

Air Injection System

OPERATION

The function of this system is to burn more completely the spent exhaust gases while they are still in the exhaust system. This is accomplished by injecting fresh air into the exhaust port or main converter. the system is comprised of reed valves, air control valves and an Air Control Valve (ACV) solenoid which is controlled by the engine control unit.

SERVICE

Reed Valve

1. Run the engine until normal operating temperature is reached. Shut off the engine. Remove the air cleaner cover and filter element.
2. Start and run engine at idle speed. Place a piece of paper over the intake ports **B** and **C** of the reed valve. Air should be sucked into the valve.
3. Increase speed to 1,500 rpm to verify that air is being sucked in.
4. Increase the engine speed to about 3000 rpm and verify that no exhaust gas is leaking from the air inlet port. Replace the reed valve if not as specified.
5. Disconnect the vacuum line from the No.2 air control valve and plug it.
6. Place a thin piece of paper over inlet port **D** of the reed valve. Increase the engine speed to about 1500 rpm and verify that air is being pulled in. If not, inspect No. 1 control valve.
7. Disconnect the vacuum line from the No.1 air control valve and plug it.
8. Using a hand vacuum pump apply 3.54 in Hg of vacuum to the No. 2 control valve.
9. Place a thin piece of paper over port **E** and increase the engine speed to about 1500 rpm. Verify that air is being pulled in. If not as specified, inspect No.2 air control valve and check the reed valve.
10. Turn the ignition switch OFF and disconnect the water temperature switch connector. Start the engine and verify that there is no vacuum at the No. 2 air control vacuum valve hose. Increase the engine speed to 1500 rpm and verify that vacuum is present at the hose.
11. If not as specified, ACV solenoid valve operation should be suspect.

Air Control Valve Solenoid

1. Disconnect the vacuum hoses from solenoid after labeling them for installation reference.
2. Blow air through the solenoid inlet closest to the wire connector and verify that air comes out of the valve air filter.
3. Disconnect the solenoid wire connector and apply battery voltage across the terminals.
4. Blow air through the solenoid inlet closest to the wire connector and verify the air comes out the other vacuum port.
5. If the valve does not operate as specified, replace it.

No.1 Air Control Valve

1. Remove the valve. Connect a vacuum pump to the valve.
2. Blow air into port **A** and verify that the air does not come out port **B**.
3. Apply 15.7 in. Hg of vacuum to the valve.
4. Blow air into port **A** and verify that air does come out port **B**.
5. If not as specified, replace the valve.

No.2 Air Control Valve

1. Remove the valve. Connect a vacuum pump to the valve.
2. Apply vacuum gradually and verify that the stem of the valve starts to move at approximately 1.97 in. Hg and stops at about 3.54 in. Hg vacuum.
3. If not as specified, replace the valve.

DRIVEABILITY AND EMISSION CONTROLS 4-7

Deceleration Control System

OPERATION

The deceleration control system on B2200 carburetor equipped Pickups is designed to maintain a balance air/fuel mixture during periods of engine deceleration. Although the components used vary in some years, the basic theory remains the same: To more thoroughly burn or dilute the initial rich mixture formed when throttle is suddenly closed, and to smooth out the transition to a lean mixture by enriching the mixture slightly after the throttle has closed. Although the process may seem contradictory, they act in sequence to provide an overall ideal mixture.

The deceleration control system used on carburetor equipped B2600 models and on the MPV reduces the fuel flow to decrease HC emissions and to improve fuel economy during deceleration and cut the fuel flow when the ignition switch is OFF to prevent "run-on".

SERVICE

2.2L Carbureted Engine

SLOW FUEL CUT SYSTEM TESTING

1. Warm the engine to normal operating temperature.
2. Disconnect the neutral switch or inhibitor switch wire connectors.
3. Remove the air cleaner assembly.
4. Connect a voltmeter to the F (LG) terminal of the carburetor connector.
5. Increase the engine speed to about 300 rpm.
6. Lift the idle switch arm.
7. Verify the voltmeter reading is about 12 volts when the engine speed is above 2500 rpm.
8. Verify that the voltmeter reading is less than 1.5 volts when the engine speed is less than 2500 rpm.
9. If not as specified, check the 2D terminal of the emission control unit and the slow cut solenoid valve.
10. Start the engine and run at idle.
11. Disconnect the carburetor wire connector.
12. Verify that the engine stops running.
13. If not as specified, inspect the carburetor ports or replace the solenoid valve.

RICHER SYSTEM TESTING

1. Warm the engine to normal operating temperature.
2. Disconnect the neutral switch or inhibitor switch wire connectors.
3. Remove the air cleaner assembly.
4. Connect a voltmeter to the H (BR/B) terminal of the carburetor connector.
5. Increase the engine speed to about 3000 rpm.
6. Lift the idle switch arm.
7. Verify the voltmeter reading is about 12 volts when the engine speed is above 2500 rpm.
8. Verify the voltmeter reading is less than 1.5 volts, one second after engine speed is 1400-2500 rpm.
9. Verify that the voltmeter reading is about 12 volts when the engine speed is below 1400 rpm.
10. If not as specified, inspect terminal 2H of the emission control unit and the coasting richer solenoid valve.
11. Start the engine and run at idle.
12. Ground the H (BR/B) terminal of the carburetor connector with a jumper wire.
13. Verify that the engine speed increases.
14. If not as specified, inspect the carburetor ports or replace the solenoid valve.

ADVANCE SYSTEM TESTING

1. Warm the engine to normal operating temperature.
2. Disconnect the neutral switch or inhibitor switch wire connectors.
3. Remove the air cleaner assembly.
4. Connect a voltmeter to the W/G terminal of the coasting advance solenoid wire connector.
5. Increase the engine speed to about 3000 rpm.
6. Lift the idle switch arm.
7. Verify that the voltmeter reading is about 12 volts when the engine speed is above 2500 rpm.
8. Verify that the voltmeter reading is less than 1.5 volts when the engine speed is 1700-2500 rpm.
9. Verify the voltmeter reading is about 12 volts when the engine speed is below 1700 rpm.
10. If not as specified, inspect terminal 1S of the emission control unit and the vacuum solenoid valve.
11. Remove the vacuum solenoid valve.
12. Connect vacuum hoses to the valve.
13. Blow air through the port closest to the solenoid wire connector.
14. Verify that air comes out of the valve air filter.
15. Apply battery voltage to the terminals of the solenoid valve.
16. Blow air through the solenoid port closest to the wire connector and verify that air comes out the opposite vacuum port.
17. If not as specified, replace the vacuum solenoid valve.

MIXTURE CONTROL VALVE TESTING

1. Start the engine.
2. Block the intake port of the mixture control valve and verify that engine rpm does not increase.
3. Increase the engine speed and quickly decelerate.
4. Verify that air is pulled into the intake port of the mixture control valve for about one to two seconds after the accelerator is released.
5. If not as specified, replace the mixture control valve.

DASHPOT TESTING

1. Push the dashpot rod in, making sure the rod goes into the dashpot slowly.
2. Release the rod and make sure it comes out quickly.
3. Replace the dashpot if it does not perform as specified.

DASHPOT ADJUSTMENT

1. Run the engine until normal operating temperature is reached.
2. Connect a suitable tachometer to the engine.
3. Raise the engine speed to above 3500 rpm.
4. Slowly decrease the engine speed making sure the dashpot arm touches the lever at about 2700-2900 rpm.
5. Adjust the dashpot as required, for correct curb idle speed, by loosening the locknut and turning the dashpot in or out as necessary. Tighten the locknut after adjustment.

B2600

SLOW FUEL CUT SYSTEM TESTING

1. Warm the engine to normal operating temperature.
2. Connect a tachometer to the engine.
3. Connect a voltmeter to the solenoid valve connector (e) terminal (YL).
4. Check the voltage at the following conditions. At idle 0-0.6 volts. Accelerate to 4000 rpm and decelerate quickly, momentary voltage should be 13-15 volts.
5. If not to specifications, check the deceleration vacuum switch, feedback control unit and wiring of the slow fuel cut system.

SLOW FUEL CUT SOLENOID VALVE TEST

1. Run the engine at idle.
2. Disconnect the connector for the solenoid valves.
3. Check that the engine stops.

DECELERATION VACUUM SWITCH TESTING

1. Turn the ignition ON.
2. Connect a hand vacuum pump to the deceleration vacuum switch.
3. Apply 10.2-11.- in. Hg vacuum to the switch and check the voltage at (e) terminal (GB).
4. Voltages should be: with vacuum applied 0 volts. Vacuum not applied 12 volts.
5. If the voltages are not correct check the deceleration vacuum switch.
6. Turn the ignition OFF.

4-8 DRIVEABILITY AND EMISSION CONTROLS

7. Connect the vacuum pump to the deceleration vacuum switch.
8. Disconnect the vacuum switch connector. Apply 10.2-11.- in. Hg of vacuum and check continuity between (e) and (f) terminals (GB) and (B). There should be continuity when vacuum is applied. If not correct, replace the deceleration vacuum switch.

DASHPOT TESTING

Move the throttle lever to the full throttle position, then release the throttle lever and check that the dashpot rod extends slowly.

DASHPOT ADJUSTMENT

1. Warm the engine to normal operating temperature.
2. Connect a tachometer to the engine.
3. Push the dashpot pushrod up into the pot.
4. Loosen the dashpot adjust screw.
5. Turn the adjust screw clockwise and set the curb idle speed to specification.

MPV

DASHPOT ADJUSTMENT

1. The dashpot is located at the throttle body. Run the engine until it reaches normal operating temperature and then let it idle.
2. Connect a suitable tachometer to the engine.
3. Raise the engine speed to above 4000 rpm.
4. Slowly decrease the engine speed making sure the dashpot rod touches the lever at about 3200-3800 rpm.
5. To adjust, loosen the locknut and turn the dashpot. Secure the locknut when adjustment is finished.

Emission Maintenance Warning Light

The emission maintenance warning light system consists of an instrument panel mounted amber lens that is electrically connected to a sensor module located under the instrument panel. The purpose of the system is to alert the driver that emission system maintenance is required.

The system actually measures accumulated vehicle ignition key on-time and is designed to continuously close an electrical circuit to the amber lens after 2000 hours of vehicle operation. Assuming an average vehicle speed of 30 mph (48 kph), the 2000 hours equates to approximately 60,000 miles (96,500 km) of vehicle operation. Actual vehicle mileage intervals will vary considerably as individual driving habits vary.

Every time the ignition is switched on, the warning light will glow for 2-5 seconds as a bulb check and to verify that the system is operating properly. When approximately 60,000 miles (96,500 km) is reached, the warning light will remain on continuously to indicate that service is required. After the required maintenance is performed, the sensor must be reset for another 60,000 mile (96,500 km) period. The sensor module is located above the right front corner of the glove box assembly.

RESETTING

1987-88 Models

1. Make sure the ignition key is OFF.
2. Locate the sensor (above the right front corner of the glove box), and lightly push a Phillips screwdriver or small rod tool through the 0.2 inch (5mm) diameter hole with the sticker labeled "RESET" and lightly press down and hold.
3. While lightly holding the screwdriver or tool down, turn the ignition switch to the RUN position. The emission warning light will then light and should remain on for as long as the screwdriver is held down. Hold the screwdriver down for approximately 5 seconds.
4. Remove the screwdriver or tool. The lamp should go out within 2-5 seconds, indicating that a reset has occurred. If the light remains on, begin again at Step 1. If the light goes out, turn the ignition off and go to the next Step.
5. Turn the ignition to the RUN position. The warning light should illuminate for 2-5 seconds and then go out. This verifies that a proper reset of the module has been accomplished. If the light remains on, repeat the reset procedure.

1989-93 Models

B SERIES PICKUP

♦ See Figures 12 and 13

On Federal and Canadian vehicles, the Malfunction Indicator Light (MIL) will come on at 60,000-80,000 mile (96,500-128,700 km) intervals to indicate the need for scheduled maintenance of the emission control system.

On California vehicles, the MIL will come on any time an engine management input device malfunctions, indicating that service is necessary.

After the required service or maintenance has been performed, the MIL can be reset by changing the connector connections.

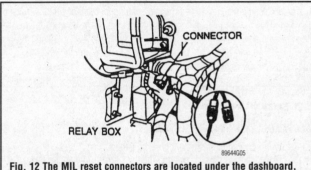

Fig. 12 The MIL reset connectors are located under the dashboard, near the realay box

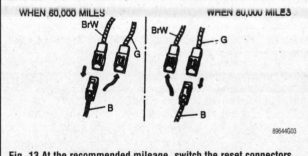

Fig. 13 At the recommended mileage, switch the reset connectors

MPV

♦ See Figure 14

Federal and Canadian vehicles are equipped with a mileage sensor which is linked to the odometer. At every 80,000 miles (128,744 km), the mileage sensor

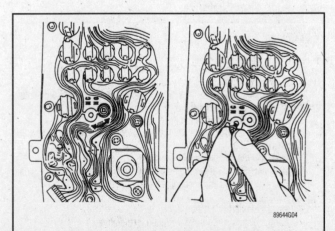

Fig. 14 After replacing the oxygen sensor, remove the instrument gauge cluster assembly and, on the back, reset the MIL by reversing the set screw position

DRIVEABILITY AND EMISSION CONTROLS 4-9

will cause the Malfunction Indicator Light (MIL) to illuminate, indicating that the oxygen sensor must be replaced.

After replacing the oxygen sensor, remove the instrument cluster and reset the MIL by reversing the position of the MIL set screw.

NAVAJO

All vehicles are equipped with a "CHECK ENGINE" or warning light located on the instrument cluster. This light should come on briefly when the ignition key is turned **ON**, but should turn off when the engine starts. If the light does not come ON when the ignition key is turned **ON** or if it comes ON and stays ON when the engine is running, there is a malfunction in the electronic engine control system. After the malfunction has been remedied, using the proper procedures, the "CHECK ENGINE" light will go out.

ELECTRONIC ENGINE CONTROLS

The PCM/ECM receives data from a number of sensors and other electronic components (switches, relay, etc.). Based on information received and information programmed in the PCM/ECM's memory, it generates output signals to control various relay, solenoids and other actuators. The PCM/ECM in this system has calibration modules located inside the assembly that contain calibration specifications for optimizing emissions, fuel economy and driveability. The calibration module is called a PROM.

The following are the electronic engine controls used by Mazda trucks (note that not all of these sensors may be found on one engine):

- Powertrain Control Module (PCM)
- Throttle Position (TP) sensor
- Mass Air Flow (MAF) sensor
- Intake Air Temperature (IAT) sensor
- Idle Air Control (IAC) valve
- Engine Coolant Temperature (ECT) sensor
- Oxygen Sensor or Heated Oxygen Sensor (HO2S)
- Camshaft Position (CMP) sensor
- Knock Sensor (KS)
- Vehicle Speed Sensor (VSS)
- Crankshaft Position (CKP) sensor

➡ Because of the complicated nature of this system, special tools and procedures are necessary for testing and troubleshooting.

Powertrain Control Module (PCM)/Electronic

OPERATION

The PCM/ECM performs many functions. The module accepts information from various engine sensors and computes the required fuel flow rate necessary to maintain the correct amount of air/fuel ratio throughout the entire engine operational range.

Based on the information that is received and programmed into the PCM's memory, the PCM generates output signals to control relays, actuators and solenoids.

REMOVAL & INSTALLATION

▶ See Figure 15

Pick-up and Navajo Models

1. Disconnect the negative battery cable.
2. Disengage the wiring harness connector from the PCM by loosening the connector retaining bolt, then pulling the connector from the module.
3. Remove the two nuts and the PCM cover.
4. Remove the PCM from the bracket by pulling the unit outward.

To install:

5. Install the PCM in the mounting bracket.
6. Install the PCM cover and tighten the two nuts.
7. Attach the wiring harness connector to the module, then tighten the connector retaining bolt.
8. Connect the negative battery cable.

MPV Models

▶ See Figure 16

The PCM/ECM is mounted in the vehicle's interior, under the dashboard on the right (passenger) side.

1. Disconnect the negative battery cable.
2. Remove the right side scuff plate and right front side trim.
3. Lift up the front mat.
4. Remove the protector cover.
5. Unplug the wiring harness connector from the control module.
6. Loosen the mounting fasteners and remove the PCM/ECM from the vehicle.

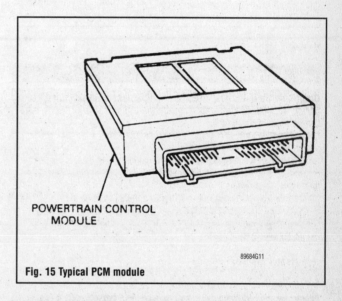

Fig. 15 Typical PCM module

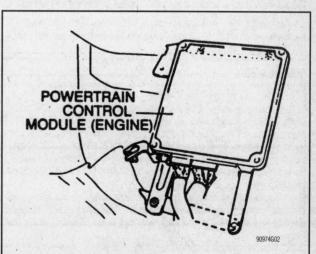

Fig. 16 Location of the control module on the front right side of the interior

4-10 DRIVEABILITY AND EMISSION CONTROLS

To install:

7. Place the PCM into the vehicle in correct position.
8. Install the mounting fasteners and tighten to 70–95 inch lbs. (8–10 Nm).
9. Plug the wiring harness connector into the PCM/ECM.
10. Install the protector cover and place the mat back in proper position.
11. Install the right front side trim and the right side scuff plate.
12. Connect the negative battery cable.

Oxygen Sensor

OPERATION

The oxygen sensor supplies the computer with a signal which indicates a rich or lean condition during engine operation. The input information assists the computer in determining the proper air/fuel ratio. A low voltage signal from the sensor indicates too much oxygen in the exhaust (lean condition) and, conversely, a high voltage signal indicates too little oxygen in the exhaust (rich condition).

The oxygen sensors are threaded into the exhaust manifold and/or exhaust pipes on all vehicles. Heated oxygen sensors are used on some models to allow the engine to reach the closed loop faster.

TESTING

Heated Sensor

⁂ **WARNING**

Do not pierce the wires when testing this sensor; this can lead to wiring harness damage. Back probe the connector to properly read the voltage of the HO2S.

1. Disconnect the HO2S.
2. Measure the resistance between PWR and GND terminals of the sensor. If the reading is approximately 6 ohms at 68°F (20°C), the sensor's heater element is in good condition.
3. With the HO2S connected and engine running, measure the voltage with a Digital Volt-Ohmmeter (DVOM) between terminals **HO2S** and **SIG RTN** (GND) of the oxygen sensor connector. If the voltage readings are swinging rapidly between 0.01–1.1 volts, the sensor is probably okay.

Non-Heated Sensor

▶ See Figure 17

⁂ **WARNING**

Do not pierce the wires when testing this sensor; this can lead to wiring harness damage. Back probe the connector to properly read the voltage of the O2S.

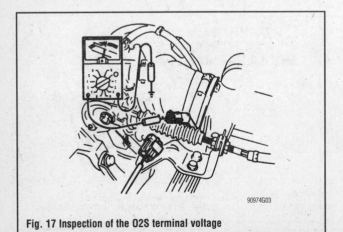

Fig. 17 Inspection of the O2S terminal voltage

1. Warm up the engine and allow it to run at idle.
2. Disconnect the O2S.
3. Connect a voltmeter between the O2S and a good ground.
4. Run the engine at 4,500 rpm until the voltmeter reading indicates approximately 0.7V.
5. Increase and decrease the engine speed suddenly several times. Check to see that when the speed is increased, the voltmeter reads between 0.5V–1.0V and when the speed is decreased, the reading is between 0V–0.4V.
6. If the actual readings are out of specifications, replace the O2S.

REMOVAL & INSTALLATION

▶ See Figures 18 and 19

1. Disconnect the negative battery cable.
2. Raise and safely support the vehicle on jackstands.
3. Disconnect the oxygen sensor from the engine control sensor wiring.

➡**If excessive force is needed to remove the sensors, lubricate the sensor with penetrating oil prior to removal.**

4. Unscrew the sensor.

To install:

5. Install the sensor in the mounting boss, then tighten it to 27–33 ft. lbs. (37–45 Nm).
6. Reattach the sensor electrical wiring connector to the engine wiring harness.

Fig. 18 From underneath, locate the oxygen sensor, which is screwed into an exhaust component

Fig. 19 Using the proper wrench, loosen the oxygen sensor. Ensure that the wire and connector are unplugged

DRIVEABILITY AND EMISSION CONTROLS 4-11

7. Lower the vehicle.
8. Connect the negative battery cable.

Idle Air Control (IAC) Valve

OPERATION

♦ See Figure 20

The Idle Air Control (IAC) valve controls the engine idle speed and dashpot functions. The valve is located on the side of the throttle body. This valve allows air, determined by the Powertrain Control Module (PCM) and controlled by a duty cycle signal, to bypass the throttle plate in order to maintain the proper idle speed.

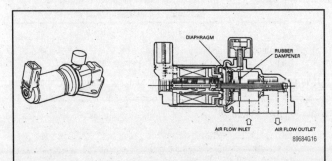

Fig. 20 Typical Idle Air Control (IAC) valve which is mounted to the throttle body—cutaway view shows air bypass direction

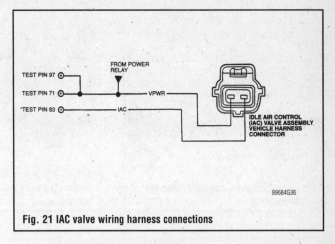

Fig. 21 IAC valve wiring harness connections

1. Turn the ignition switch to the **OFF** position.
2. Disengage the wiring harness connector from the IAC valve.
3. Using an ohmmeter, measure the resistance between the terminals of the valve.
4. The resistance should measure 10.7–12.3 ohms at 68° F (20° C) for 3.0L equipped models and 7.7–9.3 ohms at 73° F (23° C) for 2.6L engine equipped models.
5. If the resistance does not measure within specifications, replace the BAC valve.

REMOVAL & INSTALLATION

Navajo Models

♦ See Figures 23, 24, 25 and 26

1. Disconnect the negative battery cable.
2. Disengage the wiring harness connector from the IAC valve.
3. Remove the two retaining screws, then remove the IAC valve and discard the old gasket.
   To install:
4. Clean the IAC valve mounting surface on the throttle body of old gasket material.
5. Using a new gasket, position the IAC valve on the throttle body. Install and tighten the retaining screws to 71–106 inch lbs. (8–12 Nm).
6. Attach the wiring harness connector to the IAC valve.
7. Connect the negative battery cable.

Pick-up and MPV Models

♦ See Figure 27

➡ The Idle Air Control (IAC) valve is an integral component of the Bypass Air Control (BAC) valve.

TESTING

Navajo Models

♦ See Figure 21

1. Turn the ignition switch to the **OFF** position.
2. Disengage the wiring harness connector from the IAC valve.
3. Using an ohmmeter, measure the resistance between the terminals of the valve.

➡ Due to the diode in the solenoid, place the ohmmeter positive lead on the VPWR terminal and the negative lead on the ISC terminal.

4. If the resistance is not 7–13 ohms, replace the IAC valve.

Pick-up and MPV Models

♦ See Figure 22

➡ The Idle Air Control (IAC) valve is an integral component of the Bypass Air Control (BAC) valve.

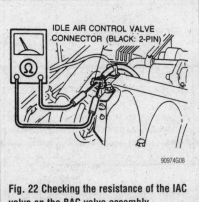

Fig. 22 Checking the resistance of the IAC valve on the BAC valve assembly

Fig. 23 To remove the IAC valve, first disconnect the negative battery cable, then the IAC wire harness plug

Fig. 24 Next, remove the two IAC valve attaching bolts . . .

4-12 DRIVEABILITY AND EMISSION CONTROLS

Fig. 25 . . . and pull the valve from the intake manifold

Fig. 26 When installing the valve, always discard the old gasket and install it using a new one

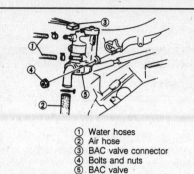

① Water hoses
② Air hose
③ BAC valve connector
④ Bolts and nuts
⑤ BAC valve

Fig. 27 Removal of the BAC/IAC valve assembly—3.0L engine shown

1. Disconnect the negative battery cable.
2. Disconnect the water hoses (if equipped) and air hose from the BAC/IAC valve.
3. Disengage the wiring harness connector from the BAC/IAC valve.
4. Remove the mounting fasteners from the valve assembly.
5. Remove the BAC/IAC valve assembly, along with the mounting gasket and O-ring from the vehicle.

To install:

6. Remove any foreign material from the contact surface.
7. Install the BAC/IAC valve assembly, with a new mounting gasket and O-ring, onto the engine in correct position.
8. Install and tighten the mounting fasteners to 14–19 ft. lbs. (19–25 Nm) on 3.0L engines, and 22–30 inch lbs. (2.5–3.4 Nm) on 2.6L engines.
9. Plug in the valve wiring harness connector.
10. Connect the air hose and water hoses (if equipped) to the valve assembly.
11. Connect the negative battery cable.
12. If equipped with water hoses, verify that the valve assembly does not leak.

Engine Coolant Temperature (ECT) Sensor

OPERATION

The engine coolant temperature sensor resistance changes in response to engine coolant temperature. The sensor resistance decreases as the surrounding temperature increases. This provides a reference signal to the PCM, which indicates engine coolant temperature.

The ECT sensor is mounted on the lower intake manifold near the water outlet/thermostat housing, or on the water outlet/thermostat housing itself.

TESTING

♦ See Figures 28 and 29

1. Disengage the engine wiring harness connector from the ECT sensor.
2. Connect an ohmmeter between the ECT sensor terminals, and set the ohmmeter scale on 200,000 ohms.
3. With the engine cold and the ignition switch in the **OFF** position, measure and note the ECT sensor resistance. Attach the engine wiring harness connector to the sensor.
4. Start the engine and allow the engine to warm up to normal operating temperature.
5. Once the engine has reached normal operating temperature, turn the engine **OFF**.
6. Once again, detach the engine wiring harness connector from the ECT sensor.
7. Measure and note the ECT sensor resistance, then compare the cold and hot ECT sensor resistance measurements with the accompanying chart.
8. Replace the ECT sensor if the readings do not approximate those in the chart, otherwise reattach the engine wiring harness connector to the sensor.

REMOVAL & INSTALLATION

♦ See Figure 30

1. Partially drain the engine cooling system until the coolant level is below the ECT sensor mounting hole.

IAT/ECT SENSOR VOLTAGE AND RESISTANCE SPECIFICATIONS

Temperature		Engine Coolant/Intake Air Temperature Sensor Values	
°C	°F	Voltage (volts)	Resistance (K ohms)
120	248	0.27	1.18
110	230	0.35	1.55
100	212	0.46	2.07
90	194	0.60	2.80
80	176	0.78	3.84
70	158	1.02	5.37
60	140	1.33	7.70
50	122	1.70	10.97
40	104	2.13	16.15
30	86	2.60	24.27
20	68	3.07	37.30
10	50	3.51	58.75

Fig. 28 IAT and ECT sensor specifications chart—Navajo models

DRIVEABILITY AND EMISSION CONTROLS

Coolant	Resistance
−20°C (−4°F)	14.5 —17.8 kΩ
20°C (68°F)	2.2 — 2.7 kΩ
80°C (176°F)	0.28— 0.35 kΩ

Fig. 29 ECT sensor specifications chart—Pick-up and MPV models

Temperature	Resistance (kΩ)
25°C (77°F)	29.7—36.3
85°C (185°F)	3.3—3.7

Fig. 31 IAT sensor specifications chart—Pick-up and MPV models

Fig. 30 Most ECT sensors can be found on the intake manifold, near the water outlet housing

2. Disconnect the negative battery cable.
3. Detach the wiring harness connector from the ECT sensor.
4. Using an open-end wrench, remove the coolant temperature sensor from the intake manifold or thermostat housing.

To install:
5. Thread the sensor into the intake manifold, or thermostat housing, by hand, then tighten it securely.
6. Connect the negative battery cable.
7. Refill the engine cooling system.
8. Start the engine, check for coolant leaks and top off the cooling system.

Intake Air Temperature (IAT) Sensor

OPERATION

The Intake Air Temperature (IAT) sensor resistance changes in response to the intake air temperature. The sensor resistance decreases as the surrounding air temperature increases. This provides a signal to the PCM indicating the temperature of the incoming air charge.

Most engines mount the IAT sensor in the air cleaner-to-throttle body supply tube. However, some earlier engines have it mounted to the upper intake manifold.

TESTING

▶ See Figures 28 and 31

Turn the ignition switch **OFF**.
1. Disengage the wiring harness connector from the IAT sensor.
2. Using a Digital Volt-Ohmmeter (DVOM), measure the resistance between the two sensor terminals.
3. Compare the resistance reading with the accompanying chart. If the reading for a given temperature is approximately that shown in the table, the IAT sensor is okay.
4. Attach the wiring harness connector to the sensor.

REMOVAL & INSTALLATION

1. Disconnect the negative battery cable.
2. Disengage the wiring harness connector from the IAT sensor.
3. Remove the sensor from the air cleaner outlet tube.

To install:
4. Wipe down the IAT sensor mounting boss to clean the sensor area of all dirt and grime.
5. Install the sensor into the air cleaner outlet tube (B Series Pick-up and Navajo) or upper intake manifold assembly securely.
6. Attach the wiring harness connector to the IAT sensor.
7. Connect the negative battery cable.

Mass Air Flow (MAF) Sensor

OPERATION

▶ See Figure 32

The Mass Air Flow (MAF) sensor directly measures the amount of the air flowing into the engine. The sensor is mounted between the air cleaner assembly and the air cleaner outlet tube.

The sensor utilizes a hot wire sensing element to measure the amount of air entering the engine. The sensor does this by sending a signal, generated by the sensor when the incoming air cools the hot wire down, to the PCM. The signal is used by the PCM to calculate the injector pulse width, which controls the air/fuel ratio in the engine. The sensor and plastic housing are integral and must be replaced if found to be defective.

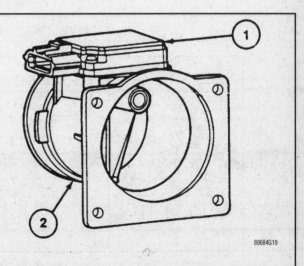

Fig. 32 The Mass Air Flow (MAF) sensor (1) is mounted to the air cleaner housing (2)

4-14 DRIVEABILITY AND EMISSION CONTROLS

TESTING

Navajo

▶ See Figure 33

1. With the engine running at idle, use a DVOM to verify there is at least 10.5 volts between terminals **A** and **B** of the MAF sensor connector. This indicates the power input to the sensor is correct. Then, measure the voltage between MAF sensor connector terminals **C** and **D**. If the reading is approximately 0.34–1.96 volts, the sensor is functioning properly.

Pick-up and MPV Models

MAF SENSOR

▶ See Figures 34 and 35

1. Remove the rubber boot from the Mass Air Flow (MAF) sensor wiring connector.
2. Using a voltmeter, measure the terminal voltages. Refer to the specifications chart.

Condition Terminal wire	Ignition switch ON	Engine running
B/W (Power supply)	B+	
G/O (Burn-off)	0V	
G/B (Airflow mass)	1.0—2.0V	1.9—5V
B/W (Ground)	0V	
B/O (Ground)	0V	

B+ : Battery positive voltage

Fig. 35 MAF sensor voltage specifications chart

3. If not as specified, inspect the wiring harness for an open or short circuit. If the wiring is okay, check the burn off operation as follows:
   a. Disconnect the negative battery cable for 20 seconds and reconnect it.
   b. Start the engine and allow it to warm up to operating temperature.
   c. Remove the rubber boot from the Mass Air Flow (MAF) sensor wiring connector.
   d. Run the engine for more than 5 seconds at approximately 2000 rpm in neutral.
   e. Turn the ignition switch OFF and check the voltage at the MAF sensor terminal wire (G/O) and terminal 2H of the powertrain control module. There should be 0 volts just after the ignition switch is turned OFF, then approximately 8 volts battery positive voltage momentarily 2–5 seconds after turning the ignition switch OFF.
4. If not as specified, replace the MAF sensor.

VOLUME AIR FLOW SENSOR

▶ See Figures 36 thru 41

➡ Testing of the volume air flow sensor also includes testing of the integral intake air temperature sensor.

1. Remove the volume air flow sensor from the vehicle.
2. Inspect the volume air flow sensor for cracks or any other signs of damage.
3. Check that the sensor shutter plate opens and closes smoothly.
4. Using an ohmmeter, open and close the sensor shutter plate while checking the resistance between the terminals.
5. Using a voltmeter with a temperature probe, an ohmmeter and a hairdryer for temperature variation, measure the resistance fluctuation of the integral intake air temperature sensor.
6. If the resistance does not measure within specifications, replace the volume air flow sensor.

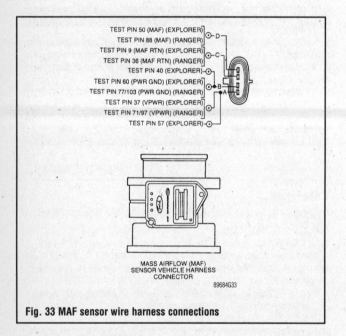

Fig. 33 MAF sensor wire harness connections

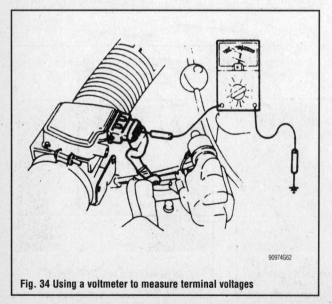

Fig. 34 Using a voltmeter to measure terminal voltages

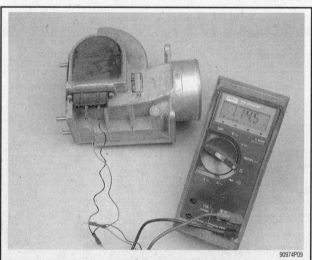

Fig. 36 Checking the resistance of the volume airflow sensor with the shutter closed

DRIVEABILITY AND EMISSION CONTROLS 4-15

Fig. 37 Checking the resistance of the volume airflow sensor while moving the shutter

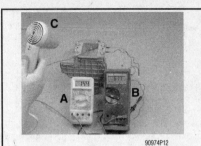

Fig. 38 Using a voltmeter with a temperature probe (A), an ohmmeter (B) and a hairdryer (C), measure the resistance fluctuation of the integral intake air temperature sensor

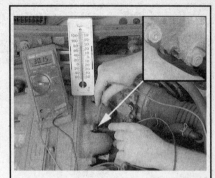

Fig. 39 Checking the resistance of the intake air temp sensor

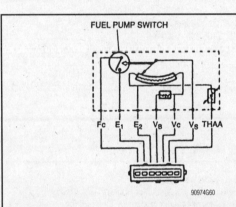

Fig. 40 Volume air flow sensor terminal identifications

Terminal	Resistance (Ω)	
	Closed throttle position	Wide open throttle
$E_2 \leftrightarrow V_S$	20—400	20—1,000
$E_2 \leftrightarrow V_C$	100—300	
$E_2 \leftrightarrow V_B$	200—400	
$E_2 \leftrightarrow THA_A$ (Intake air temperature sensor)	−20 °C (−4 °F) 20 °C (68 °F) 60 °C (140 °F)	13,600—18,400 2,210— 2,690 493— 667
$E_1 \leftrightarrow F_C$	∞	0

Fig. 41 Resistance specifications for the volume air flow sensor

REMOVAL & INSTALLATION

▶ See Figures 42, 43, 44 and 45

※※ **CAUTION**

The mass air flow sensor hot wire sensing element and housing are calibrated as a unit and must be serviced as a complete assembly. Do not damage the sensing element or possible failure of the sensor may occur.

1. Disconnect the negative battery cable.
2. Disengage the wiring harness connector from the MAF sensor, and if necessary, the IAT sensor.
3. Loosen the engine air cleaner outlet tube clamps, then remove the tube from the engine.
4. Remove the MAF sensor from the air cleaner assembly by disengaging the retaining clips.

To install:

➡ If necessary, be sure to align the arrow on the MAF sensor with the direction of the air flow.

5. Install the MAF sensor to the air cleaner assembly and ensure that the retaining clips are fully engaged.
6. Install the air cleaner outlet tube, then tighten the outlet tube clamps until snug.
7. Attach the engine wiring harness connectors to the IAT and MAF sensors.
8. Connect the negative battery cable.

Throttle Position (TP) Sensor

OPERATION

The Throttle Position (TP) sensor is a potentiometer that provides a signal to the PCM that is directly proportional to the throttle plate position. The TP sen-

Fig. 42 To remove the MAF sensor, first disconnect the wire harness plug . . .

Fig. 43 . . . then remove the air cleaner-to-throttle body air tube

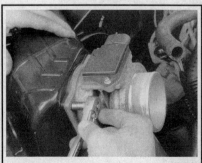

Fig. 44 Remove the four MAF sensor attaching screws . . .

4-16 DRIVEABILITY AND EMISSION CONTROLS

Fig. 45 . . . then remove the sensor from the air cleaner housing

sor is mounted on the side of the throttle body and is connected to the throttle plate shaft. The TP sensor monitors throttle plate movement and position, and transmits an appropriate electrical signal to the PCM. These signals are used by the PCM to adjust the air/fuel mixture, spark timing and EGR operation according to engine load at idle, part throttle, or full throttle. The TPS is not adjustable.

TESTING

Navajo Models

♦ See Figure 46

1. Disconnect the negative battery cable.
2. Disengage the wiring harness connector from the TP sensor.
3. Using a Digital Volt-Ohmmeter (DVOM) set on ohmmeter function, probe the terminals, which correspond to the Brown/White and the Gray/White connector wires, on the TP sensor. Do not measure the wiring harness connector terminals, rather the terminals on the sensor itself.
4. Slowly rotate the throttle shaft and monitor the ohmmeter for a continuous, steady change in resistance. Any sudden jumps, or irregularities (such as jumping back and forth) in resistance indicates a malfunctioning sensor.
5. Reconnect the negative battery cable.
6. Turn the DVOM to the voltmeter setting.

※※ WARNING

Ensuring the DVOM is on the voltmeter function is vitally important, because if you measure circuit resistance (ohmmeter function) with the battery cable connected, your DVOM will be destroyed.

7. Detach the wiring harness connector from the PCM (located behind the lower right-hand kick panel in the passengers' compartment), then install a break-out box between the wiring harness connector and the PCM connector.
8. Turn the ignition switch **ON** and using the DVOM on voltmeter function, measure the voltage between terminals 89 and 90 of the breakout box. The specification is 0.9 volts.
9. If the voltage is outside the standard value or if it does not change smoothly, inspect the circuit wiring and/or replace the TP sensor.

Pick-up and MPV Models

♦ See Figures 47 and 48

1. Verify that the throttle valve is at the closed throttle position.
2. Disengage the wiring harness connector from the TP sensor.
3. Using a Digital Volt-Ohmmeter (DVOM) set on ohmmeter function, probe terminals C and D on the TP sensor.
4. Insert a 0.020 inch (0.50mm) feeler gauge between the throttle adjusting screw and the throttle lever. Verify that there is no continuity.
5. If there is no continuity, adjust the TPS as follows:
   a. Loosen the TPS mounting screws.
   b. Insert a feeler gauge between the throttle adjusting screw and the throttle lever.
   c. There should be continuity when inserting a 0.006 inch (0.15mm) feeler gauge and there should be NO continuity when inserting a 0.020 inch (0.50mm) feeler gauge.
   d. Tighten the mounting screws.
6. Replace the TPS if not as specified.

REMOVAL & INSTALLATION

1. Disconnect the negative battery cable.
2. Disengage the wiring harness connector from the TP sensor.
3. Remove the two TP sensor mounting screws, then pull the TP sensor out of the throttle body housing.

To install:

4. Position the TP sensor against the throttle body housing, ensuring that the mounting screw holes are aligned. When positioning the TP sensor against the throttle body, slide the sensor straight onto the housing.
5. Install and tighten the sensor mounting screws until snug.
6. Attach the wiring harness connector to the sensor, then connect the negative battery cable.

Camshaft Position (CMP) Sensor

OPERATION

The CMP sensor provides the camshaft position information, called the CMP signal, which is used by the Powertrain Control Module (PCM) for fuel synchronization.

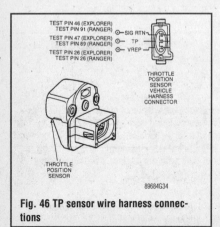

Fig. 46 TP sensor wire harness connections

Fig. 47 Checking the resistance of the ignition switch within the TPS housing using a feeler gauge (A) and an ohmmeter on terminals C and D (B)

Fig. 48 Place a feeler gauge between the throttle lever and throttle lever stop

DRIVEABILITY AND EMISSION CONTROLS 4-17

TESTING

Navajo

THREE WIRE SENSORS

♦ See Figure 49

1. With the ignition **OFF**, disconnect the CMP sensor. With the ignition **ON** and the engine **OFF**, measure the voltage between sensor harness connector **VPWR** and **PWR GND** terminals (refer to the accompanying illustration). If the reading is greater than 10.5 volts, the power circuit to the sensor is okay.

2. With the ignition **OFF**, install break-out box between the CMP sensor and the PCM. Using a Digital Volt-Ohmmeter (DVOM) set to the voltage function (scale set to monitor less than 5 volts), measure voltage between break-out box terminals **24** and **40** with the engine running at varying RPM. If the voltage reading varies more than 0.1 volt, the sensor is okay.

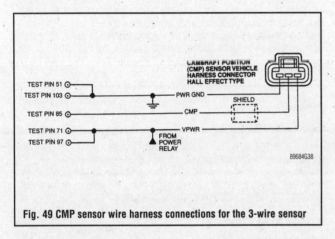

Fig. 49 CMP sensor wire harness connections for the 3-wire sensor

TWO WIRE SENSORS

♦ See Figure 50

1. With the ignition **OFF**, install a break-out box between the CMP sensor and PCM.
2. Using a Digital Volt-Ohmmeter (DVOM) set to the voltage function (scale set to monitor less than 5 volts), measure the voltage between break-out box terminals **24** and **46** with the engine running at varying RPM. If the voltage reading varies more than 0.1 volt AC, the sensor is okay.

Pick-up and MPV Models

♦ See Figures 51 and 52

1. Disconnect the negative battery cable.
2. Remove the distributor assembly.
3. Unplug the fuel injector connector.

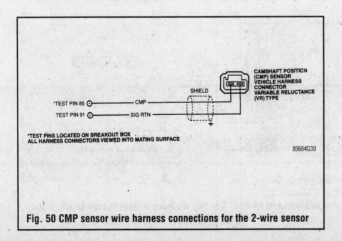

Fig. 50 CMP sensor wire harness connections for the 2-wire sensor

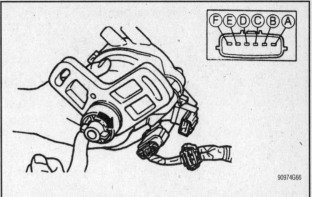

Fig. 51 Rotate the distributor drive and check the output signal

Signal	Terminal	Voltage
SGC	4G	Approx. 5 V (1 pulse/rev)

Fig. 52 Output signal specification

4. Connect ONLY the distributor 6-pin connector. Do NOT connect the 3-pin connector.
5. Turn the ignition switch to the ON position.
6. Turn the distributor drive by hand and check the output signal.
7. If not as specified, inspect the following:
• Harness continuity between PCM terminal 4G and distributor 6-pin connector terminal D
• Harness continuity between PCM terminal 4A and distributor 6-pin connector terminal A
• Harness continuity between main relay terminal D and distributor 6-pin connector terminal B
• Inspect and measure for battery positive voltage at distributor 6-pin connector terminal B
8. If there is incorrect terminal voltage or harness continuity, replace the distributor assembly.

REMOVAL & INSTALLATION

Pick-up and MPV Models

Refer to Section 2 in this manual for distributor removal and installation.

Navajo

♦ See Figures 53 and 54

➡If the camshaft position sensor housing does not contain a plastic locator cover tool, a special service tool such as T89P-12200-A, or equivalent, must be obtained prior to installation. Failure to follow this procedure may result in improper stator alignment. This will result in the fuel system being out of time with the engine, possibly causing engine damage.

1. Disconnect the negative battery cable.
2. Remove the ignition coil, radio capacitor and ignition coil bracket.
3. Disengage the wiring harness connector from the CMP sensor.

4-18 DRIVEABILITY AND EMISSION CONTROLS

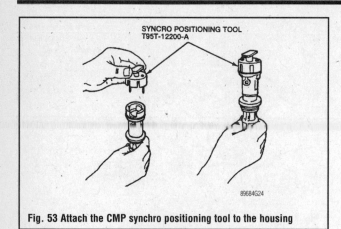

Fig. 53 Attach the CMP synchro positioning tool to the housing

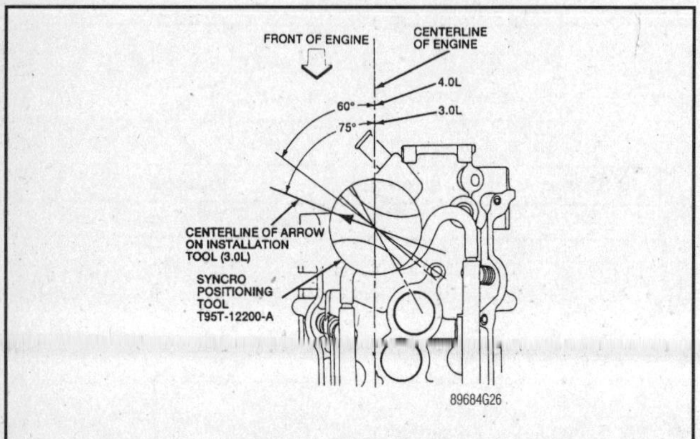

Fig. 54 CMP sensor positioning

→Prior to removing the camshaft position sensor, set the No. 1 cylinder to 10° After Top Dead Center (ATDC) of the compression stroke. Note the position of the sensor electrical connection. When installing the sensor, the connection must be in the exact same position.

4. Position the No. 1 cylinder at 10° ATDC, then matchmark the CMP sensor terminal connector position with the engine assembly.
5. Remove the camshaft position sensor retaining screws and sensor.
6. Remove the retaining bolt and hold-down clamp.

→The oil pump intermediate shaft should be removed with the camshaft sensor housing.

7. Remove the CMP sensor housing from the front engine cover.

To install:

8. If the plastic locator cover is not attached to the replacement camshaft position sensor, attach a synchro positioning tool, such as Ford Tool T89P-12200-A or equivalent. To do so, perform the following:
   a. Engage the sensor housing vane into the radial slot of the tool.
   b. Rotate the tool on the camshaft sensor housing until the tool boss engages the notch in the sensor housing.

→The cover tool should be square and in contact with the entire top surface of the camshaft position sensor housing.

9. Transfer the oil pump intermediate shaft from the old camshaft position sensor housing to the replacement sensor housing.
10. Install the camshaft sensor housing so that the drive gear engagement occurs when the arrow on the locator tool is pointed approximately 30° counter-clockwise (the sensor terminal connector should be aligned with its matchmarks) from the face of the cylinder block.
11. Install the hold-down clamp and bolt, then tighten the bolt to 15–22 ft. lbs. (20–30 Nm).
12. Remove the synchro positioning tool.

※※ CAUTION

If the sensor connector is positioned correctly, DO NOT reposition the connector by rotating the sensor housing. This will result in the fuel system being out of time with the engine. This could possibly cause engine damage. Remove the sensor housing and repeat the installation procedure beginning with step one.

13. Install the sensor and retaining screws, tighten the screws to 22–31 inch lbs. (2–4 Nm).
14. Attach the engine control sensor wiring connector to the sensor.
15. Install the ignition coil bracket, radio ignition capacitor and ignition coil.
16. Connect the negative battery cable.

Crankshaft Position (CKP) Sensor

OPERATION

The Crankshaft Position (CKP) sensor, located on the front cover (near the crankshaft pulley) is used to determine crankshaft position and crankshaft rpm. The CKP sensor is a reluctance sensor which senses the passing of teeth on a sensor ring because the teeth disrupt the magnetic field of the sensor. This disruption creates a voltage fluctuation, which is monitored by the PCM.

TESTING

Navajo

▶ See Figure 55

Using a DVOM set to the DC scale to monitor less than 5 volts, measure the voltage between the sensor Cylinder Identification (CID) terminal and ground by backprobing the sensor connector. If the connector cannot be backprobed, fabricate or purchase a test harness. The sensor is okay if the voltage reading varies more than 0.1 volt with the engine running at varying RPM.

Pick-up and MPV Models

▶ See Figures 56 and 57

1. Disconnect the negative battery cable.
2. Remove the distributor assembly.
3. Plug in the distributor connector.

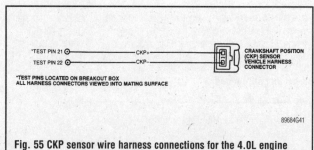

Fig. 55 CKP sensor wire harness connections for the 4.0L engine

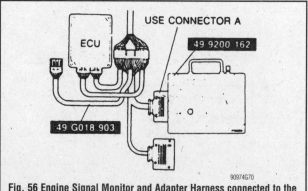

Fig. 56 Engine Signal Monitor and Adapter Harness connected to the engine control module

DRIVEABILITY AND EMISSION CONTROLS

Engine	Signal	Terminal	Voltage
JE	NE	2E	Approx. 5V (6 pulses)
	G1	2G	Approx. 5V (1 pulse)
	G2	2H	Approx. 5V (1 pulse)
G6	NE	2E	Approx. 5V (4 pulses)
	G	2G	Approx. 5V (1 pulse)

Fig. 57 Output signal specifications chart (JE—3.0L / G6—2.6L)

4. Unplug the fuel injector connector.
5. Turn the ignition switch to the **ON** position.
6. Connect the Engine Signal Monitor and Adapter Harness to the engine control module as illustrated.
7. Set the Engine Signal Monitor.
8. Turn the distributor drive by hand and measure the output voltage.
9. If not as specified, replace the distributor assembly.

REMOVAL & INSTALLATION

Navajo

▶ See Figures 58 and 59

1. Disconnect the negative battery cable.
2. Disengage the wiring harness connector from the CKP sensor.
3. Loosen the CKP sensor mounting stud/bolts, then separate the sensor form the engine front cover.

To install:

→When installing a new sensor on the 4.0L engine, position the sensor against the crankshaft damper. There are small rub tabs which wear off and allow the sensor to be perfectly spaced from the damper.

4. Position the sensor against the engine front cover, then install the mounting stud/bolts. Tighten them until snug.
5. Install the CKP sensor cover and retaining nuts; tighten the nuts until snug.
6. Reattach the wiring harness connector to the CKP sensor.
7. Connect the negative battery cable.

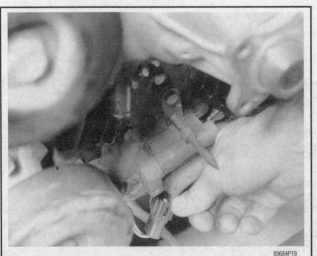

Fig. 58 To remove the CKP sensor, first disconnect the wire harness plug . . .

Fig. 59 . . . then unbolt the sensor and pull it away from the engine—crankshaft damper removed for clarity

Pick-up and MPV Models

Refer to Section 2 in this manual for distributor removal and installation.

Vehicle Speed Sensor (VSS)

OPERATION

The Vehicle Speed Sensor (VSS) is a magnetic pick-up that sends a signal to the Powertrain Control Module (PCM). The sensor measures the rotation of the transmission and the PCM determines the corresponding vehicle speed.

TESTING

▶ See Figure 60

1. Turn the ignition switch to the **OFF** position.
2. Disengage the wiring harness connector from the VSS.
3. Using a Digital Volt-Ohmmeter (DVOM), measure the resistance (DVOM ohmmeter function) between the sensor terminals. If the resistance is 190–250 ohms, the sensor is okay.

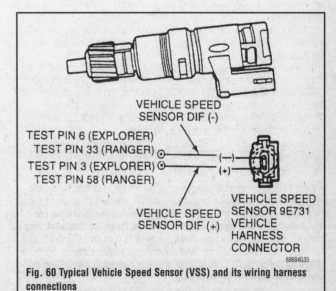

Fig. 60 Typical Vehicle Speed Sensor (VSS) and its wiring harness connections

4-20 DRIVEABILITY AND EMISSION CONTROLS

REMOVAL & INSTALLATION

The VSS is located half-way down the right-hand side of the transmission assembly.

1. Apply parking brake, block the rear wheels, then raise and safely support the front of the vehicle on jackstands.
2. From under the right-hand side of the vehicle, disengage the wiring harness connector from the VSS.
3. Loosen the VSS hold-down bolt, then pull the VSS out of the transmission housing.

To install:

4. If a new sensor is being installed, transfer the driven gear retainer and gear to the new sensor.
5. Ensure that the O-ring is properly seated in the VSS housing.
6. For ease of assembly, engage the wiring harness connector to the VSS, then insert the VSS into the transmission assembly.
7. Install and tighten the VSS hold-down bolt to 62–88 inch lbs. (7–10 Nm).
8. Lower the vehicle and remove the wheel blocks.

TROUBLE CODES

General information

Mazda Navajo, B Series Pick-up and MPV vehicles employ an electronic engine control system, to manage fuel, ignition and emissions on vehicle engines.

The Powertrain Control Module (PCM) is given responsibility for the operation of the emission control devices, ignition and advance and in some cases, automatic transmission functions. Because the control system oversees both the ignition timing and the fuel injector operation, a precise air/fuel ratio will be maintained under all operating conditions. The PCM is a microprocessor or small computer which receives electrical inputs from several sensors, switches and relays on and around the engine.

Based on combinations of these inputs, the PCM controls outputs to various devices concerned with engine operation and emissions. The engine control assembly relies on the signals to form a correct picture of current vehicle operation. If any of the input signals is incorrect, the PCM reacts to whatever picture is painted for it. For example, if the coolant temperature sensor is inaccurate and reads too low, the PCM may see a picture of the engine never warming up. Consequently, the engine settings will be maintained as if the engine were cold. Because so many inputs can affect one output, correct diagnostic procedures are essential on these systems.

One part of the PCM is devoted to monitoring both input and output functions within the system. This ability forms the core of the self-diagnostic system. If a problem is detected within a circuit, the controller will recognize the fault, assign it an identification code, and store the code in a memory section. Depending on the year and model, the fault code(s) may be represented by two or three digit numbers. The stored code(s) may be retrieved during diagnosis.

While the electronic engine control system is capable of recognizing many internal faults, certain faults will not be recognized. Because the computer system sees only electrical signals, it cannot sense or react to mechanical or vacuum faults affecting engine operation. Some of these faults may affect another component which will set a code. For example, the PCM monitors the output signal to the fuel injectors, but cannot detect a partially clogged injector. As long as the output driver responds correctly, the computer will read the system as functioning correctly. However, the improper flow of fuel may result in a lean mixture. This would, in turn, be detected by the oxygen sensor and noticed as a constantly lean signal by the PCM. Once the signal falls outside the pre-programmed limits, the engine control assembly would notice the fault and set an identification code.

Additionally, the electronic control system employs adaptive fuel logic. This process is used to compensate for normal wear and variability within the fuel system. Once the engine enters steady-state operation, the engine control assembly watches the oxygen sensor signal for a bias or tendency to run slightly rich or lean. If such a bias is detected, the adaptive logic corrects the fuel delivery to bring the air/fuel mixture towards a centered or 14.7:1 ratio. This compensating shift is stored in a non-volatile memory which is retained by battery power even with the ignition switched **OFF**. The correction factor is then available the next time the vehicle is operated.

➡If the battery cable(s) is disconnected for longer than 5 minutes, the adaptive fuel factor will be lost. After repair it will be necessary to drive the vehicle at least 10 miles to allow the processor to relearn the correct factors. The driving period should include steady-throttle open road driving if possible. During the drive, the vehicle may exhibit driveability symptoms not noticed before. These symptoms should clear as the PCM computes the correction factor. The PCM will also store a code that indicates loss of power to the controller.

MALFUNCTION INDICATOR LAMP (MIL)

The CHECK ENGINE or SERVICE ENGINE SOON dashboard warning lamp is referred to as the Malfunction Indicator Lamp (MIL). The lamp is connected to the engine control assembly and will alert the driver to certain malfunctions within the EEC system. When the lamp is lit, the PCM has detected a fault and stored an identity code in memory. The engine control system will usually enter either a failure or limited operation mode and driveability will be impaired.

The light will stay on as long as the fault causing it is present. Should the fault self-correct, the MIL will extinguish, but the stored code will remain in memory.

Under normal operating conditions, the MIL should light briefly when the ignition key is turned **ON**. As soon as the PCM receives a signal that the engine is cranking, the lamp will be extinguished. The dash warning lamp should remain out during the entire operating cycle.

Reading Codes

PICKUP AND MPV

Self Diagnosis Checker

A self-diagnosis checker (such as 49-H018-9A1) is used to retrieve code numbers of malfunctions which have happened and were memorized or are continuing. The malfunction is indicated by a code number.

If there is more than one malfunction, the code numbers will display on the self-diagnosis checker one by one in numerical order. In the case of malfunctions, 09, 19, 01, the code numbers are displayed in order of 01, 09 and then 13.

The memory of malfunctions is canceled by disconnecting the negative battery cable.

The ECU had a fail-safe mechanism for the main input sensors. If a malfunction occurs, the emission control unit will substitute values. This will slightly affect the driving performance, but the vehicle may still be driven.

The ECU continuously checks for malfunctions of the input devices during operation. But, the ECU checks for malfunctions of the output devices within three seconds after turning the ignition switch to the ON position and the test connector is grounded.

The malfunction indicator light (MIL) indicates a pattern the same as a self-diagnosis checker when the self-diagnosis check connector is grounded. When the self-diagnosis check connector is not grounded, the lamp illuminates steady while malfunction of the main input sensor occurs and goes out if the malfunction recovers. However the malfunction code is memorized in the emission control unit.

Testing Procedure

SELF-DIAGNOSIS CHECKER

1. Connect the Self-Diagnosis Checker to the check connector. The check connector is usually located above the right side wheel housing.
2. Set the selector switch on the tester to the A position.
3. Ground the green test connector using a suitable jumper wire.
4. Turn the ignition switch to the ON position. Check that number "88" flashes on the digital display and the buzzer sounds for three seconds after turning the ignition switch ON.
5. If the number "88" does not flash, check the main power relay, power supply circuit and check the connector wiring.

DRIVEABILITY AND EMISSION CONTROLS

6. If the number "88" flashes and the buzzer sounds continuously for more than 20 seconds, replace the ECU and perform steps 3 and 4 again.

7. Note the code numbers and check the causes. Repair as necessary. Be sure to recheck for code numbers after repairing.

MALFUNCTION LIGHT (MIL)

1. During the self-test a service code is reported by the malfunction indicator light. It will represent itself as a flash on the CHECK ENGINE or SERVICE ENGINE SOON light on the dash panel. A single digit number of 3 will be reported by 3 flashes. However, if a service code is represented by a 3-digit number such as 111, the code will appear on the MIL light as three flashes, then after a two second pause, the light will flash three times.

2. Start the engine and allow it to reach normal operating temperatures.

3. Turn the engine off. Connect a jumper wire from the STI to the SIG RTN at the self test connectors and then turn the ignition switch to the ON position. Service codes will be flashed on the MIL light.

NAVAJO

The EEC-IV system may be interrogated for stored codes using the Quick Test procedures. These tests will reveal faults immediately present during the test as well as any intermittent codes set within the previous 80 warm up cycles. If a code was set before a problem self-corrected (such as momentarily loose connector), the code will be erased if the problem does not reoccur within 80 warm-up cycles.

In 1991 the EEC-VI system check from 2 digit codes to 3 digit codes. These codes are obtained the same way and the system functions the same. All of the test procedures apply to both versions of the ECA.

The Quick Test procedure is divided into 2 sections, Key On Engine Off (KOEO) and Key On Engine Running (KOER). These 2 procedures must be performed correctly if the system is to run the internal self-checks and provide accurate fault codes. Codes will be output and displayed as numbers on the hand scan tool, i.e. 23. If the codes are being read through the dashboard warning lamp, the codes will be displayed as groups of flashes separated by pauses. Code 23 would be shown as two flashes, a pause and three more flashes. A longer pause would occur between codes. If the codes are being read on an analog voltmeter, the needle sweeps indicate the code digits in the same manner as the lamp flashes.

In all cases, the codes 11 or 111 are used to indicate PASS during testing. Note that the PASS code may appear, followed by other stored codes. These are codes from the continuous memory and may indicate intermittent faults, even though the system does not presently contain the fault. The PASS designation only indicates the system passes all internal tests at the moment.

✳✳ CAUTION

To prevent injury and/or property damage, always block the drive wheels, firmly apply the parking brake, place the transmission in P or N and turn all the electrical loads off before performing the Quick Test procedures.

Super Star II Tester

KEY ON ENGINE OFF (KOEO)

1. Specific instructions for the use of the tester are included with the tester.

2. Plug the tester connectors into the vehicle self-test connectors.

3. Set the switches on the tester to the appropriate settings for type of EEC system and for fast or slow mode. The same information is contained in fast or slow mode, but the fast mode is 100 times faster than the slow and is for use with the Super Star Tester.

4. Turn on the tester power and depress the test button. Turn ON the vehicle ignition key.

5. The tester will read the codes as they exist.

KEY ON ENGINE RUNNING (KOER)

1. Connect the tester as in the KOEO test and start and warm the engine to normal operating temperature.

2. Turn the engine off and depress the test button so it stays in the test position.

3. Restart the engine and the tester will read any existing codes. The tester will display an engine I.D. code, then on some vehicles a Dynamic Response code. The final display will be the service codes.

Analog Voltmeter

KEY ON ENGINE OFF (KOEO) AND CONTINUOUS CODES

1. Turn the vehicle ignition switch to OFF. Set the range of the analog voltmeter to 0-15 volts.

2. Connect the positive lead of an analog voltmeter to the positive of the battery.

3. Connect the negative lead of the voltmeter to the Self-Test Output pin of the Self-Test connector.

4. Connect a jumper wire between the Self-Test Input connector and the Signal Return pin of the Self-Test connector.

5. The needle will sweep 3 times for a code digit of 3, for example. If the code is 123, the needle will sweep once, pause for 2 seconds, sweep twice, pause for 2 seconds, then sweep three times. There will be a 4 second pause between codes. The method of readout of 2 digit codes is similar to the three digit code readout, but only 2 sets of sweeps will occur.

6. The continuous memory codes will be displayed first, then there will be a 6 second pause, a single sweep, anther 6 second pause then the KOEO codes will be displayed.

Malfunction Indicator or Check Engine Light

KEY ON ENGINE OFF (KOEO) AND CONTINUOUS CODES

1. Connect a jumper wire between the Self-Test Input (STI) connector and the Signal Return pin of the Self-Test connector.

2. This will output the codes to the MIL or Check Engine light on the instrument panel.

3. The light will flash 3 times for a code digit of 3, for example. If the code is 123, the light will flash once, pause for 2 seconds, flash twice, pause for 2 seconds, then flash three times. There will be a 4 second pause between codes. The method of readout for 2 digit code is similar to the three digit code readout, but only 2 sets of sweeps will occur.

4. The continuous memory codes will be displayed first, then there will be a 6 second pause, a single flash, another 6 second pause, then the KOEO codes will be displayed.

Other Test Modes

CONTINUOUS MONITOR OR WIGGLE TEST MODE

Once entered, this mode allows the technician to attempt to recreate the intermittent faults by wiggling or tapping components, wiring or connectors. The test may by performed during either KOEO or KOER procedures. The test requires the use of either an analog voltmeter or a hard scan tool.

To enter the continuous monitor mode during KOEO testing, turn the ignition switch ON. Activate the test, wait 10 seconds, then deactivate and reactivate the test; the system will enter the continuous monitor mode. Tap, move, or wiggle the harness, component, or connector suspected of causing the problem; if a fault is detected, the code will store in the memory. When the fault occurs, the dash warning lamp will illuminate, the STAR tester will light a red indicator (and possible beep) and the analog monitor needle will sweep once.

To enter this mode in the KOER test:

1. Start the engine and run it at 2000 rpm for two minutes. This action warms up the oxygen sensor.

2. Turn the ignition switch to OFF for 10 seconds.

3. Start the engine.

4. Activate the test, wait 10 seconds, then deactivate and reactivate the test; the system will enter the continuous monitor mode.

5. Tap, move, or wiggle the harness, component, or connector suspected of causing the problem; if a fault is detected, the code will store in the memory.

6. When the fault occurs, the dash warning lamp will illuminate, the STAR tester will light the red indicator (and possibly beep) and the analog meter needle will sweep once.

OUTPUT STATE CHECK

This testing mode allows the operator to energize the de-energize most of the outputs controlled by the EEC-IV system. Many of the outputs may be checked at the component by listening for a click or feeling the item move or engage by a hand placed on the case. To enter this check:

1. Enter the KOEO test mode.

2. When all codes have been transmitted, depress the accelerator all the way to the floor and release it.

3. The output actuators are now all ON. Depressing the throttle pedal to the floor again switches all the actuator outputs OFF.

4-22 DRIVEABILITY AND EMISSION CONTROLS

4. This test may be performed as often as necessary, switching between ON and OFF by depressing the throttle.
5. Exit the test by turning the ignition switch OFF, disconnecting the jumper at the diagnostic connector or releasing the test button on the scan tool.

Clearing Codes

Pickup and MPV

1. Cancel the memory of malfunctions by disconnecting the negative battery cable and depressing the brake pedal for at least twenty seconds. Reconnect the negative battery cable.
2. Connect the Self-Diagnosis Checker at the check connector. Ground the green one pin test connector using a suitable jumper wire.
3. Turn the ignition switch to the ON position, but do not start the engine for at least six seconds.
4. Start the engine and allow it to reach normal operating temperatures. Run the engine at 2000 rpm for two minutes. Check that no codes are displayed.

Navajo

CONTINUOUS CODES

1. Run the Self-Test for the output system being used: Super Star II, analog voltmeter, or dash light.
2. When the service codes start to be displayed the codes will be cleared when the Test button is unlatched on the Super Star II tester. For the other retrieval methods the codes will be cleared when the jumper is removed between the Self-Test Input connector and the Signal Return pin of the Self-Test connector.
3. Disconnect the Super Star II tester or voltmeter.

KEEP MEMORY ALIVE

The Keep Alive Memory will be erased when the negative battery cable is disconnected. Keep the cable off for a minimum of 5 minutes. When the cable is connected and the vehicle driven, it may exhibit some drivability problems for the first 10 miles (16 km) or so. This is due to the relearning process where the ECA memorizes values needed for optimum drivability.

Code Description

PICKUP AND MPV

Except Feedback Carbureted Models

- Code 01—Ignition Pulse
- Code 02—Crank Angle Sensor
- Code 06—Vehicle Speed Sensor
- Code 08—Air Flow Meter
- Code 09—Water Temperature Sensor
- Code 10—Intake Air Temperature at Airflow Meter
- Code 11—Intake Air Temperature at Dynamic Chamber
- Code 12—Throttle Sensor
- Code 14—ECA Atmospheric Pressure Sensor
- Code 15—Oxygen Sensor or Circuit
- Code 16—EGR Position Sensor
- Code 17—Oxygen Sensor Feedback System
- Code 23—Right Side Oxygen Sensor
- Code 24—Right Side Oxygen Sensor Feedback System
- Code 25—Pressure Regulator Solenoid Valve
- Code 26—Purge Control Solenoid Valve
- Code 28—EGR Vacuum Solenoid Valve
- Code 29—EGR Vent Solenoid Valve
- Code 30—Cold Start Injector Relay
- Code 34—Idle Speed Control Valve
- Code 55—Pulse Generator

Feedback Carbureted Models

- Code 01—Ignition Pulse
- Code 09—Water Temperature Sensor
- Code 13—Vacuum Sensor
- Code 14—Atmospheric Pressure Sensor
- Code 15—Oxygen Sensor Open or Short
- Code 17—Oxygen Sensor Feedback System
- Code 18—Air/Fuel Solenoid Valve
- Code 22—Slow Fuel Cut Solenoid Valve
- Code 23—Coasting Richer Solenoid Valve
- Code 26—Purge Solenoid Valve
- Code 28—Duty Solenoid Vacuum Valve
- Code 29—Duty Solenoid Vent Valve
- Code 30—Air Control Valve Solenoid
- Code 34—Idle-Up Solenoid Valve for AC
- Code 35—Idle-Up Solenoid Valve for AT
- Code 45—Vacuum Solenoid Valve

NAVAJO

2 Digit Codes

- Code 11—System pass
- Code 12—Cannot control rpm during high rpm check
- Code 13—Cannot control rpm during low rpm check
- Code 14—PIP circuit failure
- Code 15—ECA keep alive only memory (KAM) test failed
- Code 15—ECA read only memory (ROM) test failed
- Code 16—IDM signal not received
- Code 18—Loss of IDM input or SAW circuit open
- Code 19—Failure in ECA internal voltage
- Code 21—ECT out of range
- Code 22—Barometric Pressure (BP) sensor
- Code 23—TP sensor out of self-test range
- Code 24—ACT sensor out of self-test range
- Code 26—MAF out of self-test range
- Code 29—Insufficient input from VSS
- Code 41—Fuel system adaptive limits, no HEGO switch
- Code 41—Lack of HEGO switches, indicates lean
- Code 42—Lack of HEGO switches, indicates rich
- Code 45—Coil 1, 2, or 3 failure
- Code 51—ECT indicated—40°F (4.4°C) open circuit
- Code 53—TP sensor circuit above maximum voltage
- Code 54—ACT indicated—40°F (4.4°C) open circuit
- Code 56—MAF circuit above maximum voltage
- Code 61—ECT indicated 254°F (123°C) grounded circuit
- Code 63—TP sensor voltage below minimum voltage
- Code 64—ACT indicated 254°F (123°C) grounded circuit
- Code 66—MAF below minimum voltage
- Code 67—Clutch switch circuit failure
- Code 67—Vehicle not in Park during KOEO
- Code 72—Insufficient MAF change during test
- Code 73—Insufficient TP change dynamic response test
- Code 74—BOO circuit failure/not actuated during self-test
- Code 77—Operator error during test
- Code 79—A/C ON/ Defroster ON/ during self-test
- Code 86—Shift solenoid circuit failure
- Code 87—Primary fuel pump circuit failure
- Code 89—Clutch Converter Override (CCO) circuit failure
- Code 95—Fuel pump circuit open—ECA to motor ground
- Code 96—Fuel pump circuit open—ECA to battery
- Code 98—Hard fault present
- Code not listed

3 Digit Codes

- Code 111—System pass
- Code 112—ACT indicated 254°F (123°C) grounded circuit
- Code 113—ACT indicated—40°F (4.4°C) open circuit
- Code 114—ACT sensor out of self-test range
- Code 116—ECT out of range
- Code 117—ECT indicated 254°F (123°C) grounded circuit
- Code 118—ECT indicated—40°F (4.4°C) open circuit
- Code 121—TP sensor out of self-test range
- Code 122—TP sensor circuit below minimum voltage

DRIVEABILITY AND EMISSION CONTROLS 4-23

- Code 123—TP sensor circuit above maximum voltage
- Code 126—Barometric Pressure (BP) sensor
- Code 128—Barometric Pressure (BP) sensor
- Code 129—Insufficient MAF change during test
- Code 136—Fuel control
- Code 137—Fuel control
- Code 139—Fuel control
- Code 144—Fuel control
- Code 157—MAF below minimum voltage
- Code 158—MAF circuit above maximum voltage
- Code 159—MAF out of self-test range
- Code 167—Insufficient TP change dynamic response test
- Code 171—Fuel system at adaptive limits, no HEGO switch
- Code 172—Lack of HEGO switches, indicates lean
- Code 173—Lack of HEGO switches, indicates rich
- Code 174—Fuel control
- Code 175—Fuel control
- Code 176—Fuel control
- Code 177—Fuel control
- Code 178—Fuel control
- Code 179—Lean adaptive limit at part throttle, system rich
- Code 181—Rich adaptive limit at part throttle, system lean
- Code 182—Fuel control
- Code 183—Fuel control
- Code 188—Fuel control
- Code 189—Fuel control
- Code 191—Fuel control
- Code 192—Fuel control
- Code 194—Fuel control
- Code 195—Fuel control
- Code 211—PIP circuit failure
- Code 212—Loss of IDM input or SAW circuit grounded
- Code 213—SAW circuit open
- Code 215—Ignition system
- Code 216—Ignition system
- Code 217—Ignition system
- Code 218—Ignition system
- Code 222—Ignition system
- Code 223—IDM signal not received
- Code 224—IDM signal not received
- Code 226—IDM signal
- Code 232—Coil 1, 2, or 3 failure
- Code 238—Ignition system
- Code 338—Temperature sensor
- Code 339—Temperature sensor
- Code 341—Octane adjust circuit open
- Code 411—Cannot control rpm during low rpm check
- Code 412—Cannot control rpm during high rpm check
- Code 452—Insufficient input from VSS
- Code 511—ECA read only memory (ROM) test failed
- Code 512—ECA keep alive only memory (KAM) test failed
- Code 513—Failure in ECA internal voltage
- Code 522—Vehicle not in P or N during KOEO
- Code 525—Vehicle not in P or N durung KOEO
- Code 528—Clutch switch circuit failure
- Code 529—Check Engine light
- Code 533—Check Engine light
- Code 536—BOO circuit failure/not actuated during self-test
- Code 538—Operator error during test
- Code 539—A/C ON/ Defroster ON/ during self-test
- Code 542—Fuel pump circuit open—ECA to motor guard
- Code 542—Fuel pump circuit open—ECA to battery
- Code 556—Primary fuel pump circuit failure
- Code 565—Canister purge (CANP) circuit failure
- Code 566—Shift solenoid circuit failure
- Code 569—Canister purge (CANP) circuit failure
- Code 629—Clutch Converter Override (CCO) circuit failure
- Code 998—Hard fault present
- Code not listed
- 600 series codes are transmission codes

COMPONENT LOCATIONS

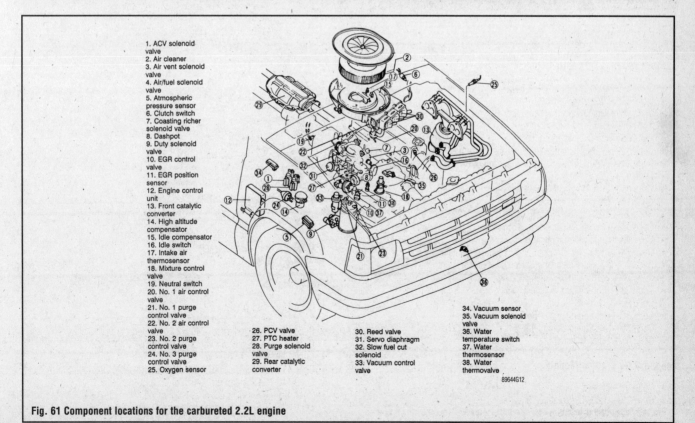

1. ACV solenoid valve
2. Air cleaner
3. Air vent solenoid valve
4. Air/fuel solenoid valve
5. Atmospheric pressure sensor
6. Clutch switch
7. Coasting richer solenoid valve
8. Dashpot
9. Duty solenoid valve
10. EGR control valve
11. EGR position sensor
12. Engine control unit
13. Front catalytic converter
14. High altitude compensator
15. Idle compensator
16. Idle switch
17. Intake air thermosensor
18. Mixture control valve
19. Neutral switch
20. No. 1 air control valve
21. No. 1 purge control valve
22. No. 2 air control valve
23. No. 2 purge control valve
24. No. 3 purge control valve
25. Oxygen sensor
26. PCV valve
27. PTC heater
28. Purge solenoid valve
29. Rear catalytic converter
30. Reed valve
31. Servo diaphragm
32. Slow fuel cut solenoid
33. Vacuum control valve
34. Vacuum sensor
35. Vacuum solenoid valve
36. Water temperature switch
37. Water thermosensor
38. Water thermovalve

Fig. 61 Component locations for the carbureted 2.2L engine

4-24 DRIVEABILITY AND EMISSION CONTROLS

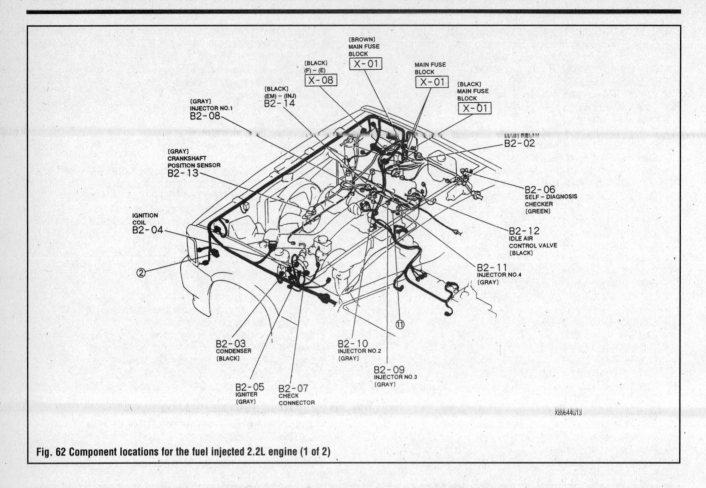

Fig. 62 Component locations for the fuel injected 2.2L engine (1 of 2)

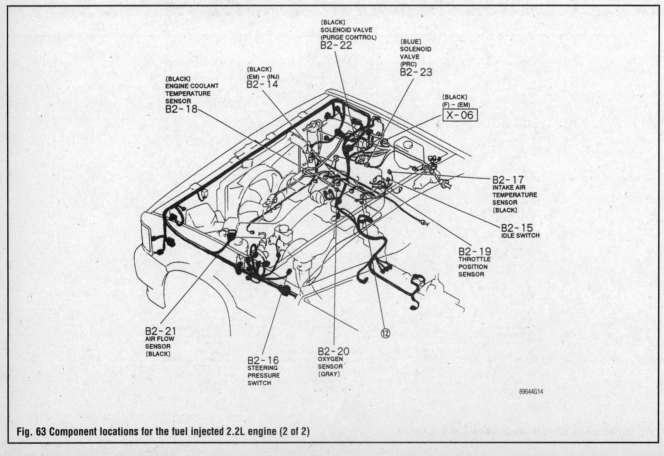

Fig. 63 Component locations for the fuel injected 2.2L engine (2 of 2)

DRIVEABILITY AND EMISSION CONTROLS 4-25

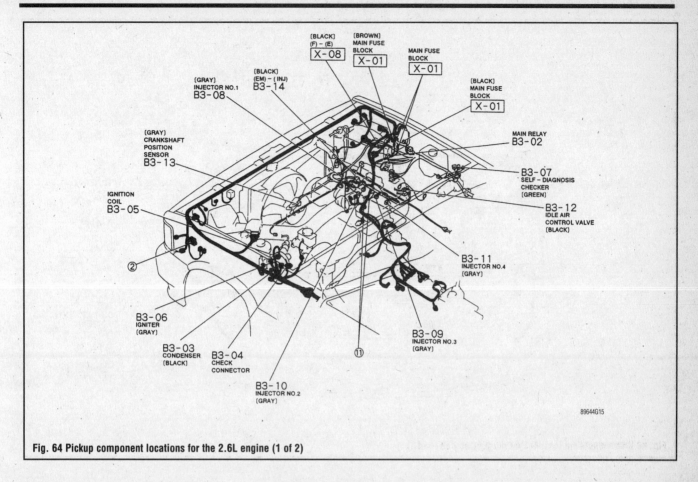

Fig. 64 Pickup component locations for the 2.6L engine (1 of 2)

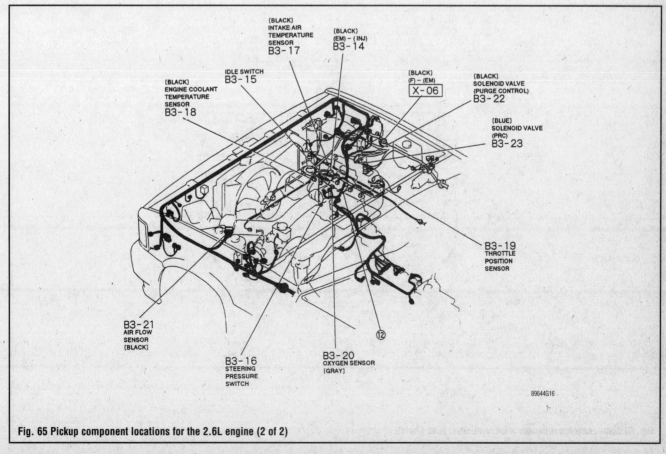

Fig. 65 Pickup component locations for the 2.6L engine (2 of 2)

4-26 DRIVEABILITY AND EMISSION CONTROLS

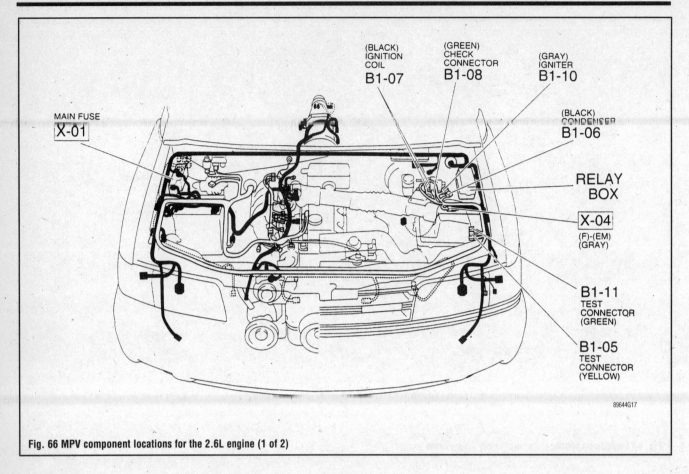

Fig. 66 MPV component locations for the 2.6L engine (1 of 2)

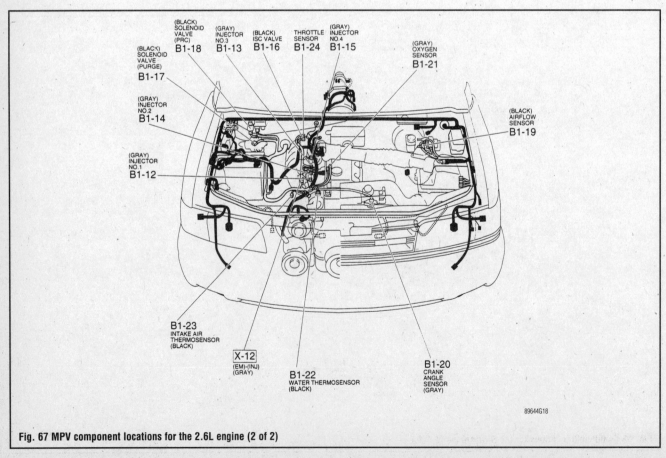

Fig. 67 MPV component locations for the 2.6L engine (2 of 2)

DRIVEABILITY AND EMISSION CONTROLS

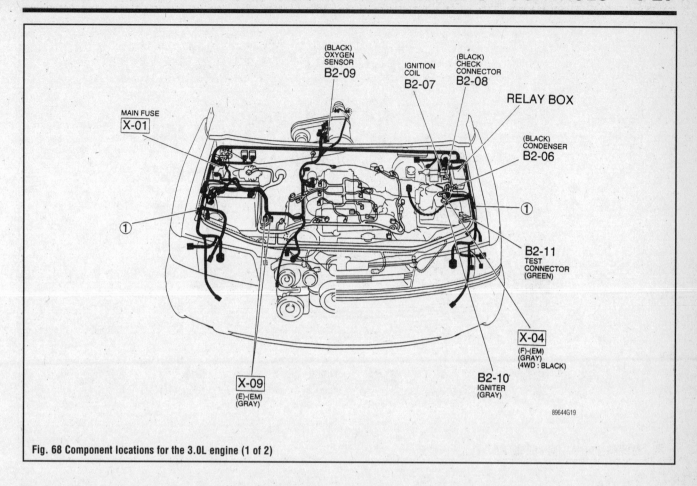

Fig. 68 Component locations for the 3.0L engine (1 of 2)

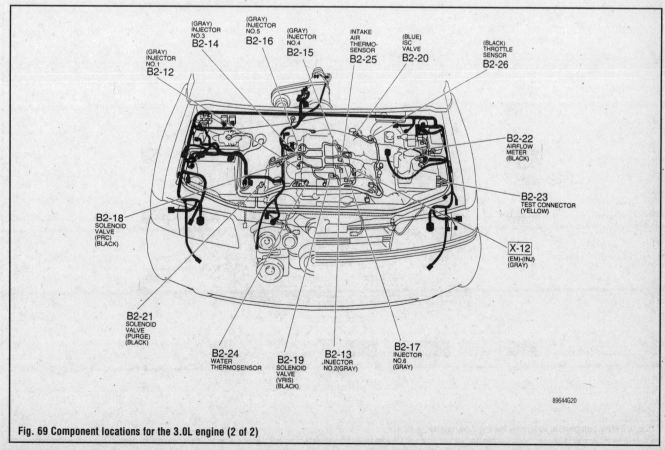

Fig. 69 Component locations for the 3.0L engine (2 of 2)

4-28 DRIVEABILITY AND EMISSION CONTROLS

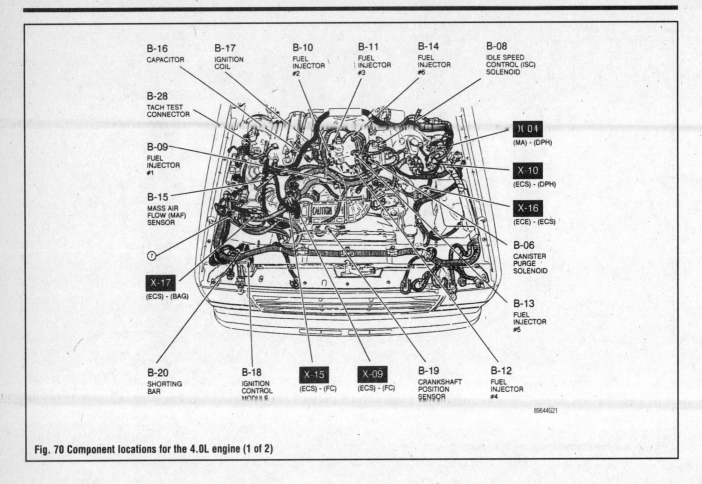

Fig. 70 Component locations for the 4.0L engine (1 of 2)

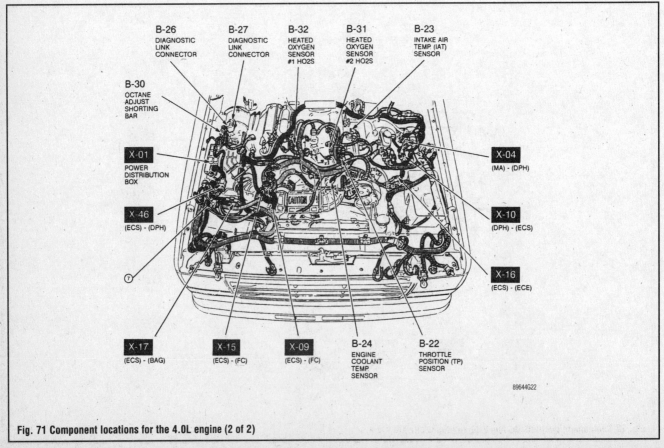

Fig. 71 Component locations for the 4.0L engine (2 of 2)

DRIVEABILITY AND EMISSION CONTROLS   4-29

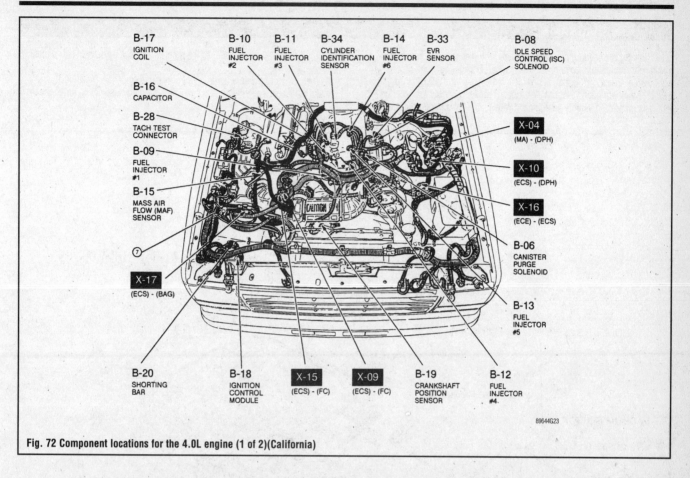

Fig. 72 Component locations for the 4.0L engine (1 of 2)(California)

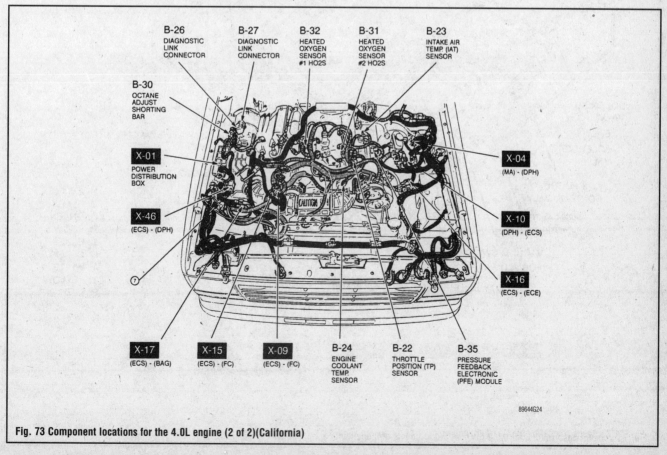

Fig. 73 Component locations for the 4.0L engine (2 of 2)(California)

4-30 DRIVEABILITY AND EMISSION CONTROLS

VACUUM DIAGRAMS

Following are vacuum diagrams for most of the engine and emissions package combinations covered by this manual. Because vacuum circuits will vary based on various engine and vehicle options, always refer first to the vehicle emission control information label, if present. Should the label be missing, or should vehicle be equipped with a different engine from the vehicle's original equipment, refer to the diagrams below for the same or similar configuration.

If you wish to obtain a replacement emissions label, most manufacturers make the labels available for purchase. The labels can usually be ordered from a local dealer.

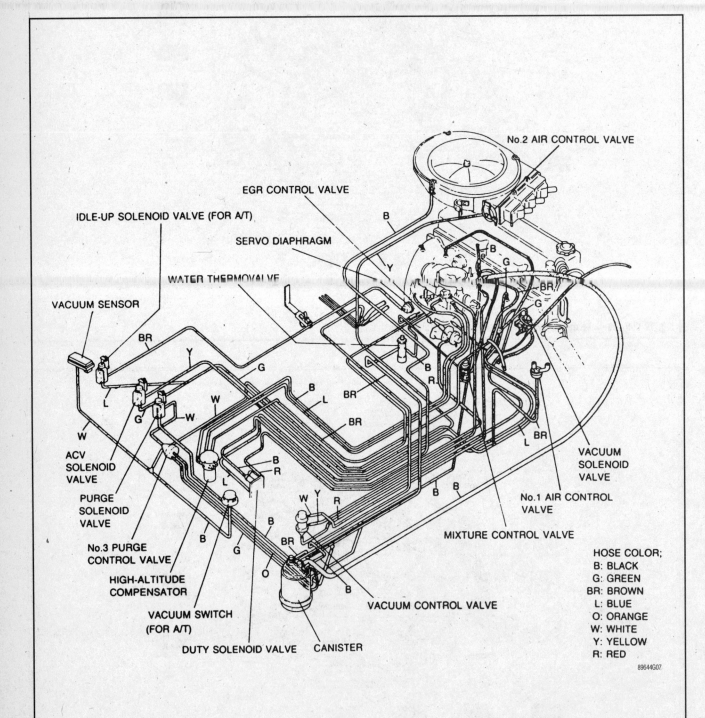

Fig. 74 Vacuum hose routing for the B Series Pickup models equipped with 2.6 L engine

DRIVEABILITY AND EMISSION CONTROLS 4-31

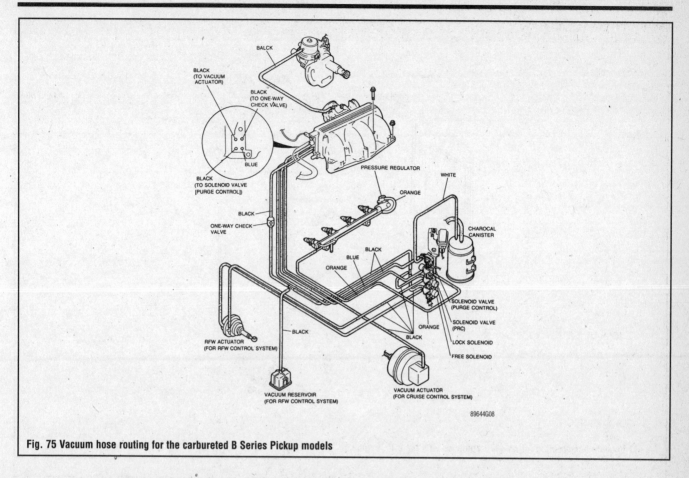

Fig. 75 Vacuum hose routing for the carbureted B Series Pickup models

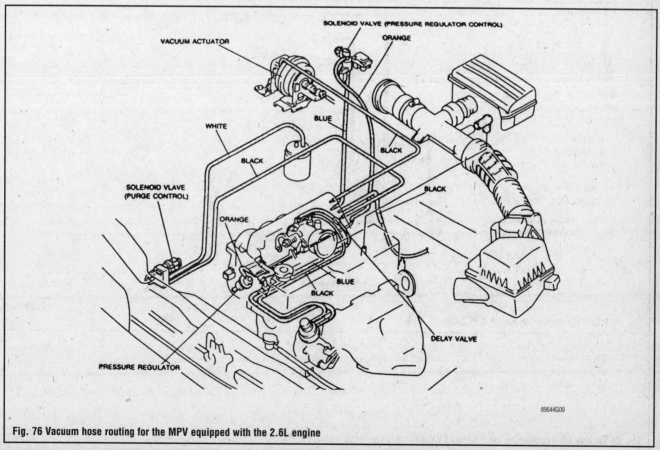

Fig. 76 Vacuum hose routing for the MPV equipped with the 2.6L engine

4-32 DRIVEABILITY AND EMISSION CONTROLS

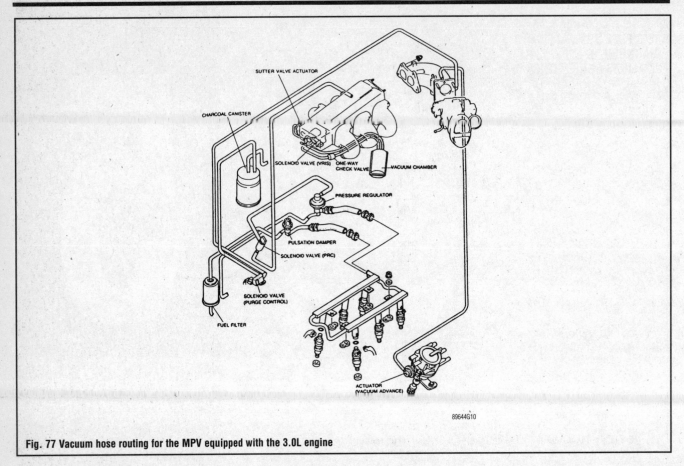

Fig. 77 Vacuum hose routing for the MPV equipped with the 3.0L engine

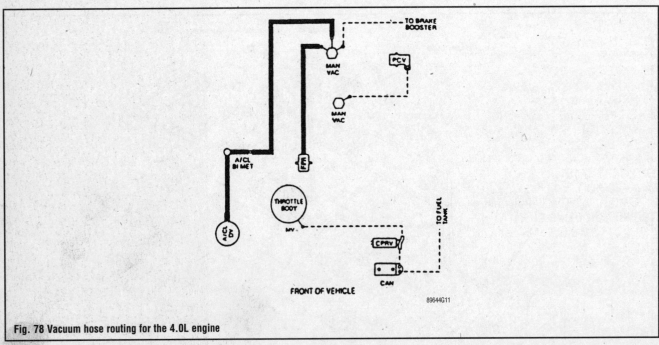

Fig. 78 Vacuum hose routing for the 4.0L engine

**BASIC FUEL SYSTEM
   DIAGNOSIS 5-2**
FUEL LINES AND FITTINGS 5-2
GENERAL INFORMATION 5-2
HAIRPIN CLIP FITTING 5-2
   REMOVAL & INSTALLATION 5-2
DUCKBILL CLIP FITTING 5-2
   REMOVAL & INSTALLATION 5-2
SPRING LOCK COUPLING 5-2
   REMOVAL & INSTALLATION 5-2
CARBURETED FUEL SYSTEM 5-3
MECHANICAL FUEL PUMP 5-3
   TESTING 5-3
   REMOVAL & INSTALLATION 5-3
ELECTRIC FUEL PUMP 5-3
   TESTING 5-3
   REMOVAL & INSTALLATION 5-3
CARBURETOR 5-3
   REMOVAL & INSTALLATION 5-3
   OVERHAUL 5-3
   ADJUSTMENTS 5-3
FUEL INJECTION SYSTEM 5-7
GENERAL INFORMATION 5-7
   FUEL SYSTEM SERVICE
      PRECAUTIONS 5-7
RELIEVING FUEL SYSTEM
   PRESSURE 5-7
   PICKUP & MPV 5-7
   NAVAJO 5-7
FUEL PUMP 5-8
   REMOVAL & INSTALLATION 5-8
   TESTING 5-9
THROTTLE BODY 5-10
   REMOVAL & INSTALLATION 5-10
FUEL INJECTORS 5-11
   REMOVAL & INSTALLATION 5-11
   TESTING 5-13
FUEL CHARGING ASSEMBLY 5-13
   REMOVAL & INSTALLATION 5-13
FUEL PRESSURE REGULATOR 5-14
   REMOVAL & INSTALLATION 5-14
FUEL TANK 5-15
TANK ASSEMBLY 5-15
   REMOVAL & INSTALLATION 5-15
TROUBLESHOOTING CHART
   FUEL SYSTEM PROBLEMS 5-16

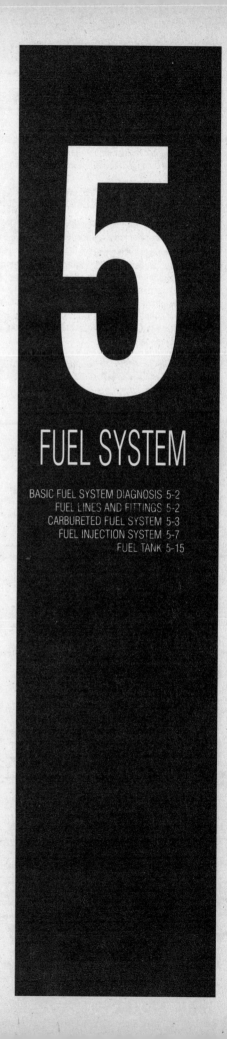

5
FUEL SYSTEM

BASIC FUEL SYSTEM DIAGNOSIS 5-2
FUEL LINES AND FITTINGS 5-2
CARBURETED FUEL SYSTEM 5-3
FUEL INJECTION SYSTEM 5-7
FUEL TANK 5-15

5-2 FUEL SYSTEM

BASIC FUEL SYSTEM DIAGNOSIS

When there is a problem starting or driving a vehicle, two of the most important checks involve the ignition and the fuel systems. The questions most mechanics attempt to answer first, "is there spark?" and "is there fuel?" will often lead to solving most basic problems. For ignition system diagnosis and testing, please refer to the information on engine electrical components and ignition systems found earlier in this manual. If the ignition system checks out (there is spark), then you must determine if the fuel system is operating properly (is there fuel?).

FUEL LINES AND FITTINGS

General Information

➥Quick-connect (push type) fuel line fittings must be disconnected using proper procedure or the fitting may be damaged. There are two types of retainers used on the push connect fittings. Line sizes of ⅜ and ⁵⁄₁₆ inch diameter use a hairpin clip retainer. The ¼ inch diameter line connectors use a duck-bill clip retainer. In addition, some engines use spring-lock connections, secured by a garter spring, which require Special Spring Lock Coupling Tools 49 UN01 051 and 49 UN01 052 (or equivalent) for removal.

Hairpin Clip Fitting

REMOVAL & INSTALLATION

♦ See Figures 1 and 2

1. Clean all dirt and grease from the fitting. Spread the two clip legs about ⅛ inch (3mm) each to disengage from the fitting and pull the clip outward from the fitting. Use finger pressure only; do not use any tools.
2. Grasp the fitting and hose assembly and pull away from the steel line. Twist the fitting and hose assembly slightly while pulling, if the assembly sticks.
3. Inspect the hairpin clip for damage, replacing the clip if necessary. Reinstall the clip in position on the fitting.
4. Inspect the fitting and inside of the connector to ensure freedom from dirt or obstruction. Install the fitting into the connector and push together. A click will be heard when the hairpin snaps into the proper connection. Pull on the line to insure full engagement.

Duckbill Clip Fitting

REMOVAL & INSTALLATION

♦ See Figures 2 and 3

1. Special tools are available from Mazda and other manufacturers for removing retaining clips. Use Mazda Tools 49-UN01-053 and 49-UN01-054 or equivalent. If the tool is not on hand, go onto step 2. Align the slot on the push connector disconnect tool with either tab on the retaining clip. Pull the line from the connector.
2. If the special clip tool is not available, use a pair of narrow 6-inch slip-jaw pliers with a jaw width of 0.2 inch (5mm) or less. Align the jaws of the pliers with the openings of the fitting case and compress the part of the retaining clip that engages the case. Compressing the retaining clip will release the fitting, which may be pulled from the connector. Both sides of the clip must be compressed at the same time to disengage.
3. Inspect the retaining clip, fitting end and connector. Replace the clip if any damage is apparent.
4. Push the line into the steel connector until a click is heard, indicating the clip is in place. Pull on the line to check engagement.

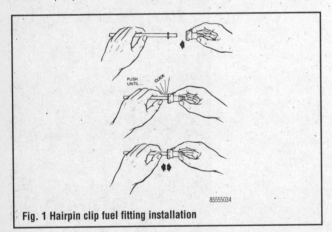

Fig. 1 Hairpin clip fuel fitting installation

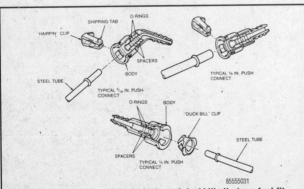

Fig. 2 Exploded views of the hairpin and duckbill clip type fuel fittings

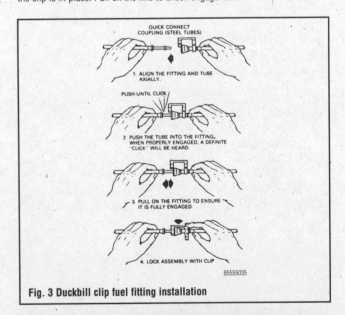

Fig. 3 Duckbill clip fuel fitting installation

Spring Lock Coupling

REMOVAL & INSTALLATION

♦ See Figures 4 and 5

The spring lock coupling is held together by a garter spring inside a circular cage. When the coupling is connected together, the flared end of the female fitting

FUEL SYSTEM 5-3

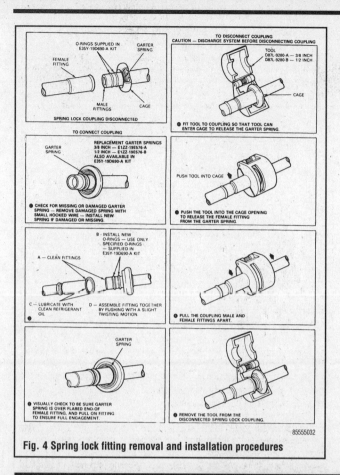

Fig. 4 Spring lock fitting removal and installation procedures

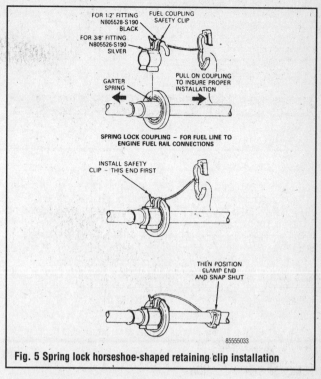

Fig. 5 Spring lock horseshoe-shaped retaining clip installation

slips behind the garter spring inside the cage of the male fitting. The garter spring and cage then prevent the flared end of the female fitting from pulling out of the cage. As an additional locking feature, most vehicles have a horseshoe-shaped retaining clip that improves the retaining reliability of the spring lock coupling.

CARBURETED FUEL SYSTEM

Mechanical Fuel Pump

1987-88 2.2L engine with manual transmission use a mechanically driven fuel pump, mounted on the right side of the cylinder head, driven by the camshaft.

TESTING

♦ See Figure 6

1. Disconnect the negative battery cable.
2. Disconnect the fuel line from the carburetor.
3. Connect a fuel pressure gauge to the fuel line.
4. Disconnect the fuel return hose from the fuel pump and plug the fuel pump return outlet.
5. Connect the negative battery cable and start the engine.
6. Check the fuel pressure while the engine is idling. The fuel pressure should be 3.7–4.7 psi.
7. Replace the pump if the pressure is not as specified.
8. Shut off the engine and disconnect the negative battery cable. Remove the fuel pressure gauge and reconnect the fuel lines.
9. Connect the negative battery cable.

REMOVAL & INSTALLATION

♦ See Figure 7

1. Disconnect the negative battery cable.
2. Disconnect and plug the inlet, outlet and return hoses at the fuel pump.
3. Remove the fuel pump mounting bolts and remove the fuel pump, insulator and gaskets.

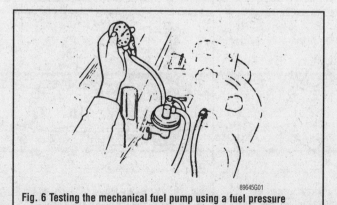

Fig. 6 Testing the mechanical fuel pump using a fuel pressure gauge

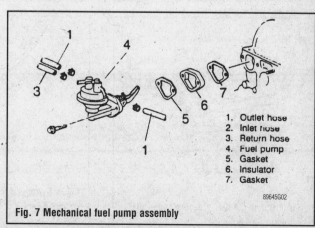

1. Outlet hose
2. Inlet hose
3. Return hose
4. Fuel pump
5. Gasket
6. Insulator
7. Gasket

Fig. 7 Mechanical fuel pump assembly

5-4 FUEL SYSTEM

To install:
4. Clean all gasket mating surfaces.
5. Install new gaskets and the insulator and fuel pump. Install the mounting bolts and tighten to 14–19 ft. lbs. (19–25 Nm).
6. Unplug and connect the fuel lines.
7. Connect the negative battery cable, start the engine and check for leaks.

Electric Fuel Pump

TESTING

▶ See Figures 8 and 9

1. Turn the ignition switch **OFF** and disconnect the negative battery cable.
2. Disconnect the main fuel hose and connect a pressure gauge to it.
3. Connect the negative battery cable. Connect a jumper wire between the **B** and **D** terminals of the fuel pump control unit.
4. Turn the ignition switch **ON** and check the fuel pressure. It should be 2.8–3.6 psi.
5. If the fuel pressure is not as specified, replace the fuel pump.
6. Turn the ignition switch **OFF** and disconnect the negative battery cable.
7. Remove the fuel pressure gauge and reconnect the main fuel hose. Remove the jumper wire from the fuel pump control unit.
8. Connect the negative battery cable.

➡ The fuel pump is located inside the fuel tank, attached to the tank sending unit assembly.

REMOVAL & INSTALLATION

▶ See Figure 10

1. Disconnect the negative battery cable.
2. Remove the fuel tank.
3. Remove any dirt that has accumulated around the sending unit/fuel pump assembly so it will not enter the fuel tank during removal and installation.
4. Remove the attaching screws and remove the sending unit/fuel pump assembly.
5. If necessary, disconnect the electrical connectors and the fuel hose and remove the pump from the sending unit assembly.
6. Installation is the reverse of the removal procedure. Be sure to install a new seal rubber gasket.

Carburetor

REMOVAL & INSTALLATION

1. Remove the air cleaner and duct.
2. Disconnect the accelerator shaft from the throttle lever.
3. Disconnect and plug the fuel supply and return lines.
4. Disconnect the leads from the throttle solenoid and deceleration valve at the quick—disconnects.
5. Disconnect the carburetor-to-distributor vacuum line.
6. Disconnect the throttle return spring.
7. Disconnect the choke cable, and, if equipped, the cruise control cable.
8. Remove the carburetor attaching nuts from the intake manifold studs and remove the carburetor. The attaching nuts are tucked underneath the carburetor body and are difficult to reach; a small socket with an **L** shaped hex drive, or a short, thin wrench sold for work on ignition systems will make removal easier.

To install:
9. Install a new carburetor gasket on the manifold.
10. Install the carburetor and tighten the carburetor attaching nuts.
11. Connect the throttle return spring.
12. Connect the accelerator shaft to the throttle shaft.
13. Connect the electrical leads to the throttle solenoid and deceleration valve.
14. Connect the distributor vacuum line.
15. Connect the fuel supply and fuel return lines.
16. Connect and adjust the choke cable and, if equipped, the cruise control cable.
17. Install the air cleaner and duct.
18. Start the engine and check for fuel leaks.

OVERHAUL

The following instructions are general overhaul procedures. Most good carburetor rebuilding kits come complete with exploded views and specific instructions.

Efficient operation depends greatly on careful cleaning and inspection during overhaul, since dirt, gum, water, or varnish in or on the carburetor parts are often responsible for poor performance.

Overhaul your carburetor in a clean, dust-free area. Carefully disassemble the carburetor, referring often to the exploded views. Keep all similar and look alike parts segregated during disassembly and cleaning to avoid accidental interchange during assembly. Make a note of all jet sizes.

When the carburetor is disassembled, wash all parts (except diaphragms, electric choke units, pump plunger, and any other plastic, leather, fiber or rubber parts) in clean carburetor solvent. Do not leave parts in the solvent any longer than is necessary to sufficiently loosen the deposits. Excessive cleaning may remove the special finish from the float bowl and choke valve bodies, leaving these parts unfit for service.

Rinse all parts in clean solvent and blow them dry with compressed air or allow them to air dry. Wipe clean all cork, plastic, leather, and fiber parts with a clean, lint-free cloth.

Blow out all passages and jets with compressed air and be sure that there are no restrictions or blockages. Never use wire or similar tools to clean jets, fuel passages, or air bleeds. Clean all jets and valves separately to avoid accidental interchange.

Check all parts for wear or damage. If wear or damage is found, replace the defective parts. Especially check the following:
1. Check the float needle and seat for wear. If wear is found, replace the complete assembly.
2. Check the float hinge pin for wear and the float(s) for dents or distortion. Replace the float if fuel has leaked into it.

Fig. 8 Connect a fuel pressure gauge to the main fuel hose (A)

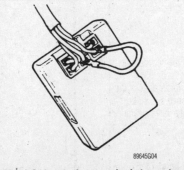

Fig. 9 Connect a jumper wire between terminals B and D of the fuel pump control unit

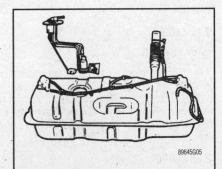

Fig. 10 The electric fuel is mounted on the fuel gauge assembly which is located in the fuel tank

FUEL SYSTEM 5-5

3. Check the throttle and choke shaft bores for wear or an out-of-round condition. Damage or wear to the throttle arm, shaft, or shaft bore will often require replacement of the throttle body. These parts require a close tolerance of fit; wear may allow air leakage, which could affect starting and idling.

➡ **Throttle shafts and bushings are not included in overhaul kits. They can be purchased separately.**

4. Inspect the idle moisture adjusting needles for burrs or grooves. Any such condition requires replacement of the needle, since you will not be able to obtain a satisfactory idle.
5. Test the accelerator pump check valves. They should pass air one way but not the other. Test for proper seating by blowing and sucking on the valve. Replace the valve if necessary. If the valve is satisfactory, wash the valve again to remove breath moisture.
6. Check the bowl cover for warped surfaces with a straight edge.
7. Closely inspect the valves and seats for wear or damage, replacing as necessary.
8. After the carburetor is assembled, check the choke valve for freedom of operation.

Carburetor overhaul kits are recommended for each overhaul. These kits contain all gaskets and new parts to replace those that deteriorate most rapidly. Failure to replace all parts supplied with the kit (especially gaskets) can result in poor performance later.

Some carburetor manufacturers supply overhaul kits of three basic types: minor repair; major repair; and gasket kits. Basically, they contain the following:

Minor Repair Kits:
- All gaskets
- Float needle valve
- Volume control screw
- All diaphragms
- Spring for the pump diaphragm

Major Repair Kits:
- All jet and gaskets
- All diaphragms
- Float needle valve
- Volume control screw
- Pump ball valve
- Main jet carrier
- Float
- Complete intermediate rod
- Intermediate pump lever
- Complete injector tube
- Some cover hold-down screws and washers

Gasket Kits:
- All gaskets

After cleaning and checking all components, reassemble the carburetor, using new parts and referring to the exploded view. When reassembling, make sure that all screws and jets are tight in their seats, but do not overtighten as the tips will be distorted. Tighten all screws gradually, in rotation. Do not tighten needle valves into their seats; uneven jetting will result. Always use new gaskets. Be sure to adjust the float lever when reassembling.

ADJUSTMENTS

Fast Idle

2.2L ENGINE

♦ See Figure 11

1. Set the fast idle cam so that the fast idle lever rests on the second step of the cam.
2. Adjust the clearance between the air horn wall and the lower edge of the throttle plates, by turning the fast idle adjusting screw. Clearance should be 0.0331–0.0409 inch (0.84–1.04mm). Make sure that the choke valve clearance hasn't changed.

2.6L ENGINE

➡ **Before making the adjustment, the carburetor should be at 73°F (23°C) for at least 1 hour.**

1. With the carburetor off the engine, invert it and check the opening between the throttle plate and the air horn wall. The clearance should be 0.0279 inch (0.71mm) if equipped with manual transmission; 0.0315 inch (0.80mm) if equipped with automatic transmission.
2. If not, adjust the opening by turning the fast idle adjustment screw.

Idle Speed

Make sure the ignition timing, spark plugs, and carburetor float level are all in normal operating condition. Turn off all lights and other unnecessary electrical loads.

1. Connect a tachometer to the engine.
2. Start the engine and allow it to reach normal operating temperature. Make sure the choke valve has fully opened.
3. Check the idle speed. If necessary, adjust it to specification by turning the throttle adjusting screw. The idle speed should be 800–850 rpm with manual transmission in neutral or automatic transmission in **P**.

Idle Mixture

2.2L ENGINE

1. Start the engine and allow it to reach normal operating temperature. Let the engine run at idle.
2. Connect a dwellmeter (90°, 4-cylinder) to the air/fuel check connector (Br/Y).
3. Check the idle mixture at the specified idle speed. The idle mixture should be 20–70°. If the idle mixture is not as specified, adjust as follows:
   a. Remove the carburetor and knock out the spring pin. Install the carburetor.
   b. Install the air cleaner and make sure the idle compensator is closed. Make sure all vacuum hoses are properly connected.
   c. Connect a tachometer to the engine.
   d. Warm up the engine and run it at idle. Make sure the idle speed is correct.
   e. Reconnect the dwell meter to the air/fuel check connector.
   f. Adjust the idle mixture to 27–45° by turning the mixture adjust screw.
   g. Tap in the spring pin.

2.6L ENGINE

1. Start the engine and allow it to reach normal operating temperature. Let the engine run at idle.
2. Connect a dwellmeter (90°, 4-cylinder) to the jet mixture solenoid valve connector (Y/G) wire. The check connector is located near the No. 3 ACV and reed valve case assembly.
3. Check the idle mixture at the specified idle speed. The idle mixture should be 27–45°. If the idle mixture is not as specified, check the oxygen sensor, feedback control unit and the wiring between the sensor and control unit.
4. If the oxygen sensor, feedback control unit and wiring are okay, adjust the idle mixture as follows:
   a. Remove the carburetor and mount it in a vise with the mixture adjust screws facing up. Protect the gasket surface from the vise jaws.

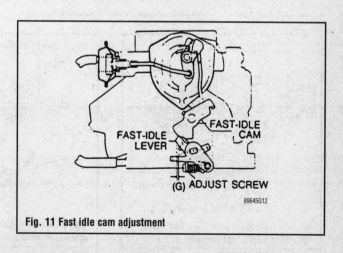

Fig. 11 Fast idle cam adjustment

5-6 FUEL SYSTEM

b. Drill a 0.079 inch (2mm) pilot hole in the casting toward the mixture adjust screw, then redrill the hole to 0.118 inch (3mm).

c. Insert a blunt punch into the hole and drive out the antitamper plug. Reinstall the carburetor.

d. Connect a tachometer to the engine. Start the engine and allow it to reach normal operating temperature. Let the engine run at idle. Make sure the idle speed is correct.

e. Reconnect the dwell meter to the jet mixture solenoid valve connector. Adjust the idle mixture to 30° ±plusmn; 9° by turning the mixture adjust screw.

f. Press in a new antitamper plug.

Float and Fuel Level

2.2L ENGINE

♦ See Figures 12 and 13

1. Remove the air horn from the carburetor.
2. Turn the air horn upside down on a level surface. Allow the float to hang under its own weight.
3. Measure the clearance between the float and the air horn gasket surface. The gap should be 0.4567–0.4960 inch (11.6–12.6mm) for vehicles with manual transmission; 0.4212–0.4606 inch (10.7–11.7mm) for vehicles with automatic transmission. If not, bend the float seat lip to obtain the correct gap.
4. Turn the air horn right side up and allow the float to hang under its own weight.
5. Measure the gap between the BOTTOM of the float and the air horn gasket surface. The gap should be 1.811–1.850 inch (46.0–47.0mm). If not, bend the float stopper until it is.

2.6L ENGINE

1. Remove air horn from carburetor main body.
2. Remove air horn gasket and invert air horn.

3. Using a gauge measure the distance from bottom of float to air horn surface. The distance should be 0.787 inch ± 0.0394 inch (20mm ± 1mm). If distance is not within specification, the shim under the needle and seat must be changed. A shim pack is available which has three shims: 0.008 inch (0.2mm), 0.012 inch (0.3mm), or 0.021 inch (0.5mm). Adding or removing one shim will change the float level three times its thickness.

Accelerator Linkage

1. Check the cable deflection at the carburetor. Deflection should be 0.0394–0.1181 inch (1.0–3.0mm). If not, adjust it by turning the adjusting nut at the bracket near the carburetor.
2. Depress the accelerator pedal to the floor. The throttle plates should be vertical. If not, adjust their position by turning the adjusting nut on the accelerator pedal bracket.

Secondary Throttle Valve

♦ See Figure 14

The clearance between the primary throttle valve and the air horn wall, when the secondary throttle valve just starts to open should be 0.2894–0.3249 inch (7.35–8.25mm). If not, bend the tab.

Choke

2.2L ENGINE

Three choke related adjustments are performed on these units.

CHOKE DIAPHRAGM

♦ See Figure 15

1. Disconnect the vacuum line from the choke diaphragm unit.
2. Using a vacuum pump, apply 15.7 inch Hg to the diaphragm.
3. Using light finger pressure, close the choke valve. Check the clearance at the upper edge of the choke valve. Clearance should be 0.067–0.085 inch (1.70–2.16mm).
4. If not, bend the tab on the choke lever to adjust it.

CHOKE VALVE CLEARANCE

♦ See Figure 16

1. Position the fast idle lever on the second step of the fast idle cam.
2. The leading edge of the choke valve should be 0.024–0.045 inch (0.60–1.14mm) from fully closed. If not, bend either the tab on the cam or the choke rod to adjust. The tab will give smaller adjustment increments.

CHOKE UNLOADER

♦ See Figure 17

1. Open the primary throttle valve all the way and hold it in this position.
2. The leading edge of the choke valve should be 0.110–0.143 inch (2.80–3.62mm) from fully closed. If not, bend the tab on the throttle lever.

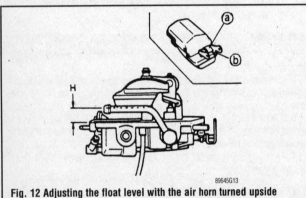

Fig. 12 Adjusting the float level with the air horn turned upside down

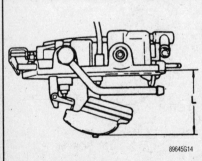

Fig. 13 Adjusting the float level with the air horn right side up and the float hanging under its own weight

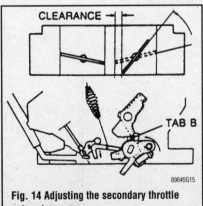

Fig. 14 Adjusting the secondary throttle valve clearance

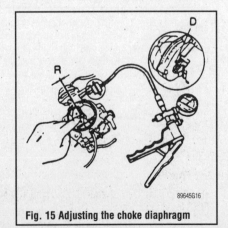

Fig. 15 Adjusting the choke diaphragm

FUEL SYSTEM 5-7

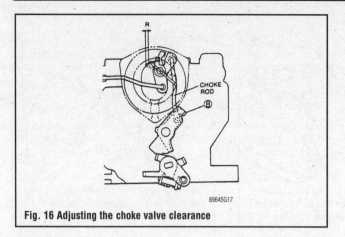

Fig. 16 Adjusting the choke valve clearance

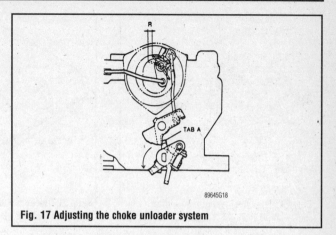

Fig. 17 Adjusting the choke unloader system

Idle Switch

1. The idle switch adjustment is not necessary unless it has been replaced or tampered with.
2. Start the engine and allow it to reach normal operating temperature. Stop the engine and connect a suitable tachometer.
3. Connect a voltmeter to the idle switch terminal (LgR).
4. Start the engine and increase the speed to above 2000 rpm and decelerate gradually.
5. Take a reading on the voltmeter. At idle, there should be approximately 12 volts and above 100–1200 rpm it should be below 1.5 volts.
6. If the readings do not agree with these specifications, turn the adjusting screw to obtain the correct reading.

FUEL INJECTION SYSTEM

General Information

The fuel injection system supplies the fuel necessary for combustion to the injectors at a constant pressure. Fuel is metered and injected into the intake manifold according to the injection control signals from the Engine Control Unit (ECU) or Engine Control Assembly (ECA). Most fuel injection systems consist of the following components: a fuel filter, distribution pipe, pulsation dampener, pressure regulator, injectors, fuel pump switch located in the airflow meter or mass airflow sensor for Navajo and the electric fuel pump usually located in the fuel tank in order to keep operating noise to a minimum.

The ECU or ECA, through various input sensors, monitors battery voltage, engine rpm, amount of air intake, cranking signals, intake temperature, coolant temperature, oxygen concentration in the exhaust gases, throttle opening, atmospheric or barometric pressure, gearshift position, clutch engagement, braking, power steering operation and air conditioner compressor operation.

The ECU or ECA controls operation of the fuel injection system, idle up system, fuel evaporation system and ignition timing. The control assembly has a built in fail safe mechanism. If a fault occurs while driving, the control assembly will substitute pre-programmed values. Driving performance will be affected but the vehicle will still, in most cases be drivable.

FUEL SYSTEM SERVICE PRECAUTIONS

Safety is the most important factor when performing not only fuel system maintenance but any type of maintenance. Failure to conduct maintenance and repairs in a safe manner may result in serious personal injury or death. Maintenance and testing of the vehicle's fuel system components can be accomplished safely and effectively by adhering to the following rules and guidelines.

- To avoid the possibility of fire and personal injury, always disconnect the negative battery cable unless the repair or test procedure requires that battery voltage be applied.
- Always relieve the fuel system pressure prior to disconnecting any fuel system component (injector, fuel rail, pressure regulator, etc.), fitting or fuel line connection. Exercise extreme caution whenever relieving fuel system pressure to avoid exposing skin, face and eyes to fuel spray. Please be advised that fuel under pressure may penetrate the skin or any part of the body that it contacts.
- Always place a shop towel or cloth around the fitting or connection prior to loosening to absorb any excess fuel due to spillage. Ensure that all fuel spillage (should it occur) is quickly removed from engine surfaces. Ensure that all fuel soaked cloths or towels are deposited into a suitable waste container.
- Always keep a dry chemical (Class B) fire extinguisher near the work area.
- Do not allow fuel spray or fuel vapors to come into contact with a spark or open flame.
- Always use a backup wrench when loosening and tightening fuel line connection fittings. This will prevent unnecessary stress and torsion to fuel line piping. Always follow the proper torque specifications.
- Always replace worn fuel fitting O-rings with new. Do not substitute fuel hose or equivalent where fuel pipe is installed.

Relieving Fuel System Pressure

Fuel lines on fuel injected vehicles will remain pressurized after the engine is shut off. This residual pressure must be relieved before any fuel lines or components are disconnected.

PICKUP & MPV

♦ See Figures 18 thru 23

1. Start the engine.
2. Disconnect the circuit opening relay connector, volume airflow meter connector or fuel pump connector.
3. After the engine stalls, turn **OFF** the ignition switch.
4. Reconnect the electrical connector.

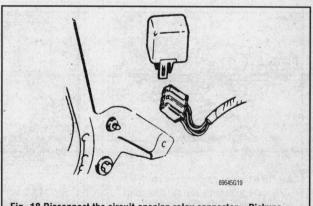

Fig. 18 Disconnect the circuit opening relay connector—Pickups

5-8 FUEL SYSTEM

Fig. 19 Disconnect the volume airflow meter connector—MPV with 3.0L engine

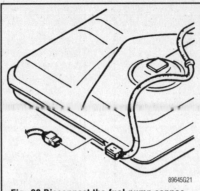

Fig. 20 Disconnect the fuel pump connector—MPV with 2.6L engine

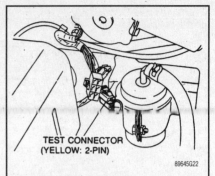

Fig. 21 Connect the terminals of the yellow 2-pin test connector with a jumper wire—Pickups

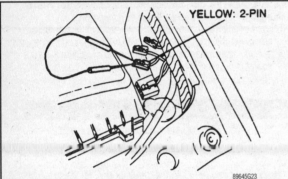

Fig. 22 Connect the terminals of the yellow 2-pin test connector with a jumper wire—MPV with 3.0L engine

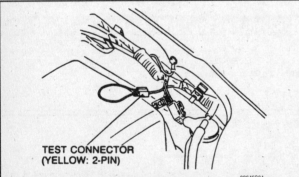

Fig. 23 Connect the terminals of the yellow 2-pin test connector with a jumper wire—MPV with 2.6L engine

➥After releasing fuel system pressure, the system must be primed before starting the engine to avoid excessive cranking. To prime the system, connect the terminals of the yellow 2-pin test connector with a jumper wire and turn the ignition switch ON for approximately 10 seconds. Check for fuel leaks, then turn the ignition switch OFF and remove the jumper wire.

NAVAJO

♦ See Figures 24 and 25

1. Disconnect the negative battery cable and remove the fuel filler cap.
2. Remove the cap from the pressure relief valve on the fuel supply manifold. Install pressure gauge 49 UN01 010 or equivalent, to the pressure relief valve.
3. Direct the gauge drain hose into a suitable container and depress the pressure relief button.
4. Remove the gauge and replace the cap on the pressure relief valve.

➥As an alternate method, disconnect the inertia switch and crank the engine for 15–20 seconds until the pressure is relieved.

Fuel Pump

REMOVAL & INSTALLATION

➥The fuel pump is located inside the fuel tank, attached to the tank sending unit assembly.

Pickup

♦ See Figure 26

1. Relieve the fuel system pressure and disconnect the negative battery cable.
2. Remove the fuel tank.

Fig. 24 The fuel pressure relief valve is located on the fuel supply manifold

Fig. 25 Fuel pressure can be checked and/or released using an inexpensive pressure/vacuum gauge

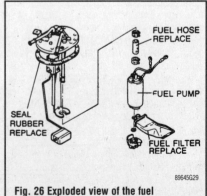

Fig. 26 Exploded view of the fuel pump/sending unit assembly

FUEL SYSTEM 5-9

> ✱✱ **CAUTION**
>
> Observe all applicable safety precautions when working around fuel. Whenever servicing the fuel system, always work in a well ventilated area. Do not allow fuel spray or vapors to come in contact with a spark or open flame. Keep a dry chemical fire extinguisher near the work area. Always keep fuel in a container specifically designed for fuel storage; also, always properly seal fuel containers to avoid the possibility of fire or explosion.

3. Remove any dirt that has accumulated around the sending unit/fuel pump assembly so it will not enter the fuel tank during removal and installation.
4. Remove the attaching screws and remove the sending unit/fuel pump assembly.
5. If necessary, disconnect the electrical connectors and the fuel hose and remove the pump from the sending unit assembly.
6. Installation is the reverse of the removal procedure. Be sure to install a new seal rubber gasket.

MPV

▶ See Figure 27

1. Relieve the fuel system pressure and disconnect the negative battery cable.

> ✱✱ **CAUTION**
>
> Observe all applicable safety precautions when working around fuel. Whenever servicing the fuel system, always work in a well ventilated area. Do not allow fuel spray or vapors to come in contact with a spark or open flame. Keep a dry chemical fire extinguisher near the work area. Always keep fuel in a container specifically designed for fuel storage; also, always properly seal fuel containers to avoid the possibility of fire or explosion.

2. Remove the rear seat and lift up the rear floormat. Remove the fuel pump cover.
3. Disconnect the sending unit/fuel pump assembly electrical connector and the fuel lines.
4. Remove any dirt that has accumulated around the sending unit/fuel pump assembly so it will not enter the fuel tank during removal and installation.
5. Remove the attaching screws and remove the sending unit/fuel pump assembly.
6. If necessary, disconnect the electrical connectors and the fuel hose and remove the pump from the sending unit assembly.
7. Installation is the reverse of the removal procedure. Be sure to install a new seal rubber gasket.

Navajo

1. Disconnect the negative battery cable and relieve the fuel system pressure.
2. Raise and safely support the vehicle.

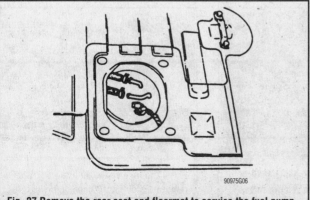

Fig. 27 Remove the rear seat and floormat to service the fuel pump

> ✱✱ **CAUTION**
>
> Observe all applicable safety precautions when working around fuel. Whenever servicing the fuel system, always work in a well ventilated area. Do not allow fuel spray or vapors to come in contact with a spark or open flame. Keep a dry chemical fire extinguisher near the work area. Always keep fuel in a container specifically designed for fuel storage; also, always properly seal fuel containers to avoid the possibility of fire or explosion.

3. Remove the fuel tank.
4. Remove any dirt that has accumulated around the fuel pump attaching flange so it will not enter the fuel tank during removal and installation.

> ✱✱ **CAUTION**
>
> If utilizing a hammer and drift (punch) tool to loosen or tighten a fuel pump lock ring, be sure that it is a brass drift. A metal, non-brass drift will most likely produce a spark when struck with or against metal, which can result a fire or explosion.

5. Turn the fuel pump locking ring counterclockwise using a hammer and brass drift or a spanner tool and ratchet. Remove the locking ring.
6. Remove the fuel pump and discard the seal ring. Separate the fuel pump from the sending unit, if required.

To install:

7. Clean the fuel pump mounting flange and tank mounting surface and seal ring groove.
8. Apply a light coating of Molybdenum grease on a new seal ring and install it in the groove.
9. Install the fuel pump to the sending unit, if removed. Install the fuel pump assembly in the tank, making sure the locating keys are in the keyways and the seal ring is in place.
10. Hold the fuel pump assembly and the seal ring in place and install the locking ring. Rotate the ring clockwise using a suitable tool. If using a hammer and brass drift, carefully tap the ring until it has achieved a tight seal. If using an adjustable spanner tool, attach a torque wrench to the spanner and tighten the locking ring to 40–45 ft. lbs. (54–61 Nm).
11. Install the fuel tank in the vehicle.
12. Lower the vehicle and fill the fuel tank with at least 10 gallons of fuel. Connect the negative battery cable. Turn the ignition key to **RUN** for 3 seconds repeatedly, 5–10 times, to pressurize the system. Check for leaks.
13. Start the engine and check for leaks.

TESTING

Pickup and MPV

▶ See Figures 21, 22, 23 and 28

1. Relieve the fuel system pressure and disconnect the negative battery cable.
2. Disconnect the fuel line from the fuel filter outlet. Connect a fuel pressure gauge to the fuel filter outlet.
3. Connect the negative battery cable. Connect the terminals of the yellow 2-pin test connector with a jumper wire.

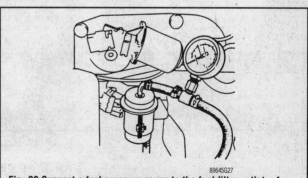

Fig. 28 Connect a fuel pressure gauge to the fuel filter outlet—4-cylinder engine shown, 6-cylinder engine similar

5-10 FUEL SYSTEM

4. Turn the ignition switch **ON** to operate the fuel pump and check the fuel pressure. It should be 64–85 psi.
5. If the fuel pressure is not as specified, replace the fuel pump.
6. Turn the ignition switch **OFF** and disconnect the negative battery cable. Remove the jumper wire from the test connector.
7. Remove the fuel pressure gauge and reconnect the fuel line to the fuel filter outlet.
8. Connect the negative battery cable.

Navajo

1. Make sure there is an adequate fuel supply.
2. Relieve the fuel system pressure.
3. Turn the ignition key **OFF**.
4. Connect a suitable fuel pressure gauge to the Schrader valve on the fuel rail.
5. Install a test lead to the **FP** terminal on the VIP test connector.
6. Turn the ignition key to the **RUN** position, then ground the test lead to run the fuel pump.
7. Observe the fuel pressure reading on the pressure gauge. The fuel pressure should be 35–45 psi.
8. Relieve the fuel system pressure and turn the ignition key **OFF**. Remove the fuel pressure gauge and the test lead.

Throttle Body

REMOVAL & INSTALLATION

Pickup

▶ See Figure 29

1. Disconnect the negative battery ground cable.
2. Disconnect the air hose.
3. Disconnect the ventilation hose.
4. Remove the air pipe and resonance chamber.
5. Disconnect the accelerator cable from the throttle lever.
6. Drain the engine coolant.

✲✲ CAUTION

When draining the coolant, keep in mind that cats and dogs are attracted by the ethylene glycol antifreeze, and are quite likely to drink any that is left in an uncovered container or in puddles on the ground. This will prove fatal in sufficient quantity. Always drain the coolant into a sealable container. Coolant should be reused unless it is contaminated or several years old.

7. Disconnect the coolant lines at the manifold.
8. Tag and disconnect all vacuum hoses.
9. Tag and disconnect all wiring.
10. Unbolt and remove the throttle body.
11. Always use a new mounting gasket. Place the new gasket and throttle body in position. Install and torque the fasteners to 15 ft. lbs. Connect the vacuum lines, hoses and intake system components. Fill the cooling system.

MPV 2.6L Engine

▶ See Figure 29

1. Disconnect the battery ground.
2. Disconnect the air hose.
3. Disconnect the ventilation hose.
4. Remove the air pipe and resonance chamber.
5. Disconnect the accelerator cable from the throttle lever.
6. Drain the engine coolant.

✲✲ CAUTION

When draining the coolant, keep in mind that cats and dogs are attracted by the ethylene glycol antifreeze, and are quite likely to drink any that is left in an uncovered container or in puddles on the ground. This will prove fatal in sufficient quantity. Always drain the coolant into a sealable container. Coolant should be reused unless it is contaminated or several years old.

7. Disconnect the coolant lines at the manifold.
8. Tag and disconnect all vacuum hoses.
9. Tag and disconnect all wiring.
10. Unbolt and remove the throttle body.
11. Always use a new mounting gasket. Place the new gasket and throttle body in position. Install and torque the fasteners to 15 ft. lbs. (20 Nm). Connect the vacuum lines, hoses and intake system components. Fill the cooling system.

3.0L Engine

▶ See Figure 30

The throttle body is located at the left end of the air intake plenum. The throttle body contains the throttle position sensor and the automatic idle speed motor.

1. Disconnect battery negative cable.
2. Loosen air cleaner to throttle body hose clamps and remove hose.
3. Remove accelerator cable and transaxle kickdown linkage.
4. Remove harness connector from throttle position sensor (TPS), and automatic idle speed (AIS) motor.
5. Label and remove vacuum hoses necessary.
6. Remove throttle body mounting nuts and remove throttle body and gasket.
7. Install throttle body using a new gasket on air intake plenum. Secure with mounting nuts. Tighten the mounting nuts to 14–18 ft. lbs. (19–25 Nm).
8. Reconnect vacuum hoses, TPS and ASI electrical connectors.
9. Reconnect accelerator cable and transaxle linkage.
10. Install air cleaner to throttle body hose and tighten clamps.
11. Reconnect battery negative cable.

4.0L Engine

▶ See Figures 31 thru 37

1. Disconnect the negative battery cable. Remove the air cleaner intake hose.
2. Remove the linkage shield.
3. Disconnect the throttle cable at the ball stud.
4. Disconnect the canister purge hose from under the throttle body.

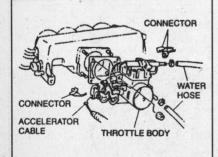

Fig. 29 Removal of throttle body components

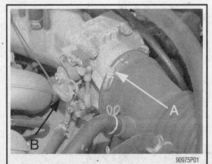

Fig. 30 Loosen the air cleaner-to-throttle body hose clamps (A) and remove hose, then disconnect the accelerator cable (B)

Fig. 31 To remove the throttle body, first remove the air intake hose, then the linkage shield

FUEL SYSTEM 5-11

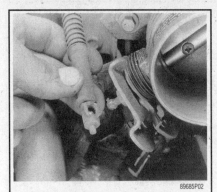
Fig. 32 Next, disconnect the throttle control cable(s) from the throttle body . . .

Fig. 33 . . . as well as any electrical connections. . .

Fig. 34 . . . and vacuum fittings. Make sure to label them to assure proper installation

Fig. 35 Remove the throttle body-to-upper intake attaching bolts . . .

Fig. 36 . . . then remove the throttle body assembly

Fig. 37 Remove the old gasket and thoroughly clean the mating surfaces of both pieces

5. Disconnect the wiring harness at the throttle position sensor.
6. Remove the 4 retaining bolts and lift the throttle body assembly off the upper intake manifold.
7. Remove and discard the gasket.

To install:

8. Clean and inspect the mounting faces of the throttle body assembly and the upper intake manifold. Both surfaces must be clean and flat.
9. Install a new gasket on the manifold.
10. Install the air throttle body assembly on the intake manifold.
11. Install the bolts finger tight, then tighten them evenly to 76–106 inch lbs. (8–12 Nm).
12. Connect the wiring harness to the throttle position sensor.
13. Install the canister purge hose.
14. Install the snow shield and air cleaner outlet tube. Reconnect the battery cable. Start engine, check for proper operation.

Fuel Injectors

REMOVAL & INSTALLATION

4-Cylinder Engines

▶ See Figure 38

1. Relieve the fuel system pressure.
2. Remove the air chamber.
3. Disconnect the vacuum hose.
4. Disconnect the fuel lines.
5. Remove the pressure regulator and fuel rail assembly.
6. Disconnect the injector wiring.
7. Pull off the injectors. Discard the grommets and O-rings.
8. Installation is the reverse of removal. Tighten the fuel rail bolts to 15 ft. lbs. (20 Nm). Always use new O-rings coated with clean engine oil.

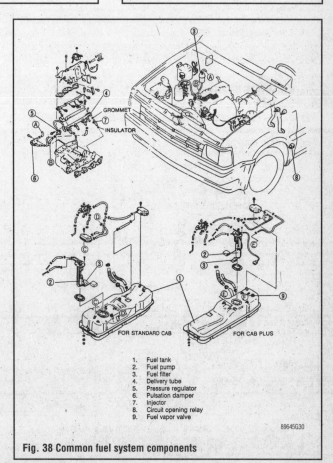
Fig. 38 Common fuel system components

1. Fuel tank
2. Fuel pump
3. Fuel filter
4. Delivery tube
5. Pressure regulator
6. Pulsation damper
7. Injector
8. Circuit opening relay
9. Fuel vapor valve

5-12 FUEL SYSTEM

3.0L Engine

♦ See Figures 39 and 40

Removal of the injectors requires removal of the fuel injector rail assembly. Refer to Fuel Charging Assembly removal procedure in this section.

1. Disconnect injector harness from injectors.
2. Invert the fuel charginging (fuel rail) assembly.
3. Remove lock rings securing injectors to fuel rail receiver cups. Pull injectors upward from receiver cups.
4. If injectors are to be reused, place a protective cap on injector nozzle to prevent dirt or other damage.
5. Lubricate the new O-ring of each injectors with a clean drop of engine oil prior to installation.
6. Assemble each injectors into fuel rail receiver cups. Be careful not to damage O-rings.
7. Install lock ring between receiver cup ridge and injector slot.

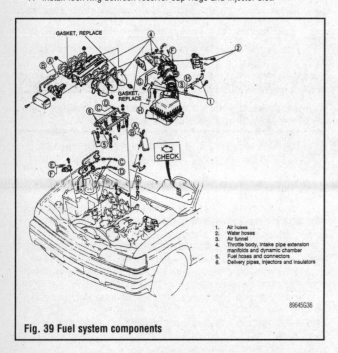

Fig. 39 Fuel system components

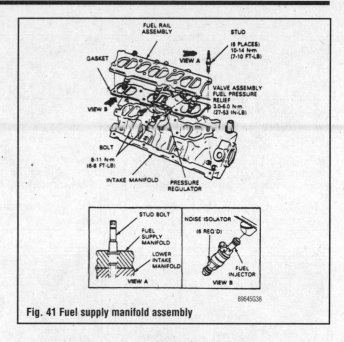

Fig. 41 Fuel supply manifold assembly

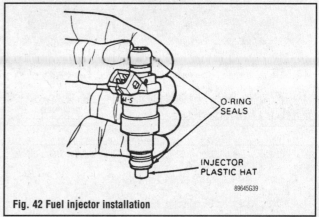

Fig. 42 Fuel injector installation

4. Remove the injector retaining clips and remove the injectors from the manifold by grasping the injector body and pulling up while rocking the injector from side-to-side.
5. Remove and discard the injector O-rings.
6. Inspect the injector plastic pintle protection cap and washer for signs of deterioration. Replace the complete injector as required. If the plastic pintle protection cap is missing, look for it in the intake manifold.

➡The plastic pintle protection cap is not available as a separate part.

To install:

7. Lubricate new O-rings with clean light grade oil and install 2 on each injector.

➡Never use silicone grease at it will clog the injectors.

8. Install the injectors, using a light, twisting, pushing motion.
9. Install the fuel supply manifold, pushing down to make sure all the fuel injector O-rings are fully seated in the fuel supply manifold cups and intake manifold.
10. Install the upper intake manifold and throttle body assembly.
11. After the upper intake manifold has been installed and before the fuel injector wire connectors have been connected, connect the negative battery cable and turn the ignition switch **ON**. This will cause the fuel pump to run for 2–3 seconds and pressurize the system.
12. Check for fuel leaks where the fuel injector is installed into the fuel supply manifold.
13. Turn the ignition switch **OFF** and disconnect the negative battery cable.
14. Connect the injector wire connectors and the vacuum line to the regulator.

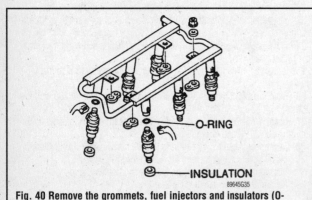

Fig. 40 Remove the grommets, fuel injectors and insulators (O-rings) from the fuel rail assembly

4.0L Engine

♦ See Figures 41 and 42

1. Disconnect the negative battery cable and relieve the fuel system pressure.
2. Remove the upper intake manifold and throttle body assembly.
3. Remove the fuel supply manifold as outlined in the Fuel Charging Assembly procedure later in this section.

FUEL SYSTEM 5-13

15. Install the air inlet tube from the throttle body to the air cleaner.
16. Connect the negative battery cable, start the engine and let it idle for 2 minutes.
17. Turn the engine **OFF** and check for fuel leaks.

TESTING

♦ See Figures 43 and 44

The fuel injectors can be tested with a Digital Volt-Ohmmeter (DVOM). To test an injector, detach the engine wiring harness connector from it. This may require removing the upper intake manifold or other engine components.

Once access to the injector is gained and the wiring is disconnected from it, set the DVOM to measure resistance (ohms). Measure the resistance of the injector by probing one terminal with the positive DVOM lead and the other injector terminal with the negative lead. The resistance measured should be between 12–16 ohms. If the resistance is not within this range, the fuel injector is faulty and must be replaced with a new one.

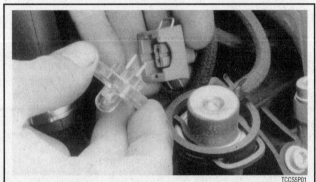

Fig. 43 A noid light can be attached to the fuel injector harness in order to test for injector pulse

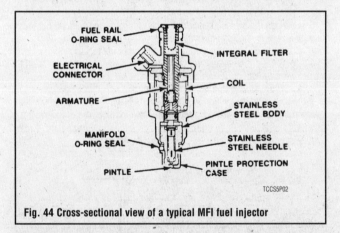

Fig. 44 Cross-sectional view of a typical MFI fuel injector

Fuel Charging Assembly

REMOVAL & INSTALLATION

4-Cylinder Engines

To remove the fuel charging (fuel rail) assembly, refer to the fuel injector removal procedure earlier in this section.

3.0L Engine

♦ See Figure 39

1. Perform fuel system pressure release procedure. Refer to fuel system pressure release procedure in this section.
2. Disconnect the negative battery cable.
3. Loosen clamps securing air cleaner to throttle body hose and remove hose.
4. Remove the throttle cable and transaxle kickdown linkage.
5. Remove harness connector from throttle position sensor (TPS), and automatic idle speed (AIS) motor.
6. Label and remove vacuum hoses from throttle body. Remove PCV and brake booster hoses from air intake plenum.
7. Remove EGR tube to intake plenum.
8. Remove electrical connection from charge temperature and coolant temperature sensor.
9. Remove vacuum connection from pressure regulator and air intake connection from manifold.
10. Remove fuel hoses to fuel rail connection.
11. Remove air intake plenum to manifold bolts (8) and remove air intake plenum and gaskets.

✱✱ WARNING

Whenever air intake plenum is remove, cover intake manifold properly to avoid objects from entering cylinder head.

12. Disconnect fuel injector wiring harness from engine wiring harness.
13. Remove pressure regulator attaching bolts and remove pressure regulator from rail.
14. Remove fuel rail attaching bolts and remove fuel rail.

To install:
15. Make certain injector are properly seated in receiver cup with lock rings in place and injector discharge holes are clean.
16. Lubricate injector O-rings with a clean drop of engine oil.
17. Install injector rail assembly making sure each injector seats in their respective ports. Tighten the fuel rail attaching bolts to 115 inch lbs. (13 Nm).
18. Lubricate pressure regulator O-ring with a drop of clean engine oil and install regulator to fuel rail. Tighten the nuts to 77 inch lbs. (9 Nm).
19. Install hold down bolts on fuel supply and return tube, and vacuum crossover tube.
20. Install and tighten the fuel pressure regulator hose clamps to 10 inch lbs. (1 Nm).
21. Reconnect injector wiring harness.
22. Reconnect vacuum hoses to fuel pressure regulator and fuel rail.
23. Set the air intake plenum gasket in place with beaded sealer in the **up** position.
24. Install air intake plenum and tighten (8) attaching screws to 115 inch lbs. (13 Nm).
25. Reconnect fuel line to fuel rail and tighten clamps to 10 inch lbs. (1 Nm).
26. Reconnect EGR tube to intake plenum and torque nuts to 200 inch lbs. (23 Nm).
27. Reconnect electrical wiring to charge temperature sensor, coolant temperature sensor, TPS and AIS motor.
28. Reconnect vacuum connection to throttle body and air intake plenum.
29. Install accelerator cable and transaxle kickdown cable.
30. Install air cleaner to throttle body hose and tighten clamps.
31. Reconnect battery negative cable.

4.0L Engine

♦ See Figure 41

1. Disconnect the battery ground cable.
2. Remove the air cleaner and intake duct.

5-14 FUEL SYSTEM

3. Remove the linkage shield.
4. Disconnect the throttle cable and bracket.
5. Tag and disconnect all vacuum lines connected to the manifold.
6. Tag and disconnect all electrical wires connected to the manifold assemblies.
7. Relieve the fuel system pressure.

✱✱ CAUTION

The fuel system is under pressure. Release pressure slowly and contain spillage. Observe no smoking/no open flame precautions. Have a Class B–C (dry powder) fire extinguisher within arm's reach at all times.

8. Tag and remove the spark plug wires.
9. Remove the EDIS ignition coil and bracket.
10. Remove the throttle body and discard the gasket.
11. Remove the 6 attaching nuts and lift off the upper manifold.
12. Disconnect the fuel supply line at the fuel supply manifold.
13. Disconnect the fuel return line at the pressure regulator as follows:
   a. Disengage the locking tabs on the connector retainer and separate the retainer halves.
   b. Check the visible, internal portion of the fitting for dirt. Clean the fitting thoroughly.
   c. Push the fitting towards the regulator, insert the fingers on Fuel Line Coupling Key T90P–9550–A, or equivalent, into the slots in the coupling. Using the tool, pull the fitting from the regulator. The fitting should slide off easily, if properly disconnected.
14. Remove the 6 Torx® head stud bolts retaining the manifold and remove the manifold.
15. Remove the electrical harness connector from each injector.
16. If required, remove the fuel injectors.

To install:
17. If removed, install the fuel injectors into the fuel supply manifold.
18. Install the retainers and electrical harness connectors.
19. Position the fuel supply manifold and press it down firmly until the injectors are fully seated in the fuel supply manifold and lower intake manifold.
20. Install the 6 Torx® head bolts and tighten them to 7–10 ft. lbs. (9–14 Nm).
21. Install the fuel supply line and tighten the fitting to 15–18 ft. lbs. (20–24 Nm).
22. Install the fuel return line on the regulator by pushing it onto the fuel pressure regulator line of to the shoulder.

✱✱ WARNING

The connector should grip the line securely!

23. Install the connector retainer and snap the two halves of the retainer together.
24. Install the upper manifold. Tighten the nuts to 18 ft. lbs. (24 Nm).
25. Install the EDIS coil.
26. Connect the fuel and return lines.
27. Ensure that the mating surfaces of the throttle body and upper manifold are clean and free of gasket material.
28. Install the throttle body assembly.
29. Connect all wires.
30. Connect all vacuum lines.
31. Connect the throttle linkage.
32. Install the linkage shield.
33. Install the air cleaner and duct.
34. Fill and bleed the cooling system.
35. Connect the battery ground.
36. Run the engine and check for leaks.

Fuel Pressure Regulator

REMOVAL & INSTALLATION

4-Cylinder Engines

1. Relieve the fuel system pressure.
2. Disconnect the vacuum hose.
3. Disconnect the fuel return hose.
4. Unbolt and remove the unit.
5. Install in the reverse order. Tighten the bolts to 95 inch lbs. (11 Nm).

3.0L Engine

Refer to Fuel Charging Assembly Removal and Installation procedure above.

4.0L Engine

▶ See Figure 45

1. Depressurize the fuel system.

✱✱ CAUTION

The fuel system is under pressure. Release pressure slowly and contain spillage. Observe no smoking/no open flame precautions. Have a Class B–C (dry powder) fire extinguisher within arm's reach at all times.

2. Remove the vacuum and fuel lines at the pressure regulator.
3. Remove the 2 or 3 Allen retaining screws from the regulator housing.
4. Remove the pressure regulator assembly, gasket and O-ring. Discard the gasket and check the O-ring for signs of cracks or deterioration.

To install:
5. Clean the gasket mating surfaces. If scraping is necessary, be careful not to damage the fuel pressure regulator or supply line gasket mating surfaces.
6. Lubricate the pressure regulator O-ring with light engine oil. Do not use silicone grease; it will clog the injectors.
7. Install the O-ring and a new gasket on the pressure regulator.
8. Install the pressure regulator on the fuel manifold and tighten the retaining screws to 6–8 ft. lbs. (8–11 Nm).
9. Install the vacuum and fuel lines at the pressure regulator. Build up fuel pressure by turning the ignition switch ON and OFF at least 6 times, leaving the ignition on for at least 5 seconds each time. Check for fuel leaks.

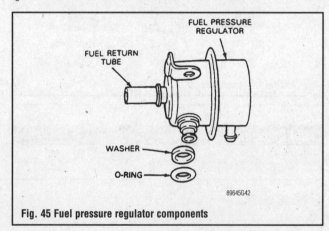

Fig. 45 Fuel pressure regulator components

FUEL SYSTEM 5-15

FUEL TANK

Tank Assembly

REMOVAL & INSTALLATION

♦ See Figure 46

Pickup

1. Relieve the fuel system pressure and disconnect the negative battery cable.
2. Remove the fuel filler cap. Raise and safely support the vehicle.
3. Position a suitable container under the fuel tank. Remove the drain plug and drain the tank.
4. Disconnect the electrical connector from the sending unit or sending unit/fuel pump assembly.
5. Disconnect the fuel filler hose, evaporative hoses, breather hose and fuel lines.
6. Position a jack under the fuel tank and remove the tank attaching nuts. Lower the tank from the vehicle.

To install:

7. Raise the tank into position and install the attaching nuts. Remove the jack.
8. Connect the fuel lines and evaporative hoses, making sure they are pushed onto the fuel tank fittings at least 1 inch (25mm). Connect the breather hose.
9. Connect the fuel filler hose, making sure the hose is pushed onto the fuel tank pipe and filler pipe at least 1.4 inch (35mm).
10. Connect the electrical connector to the sending unit or sending unit/fuel pump assembly.
11. Install the drain plug and lower the vehicle.
12. Fill the fuel tank and install the filler cap. Check for leaks.
13. Start the engine and check for leaks.

MPV

1. Relieve the fuel system pressure and disconnect the negative battery cable.
2. Remove the fuel filler cap. Raise and safely support the vehicle.
3. Position a suitable container under the fuel tank. Remove the drain plug and drain the tank.
4. Disconnect the fuel pump electrical connector.
5. Disconnect the fuel lines, evaporative hoses, breather hose and fuel filler hose.
6. Support the tank with a jack. Remove the retaining bolts and the fuel tank straps.
7. Lower the fuel tank from the vehicle.

To install:

8. Raise the fuel tank into position and install the straps and retaining bolts. Tighten to 32–44 ft. lbs. (43–61 Nm). Remove the jack.
9. Connect the fuel lines and evaporative hoses, making sure they are pushed onto the fuel tank fittings at least 1 inch (25mm). Connect the breather hose.
10. Connect the fuel filler hose, making sure the hose is pushed onto the fuel tank pipe and filler pipe at least 1.4 inch (35mm).
11. Connect the fuel pump electrical connector.
12. Install the drain plug and lower the vehicle.
13. Fill the fuel tank and install the filler cap. Check for leaks.
14. Start the engine and check for leaks.

Navajo

1. Disconnect the negative battery cable and relieve the fuel system pressure.
2. Raise and safely support the vehicle.
3. Drain the fuel from the fuel tank.
4. Remove the shield, skid plate and fuel tank front strap.
5. Support the tank with a jack and remove the bolt from the fuel tank rear strap.
6. Disconnect the filler pipe and vent pipe and lower the tank. Disconnect the vapor hose, fuel lines and electrical connector.
7. Lower the tank from the vehicle.

To install:

8. Raise the fuel tank and connect the electrical connector, fuel lines and vapor hose.
9. Connect the filler pipe and vent pipe. Attach the rear fuel tank strap.
10. Install the shield, skid plate and front strap.
11. Remove the jack and lower the vehicle.
12. Fill the fuel tank and check for leaks. Connect the negative battery cable.

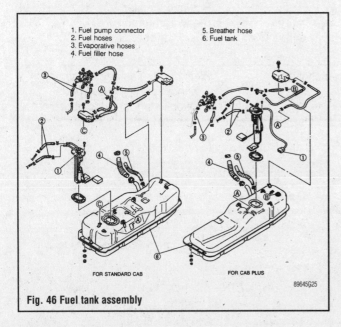

1. Fuel pump connector
2. Fuel hoses
3. Evaporative hoses
4. Fuel filler hose
5. Breather hose
6. Fuel tank

FOR STANDARD CAB   FOR CAB PLUS

Fig. 46 Fuel tank assembly

Troubleshooting Basic Fuel System Problems

Problem	Cause	Solution
Engine cranks, but won't start (or is hard to start) when cold	• Empty fuel tank • Incorrect starting procedure • Defective fuel pump • No fuel in carburetor • Clogged fuel filter • Engine flooded • Defective choke	• Check for fuel in tank • Follow correct procedure • Check pump output • Check for fuel in the carburetor • Replace fuel filter • Wait 15 minutes; try again • Check choke plate
Engine cranks, but is hard to start (or does not start) when hot— (presence of fuel is assumed)	• Defective choke	• Check choke plate
Rough idle or engine runs rough	• Dirt or moisture in fuel • Clogged air filter • Faulty fuel pump	• Replace fuel filter • Replace air filter • Check fuel pump output
Engine stalls or hesitates on acceleration	• Dirt or moisture in the fuel • Dirty carburetor • Defective fuel pump • Incorrect float level, defective accelerator pump	• Replace fuel filter • Clean the carburetor • Check fuel pump output • Check carburetor
Poor gas mileage	• Clogged air filter • Dirty carburetor • Defective choke, faulty carburetor adjustment	• Replace air filter • Clean carburetor • Check carburetor
Engine is flooded (won't start accompanied by smell of raw fuel)	• Improperly adjusted choke or carburetor	• Wait 15 minutes and try again, without pumping gas pedal • If it won't start, check carburetor

UNDERSTANDING AND
   TROUBLESHOOTING ELECTRICAL
   SYSTEMS 6-2
BASIC ELECTRICAL THEORY 6-2
   HOW DOES ELECTRICITY WORK: THE
      WATER ANALOGY 6-2
   OHM'S LAW 6-2
ELECTRICAL COMPONENTS 6-2
   POWER SOURCE 6-2
   GROUND 6-3
   PROTECTIVE DEVICES 6-3
   SWITCHES & RELAYS 6-3
   LOAD 6-4
   WIRING & HARNESSES 6-4
   CONNECTORS 6-4
TEST EQUIPMENT 6-5
   JUMPER WIRES 6-5
   TEST LIGHTS 6-5
   MULTIMETERS 6-5
TROUBLESHOOTING ELECTRICAL
   SYSTEMS 6-6
TESTING 6-6
   OPEN CIRCUITS 6-6
   SHORT CIRCUITS 6-6
   VOLTAGE 6-7
   VOLTAGE DROP 6-7
   RESISTANCE 6-7
WIRE AND CONNECTOR REPAIR 6-8
BATTERY CABLES 6-8
DISCONNECTING THE CABLES 6-8
HEATING AND AIR
   CONDITIONING 6-8
BLOWER MOTOR 6-8
   REMOVAL & INSTALLATION 6-8
HEATER CORE 6-10
   REMOVAL & INSTALLATION 6-10
HEATER CASE ASSEMBLY 6-12
   REMOVAL & INSTALLATION 6-12
AIR CONDITIONING COMPONENTS 6-12
   REMOVAL & INSTALLATION 6-12
CONTROL PANEL 6-13
   REMOVAL & INSTALLATION 6-13
CRUISE CONTROL 6-14
ENTERTAINMENT SYSTEMS 6-16
RADIO RECIEVER/TAPE PLAYER 6-16
   REMOVAL & INSTALLATION 6-16
SPEAKERS 6-18
   REMOVAL & INSTALLATION 6-18
WINDSHIELD WIPERS 6-20
WINDSHIELD WIPER BLADE AND
   ARM 6-20
   REMOVAL & INSTALLATION 6-20
WINDSHIELD WIPER MOTOR 6-21
   REMOVAL & INSTALLATION 6-21
WINDSHIELD WASHER MOTOR 6-23
   REMOVAL & INSTALLATION 6-23
INSTRUMENTS AND
   SWITCHES 6-24
INSTRUMENT CLUSTER 6-24
   REMOVAL & INSTALLATION 6-24
FUEL, OIL PRESSURE, VOLTAGE AND
   COOLANT TEMPERATURE
   GAUGES 6-26
WINDSHIELD WIPER SWITCH 6-26
   REMOVAL & INSTALLATION 6-26
REAR WIPER SWITCH 6-26
   REMOVAL & INSTALLATION 6-26
HEADLIGHT SWITCH 6-26
   REMOVAL & INSTALLATION 6-26
LIGHTING 6-27
HEADLIGHTS 6-27
   REMOVAL & INSTALLATION 6-27
   AIMING THE HEADLIGHTS 6-28
SIGNAL AND MARKER LIGHTS 6-29
   REMOVAL & INSTALLATION 6-29
LIGHT BULB APPLICATIONS 6-34
TRAILER WIRING 6-34
CIRCUIT PROTECTION 6-35
FUSES 6-35
   REPLACEMENT 6-35
FUSIBLE LINKS 6-38
CIRCUIT BREAKERS 6-38
   REPLACEMENT 6-38
FLASHERS AND RELAYS 6-38
   REPLACEMENT 6-38
WIRING DIAGRAMS 6-39
TROUBLESHOOTING CHART
   CRUISE CONTROL 6-16

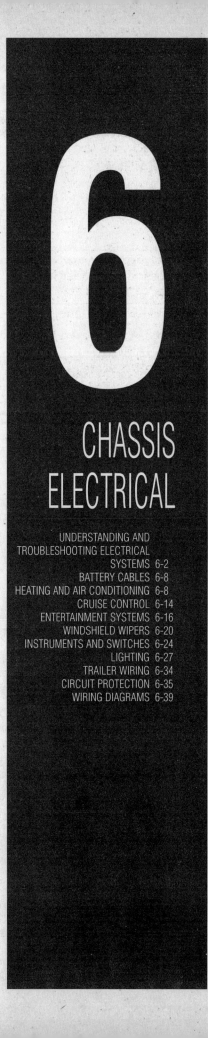

6

CHASSIS ELECTRICAL

UNDERSTANDING AND
TROUBLESHOOTING ELECTRICAL
SYSTEMS 6-2
BATTERY CABLES 6-8
HEATING AND AIR CONDITIONING 6-8
CRUISE CONTROL 6-14
ENTERTAINMENT SYSTEMS 6-16
WINDSHIELD WIPERS 6-20
INSTRUMENTS AND SWITCHES 6-24
LIGHTING 6-27
TRAILER WIRING 6-34
CIRCUIT PROTECTION 6-35
WIRING DIAGRAMS 6-39

6-2 CHASSIS ELECTRICAL

UNDERSTANDING AND TROUBLESHOOTING ELECTRICAL SYSTEMS

Basic Electrical Theory

♦ See Figure 1

For any 12 volt, negative ground, electrical system to operate, the electricity must travel in a complete circuit. This simply means that current (power) from the positive (+) terminal of the battery must eventually return to the negative (-) terminal of the battery. Along the way, this current will travel through wires, fuses, switches and components. If, for any reason, the flow of current through the circuit is interrupted, the component fed by that circuit will cease to function properly.

Perhaps the easiest way to visualize a circuit is to think of connecting a light bulb (with two wires attached to it) to the battery—one wire attached to the negative (-) terminal of the battery and the other wire to the positive (+) terminal. With the two wires touching the battery terminals, the circuit would be complete and the light bulb would illuminate. Electricity would follow a path from the battery to the bulb and back to the battery. It's easy to see that with longer wires on our light bulb, it could be mounted anywhere. Further, one wire could be fitted with a switch so that the light could be turned on and off.

The normal automotive circuit differs from this simple example in two ways. First, instead of having a return wire from the bulb to the battery, the current travels through the frame of the vehicle. Since the negative (-) battery cable is attached to the frame (made of electrically conductive metal), the frame of the vehicle can serve as a ground wire to complete the circuit. Secondly, most automotive circuits contain multiple components which receive power from a single circuit. This lessens the amount of wire needed to power components on the vehicle.

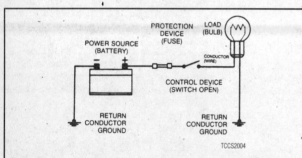

Fig. 1 This example illustrates a simple circuit. When the switch is closed, power from the positive (+) battery terminal flows through the fuse and the switch, and then to the light bulb. The light illuminates and the circuit is completed through the ground wire back to the negative (-) battery terminal. In reality, the two ground points shown in the illustration are attached to the metal frame of the vehicle, which completes the circuit back to the battery

HOW DOES ELECTRICITY WORK: THE WATER ANALOGY

Electricity is the flow of electrons—the subatomic particles that constitute the outer shell of an atom. Electrons spin in an orbit around the center core of an atom. The center core is comprised of protons (positive charge) and neutrons (neutral charge). Electrons have a negative charge and balance out the positive charge of the protons. When an outside force causes the number of electrons to unbalance the charge of the protons, the electrons will split off the atom and look for another atom to balance out. If this imbalance is kept up, electrons will continue to move and an electrical flow will exist.

Many people have been taught electrical theory using an analogy with water. In a comparison with water flowing through a pipe, the electrons would be the water and the wire is the pipe.

The flow of electricity can be measured much like the flow of water through a pipe. The unit of measurement used is amperes, frequently abbreviated as amps (a). You can compare amperage to the volume of water flowing through a pipe. When connected to a circuit, an ammeter will measure the actual amount of current flowing through the circuit. When relatively few electrons flow through a circuit, the amperage is low. When many electrons flow, the amperage is high.

Water pressure is measured in units such as pounds per square inch (psi); The electrical pressure is measured in units called volts (v). When a voltmeter is connected to a circuit, it is measuring the electrical pressure.

The actual flow of electricity depends not only on voltage and amperage, but also on the resistance of the circuit. The higher the resistance, the higher the force necessary to push the current through the circuit. The standard unit for measuring resistance is an ohm. Resistance in a circuit varies depending on the amount and type of components used in the circuit. The main factors which determine resistance are:

- Material—some materials have more resistance than others. Those with high resistance are said to be insulators. Rubber materials (or rubber-like plastics) are some of the most common insulators used in vehicles as they have a very high resistance to electricity. Very low resistance materials are said to be conductors. Copper wire is among the best conductors. Silver is actually a superior conductor to copper and is used in some relay contacts, but its high cost prohibits its use as common wiring. Most automotive wiring is made of copper.
- Size—the larger the wire size being used, the less resistance the wire will have. This is why components which use large amounts of electricity usually have large wires supplying current to them.
- Length—for a given thickness of wire, the longer the wire, the greater the resistance. The shorter the wire, the less the resistance. When determining the proper wire for a circuit, both size and length must be considered to design a circuit that can handle the current needs of the component.
- Temperature—with many materials, the higher the temperature, the greater the resistance (positive temperature coefficient). Some materials exhibit the opposite trait of lower resistance with higher temperatures (negative temperature coefficient). These principles are used in many of the sensors on the engine.

OHM'S LAW

There is a direct relationship between current, voltage and resistance. The relationship between current, voltage and resistance can be summed up by a statement known as Ohm's law.

Voltage (E) is equal to amperage (I) times resistance (R): $E = I \times R$

Other forms of the formula are $R = E/I$ and $I = E/R$

In each of these formulas, E is the voltage in volts, I is the current in amps and R is the resistance in ohms. The basic point to remember is that as the resistance of a circuit goes up, the amount of current that flows in the circuit will go down, if voltage remains the same.

The amount of work that the electricity can perform is expressed as power. The unit of power is the watt (w). The relationship between power, voltage and current is expressed as:

Power (w) is equal to amperage (I) times voltage (E): $W = I \times E$

This is only true for direct current (DC) circuits; The alternating current formula is a tad different, but since the electrical circuits in most vehicles are DC type, we need not get into AC circuit theory.

Electrical Components

POWER SOURCE

Power is supplied to the vehicle by two devices: The battery and the alternator. The battery supplies electrical power during starting or during periods when the current demand of the vehicle's electrical system exceeds the output capacity of the alternator. The alternator supplies electrical current when the engine is running. Just not does the alternator supply the current needs of the vehicle, but it recharges the battery.

The Battery

In most modern vehicles, the battery is a lead/acid electrochemical device consisting of six 2 volt subsections (cells) connected in series, so that the unit is capable of producing approximately 12 volts of electrical pressure. Each subsection consists of a series of positive and negative plates held a short distance apart in a solution of sulfuric acid and water.

The two types of plates are of dissimilar metals. This sets up a chemical reaction, and it is this reaction which produces current flow from the battery when its positive and negative terminals are connected to an electrical load.

CHASSIS ELECTRICAL 6-3

The power removed from the battery is replaced by the alternator, restoring the battery to its original chemical state.

The Alternator

On some vehicles there isn't an alternator, but a generator. The difference is that an alternator supplies alternating current which is then changed to direct current for use on the vehicle, while a generator produces direct current. Alternators tend to be more efficient and that is why they are used.

Alternators and generators are devices that consist of coils of wires wound together making big electromagnets. One group of coils spins within another set and the interaction of the magnetic fields causes a current to flow. This current is then drawn off the coils and fed into the vehicles electrical system.

GROUND

Two types of grounds are used in automotive electric circuits. Direct ground components are grounded to the frame through their mounting points. All other components use some sort of ground wire which is attached to the frame or chassis of the vehicle. The electrical current runs through the chassis of the vehicle and returns to the battery through the ground (-) cable; if you look, you'll see that the battery ground cable connects between the battery and the frame or chassis of the vehicle.

➡ It should be noted that a good percentage of electrical problems can be traced to bad grounds.

PROTECTIVE DEVICES

♦ See Figure 2

It is possible for large surges of current to pass through the electrical system of your vehicle. If this surge of current were to reach the load in the circuit, the surge could burn it out or severely damage it. It can also overload the wiring, causing the harness to get hot and melt the insulation. To prevent this, fuses, circuit breakers and/or fusible links are connected into the supply wires of the electrical system. These items are nothing more than a built-in weak spot in the system. When an abnormal amount of current flows through the system, these protective devices work as follows to protect the circuit:

• Fuse—when an excessive electrical current passes through a fuse, the fuse "blows" (the conductor melts) and opens the circuit, preventing the passage of current.

• Circuit Breaker—a circuit breaker is basically a self-repairing fuse. It will open the circuit in the same fashion as a fuse, but when the surge subsides, the circuit breaker can be reset and does not need replacement.

• Fusible Link—a fusible link (fuse link or main link) is a short length of special, high temperature insulated wire that acts as a fuse. When an excessive electrical current passes through a fusible link, the thin gauge wire inside the link melts, creating an intentional open to protect the circuit. To repair the circuit, the link must be replaced. Some newer type fusible links are housed in plug-in modules, which are simply replaced like a fuse, while older type fusible links must be cut and spliced if they melt. Since this link is very early in the electrical path, it's the first place to look if nothing on the vehicle works, yet the battery seems to be charged and is properly connected.

※※ CAUTION

Always replace fuses, circuit breakers and fusible links with identically rated components. Under no circumstances should a component of higher or lower amperage rating be substituted.

SWITCHES & RELAYS

♦ See Figures 3 and 4

Switches are used in electrical circuits to control the passage of current. The most common use is to open and close circuits between the battery and the various electric devices in the system. Switches are rated according to the amount of amperage they can handle. If a sufficient amperage rated switch is not used in a circuit, the switch could overload and cause damage.

Some electrical components which require a large amount of current to operate use a special switch called a relay. Since these circuits carry a large amount of current, the thickness of the wire in the circuit is also greater. If this large wire were connected from the load to the control switch, the switch would have to carry the high amperage load and the fairing or dash would be twice as large to accommodate the increased size of the wiring harness. To prevent these problems, a relay is used.

Relays are composed of a coil and a set of contacts. When the coil has a current passed though it, a magnetic field is formed and this field causes the contacts to move together, completing the circuit. Most relays are normally open,

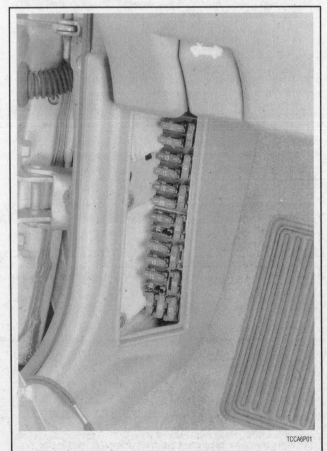

Fig. 2 Most vehicles use one or more fuse panels. This one is located on the driver's side kick panel

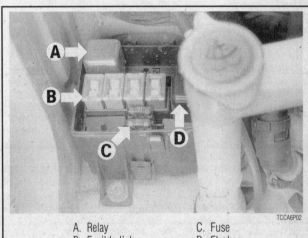

A. Relay   C. Fuse
B. Fusible link   D. Flasher

Fig. 3 The underhood fuse and relay panel usually contains fuses, relays, flashers and fusible links

6-4 CHASSIS ELECTRICAL

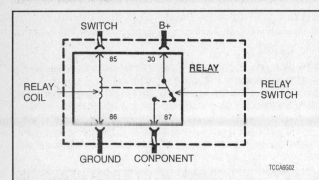

Fig. 4 Relays are composed of a coil and a switch. These two components are linked together so that when one operates, the other operates at the same time. The large wires in the circuit are connected from the battery to one side of the relay switch (B+) and from the opposite side of the relay switch to the load (component). Smaller wires are connected from the relay coil to the control switch for the circuit and from the opposite side of the relay coil to ground

preventing current from passing through the circuit, but they can take any electrical form depending on the job they are intended to do. Relays can be considered "remote control switches." They allow a smaller current to operate devices that require higher amperages. When a small current operates the coil, a larger current is allowed to pass by the contacts. Some common circuits which may use relays are the horn, headlights, starter, electric fuel pump and other high draw circuits.

LOAD

Every electrical circuit must include a "load" (something to use the electricity coming from the source). Without this load, the battery would attempt to deliver its entire power supply from one pole to another. This is called a "short circuit." All this electricity would take a short cut to ground and cause a great amount of damage to other components in the circuit by developing a tremendous amount of heat. This condition could develop sufficient heat to melt the insulation on all the surrounding wires and reduce a multiple wire cable to a lump of plastic and copper.

WIRING & HARNESSES

The average vehicle contains meters and meters of wiring, with hundreds of individual connections. To protect the many wires from damage and to keep them from becoming a confusing tangle, they are organized into bundles, enclosed in plastic or taped together and called wiring harnesses. Different harnesses serve different parts of the vehicle. Individual wires are color coded to help trace them through a harness where sections are hidden from view.

Automotive wiring or circuit conductors can be either single strand wire, multi-strand wire or printed circuitry. Single strand wire has a solid metal core and is usually used inside such components as alternators, motors, relays and other devices. Multi-strand wire has a core made of many small strands of wire twisted together into a single conductor. Most of the wiring in an automotive electrical system is made up of multi-strand wire, either as a single conductor or grouped together in a harness. All wiring is color coded on the insulator, either as a solid color or as a colored wire with an identification stripe. A printed circuit is a thin film of copper or other conductor that is printed on an insulator backing. Occasionally, a printed circuit is sandwiched between two sheets of plastic for more protection and flexibility. A complete printed circuit, consisting of conductors, insulating material and connectors for lamps or other components is called a printed circuit board. Printed circuitry is used in place of individual wires or harnesses in places where space is limited, such as behind instrument panels.

Since automotive electrical systems are very sensitive to changes in resistance, the selection of properly sized wires is critical when systems are repaired. A loose or corroded connection or a replacement wire that is too small for the circuit will add extra resistance and an additional voltage drop to the circuit.

The wire gauge number is an expression of the cross-section area of the conductor. Vehicles from countries that use the metric system will typically describe the wire size as its cross-sectional area in square millimeters. In this method, the larger the wire, the greater the number. Another common system for expressing wire size is the American Wire Gauge (AWG) system. As gauge number increases, area decreases and the wire becomes smaller. An 18 gauge wire is smaller than a 4 gauge wire. A wire with a higher gauge number will carry less current than a wire with a lower gauge number. Gauge wire size refers to the size of the strands of the conductor, not the size of the complete wire with insulator. It is possible, therefore, to have two wires of the same gauge with different diameters because one may have thicker insulation than the other.

It is essential to understand how a circuit works before trying to figure out why it doesn't. An electrical schematic shows the electrical current paths when a circuit is operating properly. Schematics break the entire electrical system down into individual circuits. In a schematic, usually no attempt is made to represent wiring and components as they physically appear on the vehicle; switches and other components are shown as simply as possible. Face views of harness connectors show the cavity or terminal locations in all multi-pin connectors to help locate test points.

CONNECTORS

▶ See Figures 5 and 6

Three types of connectors are commonly used in automotive applications—weatherproof, molded and hard shell.

• Weatherproof—these connectors are most commonly used where the connector is exposed to the elements. Terminals are protected against moisture

Fig. 5 Hard shell (left) and weatherproof (right) connectors have replaceable terminals

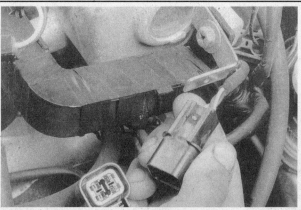

Fig. 6 Weatherproof connectors are most commonly used in the engine compartment or where the connector is exposed to the elements

CHASSIS ELECTRICAL   6-5

and dirt by sealing rings which provide a weathertight seal. All repairs require the use of a special terminal and the tool required to service it. Unlike standard blade type terminals, these weatherproof terminals cannot be straightened once they are bent. Make certain that the connectors are properly seated and all of the sealing rings are in place when connecting leads.

- Molded—these connectors require complete replacement of the connector if found to be defective. This means splicing a new connector assembly into the harness. All splices should be soldered to insure proper contact. Use care when probing the connections or replacing terminals in them, as it is possible to create a short circuit between opposite terminals. If this happens to the wrong terminal pair, it is possible to damage certain components. Always use jumper wires between connectors for circuit checking and NEVER probe through weatherproof seals.
- Hard Shell—unlike molded connectors, the terminal contacts in hard-shell connectors can be replaced. Replacement usually involves the use of a special terminal removal tool that depresses the locking tangs (barbs) on the connector terminal and allows the connector to be removed from the rear of the shell. The connector shell should be replaced if it shows any evidence of burning, melting, cracks, or breaks. Replace individual terminals that are burnt, corroded, distorted or loose.

Test Equipment

Pinpointing the exact cause of trouble in an electrical circuit is most times accomplished by the use of special test equipment. The following describes different types of commonly used test equipment and briefly explains how to use them in diagnosis. In addition to the information covered below, the tool manufacturer's instructions booklet (provided with the tester) should be read and clearly understood before attempting any test procedures.

JUMPER WIRES

※※ CAUTION

Never use jumper wires made from a thinner gauge wire than the circuit being tested. If the jumper wire is of too small a gauge, it may overheat and possibly melt. Never use jumpers to bypass high resistance loads in a circuit. Bypassing resistances, in effect, creates a short circuit. This may, in turn, cause damage and fire. Jumper wires should only be used to bypass lengths of wire or to simulate switches.

Jumper wires are simple, yet extremely valuable, pieces of test equipment. They are basically test wires which are used to bypass sections of a circuit. Although jumper wires can be purchased, they are usually fabricated from lengths of standard automotive wire and whatever type of connector (alligator clip, spade connector or pin connector) that is required for the particular application being tested. In cramped, hard-to-reach areas, it is advisable to have insulated boots over the jumper wire terminals in order to prevent accidental grounding. It is also advisable to include a standard automotive fuse in any jumper wire. This is commonly referred to as a "fused jumper". By inserting an in-line fuse holder between a set of test leads, a fused jumper wire can be used for bypassing open circuits. Use a 5 amp fuse to provide protection against voltage spikes.

Jumper wires are used primarily to locate open electrical circuits, on either the ground (-) side of the circuit or on the power (+) side. If an electrical component fails to operate, connect the jumper wire between the component and a good ground. If the component operates only with the jumper installed, the ground circuit is open. If the ground circuit is good, but the component does not operate, the circuit between the power feed and component may be open. By moving the jumper wire successively back from the component toward the power source, you can isolate the area of the circuit where the open is located. When the component stops functioning, or the power is cut off, the open is in the segment of wire between the jumper and the point previously tested.

You can sometimes connect the jumper wire directly from the battery to the "hot" terminal of the component, but first make sure the component uses 12 volts in operation. Some electrical components, such as fuel injectors or sensors, are designed to operate on about 4 to 5 volts, and running 12 volts directly to these components will cause damage.

TEST LIGHTS

▶ See Figure 7

The test light is used to check circuits and components while electrical current is flowing through them. It is used for voltage and ground tests. To use a 12 volt test light, connect the ground clip to a good ground and probe wherever necessary with the pick. The test light will illuminate when voltage is detected. This does not necessarily mean that 12 volts (or any particular amount of voltage) is present; it only means that some voltage is present. It is advisable before using the test light to touch its ground clip and probe across the battery posts or terminals to make sure the light is operating properly.

※※ WARNING

Do not use a test light to probe electronic ignition, spark plug or coil wires. Never use a pick-type test light to probe wiring on computer controlled systems unless specifically instructed to do so. Any wire insulation that is pierced by the test light probe should be taped and sealed with silicone after testing.

Like the jumper wire, the 12 volt test light is used to isolate opens in circuits. But, whereas the jumper wire is used to bypass the open to operate the load, the 12 volt test light is used to locate the presence of voltage in a circuit. If the test light illuminates, there is power up to that point in the circuit; if the test light does not illuminate, there is an open circuit (no power). Move the test light in successive steps back toward the power source until the light in the handle illuminates. The open is between the probe and a point which was previously probed.

The self-powered test light is similar in design to the 12 volt test light, but contains a 1.5 volt penlight battery in the handle. It is most often used in place of a multimeter to check for open or short circuits when power is isolated from the circuit (continuity test).

The battery in a self-powered test light does not provide much current. A weak battery may not provide enough power to illuminate the test light even when a complete circuit is made (especially if there is high resistance in the circuit). Always make sure that the test battery is strong. To check the battery, briefly touch the ground clip to the probe; if the light glows brightly, the battery is strong enough for testing.

➡A self-powered test light should not be used on any computer controlled system or component. The small amount of electricity transmitted by the test light is enough to damage many electronic automotive components.

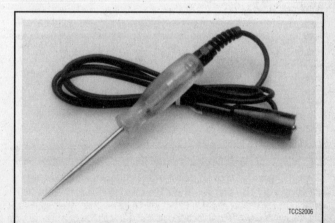

Fig. 7 A 12 volt test light is used to detect the presence of voltage in a circuit

MULTIMETERS

Multimeters are an extremely useful tool for troubleshooting electrical problems. They can be purchased in either analog or digital form and have a price range to suit any budget. A multimeter is a voltmeter, ammeter and ohmmeter (along with other features) combined into one instrument. It is often used when testing solid

6-6 CHASSIS ELECTRICAL

state circuits because of its high input impedance (usually 10 megaohms or more). A brief description of the multimeter main test functions follows:

• Voltmeter—the voltmeter is used to measure voltage at any point in a circuit, or to measure the voltage drop across any part of a circuit. Voltmeters usually have various scales and a selector switch to allow the reading of different voltage ranges. The voltmeter has a positive and a negative lead. To avoid damage to the meter, always connect the negative lead to the negative (-) side of the circuit (to ground or nearest the ground side of the circuit) and connect the positive lead to the positive (+) side of the circuit (to the power source or the nearest power source). Note that the negative voltmeter lead will always be black and that the positive voltmeter will always be some color other than black (usually red).

• Ohmmeter—the ohmmeter is designed to read resistance (measured in ohms) in a circuit or component. Most ohmmeters will have a selector switch which permits the measurement of different ranges of resistance (usually the selector switch allows the multiplication of the meter reading by 10, 100, 1,000 and 10,000). Some ohmmeters are "auto-ranging" which means the meter itself will determine which scale to use. Since the meters are powered by an internal battery, the ohmmeter can be used like a self-powered test light. When the ohmmeter is connected, current from the ohmmeter flows through the circuit or component being tested. Since the ohmmeter's internal resistance and voltage are known values, the amount of current flow through the meter depends on the resistance of the circuit or component being tested. The ohmmeter can also be used to perform a continuity test for suspected open circuits. In using the meter for making continuity checks, do not be concerned with the actual resistance readings. Zero resistance, or any ohm reading, indicates continuity in the circuit. Infinite resistance indicates an opening in the circuit. A high resistance reading where there should be none indicates a problem in the circuit. Checks for short circuits are made in the same manner as checks for open circuits, except that the circuit must be isolated from both power and normal ground. Infinite resistance indicates no continuity, while zero resistance indicates a dead short.

※※ WARNING

Never use an ohmmeter to check the resistance of a component or wire while there is voltage applied to the circuit.

• Ammeter—an ammeter measures the amount of current flowing through a circuit in units called amperes or amps. At normal operating voltage, most circuits have a characteristic amount of amperes, called "current draw" which can be measured using an ammeter. By referring to a specified current draw rating, then measuring the amperes and comparing the two values, one can determine what is happening within the circuit to aid in diagnosis. An open circuit, for example, will not allow any current to flow, so the ammeter reading will be zero. A damaged component or circuit will have an increased current draw, so the reading will be high. The ammeter is always connected in series with the circuit being tested. All of the current that normally flows through the circuit must also flow through the ammeter; if there is any other path for the current to follow, the ammeter reading will not be accurate. The ammeter itself has very little resistance to current flow and, therefore, will not affect the circuit, but it will measure current draw only when the circuit is closed and electricity is flowing. Excessive current draw can blow fuses and drain the battery, while a reduced current draw can cause motors to run slowly, lights to dim and other components to not operate properly.

Troubleshooting Electrical Systems

When diagnosing a specific problem, organized troubleshooting is a must. The complexity of a modern automotive vehicle demands that you approach any problem in a logical, organized manner. There are certain troubleshooting techniques, however, which are standard:

• Establish when the problem occurs. Does the problem appear only under certain conditions? Were there any noises, odors or other unusual symptoms? Isolate the problem area. To do this, make some simple tests and observations, then eliminate the systems that are working properly. Check for obvious problems, such as broken wires and loose or dirty connections. Always check the obvious before assuming something complicated is the cause.

• Test for problems systematically to determine the cause once the problem area is isolated. Are all the components functioning properly? Is there power going to electrical switches and motors. Performing careful, systematic checks will often turn up most causes on the first inspection, without wasting time checking components that have little or no relationship to the problem.

• Test all repairs after the work is done to make sure that the problem is fixed. Some causes can be traced to more than one component, so a careful verification of repair work is important in order to pick up additional malfunctions that may cause a problem to reappear or a different problem to arise. A blown fuse, for example, is a simple problem that may require more than another fuse to repair. If you don't look for a problem that caused a fuse to blow, a shorted wire (for example) may go undetected.

Experience has shown that most problems tend to be the result of a fairly simple and obvious cause, such as loose or corroded connectors, bad grounds or damaged wire insulation which causes a short. This makes careful visual inspection of components during testing essential to quick and accurate troubleshooting.

Testing

OPEN CIRCUITS

▶ See Figure 8

This test already assumes the existence of an open in the circuit and it is used to help locate the open portion.
1. Isolate the circuit from power and ground.
2. Connect the self-powered test light or ohmmeter ground clip to the ground side of the circuit and probe sections of the circuit sequentially.
3. If the light is out or there is infinite resistance, the open is between the probe and the circuit ground.
4. If the light is on or the meter shows continuity, the open is between the probe and the end of the circuit toward the power source.

Fig. 8 The infinite reading on this multimeter indicates that the circuit is open

SHORT CIRCUITS

➥**Never use a self-powered test light to perform checks for opens or shorts when power is applied to the circuit under test. The test light can be damaged by outside power.**

1. Isolate the circuit from power and ground.
2. Connect the self-powered test light or ohmmeter ground clip to a good ground and probe any easy-to-reach point in the circuit.
3. If the light comes on or there is continuity, there is a short somewhere in the circuit.
4. To isolate the short, probe a test point at either end of the isolated circuit (the light should be on or the meter should indicate continuity).
5. Leave the test light probe engaged and sequentially open connectors or switches, remove parts, etc. until the light goes out or continuity is broken.
6. When the light goes out, the short is between the last two circuit components which were opened.

CHASSIS ELECTRICAL 6-7

VOLTAGE

This test determines voltage available from the battery and should be the first step in any electrical troubleshooting procedure after visual inspection. Many electrical problems, especially on computer controlled systems, can be caused by a low state of charge in the battery. Excessive corrosion at the battery cable terminals can cause poor contact that will prevent proper charging and full battery current flow.

1. Set the voltmeter selector switch to the 20V position.
2. Connect the multimeter negative lead to the battery's negative (-) post or terminal and the positive lead to the battery's positive (+) post or terminal.
3. Turn the ignition switch **ON** to provide a load.
4. A well charged battery should register over 12 volts. If the meter reads below 11.5 volts, the battery power may be insufficient to operate the electrical system properly.

VOLTAGE DROP

▶ See Figure 9

When current flows through a load, the voltage beyond the load drops. This voltage drop is due to the resistance created by the load and also by small resistances created by corrosion at the connectors and damaged insulation on the wires. The maximum allowable voltage drop under load is critical, especially if there is more than one load in the circuit, since all voltage drops are cumulative.

1. Set the voltmeter selector switch to the 20 volt position.
2. Connect the multimeter negative lead to a good ground.
3. Operate the circuit and check the voltage prior to the first component (load).
4. There should be little or no voltage drop in the circuit prior to the first component. If a voltage drop exists, the wire or connectors in the circuit are suspect.
5. While operating the first component in the circuit, probe the ground side of the component with the positive meter lead and observe the voltage readings. A small voltage drop should be noticed. This voltage drop is caused by the resistance of the component.
6. Repeat the test for each component (load) down the circuit.
7. If a large voltage drop is noticed, the preceding component, wire or connector is suspect.

1. Isolate the circuit from the vehicle's power source.
2. Ensure that the ignition key is **OFF** when disconnecting any components or the battery.
3. Where necessary, also isolate at least one side of the circuit to be checked, in order to avoid reading parallel resistances. Parallel circuit resistances will always give a lower reading than the actual resistance of either of the branches.
4. Connect the meter leads to both sides of the circuit (wire or component) and read the actual measured ohms on the meter scale. Make sure the selector switch is set to the proper ohm scale for the circuit being tested, to avoid misreading the ohmmeter test value.

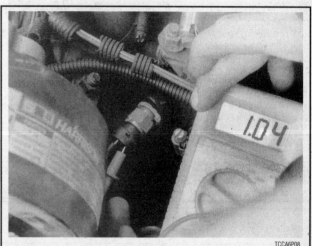

Fig. 10 Checking the resistance of a coolant temperature sensor with an ohmmeter. Reading is 1.04 kilohms

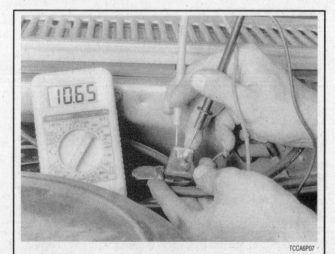

Fig. 9 This voltage drop test revealed high resistance (low voltage) in the circuit

RESISTANCE

▶ See Figures 10 and 11

✻✻ WARNING

Never use an ohmmeter with power applied to the circuit. The ohmmeter is designed to operate on its own power supply. The normal 12 volt electrical system voltage could damage the meter!

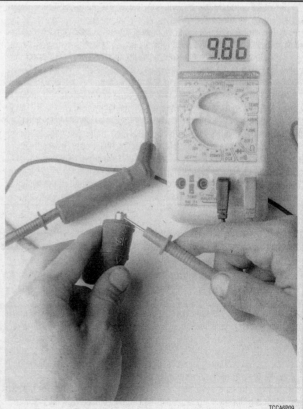

Fig. 11 Spark plug wires can be checked for excessive resistance using an ohmmeter

6-8 CHASSIS ELECTRICAL

Wire and Connector Repair

Almost anyone can replace damaged wires, as long as the proper tools and parts are available. Wire and terminals are available to fit almost any need. Even the specialized weatherproof, molded and hard shell connectors are now available from aftermarket suppliers.

Be sure the ends of all the wires are fitted with the proper terminal hardware and connectors. Wrapping a wire around a stud is never a permanent solution and will only cause trouble later. Replace wires one at a time to avoid confusion. Always route wires exactly the same as the factory.

➡ If connector repair is necessary, only attempt it if you have the proper tools. Weatherproof and hard shell connectors require special tools to release the pins inside the connector. Attempting to repair these connectors with conventional hand tools will damage them.

BATTERY CABLES

Disconnecting the Cables

When working on any electrical component on the vehicle, it is always a good idea to disconnect the negative (-) battery cable. This will prevent potential damage to many sensitive electrical components such as the Engine Control Module (ECM), radio, alternator, etc.

➡ Any time you disengage the battery cables, it is recommended that you disconnect the negative (-) battery cable first. This will prevent your accidentally grounding the positive (+) terminal to the body of the vehicle when disconnecting it, thereby preventing damage to the above mentioned components.

Before you disconnect the cable(s), first turn the ignition to the **OFF** position. This will prevent a draw on the battery which could cause arcing (electricity trying to ground itself to the body of a vehicle, just like a spark plug jumping the gap) and, of course, damaging some components such as the alternator diodes.

When the battery cable(s) are reconnected (negative cable last), be sure to check that your lights, windshield wipers and other electrically operated safety components are all working correctly. If your vehicle contains an Electronically Tuned Radio (ETR), don't forget to also reset your radio stations. Ditto for the clock.

HEATING AND AIR CONDITIONING

Blower Motor

REMOVAL & INSTALLATION

Pickup

♦ See Figure 12

1. Remove the heater assembly.
2. Remove the five screws and separate the halves of the heater assembly.
3. Loosen the fan retaining nut. Lightly tap on the nut to loosen the fan. Remove the fan and nut from the motor shaft.
4. Remove the three motor-to-case retaining screws and disconnect the bullet connector to the resistor and ground screw.
5. Rotate the motor and remove it from the case.

To install:

6. Install the motor in the case, rotating it slightly.
7. Install the retaining screws and connect the bullet connector and ground wire.
8. Install the fan on the shaft and install the nut.
9. Assemble the halves together and install the five retaining screws.
10. Install the heater in the trucks. Check the operation of the heater.

MPV

FRONT

♦ See Figure 13

1. Remove the passenger's side lower panel and lower cover.
2. Disconnect the wiring at the motor connector.
3. Remove the mounting nuts and lift out the blower motor.
4. Installation is the reverse of removal.

REAR

♦ See Figure 14

1. Disconnect the battery ground.
2. Set the rear heater temperature control knob to WARM.
3. Drain the engine coolant.

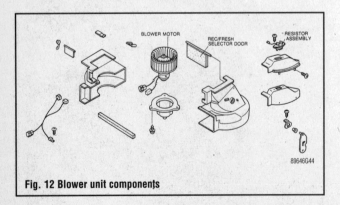

Fig. 12 Blower unit components

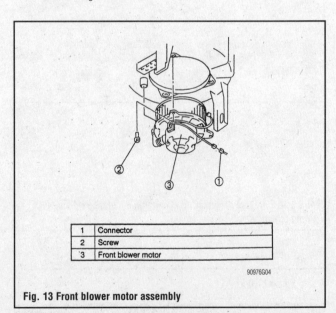

1	Connector
2	Screw
3	Front blower motor

Fig. 13 Front blower motor assembly

CHASSIS ELECTRICAL 6-9

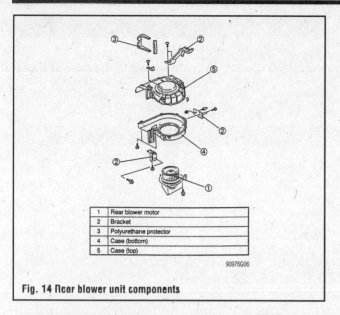

Fig. 14 Rear blower unit components

1	Rear blower motor
2	Bracket
3	Polyurethane protector
4	Case (bottom)
5	Case (top)

4. Remove the driver's seat.
5. Disconnect the heater hoses at the core tubes.
6. Disconnect the wiring at the connector.
7. Remove the mounting bolts and lift out the heater case.
8. Remove the mounting screws and lift out the blower motor.
9. Installation is the reverse of removal. Replace any damaged sealer.

Navajo

▶ See Figure 15

WITHOUT AIR CONDITIONING

▶ See Figures 16, 17, 18 and 19

1. Disconnect the negative battery cable.
2. Remove the air cleaner or air inlet duct, as necessary.
3. Remove the 2 screws attaching the vacuum reservoir to the blower assembly and remove the reservoir.
4. Disconnect the wire harness connector from the blower motor by pushing down on the connector tabs and pulling the connector off of the motor.
5. Disconnect the blower motor cooling tube at the blower motor.
6. Remove the 3 screws attaching the blower motor and wheel to the heater blower assembly.

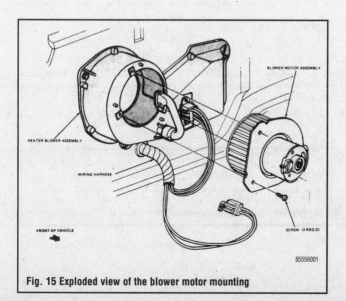

Fig. 15 Exploded view of the blower motor mounting

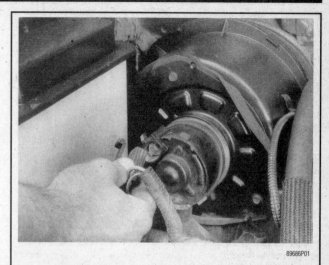

Fig. 16 To remove the blower motor, first disconnect the electrical wire harness plug . . .

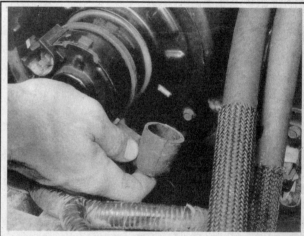

Fig. 17 . . . then remove the blower motor cooling tube from the motor

Fig. 18 Remove the blower motor attaching screws and the washer fluid reservoir . . .

6-10 CHASSIS ELECTRICAL

Fig. 19 . . . then slide the blower motor from the housing. Take care to not damage the blower wheel

7. Holding the cooling tube aside, pull the blower motor and wheel from the heater blower assembly and remove it from the vehicle.
8. Remove the blower wheel push-nut or clamp from the motor shaft and pull the blower wheel from the motor shaft.

To install:

9. Install the blower wheel on the blower motor shaft.
10. Install the hub clamp or push-nut.
11. Holding the cooling tube aside, position the blower motor and wheel on the heater blower assembly and install the 3 attaching screws.
12. Connect the blower motor cooling tube and the wire harness connector.
13. Install the vacuum reservoir on the hoses with the 2 screws.
14. Install the air cleaner or air inlet duct, as necessary.
15. Connect the negative battery cable and check the system for proper operation.

WITH AIR CONDITIONING

1. Disconnect the negative battery cable.
2. In the engine compartment, disconnect the wire harness from the motor by pushing down on the tab while pulling the connection off at the motor.
3. Remove the air cleaner or air inlet duct, as necessary.
4. Remove the solenoid box cover retaining bolts and the solenoid box cover, if equipped.
5. Disconnect the blower motor cooling tube from the blower motor.
6. Remove the 3 blower motor mounting plate attaching screws and remove the motor and wheel assembly from the evaporator assembly blower motor housing.
7. Remove the blower motor hub clamp from the motor shaft and pull the blower wheel from the shaft.

To install:

8. Install the blower motor wheel on the blower motor shaft and install a new hub clamp.
9. Install a new motor mounting seal on the blower housing before installing the blower motor.
10. Position the blower motor and wheel assembly in the blower housing and install the 3 attaching screws.
11. Connect the blower motor cooling tube.
12. Connect the electrical wire harness hardshell connector to the blower motor by pushing into place.
13. Position the solenoid box cover, if equipped, into place and install the 3 retaining screws.
14. Install the air cleaner or air inlet duct, as necessary.
15. Connect the negative battery cable and check the blower motor in all speeds for proper operation.

Heater Core

REMOVAL & INSTALLATION

Pickup

1. Remove the heater from the truck.
2. Remove the five screws and separate the halves of the case.
3. Loosen the hose clamps and slide the heater core from the case.
4. Slide the replacement core into the case. at the same time, connect the core tube to the water valve tube with the short hose and clamps.
5. Assemble the halves of the heater and install the five screws.
6. Install the heater in the truck. Check the operation of the heater.

MPV

FRONT

▶ See Figure 20

1. Drain the engine cooling system.
2. Remove the instrument panel. See the procedure, below.
3. Disconnect the heater hoses at the core tubes.
4. Unbolt and remove the instrument panel brace.
5. Support the heater case and remove the mounting nuts.
6. Carefully lift off the case and remove it from the van. Be Careful! There will be a substantial amount of coolant left in the core.
7. Lift out the heater core.
8. Installation is the reverse of removal. Use new sealer around the case and core. Refill the cooling system. Check for leaks.

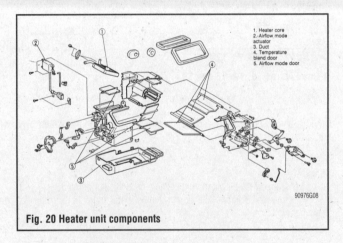

Fig. 20 Heater unit components

REAR

1. Disconnect the battery ground.
2. Set the rear heater temperature control knob to WARM.
3. Drain the engine coolant.
4. Remove the driver's seat.
5. Disconnect the heater hoses at the core tubes.
6. Disconnect the wiring at the connector.
7. Remove the mounting bolts and lift out the heater case.
8. Separate the case halves and lift out the core.
9. Installation is the reverse of removal. Replace any damaged sealer.

Navajo

▶ See Figures 21 thru 26

1. Disconnect the negative battery cable. Allow the engine to cool down. Drain the cooling system.

CHASSIS ELECTRICAL 6-11

> ✶✶ **CAUTION**
>
> When draining the coolant, keep in mind that cats and dogs are attracted by the ethylene glycol antifreeze, and are quite likely to drink any that is left in an uncovered container or in puddles on the ground. This will prove fatal in sufficient quantity. Always drain the coolant into a sealable container. Coolant should be reused unless it is contaminated or several years old.

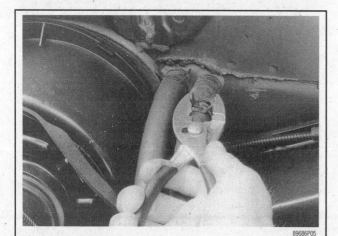

Fig. 21 From inside the engine compartment, disconnect the heater hoses from the core fittings . . .

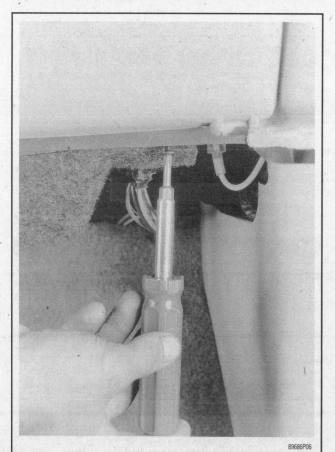

Fig. 22 . . . then, in the passenger compartment, remove the under dash cover retaining screws . . .

Fig. 23 . . . and allow the cover to drop down. If necessary, you can remove the cover if it is in your way

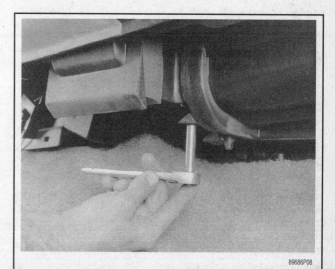

Fig. 24 Remove the heater core access panel attaching screws . . .

Fig. 25 . . . then remove the cover by pulling downward and straight back to disengage the drain tube (arrow)

6-12 CHASSIS ELECTRICAL

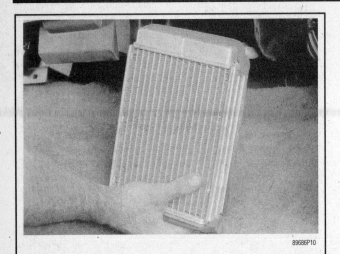

Fig. 26 Pull the heater core rearward and down to remove it from the plenum assembly

2. Disconnect the heater hoses from the heater core tubes and plug hoses.
3. In the passenger compartment, remove the five screws attaching the heater core access cover to the plenum assembly and remove the access cover.
4. Pull the heater core rearward and down, removing it from the plenum assembly.

To install:
5. Position the heater core and seal in the plenum assembly.
6. Install the heater core access cover to the plenum assembly and secure with five screws.
7. Install the heater hoses to the heater core tubes at the dash panel in the engine compartment. Do not over-tighten hose clamps.
8. Check the coolant level and add coolant as required. Connect the negative battery cable.
9. Start the engine and check the system for coolant leaks.

Heater Case Assembly

REMOVAL & INSTALLATION

Pickup

1. Disconnect the battery ground cable.
2. Drain the cooling system.

★ CAUTION

When draining the coolant, keep in mind that cats and dogs are attracted by the ethelyne glycol antifreeze, and are quite likely to drink any that is left in an uncovered container or in puddles on the ground. This will prove fatal in sufficient quantity. Always drain the coolant into a sealable container. Coolant should be reused unless it is contaminated or several years old.

3. Remove the water valve shield at the left side of the heater.
4. Disconnect the two hoses from the left side of the heater.
5. At the heat-defroster door, the water valve and the outside recirculation door, disengage the control cable housing from the mounting clip on the heater. Disconnect each of the three cable wires from the crank arms.
6. Disconnect the fan motor electrical lead.
7. Remove the glove compartment for clearance.
8. Working inside the engine compartment, remove the two retaining nuts and the single bolt and washer which hold the heater to the firewall. Later models also have a retaining bolt inside the passenger compartment which must be removed.
9. Disconnect the two defroster ducts from the heater and remove the heater.

10. Install the heater on the dash so that the heater duct indexes with the air intake duct and the two mounting studs enter their respective holes.
11. From the engine side of the firewall, install the nuts on the mounting studs. While an assistant holds the heater in position, install the mounting bolt.
12. Connect the defroster ducts.
13. Connect the heat-defrost door control cable to the door crank arm. Set the control lever (upper) in the HEAT position and turn the crank arm toward the mounting clip as far as it will go. Engage the cable housing in the clip and install the screw in the clip.
14. Connect the water valve control cable wire to the crank arm on the water valve lever. Locate the cable housing in the mounting clip. Set the control lever in the HOT position and pull the valve plunger and lever to the full outward position. This will move the lever crank arm toward the cable mounting clip as far as it will go. Tighten the clip and screw.
15. Insert the outside recirculation door control cable into the hole in the door crank arm. Bend the wire over and tighten the screw Set the center control lever in the REC position and turn the door crank arm toward the mounting clip as far as it will go. Engage the cable housing in the clip and install the screw in the clip.
16. Connect the fan motor electrical lead.
17. Connect the two hoses to the heater core tubes, at the left side of the heater, and tighten the clamp.
18. Install the water valve shield and tighten the three screws (left side of heater).
19. Refill the cooling system and connect the battery ground cable.
20. Run the engine and check for leaks. Check the operation of the heater.
21. Replace the glove compartment.

MPV

FRONT

1. Drain the engine cooling system.
2. Remove the instrument panel. See the procedure, below.
3. Disconnect the heater hoses at the core tubes.
4. Unbolt and remove the instrument panel brace.
5. Support the heater case and remove the mounting nuts.
6. Carefully lift off the case and remove it from the van. Be Careful! There will be a substantial amount of coolant left in the core.
7. Installation is the reverse of removal. Use new sealer around the case. Refill the cooling system. Check for leaks.

REAR

1. Disconnect the battery ground.
2. Set the rear heater temperature control knob to WARM.
3. Drain the engine coolant.
4. Remove the driver's seat.
5. Disconnect the heater hoses at the core tubes.
6. Disconnect the wiring at the connector.
7. Remove the mounting bolts and lift out the heater case.
8. Installation is the reverse of removal. Replace any damaged sealer.

Air Conditioning Components

REMOVAL & INSTALLATION

Repair or service of air conditioning components is not covered by this manual, because of the risk of personal injury or death, and because of the legal ramifications of servicing these components without the proper EPA certification and experience. Cost, personal injury or death, environmental damage, and legal considerations (such as the fact that it is a federal crime to vent refrigerant into the atmosphere), dictate that the A/C components on your vehicle should be serviced only by a Motor Vehicle Air Conditioning (MVAC) trained, and EPA certified automotive technician.

➡If your vehicle's A/C system uses R-12 refrigerant and is in need of recharging, the A/C system can be converted over to R-134a refrigerant (less environmentally harmful and expensive). Refer to Section 1 for additional information on R-12 to R-134a conversions, and for additional considerations dealing with your vehicle's A/C system.

CHASSIS ELECTRICAL 6-13

Control Panel

REMOVAL & INSTALLATION

Pickup

▶ See Figure 27

1. Disconnect the negative battery cable.
2. Remove the instrument panel meter hood.
3. Remove the attaching screw, knobs and nuts. Disconnect the AC and cigarette lighter connector.
4. Remove the center panel. Remove the glove compartment. Remove the attaching screws and disconnect the control head wire connectors.
5. Remove the control head assembly. Disconnect the control cables.
6. Install in reverse order. Connect the negative battery cable.

MPV

WIRE TYPE

▶ See Figures 28 thru 35

1. Disconnect the negative battery cable.
2. Remove the left and right side lower dash panels and the lower center panel. Remove the steering column cover, if required.
3. Remove the instrument cluster hood assembly. Remove the switch panel knobs, attaching nuts and washers. Remove the switch panel.
4. Remove the attaching screws and remove the temperature control assembly. Remove the airflow mode control unit attaching screws. Remove the control assembly and disconnect the wire connectors.
5. Install in reverse order.
6. Adjust the control wires.

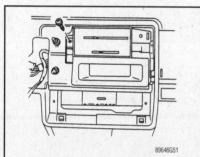

Fig. 27 Remove the attaching screws, detach the wire connectors, disconnect the cables, then remove the control head assembly

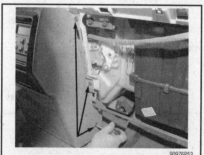

Fig. 28 Remove the three mounting screws on the right side of the lower center panel, then remove the three mounting screws on the left side

Fig. 29 Remove the lower center panel from the vehicle

Fig. 30 Remove the temperature control and fan speed knobs by pulling straight off

Fig. 31 Remove the switch panel mounting screws to access the temperature control assembly

Fig. 32 Turn the switch panel light socket counterclockwise and pull out of the switch panel housing

Fig. 33 Remove the two temperature control assembly mounting screws

Fig. 34 Remove the temperature control assembly after removing the mounting screws

Fig. 35 After disconnecting the control wires and removing the 4 mounting screws, pull out the control panel head

6-14 CHASSIS ELECTRICAL

LOGIC TYPE

▶ See Figures 28 and 29

1. Remove the instrument cluster hood.
2. Remove the left and right side lower dash panels and the lower center panel.
3. Remove the control unit mounting screws and pull the unit out enough to disengage the wiring harness connectors.
4. Installation is the reverse of the removal procedure.

Navajo

▶ See Figure 36

1. Disconnect the negative battery cable.
2. Open the ash tray and remove the 2 screws that hold the ash tray drawer slide to the instrument panel. Remove the ash tray and drawer slide bracket from the instrument panel.
3. Gently pull the finish panel away from the instrument panel and the cluster. The finish panel pops straight back for approximately 1 in., then up to remove. Be careful not to trap the finish panel around the steering column.

➡ If equipped with the electronic 4x4 shift-on-the-fly module, disconnect the wire from the rear of the 4x4 transfer switch before trying to remove the finish panel from the instrument panel.

4. Remove the 4 screws attaching the control assembly to the instrument panel.
5. Pull the control through the instrument panel opening far enough to allow removal of the electrical connections from the blower switch and control assembly illumination lamp. Using a suitable tool, remove the 2 hose vacuum harness from the vacuum switch on the side of the control.
6. At the rear of the control, using a suitable tool, release the temperature and function cable snap-in flags from the white control bracket.
7. On the bottom side of the control, remove the temperature cable from the control by rotating the cable until the T-pin releases the cable. The temperature cable is black with a blue snap-in flag.
8. Pull enough cable through the instrument panel opening until the function cable can be held vertical to the control, then remove the control cable from the function lever. The function cable is white with a black snap-in flag.
9. Remove the control assembly from the instrument panel.

To install:

10. Pull the control cables through the control assembly opening in the instrument panel for a distance of approximately 8 inch (203mm).
11. Hold the control assembly up to the instrument panel with it's face directed toward the floor of the vehicle. This will locate the face of the control in a position that is 90° out of it's installed position.
12. Carefully bend and attach the function cable that has a white color code and a black snap-in terminal to the white plastic lever on the control assembly. Rotate the control assembly back to it's normal position for installation, then snap the black cable flag into the control assembly bracket.

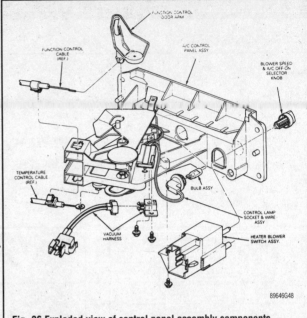

Fig. 36 Exploded view of control panel assembly components

13. On the opposite side of the control assembly, attach the black temperature control cable with the blue plastic snap-in flag to the blue plastic lever on the control. Make sure the end of the cable is seated securely with the T-top pin on the control. Rotate the cable to it's operating position and snap the blue cable flag into the control assembly bracket.
14. Connect the wiring harness to the blower switch and the illumination lamp to it's receptacle on the control assembly. Connect the dual terminal on the vacuum hose to the vacuum switch on the control assembly.
15. Position the control assembly into the instrument panel opening and install the 4 mounting screws.
16. If equipped, reconnect the 4×4 electric shift harness on the rear of the cluster finish panel.
17. Install the cluster finish panel with integral push-pins. Make sure that all pins are fully seated around the rim of the panel.
18. Reinsert the ash tray slide bracket and reconnect the illumination connection circuit. Reinstall the 2 screws that retain the ash tray retainer bracket and the finish panel.
19. Replace the ash tray and reconnect the cigarette lighter.
20. Connect the negative battery cable and check the heater system for proper control assembly operation.

CRUISE CONTROL

▶ See Figures 37, 38 and 39

The vacuum controlled cruise control system consists of the following components:
- Control switches
- Servo or vacuum actuator (throttle actuator)
- Speed sensor
- Stoplamp/brake switch
- Actuator cable
- Cruise control unit
- Clutch switch (manual transmissions)
- Vacuum dump valve (Navajo)
- Amplifier assembly (Navajo)
- Transmission range switch (Pickup and MPV)

The throttle actuator is mounted in the engine compartment and is connected to the throttle linkage with an actuator cable. The speed control amplifier regulates the throttle actuator to keep the requested speed. When the brake pedal is depressed, an electrical signal from the stoplamp switch returns the system to stand-by mode. The vacuum dump valve also mechanically releases the vacuum in the throttle actuator, thus releasing the throttle independently of the amplifier control. This feature is used as a safety backup.

CHASSIS ELECTRICAL 6-15

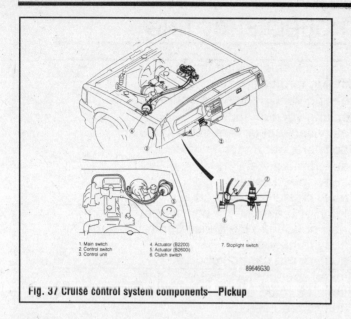

Fig. 37 Cruise control system components—Pickup

Fig. 39 The cruise control actuator unit is mounted on top of the right side strut tower in the engine compartment of the MPV

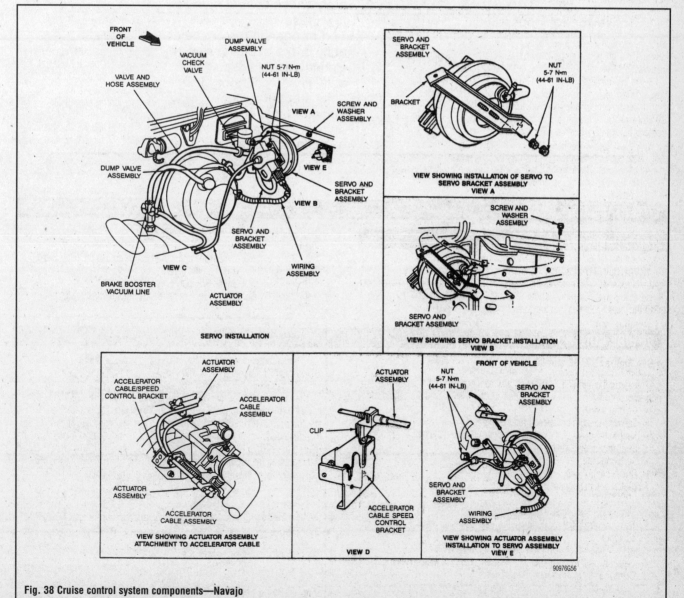

Fig. 38 Cruise control system components—Navajo

CHASSIS ELECTRICAL

CRUISE CONTROL TROUBLESHOOTING

Problem	Possible Cause
Will not hold proper speed	Incorrect cable adjustment Binding throttle linkage Leaking vacuum servo diaphragm Leaking vacuum tank Faulty vacuum or vent valve Faulty stepper motor Faulty transducer Faulty speed sensor Faulty cruise control module
Cruise intermittently cuts out	Clutch or brake switch adjustment too tight Short or open in the cruise control circuit Faulty transducer Faulty cruise control module
Vehicle surges	Kinked speedometer cable or casing Binding throttle linkage Faulty speed sensor Faulty cruise control module
Cruise control inoperative	Blown fuse Short or open in the cruise control circuit Faulty brake or clutch switch Leaking vacuum circuit Faulty cruise control switch Faulty stepper motor Faulty transducer Faulty speed sensor Faulty cruise control module

Note: Use this chart as a guide. Not all systems will use the components listed.

TCCA6C01

ENTERTAINMENT SYSTEMS

Radio Reciever/Tape Player

REMOVAL & INSTALLATION

Pickup

♦ See Figure 40

1. Disconnect the battery ground.
2. Remove the front console assembly.
3. Remove the ashtray from the instrument panel.
4. Remove the audio box from the instrument panel.
5. Disconnect the antenna lead, wiring and speaker connectors behind the radio.
6. Remove the radio mounting bracket.
7. Remove the radio pod mounting screws and lift out the unit.
8. Installation is the reverse of removal.

MPV

♦ See Figures 41 thru 47

1. Disconnect the negative battery cable.
2. Remove the ashtray by pushing down on the inside metal flap and then pulling straight out.
3. Remove the two lower center bezel mounting screws.
4. Carefully pull out the center bezel, disengaging the metal retaining clips.
5. Remove the five radio mounting screws.
6. Carefully pull the radio out of the instrument panel far enough to unplug the antenna leads and wiring connectors from the radio.
7. Remove the radio from the vehicle.

To install:

8. Connect the antenna leads and wiring connectors to the back of the radio, then slide into the instrument panel.
9. Install and tighten the five mounting screws.
10. Install the center bezel and tighten the two mounting screws.

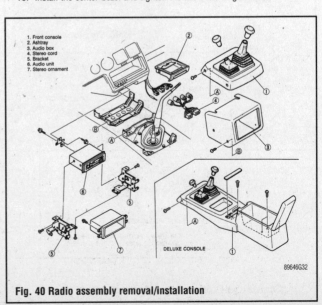

Fig. 40 Radio assembly removal/installation

CHASSIS ELECTRICAL 6-17

Fig. 41 Push down on the upper retaining flap and pull out the ashtray

Fig. 42 Remove the two center bezel retaining screws located at the bottom edge of the bezel

Fig. 43 Pull the bezel away from the dashboard disengaging the metal clips around the bezel

Fig. 44 Remove the five radio unit mounting screws

Fig. 45 Carefully pull the radio unit out of the center dash . . .

Fig. 46 . . . then unplug the wiring harness connectors from behind the radio . . .

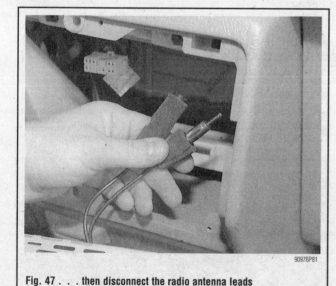

Fig. 47 . . . then disconnect the radio antenna leads

11. Install the ashtray.
12. Connect the negative battery cable.

Navajo

KNOB TYPE RADIO

1. Disconnect the battery ground cable.
2. Remove the knobs and discs from the radio control shafts.
3. Remove the two steering column shroud-to-panel retaining screws and remove the shroud.
4. Detach the cluster trim cover or appliques from the instrument panel by removing the eight screws.
5. Remove the four screws securing the mounting plate assembly to the instrument panel and remove the radio with the mounting plate and rear bracket.
6. Disconnect the antenna lead-in cable, speaker wires and the radio (power) wire.
7. Remove the nut and washer assembly attaching the radio rear support.
8. Remove the nuts and washers from the radio control shafts and remove the mounting plate from the radio.

To install:

9. Install the radio rear support using the nut and washer assembly.
10. Install the mounting plate to the radio using the two lock washers and two nuts.
11. Connect the wiring connectors to the radio and position the radio with the mounting plate to the instrument panel.

➡ Make sure that the hair pin area of the rear bracket is engaged to the instrument panel support.

12. Secure the mounting plate to the instrument panel with the four screws.

➡ Make sure the mounting plate is fully seated on the instrument panel.

13. Install the panel trim covers and steering column shroud.
14. Install the panel knobs and discs to the radio control shafts.
15. Connect the battery ground cable.

ELECTRONICALLY TUNED RADIO (ETR)

▶ See Figures 48, 49, 50, 51 and 52

1. Disconnect the negative battery cable.
2. If necessary, remove the finish panel from around the radio assembly.
3. Insert the radio removal tool 49-UN01-050 or equivalent, into the radio face.
4. Press in 1 inch (25.4mm) to release the retaining clips, then using the tool as handles, pull the radio out of the instrument panel.

6-18 CHASSIS ELECTRICAL

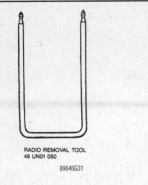

Fig. 48 Example of the special radio removal tool

Fig. 49 To remove the radio, insert the removal tool prongs into the release clip access holes (arrows)

Fig. 50 Push in approximately 1 inch (25.4mm) to release the retainer clips, then pull straight out to remove

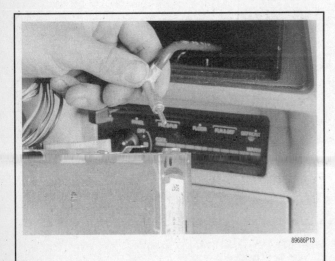

Fig. 51 Disconnect the antenna . . .

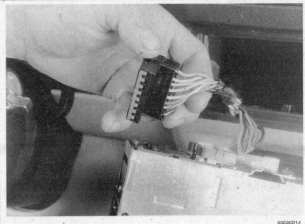

Fig. 52 . . . and the electrical wire harness plugs from the back of the radio

5. Detach the antenna and wiring connectors from the radio.

To install:

6. Connect the wiring and slide the radio into the instrument panel. Ensure that the rear mounting bracket is engaged on the mounting track in the panel.
7. If removed, install the finish panel.
8. Connect the battery cable.
9. Check the operation of the radio.

Speakers

REMOVAL & INSTALLATION

※※ WARNING

Never operate the radio with one of the speaker disconnected. Damage to the radio can occur.

Door Mounted Speakers

♦ See Figures 53, 54 and 55

1. Remove the door trim panel.
2. If necessary, peel away the plastic inner liner.
3. Remove the speaker attaching screws.
4. Pull the speaker out from the door frame and disconnect the wire harness plug.
5. Installation is the reverse of the removal procedure.

Fig. 53 Carefully peel away the inner door liner to access the speaker mounting bolts

CHASSIS ELECTRICAL 6-19

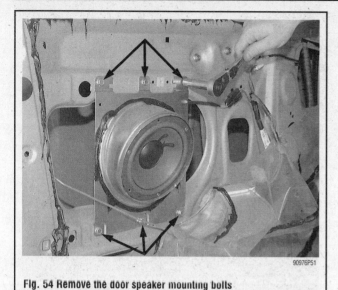

Fig. 54 Remove the door speaker mounting bolts

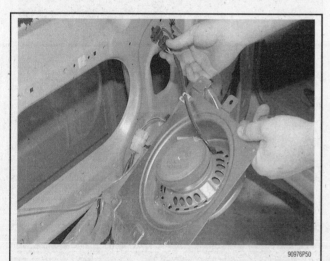

Fig. 55 Disengage the door mounted speaker wiring connector and remove from the vehicle

Body Mounted Rear Speakers

NAVAJO

▶ See Figure 56

1. Gently pry the speaker grille from the trim panel.
2. Remove the 4 screws that mount the speaker unit to the mounting bracket.
3. Pull the speaker out from the trim panel and disconnect the wire harness plug.
4. Installation is the reverse of the removal procedure.

MPV MODELS

▶ See Figure 57

1. Remove the rear interior side trim.
2. Remove the three speaker mounting screws.
3. Carefully pull the speaker out and disconnect the wire harness plug.
4. Installation is the reverse of the removal procedure.

PICKUP (STANDARD)

1. Remove the seat belt upper anchor retaining bolt.
2. Remove the back upper garnish and B pillar trim.

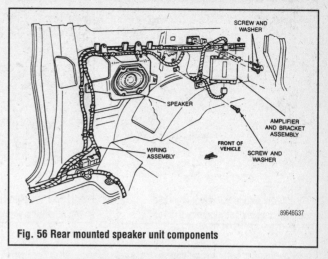

Fig. 56 Rear mounted speaker unit components

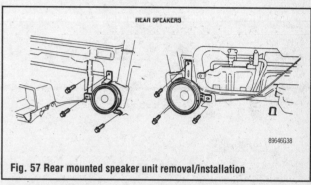

Fig. 57 Rear mounted speaker unit removal/installation

3. Remove the 4 speaker mounting screws.
4. Carefully pull the speaker out and disconnect the wire harness plug.
5. Installation is the reverse of the removal procedure.

PICKUP (CAB PLUS)

1. Remove the seat belt upper anchor retaining bolt.
2. Remove the quarter window glass.
3. Remove the back upper garnish, B pillar upper and lower trim.
4. Remove the 3 speaker mounting screws.
5. Carefully pull the speaker out and disconnect the wire harness plug.
6. Installation is the reverse of the removal procedure.

Dash Mounted Speakers

MPV

▶ See Figures 58, 59, 60, 61 and 62

➡This procedure applies to both right and left side dash mounted speakers.

1. Remove the dashboard end panel by carefully prying outward to disengage the retaining clips.
2. Remove all of the dashboard lower panel mounting screws.
3. Carefully pull the lower panel outward and unplug the speaker wiring connector.
4. Remove the speaker mounting fasteners and remove from the vehicle.
5. Installation is the reverse of the removal procedure.

PICKUP

➡This procedure applies to both right and left side dash mounted speakers.

1. Remove the speaker grille mounting screws. Remove the speaker grille.
2. Remove the speaker unit mounting screws.
3. Carefully pull the speaker unit out just enough to unplug the wiring harness.
4. Remove the speaker from the vehicle.
5. Installation is the reverse of the removal procedure.

6-20 CHASSIS ELECTRICAL

Fig. 58 Disengage the metal clips and remove the side dashboard end cover . . .

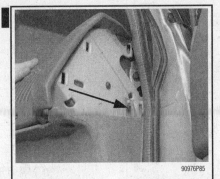

Fig. 59 . . . then remove the lower cover end mounting screw

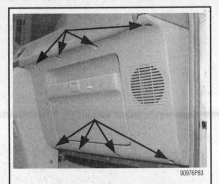

Fig. 60 Remove the lower dash panel cover mounting screws

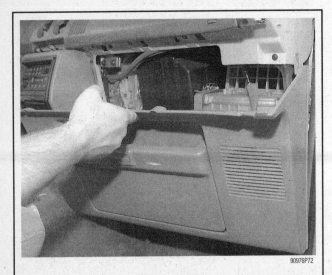

Fig. 61 Carefully pull away the lower dash cover . . .

Fig. 62 . . . then unplug the wire from the lower dash cover speaker, loosen the mounting fasteners and remove from the vehicle

WINDSHIELD WIPERS

Windshield Wiper Blade and Arm

REMOVAL & INSTALLATION

Front Wipers

PICKUP AND MPV

♦ See Figures 63 and 64

1. To remove the blade and arm, remove cap, unscrew the retaining nut and pry the blade and arm from the pivot shaft. The shaft and arm are serrated to provide for adjustment of the wiper pattern on the glass.
2. To set the arms back in the proper park position, turn the wiper switch on and allow the motor to cycle three or four times. Then turn off the wiper switch (do not turn off the wiper motor with the ignition key). This will place the wiper shafts in the proper park position.

➡Using a wire brush, clean the wiper arm connector shafts before installing the wiper arms.

3. Install the blade and arm on the shaft and install the retaining nut. Install the caps. The blades and arms should be positioned according to the illustration.

NAVAJO

To remove the arm and blade assembly, raise the blade end of the arm off of the windshield and move the slide latch away from the pivot shaft. The wiper

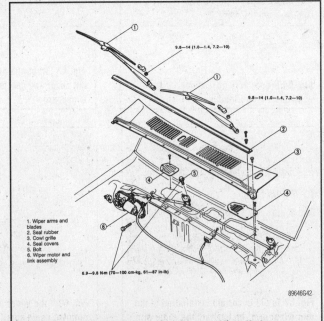

Fig. 63 Pickup models wiper assembly removal/installation

CHASSIS ELECTRICAL 6-21

arm can now be removed from the shaft without the use of any tools. To install, push the main head over the pivot shaft. Be sure the wipers are in the parked position, and the blade assembly is in its correct position. Hold the main arm head onto the pivot shaft while raising the blade end of the wiper arm and push the slide latch into the lock under the pivot shaft head. Then, lower the blade to the windshield. If the blade does not lower to the windshield, the slide latch is not completely in place.

Rear Wiper

MPV MODELS

▶ See Figures 65, 66, 67 and 68

1. To remove the blade and arm, unscrew the retaining nut and pry the blade and arm from the pivot shaft. The shaft and arm are serrated to provide for adjustment of the wiper pattern on the glass.
2. To set the arm back in the proper park position, turn the wiper switch on and allow the motor to cycle three or four times. Then turn off the wiper switch (do not turn off the wiper motor with the ignition key). This will place the wiper shafts in the proper park position.
3. Install the blade and arm on the shaft and install the retaining nut. The blades and arms should be positioned according to the illustration.

NAVAJO MODELS

▶ See Figure 69

❊❊ WARNING

Use a towel or similar device to protect the vehicle finish when performing this procedure.

1. Raise the windshield wiper blade/arm off of the glass and place it into the service position.

2. Using a small, flat bladed prytool, release the retaining clip at the base of the windshield wiper pivot arm.
3. Carefully pry the wiper arm from the shaft.

To install:
4. Ensure that the wiper motor pivot shaft is in the park position.
5. Position the wiper arm over the shaft and firmly push it on until it stops.
6. Lower the wiper blade/arm against the glass.

➡The blade is properly positioned when it firmly contacts the windshield wiper arm stop. If it does not, remove the arm again and reposition it so that it does.

Windshield Wiper Motor

REMOVAL & INSTALLATION

Pickup

1. Remove the wiper arm/blade assembly. Note that the arms are different. Don't confuse them.
2. Remove the rubber seal from the leading edge of the cowl.
3. Unbolt and remove the cowl.
4. Remove the access hole covers.
5. Remove the bolts holding the wiper shaft drives.
6. Matchmark the position of the wiper crank arm in relation to the face of the wiper motor. Disconnect the wiper linkage from the wiper motor crank arm.
7. Remove the wiper linkage.
8. Unbolt and remove the wiper motor. Disconnect the wiring harness.
9. Installation is the reverse of removal. Make sure that the parked height of the wiper arms, measured from the blade tips to the windshield molding is 0.787 inch (20mm). Tighten the arm retaining nuts to 8–10 ft. lbs. (11–14 Nm).

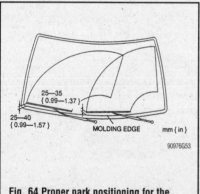

Fig. 64 Proper park positioning for the windshield wiper blades—MPV shown

Fig. 65 Lift up the plastic wiper arm pivot nut cover and carefully remove from the wiper arm

Fig. 66 Using a 10mm box wrench, remove the rear wiper arm pivot nut

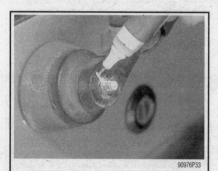

Fig. 67 To aid in correct installation of the rear wiper arm, matchmark the wiper arm to the pivot shaft

Fig. 68 If the wiper arm is difficult to remove, use a small puller tool such as the one shown

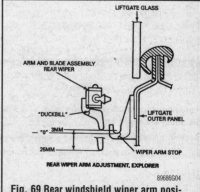

Fig. 69 Rear windshield wiper arm positioning

6-22 CHASSIS ELECTRICAL

MPV

FRONT

1. Remove the wiper arms.
2. Remove the drive link nuts from the top of the cowl.
3. Working under the hood, disconnect the battery ground cable.
4. Remove the motor and linkage mounting bolts and lift out the assembly.
5. After removal, you can separate the linkage from the motor. If you don't have to, don't remove the motor arm from the motor. The position of the arm on the motor shaft controls the automatic park position.
6. Installation is the reverse of removal. When installing the wiper arms, make sure that the at-rest position gives a gap of 30mm between the blade tips and the lower windshield molding.

REAR

▶ See Figures 70 thru 75

1. Disconnect the battery ground cable.
2. Remove the protective cap and wiper arm nut and pull off the wiper arm, seal and bushing.
3. Remove the liftgate inner trim panels.
4. Remove the weathershield.
5. Disconnect the wiper motor wiring.
6. Remove the motor mounting bolts and lift out the motor.
7. Installation is the reverse of removal.

Navajo

FRONT

▶ See Figure 76

1. Turn the wiper switch on. Turn the ignition switch on until the blades are straight up and then turn ignition off to keep them there.
2. Remove the right wiper arm and blade.
3. Remove the negative battery cable.
4. Remove the right pivot nut and allow the linkage to drop into the cowl.
5. Remove the linkage access cover, located on the right side of the dash panel near the wiper motor.
6. Reach through the access cover opening and unsnap the wiper motor clip.
7. Push the clip away from the linkage until it clears the nib on the crank pin. Then, push the clip off the linkage.
8. Remove the wiper linkage from motor crank pin.
9. Disconnect the wiper motor's wiring connector.
10. Remove the wiper motor's three attaching screws and remove the motor.

To install:

11. Install the motor and attach the three attaching screws. Tighten to 60–65 inch lbs. (6–7 Nm).
12. Connect the wiper motor's wiring connector.
13. Install the clip completely on the right linkage. Make sure the clip is completely on.
14. Install the left linkage on the wiper motor crank pin.
15. Install the right linkage on the wiper motor crank pin and pull the linkage on to the crank pin until it snaps.

➡The clip is properly installed if the nib is protruding through the center of the clip.

16. Reinstall the right wiper pivot shaft and nut.
17. Reconnect the battery and turn the ignition **ON**. Turn the wiper switch off so the wiper motor will park, then turn the ignition **OFF**. Replace the right linkage access cover.
18. Install the right wiper blade and arm.
19. Check the system for proper operation.

REAR

1. Disconnect the negative battery cable.
2. Remove the wiper arm and blade.
3. Remove the liftgate interior trim.
4. Remove the motor attaching bolts (3). Disconnect the electrical leads.

Fig. 70 After removing the wiper arm, pull the bushing off of the motor shaft

Fig. 71 Using an adjustable wrench, or a 29mm box wrench, loosen the motor shaft-to-tailgate retaining nut

Fig. 72 Remove the motor shaft retaining nut and shaft grommet

Fig. 73 Disengage the black rear wiper motor electrical connector

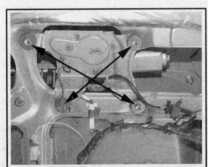

Fig. 74 Location of the rear wiper motor assembly mounting bolts. Remove these bolts using a 10mm socket

Fig. 75 Remove the rear wiper motor assembly from the tailgate door

CHASSIS ELECTRICAL 6-23

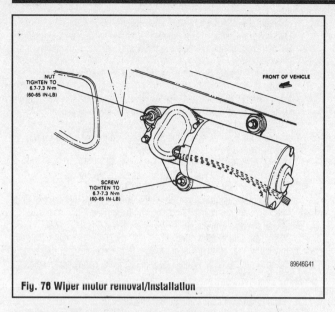

Fig. 76 Wiper motor removal/installation

5. Remove the wiper motor from the vehicle.
6. Install the wiper motor in position and connect the electrical leads.
7. Install the liftgate trim. Connect the negative battery cable.

Windshield Washer Motor

REMOVAL & INSTALLATION

Front Washer Pump Motor

WASHER MOTOR SECURED BY A RETAINING RING

▶ See Figures 77 and 78

※※ WARNING

Never operate the washer fluid pump without fluid in the reservoir. Repeatedly running the reservoir dry will ruin the pump.

➡The front washer fluid reservoir is connected to the coolant overflow tank, which will be removed as an assembly.

1. Disconnect the electrical plug and hose from the washer fluid reservoir. Also, if removing the front washer pump, disconnect the coolant overflow hose from its reservoir.
2. Remove the reservoir attaching screws and lift the assembly from the vehicle.
3. Drain and discard the contents of the washer fluid reservoir.
4. Using a small prytool, pry out the pump retaining ring.
5. Use a pair of pliers and grasp the pump on one wall around the electrical terminals. Pull out the motor and seal from the reservoir.

To install:

6. Thoroughly clean the reservoir pump chamber and ensure that there is no foreign material in it.
7. Lubricate the outside of the pump seal with a dry lubricant, such as graphite powder.

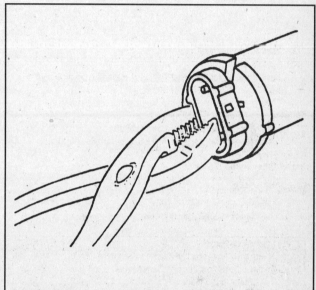

Fig. 78 Pull the washer pump from the reservoir using a pair of pliers as shown

8. If equipped, align the small tab on the motor with the notch on the reservoir.
9. Install the motor so that the seal seats against the bottom of the chamber.
10. Use a 1 inch (25.4mm) socket, preferable a 12-point socket, hand press the retaining ring securely against the motor end plate.
11. Install the reservoir assembly.
12. Connect the hoses and electrical harness plug.
13. Fill the reservoir and operate the washer system. Check for leaks.

WASHER PUMP PRESS FIT INTO RESERVOIR

▶ See Figure 79

1. Remove the windshield washer fluid reservoir mounting screws.
2. Disconnect the hose(s) from the reservoir tank.
3. Unplug the electrical connector from the washer pump.
4. Remove the washer fluid reservoir from the engine compartment.
5. Empty the washer fluid reservoir.
6. Using a small prying tool, pry out the pump, being careful not to damage the plastic housing.
7. Remove the rubber grommet and inspect for damage or debris.

To install:

8. Insert the rubber grommet, then lubricate the inside diameter of the grommet with soapy water and install the pump into the reservoir until it is firmly seated.

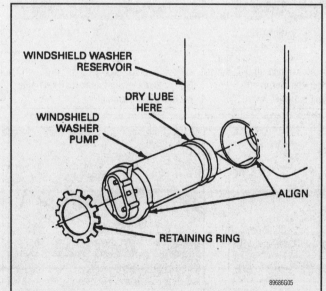

Fig. 77 Exploded view of the washer pump and retaining ring mounting

6-24 CHASSIS ELECTRICAL

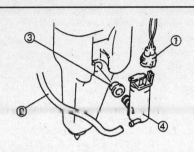

1	Connector
2	Windshield washer pipe
3	Grommet
4	Windshield washer motor

90976G50

Fig. 79 Exploded view of the press fit

grommet with soapy water and install the pump into the reservoir until it is firmly seated.
  9. Connect the hose(s), then place the reservoir into position in the engine compartment and tighten the mounting screws.
  10. Plug the wiring harness connector to the pump.
  11. Fill the reservoir with washer fluid and check pump operation.

Rear Washer Pump Motor

NAVAJO MODELS

1. Remove the left side rear quarter trim panel as follows:
   a. If equipped, remove the luggage cover from the retainer.
   b. Remove the door scuff plate from the floor.
   c. Remove the liftgate scuff plate from the floor.
   d. Remove the front seat belt attachment from the floor.
   e. Remove the rear seat belt anchor from the "B" pillar.
   f. Remove the side window latch.
   g. Remove the mounting screws and pushpins securing the quarter trim panel assembly. Remove the panel from the vehicle.
2. Disconnect the filler hose.

3. Unplug the washer pump motor electrical connector.
4. Loosen the mounting screws and remove the rear window washer reservoir.
5. Using a small prying tool, pry out the pump, being careful not to damage the plastic housing.
6. Remove the one piece seal/filter and inspect for damage or debris.

To install:
7. Insert the seal, then lubricate the inside diameter of the seal with soapy water and install the pump into the reservoir until it is firmly seated.
8. Install the washer hose to the pump discharge nipple, then place the reservoir into position on the quarter panel and tighten the mounting screws.
9. Plug the wiring harness connector to the pump.
10. Connect the filler hoses.
11. Fill the reservoir with washer fluid and check pump operation.
12. Install the left side rear quarter trim panel as follows:
   a. Position the side trim panel inside the vehicle. Insert the seat belt floor attachment through the opening in the panel, if equipped.
   b. Install the side trim panel.
   c. Install the side window latch.
   d. Install the rear seat belt anchor to the "B" pillar. Install the front seat belt attachment to the floor. Tighten the seat belt anchor and attachments to 22–30 ft. lbs. (30–40 Nm).
   e. Install the scuff plate to the floor.
   f. Install the lower garnish molding to the floor.
   g. If equipped, install the luggage cover.

MPV MODELS

1. Disconnect the negative battery cable.
2. Remove the mat set end plate.
3. Remove the right side rear trim panel.
4. Disconnect the washer fluid hose and wiring harness.
5. Loosen the mounting screws and remove the rear washer fluid reservoir.
6. Empty out the fluid from the reservoir.
7. Loosen the two retaining screws and remove the rear washer motor from the reservoir.

To install:
8. Hold the washer motor in place on the reaservoir and install the two retaining screws.
9. Hold the washer fluid reservoir in position, then connect the fluid hose and wiring harness. Install and tighten the mounting screws.
10. Connect the negative battery cable.
11. Fill the reservoir with washer fluid and check pump operation.
12. Install the right side rear trim panel.
13. Install the mat set end plate.

INSTRUMENTS AND SWITCHES

Instrument Cluster

REMOVAL & INSTALLATION

Pickup

♦ See Figure 80

1. Disconnect the battery ground cable.
2. Reach behind the cluster and disconnect the speedometer cable.
3. Remove the screws attaching the cluster hood and carefully lift the hood off.
4. Remove the screw attaching the cluster pod to the dash panel and pull the pod out toward you, gradually. Reach behind the pod and disconnect the wiring connectors.
5. Remove the trip meter knob, and, on clusters w/tachometer, the clock adjust knob.
6. Remove the screws retaining the lens cover and lift off the cover.
7. Remove the screws retaining the cluster bezel and lift off the bezel.
8. Lift out the warning light plate.
9. On clusters wo/tachometer, remove, in order:
  • fuel gauge
  • speedometer
  • temperature gauge
  • printed circuit board
10. On cluster w/tachometer, remove, in order:
  • speedometer
  • digital clock

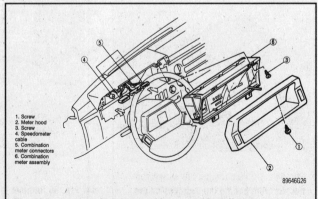

1. Screw
2. Meter hood
3. Screw
4. Speedometer cable
5. Combination meter connectors
6. Combination meter assembly

89646G26

Fig. 80 Instrument cluster assembly removal/installation

CHASSIS ELECTRICAL   6-25

- tachometer
- fuel gauge
- temperature gauge
- printed circuit board

11. Installation is the reverse of removal.

MPV

▶ See Figures 81 thru 86

1. Disconnect the negative battery cable.
2. Using a small prying tool, remove the trim panel at the front edge of the instrument cluster hood, near the bottom edge of the windshield.
3. Remove the three instrument cluster hood mounting screws underneath the trim panel.
4. Remove the two instrument cluster hood mounting screws located under the bootom edge of the hood assembly.
5. Carefully pull the instrument cluster hood assembly away from the dashboard far enough to unplug the wiring harness connectors from behind the switches on each side.
6. Remove the 4 cluster assembly retaining screws and pull the cluster towards you slowly and carefully, until you can reach behind it and unplug the electrical connectors, speedometer cable and unhook the transmission range indicator wire from the steering column.
7. Remove the cluster.

To install:

8. Hold the instrument cluster close to the dashboard and route the transmission range indicator wire through the dashboard.
9. Connect the speedometer cable to the back of the cluster assembly and plug in the electrical connectors.
10. Place the cluster into correct position, then install and tighten the four cluster retaining screws. Attach the transmission range indicator wire from the steering column.
11. Hold the instrument cluster hood assembly close to the dashboard and plug in wiring harness connectors.
12. Place the instrument cluster hood assembly onto the dashboard. Install and tighten the five mounting screws.
13. Install the trim panel cover.
14. Connect the negative battery cable. Check the operation of the instrument cluster gauges.

Navajo

▶ See Figure 87

1. Disconnect the battery ground cable.
2. Remove the two steering column shroud-to-panel retaining screws and remove the shroud.
3. Remove the lower instrument panel trim.
4. Remove the cluster trim cover from the instrument panel by removing the eight screws.
5. Remove the four instrument cluster to panel retaining screws.
6. Position the cluster slightly away from the panel for access to the back of the cluster to disconnect the speedometer.

➡ If there is not sufficient access to disengage the speedometer cable from the speedometer, it may be necessary to remove the speedometer cable at the transmission and pull cable through cowl, to allow room to reach the speedometer quick disconnect.

7. Disconnect the wiring harness connector from the printed circuit, and any bulb-and-socket assemblies from the wiring harness to the cluster assembly and remove the cluster assembly from the instrument panel.

To install:

8. Apply approximately 1/8 inch diameter ball of D7AZ-19A331–A Silicone Dielectric compound or equivalent in the drive hole of the speedometer head.
9. Position the cluster near its opening in the instrument panel.

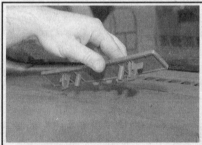

Fig. 81 Lift up the trim panel from the front edge of the instrument cluster hood assembly to access and remove the three mounting screws

Fig. 82 Remove the lower instrument cluster hood mounting screws

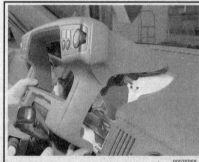

Fig. 83 Carefully pull the instrument cluster hood assembly away from the dashboard . . .

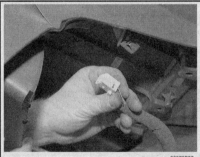

Fig. 84 . . . then unplug the wiring connectors from behind the switches on the instrument cluster hood assembly

Fig. 85 Remove the four instrument cluster assembly mounting screws

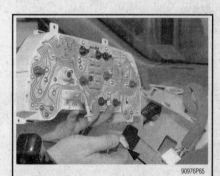

Fig. 86 When removing the instrument cluster, disconnect the transmission selector cable from the steering column

6-26 CHASSIS ELECTRICAL

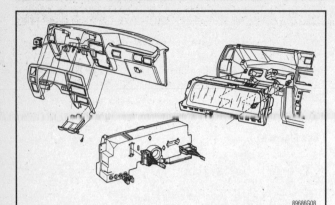

Fig. 87 Exploded view of the instrument cluster and trim panel assemblies

10. Connect the wiring harness connector to the printed circuit, and any bulb-and-socket assemblies from the wiring harness to the cluster assembly.
11. Position the cluster to the instrument panel and install the four cluster to panel retaining screws.
12. Install the panel trim covers and the steering column shroud.
13. Connect the battery ground cable.
14. Check operation of all gauges, lamps and signals.

Fuel, Oil Pressure, Voltage and Coolant Temperature Gauges

Each of the gauges can be removed in the same manner, once the instrument cluster is removed.
1. Disconnect the negative battery cable.
2. Remove the instrument cluster assembly.
3. Remove the lens from the instrument cluster.
4. If the gauge is retained by mounting screws, remove the screws. Pull the gauge from the cluster.

To install:
5. Install the gauge, by pushing it firmly into position. If necessary, secure with mounting screws.
6. Install the cluster lens and install the cluster into the instrument panel.

Windshield Wiper Switch

REMOVAL & INSTALLATION

The windshield wiper switch is part of the multi-function switch mounted in the steering column. For service procedures, see Section 8.

Rear Wiper Switch

REMOVAL & INSTALLATION

Navajo Models

1. Disconnect the negative battery cable.
2. Loosen the two mounting screws and remove the ashtray.
3. Disengage the retaining clips and remove the instrument cluster finish panel.
4. Unsnap and remove the mounting bezel that contains the rear wiper switch.
5. Disconnect the electrical lead from the switch.
6. Remove the switch from the mounting bezel by pushing on the switch from the connector side until the mounting clips unsnap.

To install:
7. Connect the wiring and install the switch in the instrument panel.
8. Install the instrument cluster finish panel.
9. Install the ashtray and tighten the two mounting screws.
10. Connect the negative battery cable. Check the operation of the switch.

1994–95 MPV Models

▶ See Figures 81, 82, 83 and 84

1. Disconnect the negative battery cable.
2. Using a small prying tool, remove the trim panel at the front edge of the instrument cluster hood, near the bottom edge of the windshield.
3. Remove the three instrument cluster hood mounting screws underneath the trim panel.
4. Remove the two instrument cluster hood mounting screws located under the bootom edge of the hood assembly.
5. Carefully pull the instrument cluster hood assembly away from the dashboard far enough to unplug the wiring harness connectors from behind the switches on each side.
6. Remove the mounting screws, then separate the rear wiper and washer switch from the left side of the cluster hood.

To install:
7. Place the rear wiper and washer switch onto the cluster hood and tighten the mounting screws.
8. Hold the instrument cluster hood assembly close to the dashboard and plug in wiring harness connectors.
9. Place the instrument cluster hood assembly onto the dashboard. Install and tighten the five mounting screws.
10. Install the trim panel cover.
11. Connect the negative battery cable. Check the operation of the switch.

Headlight Switch

REMOVAL & INSTALLATION

Pickup and MPV

The switch is part of the multi-function switch mounted in the steering column. For service procedures, see Section 8.

Navajo

1. Disconnect the battery ground cable.
2. Pull the headlight switch knob to the headlight on position.
3. Depress the shaft release button and remove the knob and shaft assembly.
4. Remove the instrument panel finish panel.
5. Unscrew the mounting nut and remove the switch from the instrument panel, then remove the wiring connector from the switch.

To install:
6. Connect the wiring connector to the headlamp switch, position the switch in the instrument panel and install the mounting nut.
7. Install the instrument panel finish panel.
8. Install the headlamp switch knob and shaft assembly by pushing the shaft into the switch until it locks into position.
9. Connect the battery ground cable, and check the operation of the headlight switch.

CHASSIS ELECTRICAL 6-27

LIGHTING

Headlights

REMOVAL & INSTALLATION

> ✲✲ **CAUTION**
>
> On models using a halogen bulb: The halogen bulb contains gas under pressure. The bulb may shatter if the glass envelope is scratched or if the bulb is dropped. Handle the bulb carefully. Grasp the bulb only by its plastic base. Avoid touching the glass envelope. Keep the bulb out of the reach of children. Energize the bulb only when installed in the headlamp.

Pickups

♦ See Figure 88

1. Remove the radiator grille attaching screws and remove the grille.
2. Remove the headlight bulb trim ring, by removing the three screws and rotating the ring clockwise. Support the headlight bulb and remove the trim ring.

➡ Do not disturb the headlight aiming screws, which are installed in the housing next to the retaining screws.

3. Pull the plug connector from the rear of the bulb and remove the bulb.
4. Connect the plug connector to the rear of a new headlight.
5. Install the headlight in the housing, and locate the bulb tabs in the slots and the housing.
6. Position the trim ring over the bulb and loosely install the retaining screws. Rotate the ring counterclockwise to lock it in position. Tighten the three attaching screws. Check the headlight operation.
7. Install the grille.
8. Have the headlight aim checked.

MPV

♦ See Figures 89, 90 and 91

1. Make sure that the headlight switch is turned **OFF**.
2. Disconnect the negative battery cable.
3. Open the vehicle's hood and secure it in an upright position.
4. Unfasten the locking ring which secures the bulb and wire harness plug assembly, then withdraw the assembly rearward.
5. Disconnect the electrical wire harness plug from the bulb and remove the locking ring as well.

To install:

6. Before connecting a light bulb to the wire harness plug, ensure that all electrical contact surfaces are free of corrosion or dirt.
7. Install the locking ring onto the bulb, then line up the replacement headlight bulb with the harness plug. Firmly push the bulb onto the plug until the spring clip latches over the bulb's projection.

> ✲✲ **WARNING**
>
> Do not touch the glass bulb with your fingers. Oil from your fingers can severely shorten the life of the bulb. If necessary, wipe off any dirt or oil from the bulb with rubbing alcohol before completing installation.

8. Position the headlight bulb and secure it with the locking ring.
9. Connect the negative battery cable.
10. To ensure that the replacement bulb functions properly, activate the applicable switch to illuminate the bulb which was just replaced. (If this is a combination low and high beam bulb, be sure to check both intensities.) If the replacement light bulb does not illuminate, either it too is faulty or there is a problem in the bulb circuit or switch. Correct if necessary.
11. Close the vehicle's hood.

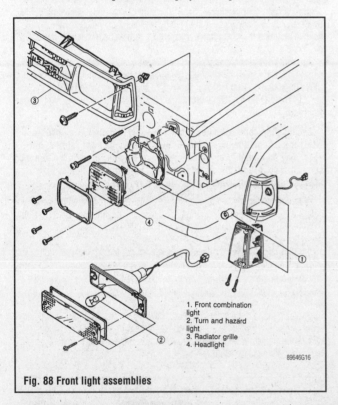

1. Front combination light
2. Turn and hazard light
3. Radiator grille
4. Headlight

Fig. 88 Front light assemblies

Fig. 89 Push down the upper retaining tab on the headlight bulb connector plug and pull straight off of the bulb

Fig. 90 Turn the headlight bulb retaining ring counterclockwise and pull off of the back of the bulb

Fig. 91 Pull the headlight bulb straight out of the headlight housing. Under any circumstance, never touch the glass of the light bulb with your fingers

6-28 CHASSIS ELECTRICAL

Navajo

♦ See Figure 92

Two aerodynamically styled headlamps are used. Each lamp uses a dual filament halogen bulb. A burned out bulb may be replaced without removing the headlamp.

1. Check that the headlamp switch is turned OFF. Lift the hood and locate the bulb installed in the rear of the headlamp body.
2. Remove the electrical connector from the bulb by squeezing the connector tabs firmly and snapping the connector rearward.
3. Remove the bulb retaining ring by rotating counterclockwise. Slide the ring off the plastic base. Keep the ring for reinstallation.
4. Carefully remove the headlamp bulb from the socket by gently pulling straight backward out of the socket. Do not rotate the bulb during removal.
5. With the flat side of the bulb facing upward, insert the glass envelope of the bulb into the socket. Turn the base slightly, if necessary, to align the grooves in the forward part of the plastic base with the corresponding locating tabs inside the socket. When the grooves are aligned, push the bulb into the socket until the mounting flange contacts the rear face of the socket.
6. Install the plastic lock ring and connect the wiring. Turn the headlamps on and check for proper operation.

➡A properly aimed headlamp need not be reaimed after installation of the bulb.

7. To remove the headlamp assembly: Disconnect the headlamp bulb connector. Remove the screw retaining the side marker assembly to the vehicle and remove the assembly.
8. Remove the fasteners retaining the grille to the grille reinforcement and remove the grille. Remove the spring clips retaining the headlamp assembly to the grille reinforcement and remove the assembly. Install in reverse order.

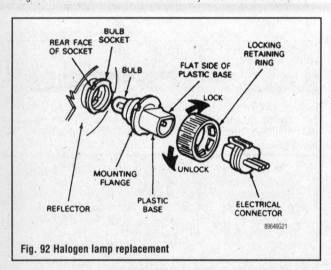

Fig. 92 Halogen lamp replacement

AIMING THE HEADLIGHTS

♦ See Figures 93 thru 98

The headlights must be properly aimed to provide the best, safest road illumination. The lights should be checked for proper aim and adjusted as necessary. Certain state and local authorities have requirements for headlight aiming; these should be checked before adjustment is made.

✱✱✱ CAUTION

About once a year, when the headlights are replaced or any time front end work is performed on your vehicle, the headlight should be accurately aimed by a reputable repair shop using the proper equipment. Headlights not properly aimed can make it virtually impossible to see and may blind other drivers on the road, possibly causing an accident. Note that the following procedure is a temporary fix, until you can take your vehicle to a repair shop for a proper adjustment.

Headlight adjustment may be temporarily made using a wall, as described below, or on the rear of another vehicle. When adjusted, the lights should not glare in oncoming car or truck windshields, nor should they illuminate the passenger compartment of vehicles driving in front of you. These adjustments are rough and should always be fine-tuned by a repair shop which is equipped with headlight aiming tools. Improper adjustments may be both dangerous and illegal.

For most of the vehicles covered by this manual, horizontal and vertical aiming of each sealed beam unit is provided by two adjusting screws which move the retaining ring and adjusting plate against the tension of a coil spring. There is no adjustment for focus; this is done during headlight manufacturing.

➡Because the composite headlight assembly is bolted into position, no adjustment should be necessary or possible. Some applications, however, may be bolted to an adjuster plate or may be retained by adjusting screws. If so, follow this procedure when adjusting the lights, BUT always have the adjustment checked by a reputable shop.

Before removing the headlight bulb or disturbing the headlamp in any way, note the current settings in order to ease headlight adjustment upon reassembly. If the high or low beam setting of the old lamp still works, this can be done using the wall of a garage or a building:

1. Park the vehicle on a level surface, with the fuel tank about ½ full and with the vehicle empty of all extra cargo (unless normally carried). The vehicle should be facing a wall which is no less than 6 feet (1.8m) high and 12 feet (3.7m) wide. The front of the vehicle should be about 25 feet from the wall.
2. If aiming is to be performed outdoors, it is advisable to wait until dusk in order to properly see the headlight beams on the wall. If done in a garage, darken the area around the wall as much as possible by closing shades or hanging cloth over the windows.

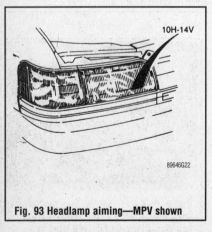

Fig. 93 Headlamp aiming—MPV shown

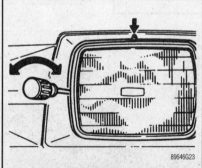

Fig. 94 Location of the aiming adjustment screws on most vehicles with sealed beam headlights—Pickup shown

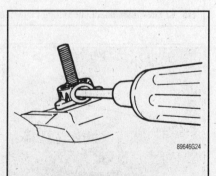

Fig. 95 Example of headlight rear adjustment screw location for composite headlamps—MPV shown

CHASSIS ELECTRICAL 6-29

3. Turn the headlights **ON** and mark the wall at the center of each light's low beam, then switch on the brights and mark the center of each light's high beam. A short length of masking tape which is visible from the front of the vehicle may be used. Although marking all four positions is advisable, marking one position from each light should be sufficient.

4. If neither beam on one side is working, and if another like-sized vehicle is available, park the second one in the exact spot where the vehicle was and mark the beams using the same-side light. Then switch the vehicles so the one to be aimed is back in the original spot. It must be parked no closer to or farther away from the wall than the second vehicle.

5. Perform any necessary repairs, but make sure the vehicle is not moved, or is returned to the exact spot from which the lights were marked. Turn the headlights **ON** and adjust the beams to match the marks on the wall.

6. Have the headlight adjustment checked as soon as possible by a reputable repair shop.

Signal and Marker Lights

REMOVAL & INSTALLATION

Front Turn Signal, Parking and Side Marker Lights

PICKUPS

The combination parking/side marker light housing is separate from the turn signal housing. However, the bulbs are replaced in the same manner.
1. Disconnect the negative battery cable.
2. Remove the lens mounting screws and pull off the lens.
3. Carefully push in on the bulb and twist it counterclockwise to remove it.
4. Installation is the reverse of removal.

MPV

▶ See Figures 99 thru 105

1. Disconnect the battery ground cable.
2. Remove the grille side moldings.

➡ The grille side molding pieces are retained by both screws and snap-clips. The snap-clips can be freed by depressing the tabs with a small screwdriver.

3. Remove the lens mounting screws and pull off the lens.
4. Turn the socket counterclockwise and pull it straight out of the back of the housing.
5. Carefully push in on the bulb and twist it counterclockwise to remove it from the socket.
6. Installation is the reverse of removal.

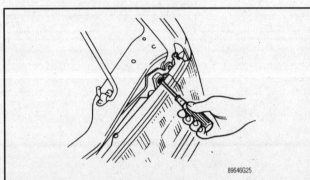

Fig. 96 Example of front facing headlight adjustment screw location for composite headlamps—MPV shown

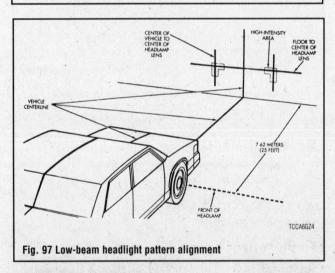

Fig. 97 Low-beam headlight pattern alignment

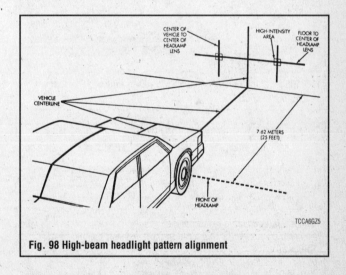

Fig. 98 High-beam headlight pattern alignment

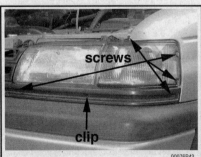

Fig. 99 To change the front turn signal/side marker bulb, five mounting screws must be removed and one clip disengaged

Fig. 100 Disengage the retaining clip in the middle of the trim panel below the headlight and turn signal housing

Fig. 101 Pull out the trim panel to aid in turn signal/side marker light housing removal

6-30 CHASSIS ELECTRICAL

Fig. 102 When removing the turn signal housing, pull firm enough to disengage the locating pin from the location hole

Fig. 103 To remove the turn signal light bulb socket from the housing, turn the socket counterclockwise

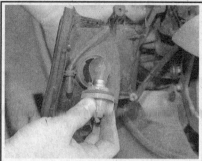

Fig. 104 Pull the socket straight out of the turn signal housing . . .

Fig. 105 . . . then press in and turn to remove the front turn signal bulb from the socket

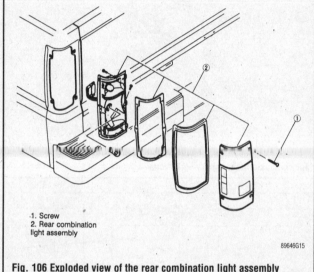

1. Screw
2. Rear combination light assembly

Fig. 106 Exploded view of the rear combination light assembly

NAVAJO

The combination parking/turn signal lamp housing is separate from the side marker lamp housing.

1. Loosen the two mounting nuts and remove the parking/turn signal lamp assembly.
2. Loosen the one mounting screw and remove the side marker lamp housing.
3. To remove the side marker, parking/turn signal bulbs and sockets from their housings, simply turn the socket(s) counterclockwise and pull straight out.
4. Remove the bulb from the socket by pulling straight out of the socket.

To install:

5. Install the new bulb straight into the socket.
6. Install the socket into the rear of the housing and turn clockwise to lock into place.
7. Install the side marker lamp housing and tighten the one mounting screw.
8. Place the parking/turn signal lamp assembly into position and tighten the two mounting nuts.
9. Check the operation of the lights.

Rear Combination Lights

PICKUPS

♦ See Figure 106

1. Remove the lens mounting screws and pull off the lens.
2. Carefully push in on the bulb and twist it counterclockwise to remove it.
3. Installation is the reverse of removal.

MPV

♦ See Figures 107 thru 117

1. Remove the lens mounting screws and pull off the lens.
2. Carefully push in on the bulb and twist it counterclockwise to remove it.
3. Installation is the reverse of removal.

NAVAJO

♦ See Figures 118, 119 and 120

1. Remove the 2 screws retaining the lamp assembly to the vehicle.
2. Remove the lamp assembly from the vehicle by pulling it outward. Make sure the 2 retainers at the bottom of the assembly release.
3. Remove the lamp sockets from the lens housing by twisting it, then pulling outward.
4. Remove the bulb from the socket by pulling it straight outward.

To install:

5. Install the bulb into the lamp socket.
6. Install the lamp socket to the lens and twist it to lock it in position.
7. Position the lens assembly to the body. Ensure that the retainers are aligned with their mounting clips, and press the assembly until fully seated.
8. Install the 2 lens assembly retaining screws.
9. Check the operation of the lights.

CHASSIS ELECTRICAL 6-31

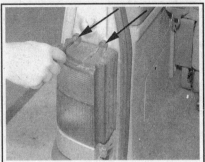

Fig. 107 Remove the two rear combination light assembly mounting screws

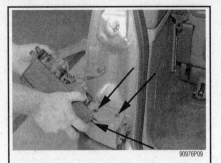

Fig. 108 When removing the rear combination light housing, be careful not to brake the locating pegs that fit into the mounting holes of the body

Fig. 109 To remove the socket from the back of the combination light housing, turn the socket counterclockwise . . .

Fig. 110 . . . then carefully pull the socket out of the opening in the housing unit

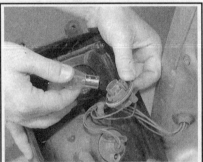

Fig. 111 Gently press in and turn the bulb, then pull it out of the socket

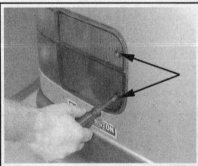

Fig. 112 Remove the two inboard combination light retaining screws . . .

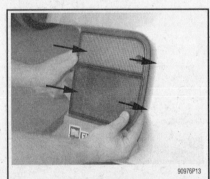

Fig. 113 . . . then gently push the assembly inward of the door to free it from the retaining tabs

Fig. 114 Part of the inboard combination light assembly is held in place by retaining tabs secured against the door metal

Fig. 115 Hold the inboard combination light assembly fimly and turn the socket counterclockwise . . .

Fig. 116 . . . then pull the bulb and socket out of the housing

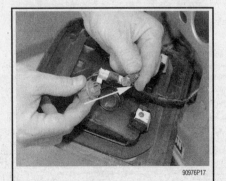

Fig. 117 While holding the socket, pull the light bulb straight out

Fig. 118 Navajo models only use two upper lens retaining screws, the bottom utilize retaining pins

6-32 CHASSIS ELECTRICAL

Fig. 119 Grasp the bulb socket then twist and pull it out of the lens assembly

Fig. 120 Remove the bulb from the socket by pulling it straight out

Fig. 121 Remove the two reverse light housing mounting screws . . .

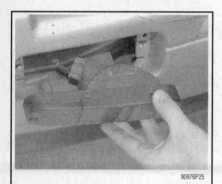

Fig. 122 . . . then pull the reverse light housing straigh out of the rear bumper

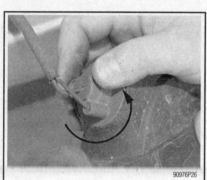

Fig. 123 Turn the reverse light bulb socket counterclockwise to release the retaining tabs . . .

Fig. 124 . . . then pull the socket straight out from behind the housing

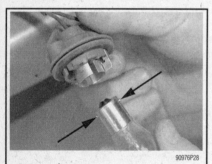

Fig. 125 When replacing bulbs (reverse light bulb shown) of any kind, be sure to note the type of bulb retaining ears that exist on each side, they can be different

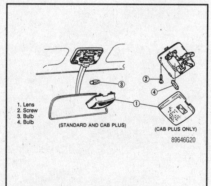

Fig. 126 Interior lamp assemblies— Pickup shown

Fig. 127 Using slight hand pressure, release the tabs . . .

Back Up Lights

▶ See Figures 121, 122, 123, 124 and 125

Only the MPV comes equipped with separately mounted back up lights.
1. Remove the back up light lens mounting screws and pull the lens away from the bumper.
2. Turn the reverse light bulb socket counterclockwise to release the retaining tabs, then pull the socket straight out from behind the housing.
3. Carefully push in on the bulb and twist it counterclockwise to remove it.
4. Installation is the reverse of removal.

High Mount Brake Light

Only the Navajo comes equipped with a high mount brake light.
1. Remove the screws retaining the lamp to the liftgate.
2. Pull the lamp away from the vehicle and disconnect the wiring connector.

3. Remove the lamp socket from the lens by twisting it and pulling outward.
4. Pull the bulb straight out from the lamp socket to remove it.

To install:
5. Install the bulb into the lamp socket.
6. Install the lamp socket to the lens and twist it to lock it in position.
7. Position the lens assembly to the body.
8. Install the lens assembly retaining screws.
9. Check the operation of the lights.

Dome Light

▶ See Figures 126, 127, 128 and 129

1. Remove the plastic cover.
2. Pull the map and/or area bulb from the dome light assembly.
3. Installation is the reverse of the removal procedure.

CHASSIS ELECTRICAL   6-33

Fig. 128 . . . then pull the dome lamp lens away from the light fixture

Fig. 129 Pull the bulb out from the contacts

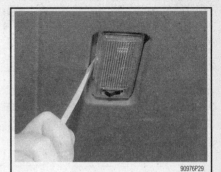

Fig. 130 Using a small prying tool, release the tabs of the tailgate inner door light lens

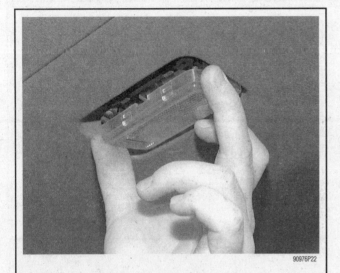

Fig. 131 Pull off the tailgate inner door light lens . . .

Fig. 132 . . . then pull the glass bulb out from between the bulb holder/connectors

Cargo or Passenger Area Lamps

♦ See Figures 126, 130, 131 and 132

The Navajo has a cargo light mounted on the roof headliner near the tailgate, the MPV has a cargo light located on the liftgate door and the Cab Plus Pickup model has an additional passenger area lamp mounted on the headliner toward the rear. To replace the bulbs, simply pry off the lens and pull out the bulb. Install the new bulb.

License Plate Lights

PICKUPS

1. Remove the lens mounting screws and pull off the lens.
2. Carefully push in on the bulb and twist it counterclockwise to remove it.
3. Installation is the reverse of removal.

MPV

1. Remove the lens mounting screws and pull off the lens.
2. Pull the assembly down and separate the lens from the mounting gasket.
3. Pull the license plate light bulb straight out of the socket.
4. Installation is the reverse of removal.

NAVAJO

1. From underneath the rear of the vehicle, grasp the lamp socket and rotate it ¼ turn.
2. Pull the lamp socket from the lens.
3. Pull the bulb from the socket
4. Installation is the reverse of the removal procedure.

6-34 CHASSIS ELECTRICAL

Light Bulb Applications

► See Figures 133, 134 and 135

LIGHT BULBS

LIGHT BULB	NUMBER OF BULBS REQUIRED	BULB TRADE NUMBER
A/C control illumination (optional)	1	161
Ashtray lamp	1	161
Back-up lamp	2	3156
Brake warning light	1	194
Cargo lamp (optional)	1	211-2
Charging system warning	1	194
Dome lamp	1	912
Door courtesy lamp	1	168
Engine coolant temperature warning	1	194
4x4 indicator light	1	194
Check engine warning light	1	194
Fasten safety belt warning light	1	194
Front parking lamp and turn signal	2	3157
Front side marker lamp	2	194
Glove compartment lamp	1	194
Headlamps	2	9004
Headlamp switch illumination	1	1015
Heater control illumination	1	161
Hi-beam indicator	1	194
Hi-mount stop lamp	2	912
Instrument panel gauge illumination	5	194
License plate lamp — rear bumper	2	194
Oil pressure indicator light	1	194
Rear tail/brake/turn lamp	2	3157
Turn signal indicator light	2	194
RABS warning light	1	194
Deluxe map reading lamp/ dome	2 1	168 906
Under hood lamp	1	906
Map lamps	2	168

Fig. 133 Light bulb chart—Navajo

Light Bulbs

Light bulb	Wattage	Bulb number
Headlights	65/55	6052
	65/35[1]	H6054[1]
Front turn signal lights	27	1156
Parking/Front side marker lights	8	67
Rear turn signal lights	27	1156
Rear side marker/Brake/Taillights	27/8	1157
Back-up lights	27	1156
License plate lights	6	—
Interior light	10	—

[1] Halogen

Fig. 134 Light bulb chart—B Series Pick-up

Light Bulbs

Light Bulb	Wattage	Bulb Trade Number
Parking lights, Side marker light, and turn signal	27/8	1157
Brake light and taillight	27/8	1157
Turn signal light	27	1156
Backup light	27	1156
Taillight	4.9	168

Fig. 135 Light bulb chart—MPV

TRAILER WIRING

Wiring the vehicle for towing is fairly easy. There are a number of good wiring kits available and these should be used, rather than trying to design your own.

All trailers will need brake lights and turn signals as well as tail lights and side marker lights. Most areas require extra marker lights for overwide trailers. Also, most areas have recently required back-up lights for trailers, and most trailer manufacturers have been building trailers with back-up lights for several years.

Additionally, some Class I, most Class II and just about all Class III and IV trailers will have electric brakes. Add to this number an accessories wire, to operate trailer internal equipment or to charge the trailer's battery, and you can have as many as seven wires in the harness.

Determine the equipment on your trailer and buy the wiring kit necessary. The kit will contain all the wires needed, plus a plug adapter set which includes the female plug, mounted on the bumper or hitch, and the male plug, wired into, or plugged into the trailer harness.

When installing the kit, follow the manufacturer's instructions. The color coding of the wires is usually standard throughout the industry. One point to note: some domestic vehicles, and most imported vehicles, have separate turn signals. On most domestic vehicles, the brake lights and rear turn signals operate with the same bulb. For those vehicles without separate turn signals, you can purchase an isolation unit so that the brake lights won't blink whenever the turn signals are operated.

One, final point, the best kits are those with a spring loaded cover on the vehicle mounted socket. This cover prevents dirt and moisture from corroding the terminals. Never let the vehicle socket hang loosely; always mount it securely to the bumper or hitch.

CHASSIS ELECTRICAL 6-35

CIRCUIT PROTECTION

Fuses

► See Figures 136 thru 145

REPLACEMENT

Fuse Panel

► See Figures 146, 147 and 148

The fuse panel is mounted in the following locations:
- Navajo and Pickup models: under the instrument panel to the left of the steering column
- MPV models: above the driver's side kick panel inside the van

1. Turn the ignition switch **OFF**.
2. If equipped, remove the fuse panel access cover.
3. If equipped, remove the fuse puller tool from the cover.
4. Grasp the push-in type fuse with the provided tool, or a pair of needle-nose pliers.
5. On pickup models only, if removing the 80 amp fuse, remove the fuse block, remove the cover, unscrew the wiring terminal and pull out the fuse.
6. Look through the side of the fuse body to determine if the fuse element is blown.

To install:

7. Check the amperage rating of the fuse which was removed and obtain a new fuse of the same rating.

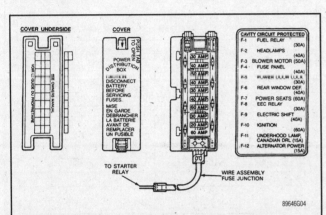

Fig. 136 Fuse panel and identification chart for the 1991–92 Navajo

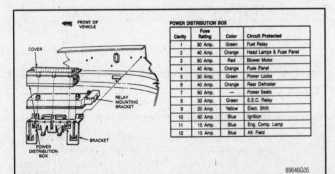

Fig. 137 Underhood power distribution box and identification chart for the 1991–92 Navajo

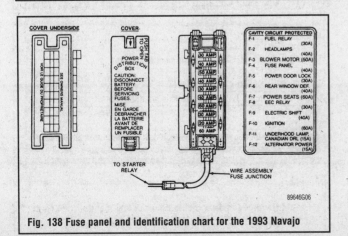

Fig. 138 Fuse panel and identification chart for the 1993 Navajo

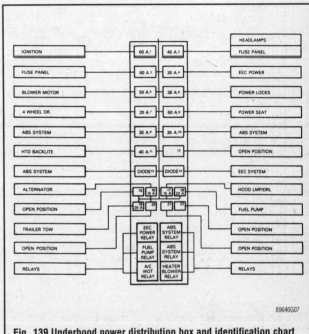

Fig. 139 Underhood power distribution box and identification chart for the 1993 Navajo

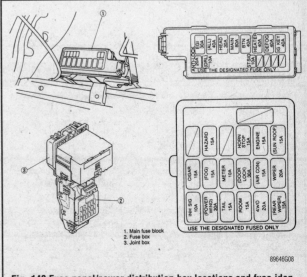

Fig. 140 Fuse panel/power distribution box locations and fuse identifications for the 1989–91 MPV

6-36 CHASSIS ELECTRICAL

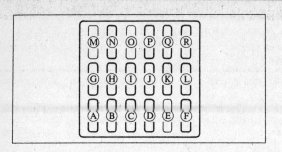

▼ Fuse panel

	DESCRIPTION	FUSE RATING	COLOR	PROTECTED COMPONENT
A	INH SIG	10A	RED	ECU, Instrument cluster
B	POWER WINDOW	30A	GREEN	Power windows
C	TAIL	15A	BLUE	Instrument panel light control, Taillights
D	ROOM	15A	BLUE	Interior lights, Key reminder
E	4WD	20A	YELLOW	4WD
F	REAR WIPER	15A	BLUE	Rear wiper and washer
G	CIGAR	15A	BLUE	Lighter, Power mirror, CPU, Clock (audio)
H	FOG	15A	BLUE	Fog lights
I	METER	10A	RED	Instrument cluster, Cruise control
J	DOOR LOCK	30A	GREEN	Power door locks
K	AIR CON	15A	BLUE	A/C switch, ALL, Rear heater main switch
L	WIPER	20A	YELLOW	Windshield wiper and washer
M	—	—	—	—
N	HAZARD	15A	BLUE	Hazard flashers
O	—	—	—	—
P	HORN STOP	15A	BLUE	Horn, Brake lights, Cruise control, ALL
Q	ENGINE	15A	BLUE	Fuel pump
R	SUNROOF*	20A	BLUE	Sunroof

Fig. 141 Fuse panel and identification chart for the 1992–93 MPV

■ Fuse Panel Description

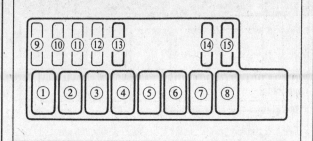

▼ Fuse block

	DESCRIPTION	FUSE RATING	COLOR	PROTECTED COMPONENT
1	IG KEY	30A	PINK	All ignition related circuits
2	DEFOG	40A	GREEN	Rear difroster
3	HEATER	40A	GREEN	Blower motor
4	BTN	40A	GREEN	Room, Horn, Stop, Door lock, Hazard and tail fuses
5	MAIN	80A	BLACK	All circuits
6	HEAD	30A	PINK	Headlights
7	ALL	40A	GREEN	ALL, ABS, Rear A/C, Rear heater
8	EGI	30A	PINK	Alternator, ECU
9	—	—	—	—
10	—	—	—	—
11	—	—	—	—
12	—	—	—	—
13	ST SIG	10A	RED	EGI-CU
14	REAR HEATER	20A	YELLOW	Rear heater
15	ANTILOCK	20A	YELLOW	Antilock brake system

Fig. 142 Underhood power distribution box and identification chart for the 1992–93 MPV

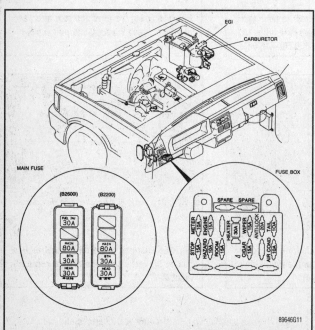

Fig. 143 Fuse panel and identification chart for the B Series Pickup

Engine compartment fuse block

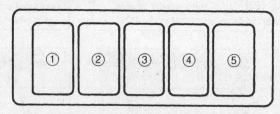

▼ Fuse block (B2200 without fuel injection)

	DESCRIPTION	FUSE RATING	COLOR	PROTECTED COMPONENT
1	HEAD	30A	PINK	Headlights
2	BTN	30A	PINK	Tail, Stop, Hazard and Room fuses
3	MAIN	80A	BLACK	For protection of all circuits
4	—	—	—	—
5	—	—	—	—

Fig. 144 Underhood power distribution box and identification chart for the B Series Pickup with 2.2L carbureted engine

CHASSIS ELECTRICAL 6-37

Engine compartment fuse block

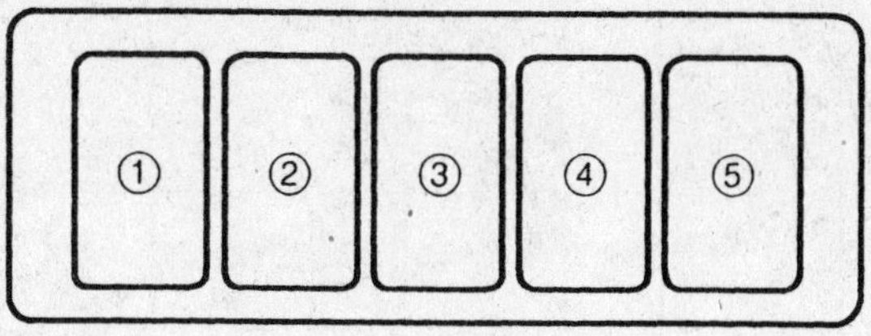

① ② ③ ④ ⑤

▼ Fuse block (B2200 with fuel injection and B2600i)

DESCRIPTION	FUSE RATING	COLOR	PROTECTED COMPONENT
1 FUEL INJ.	30A	PINK	Fuel injection system
2 —	—	—	—
3 MAIN	80A	BLACK	For protection of all circuits
4 BTN	30A	PINK	Tail, Stop, Hazard and Room fuses
5 HEAD	30A	PINK	Headlights

Fig. 145 Underhood power distribution box and identification chart for the B Series Pickup with fuel injected engines

※※ WARNING
Never replace a blown fuse with a new fuse of a higher rating. Severe electrical damage, as well as possible electrical fire could result.

8. Align the fuse with its mounting position and push it into place until fully seated in the panel.
9. If installing the 80 amp fuse on pickup models, push the fuse into place, connect the wiring terminal, install the cover and fuse block.
10. Turn the ignition switch on and operate the accessory that was protected by that fuse.

※※ WARNING
If the fuse continues to blow, inspect and test the wire harness and component or components which are protected by that fuse.

Power Distribution Box

▶ See Figures 149, 150, 151, 152 and 153

The power distribution box (also known as the main fuse box) is mounted in the following locations:
- Navajo models: in the engine compartment on the right inner fender, next to the starter relay
- MPV models: in the engine compartment on the right fender liner.
- Pickup models: in the engine compartment on the right fender apron just behind the battery on the B2200, or, on the left fender apron on B2600 models.

The power distribution box houses the fuses and relays for most of the

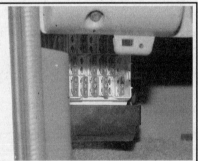

Fig. 146 Location of the interior fusebox under the driver side dashboard—MPV

Fig. 147 To access the fuses, remove the fuse panel cover. Note the spare fuses (A) and provided fuse puller tool (B)—Navajo

Fig. 148 Using the provided tool or a pair of pliers, grasp the fuse and pull it from the panel

Fig. 149 Location of the fuses are illustrated on the power distribution box top cover-right side shown

Fig. 150 Pulling out a fuse from the power distribution box located on the right side of the engine compartment

Fig. 151 Location of the power distribution box on the left side of the engine compartment

6-38 CHASSIS ELECTRICAL

Fig. 152 Release the top cover clip of the power distribution box by pressing on the clip against the cover

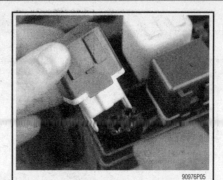

Fig. 153 Pulling out a relay from the left side power distribution box

Fig. 154 Circuit breakers can be replaced as if it were a fuse, simply grasp hold of it and pull it from the panel

under hood components which are not controlled by a dash mounted switch (such as the alternator, fuel pump, ECM, etc.). The fuses and relays are replaced in the same manner as the fuse panel inside the vehicle. The power distribution box cover is hinged on one end and utilizes a retaining latch on the other. Simply release the latch and lift the cover up to gain access to the fuses and relays.

Fusible Links

The fusible link is a short length of special, Hypalon (high temperature) insulated wire, integral with the engine compartment wiring harness and should not be confused with standard wire. It is several wire gauges smaller than the circuit which it protects. Under no circumstances should a fuse link replacement repair be made using a length of standard wire cut from bulk stock or from another wiring harness.

The fusible links are located near the starter solenoid and shares the terminal with the battery-to-starter solenoid cable.

➥Do not mistake a resistor wire for a fusible link. The resistor wire is generally longer and has print stating, "Resistor—don't cut or splice". When attaching a single No. 16, 17, 18 or 20 gauge fusible link to a heavy gauge wire, always double the stripped wire end of the fusible link before inserting and crimping it into the butt connector for positive wire retention.

Circuit Breakers

Pickup, Navajo and MPV models use circuit breakers for components which have a high start-up amperage pull (such as power windows). All breakers will automatically reset if they have been tripped. The circuit breakers are replaceable (should one go bad or not reset) and can be found on the fuse panel.

➥All circuit breakers on Navajo are located in the fuse panel except for a 4.5 amp circuit breaker for the rear wiper/washer and a 22 amp circuit breaker for the high beams which is integral with the headlight switch.

REPLACEMENT

♦ See Figure 154

The circuit breakers are replaced in the same manner as a fuse. Simply pull the breaker from the fuse panel to remove.

Flashers and Relays

REPLACEMENT

Pickups

The hazard warning flasher, on the pickup, is located to the left of the steering column, beneath the instrument panel, and is secured by a clamp and one screw. To remove it, simply unplug the electrical connector, loosen the screw, and slide the flasher out of the clamp. The turn signal flasher is located to the right of the steering column, beneath the instrument panel, and is secured in the same way as the hazard flasher.

The turn signal relay is located to the immediate right of the hazard flasher, and is secured by two screws. To remove it, unplug the electrical connector and remove the screws.

MPV

The relays and flashers are located under the dash, near the fusebox.

Navajo

Both the turn signal flasher and the hazard warning flasher are mounted on the fuse panel on the truck. To gain access to the fuse panel, remove the cover from the lower edge of the instrument panel below the steering column. First remove the two fasteners from the lower edge of the cover. Then pull the cover downward until the spring clips disengage from the instrument panel.

The turn signal flasher unit is mounted on the front of the fuse panel, and the hazard warning flasher is mounted on the rear of the fuse panel.

CHASSIS ELECTRICAL 6-39

WIRING DIAGRAMS

INDEX OF WIRING DIAGRAMS

DIAGRAM 1   Sample Diagram: How To Read & Interpret Wiring Diagrams

DIAGRAM 2   Wiring Diagram Symbols

DIAGRAM 3   1987-93 B-Series 2.2L Carbureted Engine Schematic

DIAGRAM 4   1987-88 B-Series 2.6L Carbureted Engine Schematic

DIAGRAM 5   1990-93 B-Series 2.2L Fuel Injected Engine Schematic

DIAGRAM 6   1989-93 B-Series & MPV 2.6L Fuel Injected Engine Schematic

DIAGRAM 7   1989-93 MPV 3.0L Engine Schematic

DIAGRAM 8   1991-93 Navajo 4.0L Engine Schematic

DIAGRAM 9   1987-93 B-Series Starting Chassis Schematics

DIAGRAM 10 1987-93 B-Series Charging, Fuel Pump Chassis Schematics

DIAGRAM 11 1987-93 B-Series Parking/Marker Lights, Headlights w/o DRL, Headlights w/DRL Chassis Schematics

DIAGRAM 12 1987-93 B-Series Turn/Hazard Lights, Stop Lights, Horn, Back-Up Lights Chassis Schematics

DIAGRAM 13 1989-93 MPV Starting, Charging, Fuel Pump Chassis Schematics

DIAGRAM 14 1989-93 MPV Turn/Hazard Lights, Parking/Marker Lights Chassis Schematics

DIAGRAM 15 1989-93 MPV Horn, Stop Lights, Cooling Fans Chassis Schematics

DIAGRAM 16 1989-93 MPV Headlights w/o DRL, Back-Up Lights, Headlights w/ DRL Chassis Schematics

DIAGRAM 17 1991-93 Navajo Starting, Charging, Fuel Pump Chassis Schematics

DIAGRAM 18 1991-93 Navajo Headlights w/o DRL, Back-Up Lights, Horns, Parking/Marker Lights Chassis Schematics

DIAGRAM 19 1991-93 Navajo Turn/Hazard Lights, Headlights w/ DRL Chassis Schematics

DIAGRAM 20 1987-93 B-Series, MPV, Navajo Winsheild Wipers, Washer Chassis Schematics

DIAGRAM 21 1991-93 Navajo Power Windows, Power Door Locks Chassis Schematics

DIAGRAM 22 1993 MPV Power Windows, Power Door Locks Chassis Schematics

6-40 CHASSIS ELECTRICAL

WIRING DIAGRAM SYMBOLS

SAMPLE DIAGRAM: HOW TO READ & INTERPRET WIRING DIAGRAMS

CHASSIS ELECTRICAL 6-41

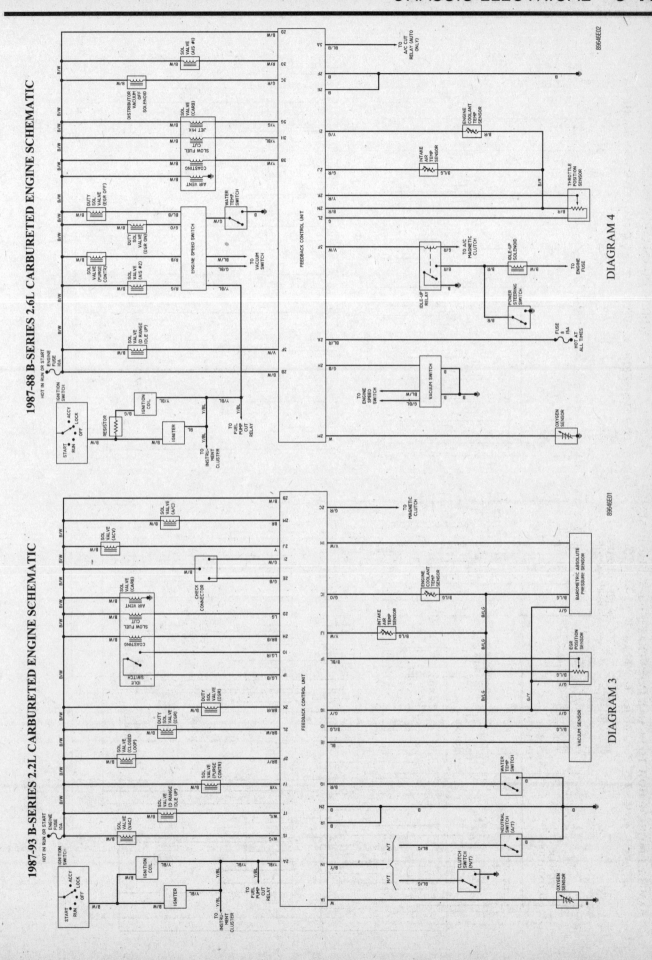

6-42 CHASSIS ELECTRICAL

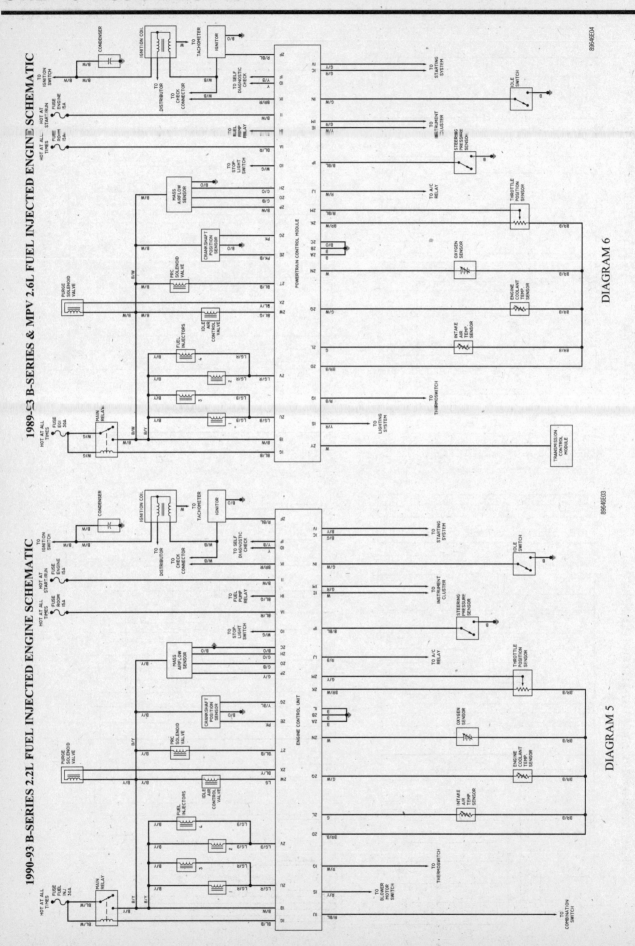

CHASSIS ELECTRICAL 6-43

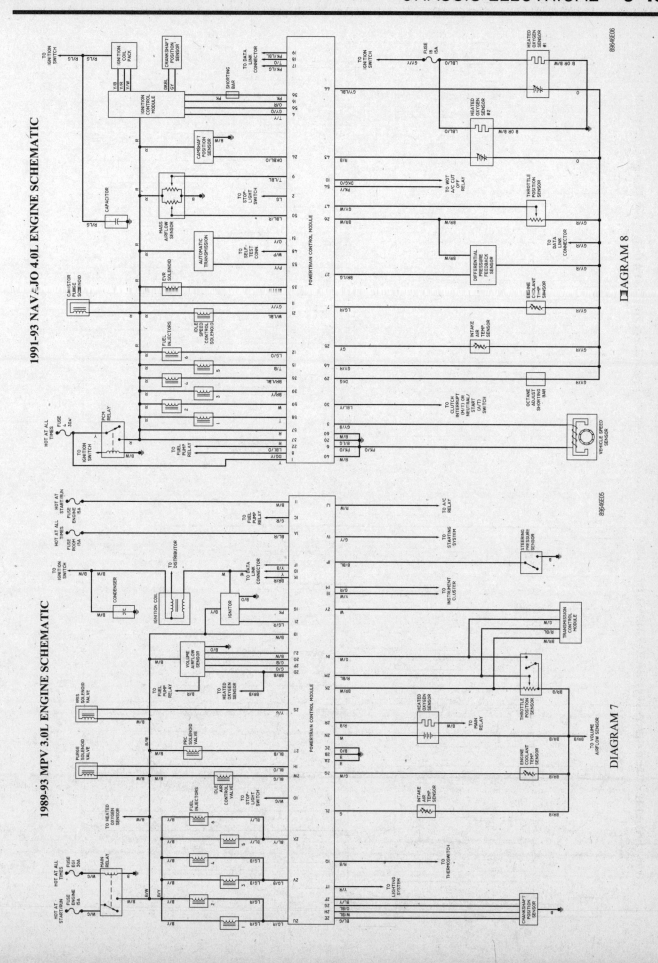

6-44 CHASSIS ELECTRICAL

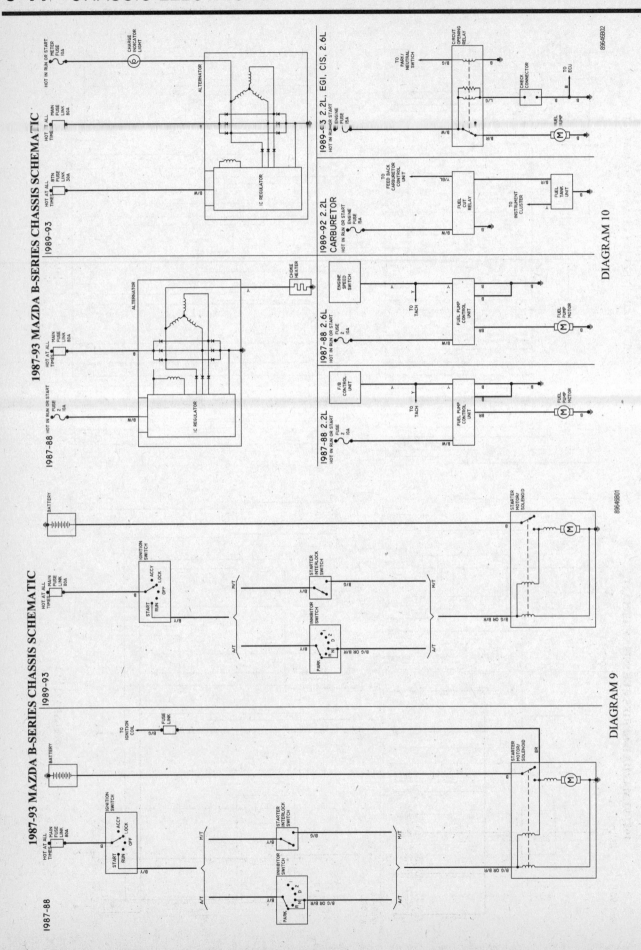

CHASSIS ELECTRICAL 6-45

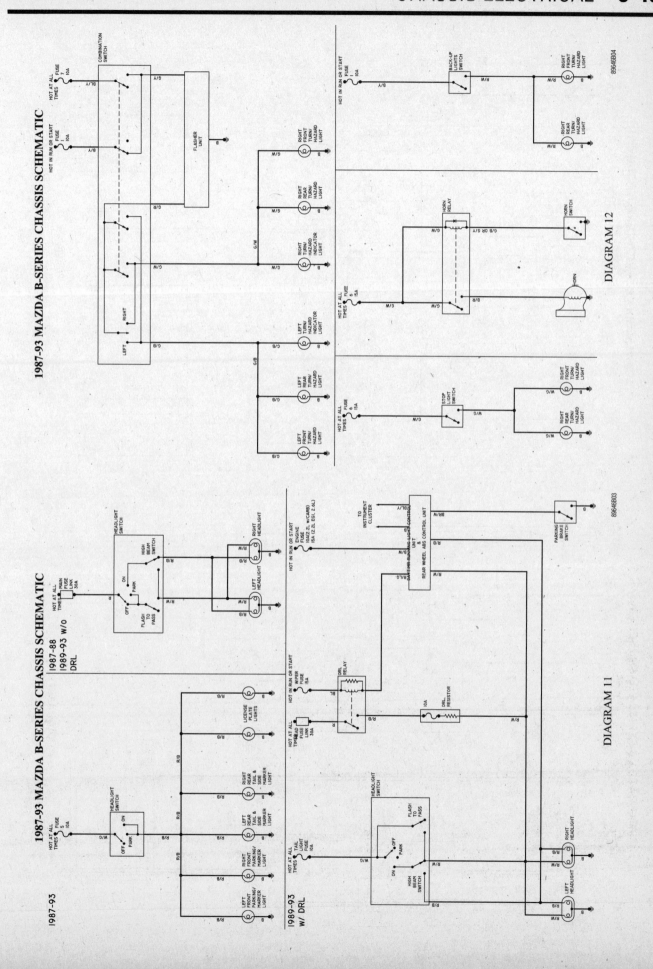

6-46 CHASSIS ELECTRICAL

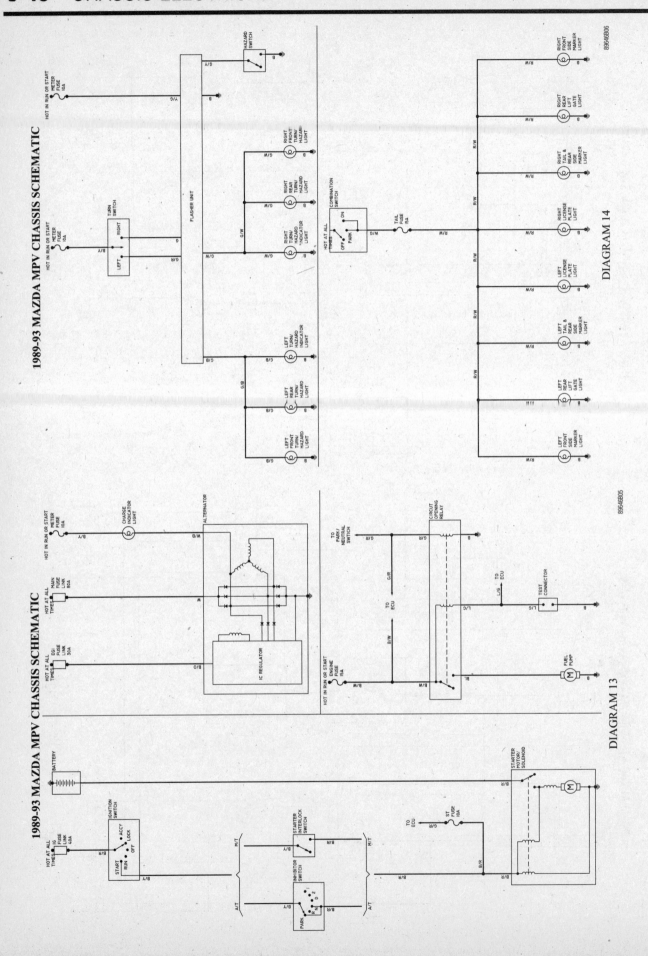

CHASSIS ELECTRICAL 6-47

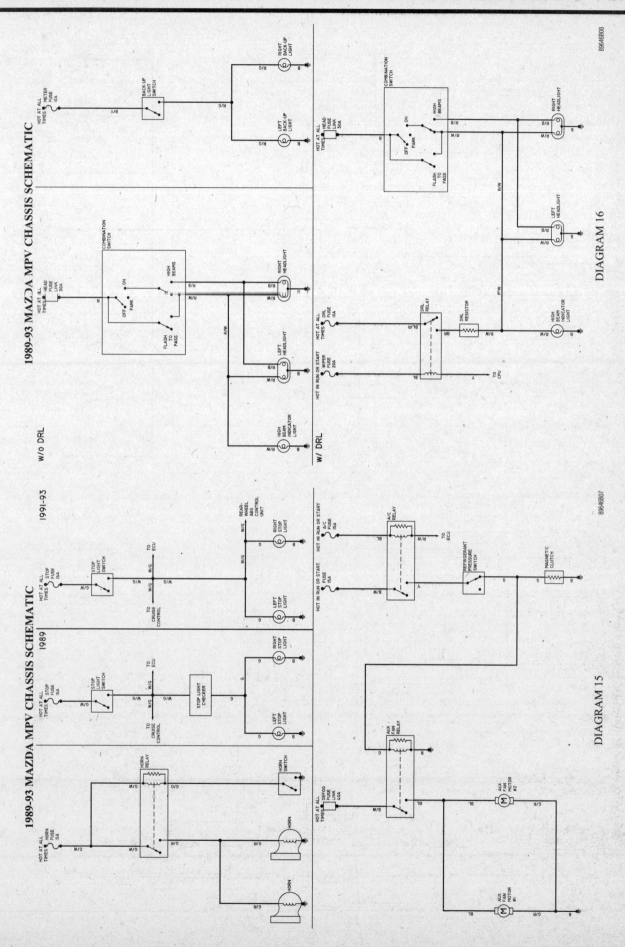

6-48 CHASSIS ELECTRICAL

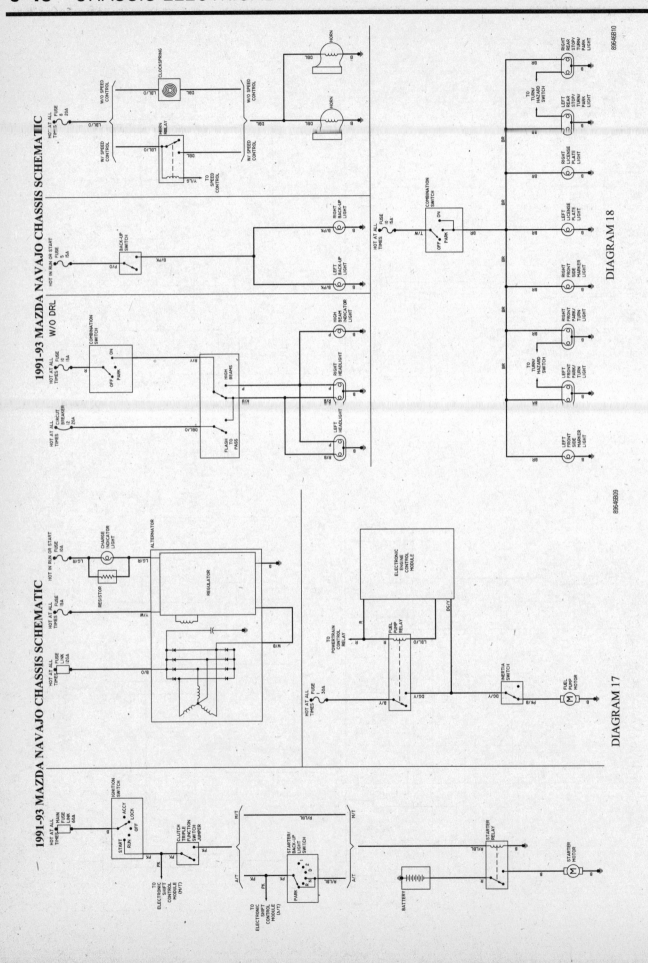

CHASSIS ELECTRICAL 6-49

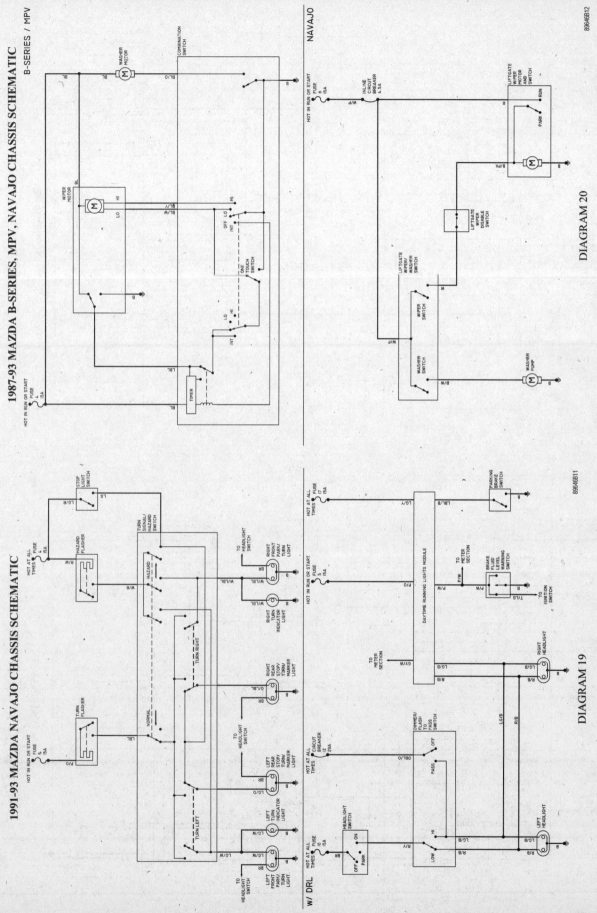

6-50 CHASSIS ELECTRICAL

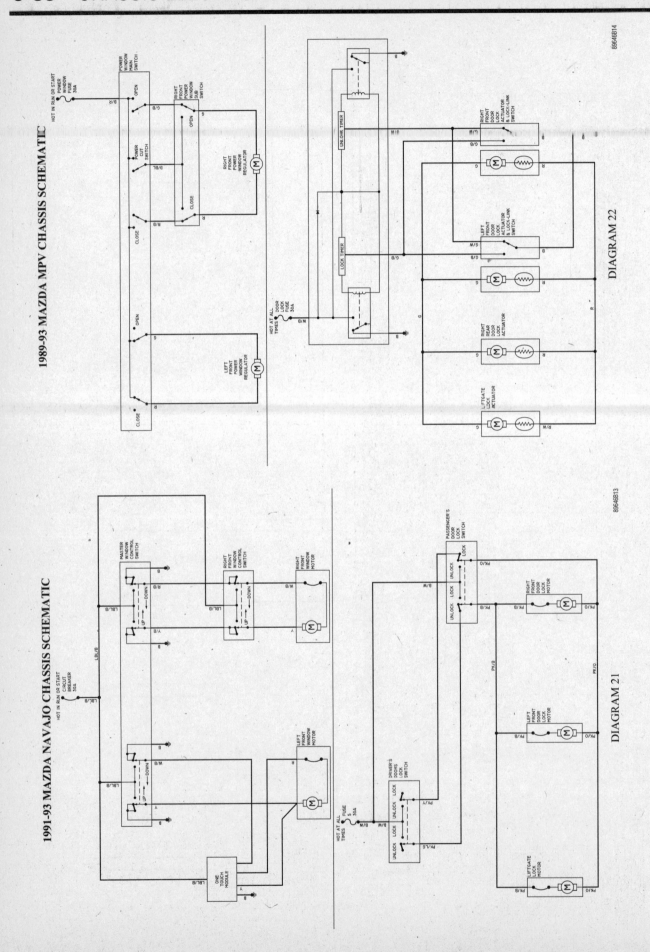

MANUAL TRANSMISSION 7-2
UNDERSTANDING THE MANUAL
  TRANSMISSION 7-2
BACK-UP LIGHT SWITCH 7-2
  REMOVAL & INSTALLATION 7-2
EXTENSION HOUSING SEAL 7-2
  REMOVAL & INSTALLATION 7-2
MANUAL TRANSMISSION
  ASSEMBLY 7-2
  REMOVAL & INSTALLATION 7-2
CLUTCH 7-4
UNDERSTANDING THE CLUTCH 7-4
DRIVEN DISC AND PRESSURE
  PLATE 7-4
  REMOVAL & INSTALLATION 7-4
  ADJUSTMENTS 7-6
CLUTCH MASTER CYLINDER 7-6
  REMOVAL & INSTALLATION 7-6
CLUTCH SLAVE CYLINDER 7-6
  REMOVAL & INSTALLATION 7-6
  BLEEDING THE HYDRAULIC
    SYSTEM 7-7
AUTOMATIC TRANSMISSION 7-7
UNDERSTANDING THE AUTOMATIC
  TRANSMISSION 7-7
NEUTRAL START SWITCH/BACK-UP
  SWITCH 7-7
  REMOVAL & INSTALLATION 7-7
  ADJUSTMENTS 7-8
EXTENSION HOUSING SEAL 7-8
  REMOVAL & INSTALLATION 7-8
AUTOMATIC TRANSMISSION
  ASSEMBLY 7-8
  REMOVAL & INSTALLATION 7-8
ADJUSTMENTS 7-10
  SHIFT LINKAGE 7-10
  KICK-DOWN SWITCH 7-10
  KICKDOWN CABLE 7-11
  MANUAL LINKAGE 7-11
TRANSFER CASE 7-12
REAR OUTPUT SHAFT SEAL 7-12
  REMOVAL & INSTALLATION 7-12
FRONT OUTPUT SHAFT SEAL 7-12
  REMOVAL & INSTALLATION 7-12
TRANSFER CASE ASSEMBLY 7-12
  REMOVAL & INSTALLATION 7-12
  ADJUSTMENTS 7-13
DRIVELINE 7-13
FRONT DRIVESHAFT AND
  U-JOINTS 7-13
  REMOVAL & INSTALLATION 7-13
  U-JOINT REPLACEMENT 7-14
REAR DRIVESHAFT 7-15
  REMOVAL & INSTALLATION 7-15
  U-JOINT REPLACEMENT 7-16
  DRIVESHAFT BALANCING 7-16
CENTER BEARING 7-17
  REPLACEMENT 7-17
FRONT DRIVE AXLE 7-18
MANUAL LOCKING HUBS 7-18
  REMOVAL & INSTALLATION 7-18
  ADJUSTMENT 7-18
AUTOMATIC LOCKING HUBS 7-19
  REMOVAL & INSTALLATION 7-19
FREEWHEEL MECHANISM 7-19
  REMOVAL & INSTALLATION 7-19
SPINDLE BEARINGS 7-20
  REMOVAL & INSTALLATION 7-20
HALFSHAFT 7-22
  REMOVAL & INSTALLATION 7-22
SPINDLE, SHAFT AND JOINT
  ASSEMBLY 7-22
  REMOVAL & INSTALLATION 7-22
PINION SEAL 7-23
  REMOVAL & INSTALLATION 7-23
DIFFERENTIAL CARRIER 7-23
  REMOVAL & INSTALLATION 7-23
REAR AXLE 7-24
AXLE SHAFT, BEARING AND SEAL 7-24
  REMOVAL & INSTALLATION 7-24
PINION SEAL 7-26
  REMOVAL & INSTALLATION 7-26
SPECIFICATION CHARTS
  TORQUE SPECIFICATIONS 7-28

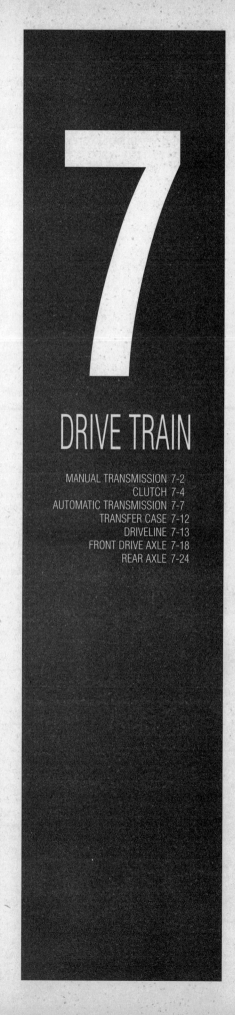

7
DRIVE TRAIN

MANUAL TRANSMISSION 7-2
CLUTCH 7-4
AUTOMATIC TRANSMISSION 7-7
TRANSFER CASE 7-12
DRIVELINE 7-13
FRONT DRIVE AXLE 7-18
REAR AXLE 7-24

7-2 DRIVE TRAIN

MANUAL TRANSMISSION

Understanding the Manual Transmission

Because of the way an internal combustion engine breathes, it can produce torque (or twisting force) only within a narrow speed range. Most overhead valve pushrod engines must turn at about 2500 rpm to produce their peak torque. Often by 4500 rpm, they are producing so little torque that continued increases in engine speed produce no power increases.

The torque peak on overhead camshaft engines is, generally, much higher, but much narrower.

The manual transmission and clutch are employed to vary the relationship between engine RPM and the speed of the wheels so that adequate power can be produced under all circumstances. The clutch allows engine torque to be applied to the transmission input shaft gradually, due to mechanical slippage. The vehicle can, consequently, be started smoothly from a full stop.

The transmission changes the ratio between the rotating speeds of the engine and the wheels by the use of gears. 4-speed or 5-speed transmissions are most common. The lower gears allow full engine power to be applied to the rear wheels during acceleration at low speeds.

The clutch driveplate is a thin disc, the center of which is splined to the transmission input shaft. Both sides of the disc are covered with a layer of material which is similar to brake lining and which is capable of allowing slippage without roughness or excessive noise.

The clutch cover is bolted to the engine flywheel and incorporates a diaphragm spring which provides the pressure to engage the clutch. The cover also houses the pressure plate. When the clutch pedal is released, the driven disc is sandwiched between the pressure plate and the smooth surface of the flywheel, thus forcing the disc to turn at the same speed as the engine crankshaft.

The transmission contains a mainshaft which passes all the way through the transmission, from the clutch to the driveshaft. This shaft is separated at one point, so that front and rear portions can turn at different speeds.

Power is transmitted by a countershaft in the lower gears and reverse. The gears of the countershaft mesh with gears on the mainshaft, allowing power to be carried from one to the other. Countershaft gears are often integral with that shaft, while several of the mainshaft gears can either rotate independently of the shaft or be locked to it. Shifting from one gear to the next causes one of the gears to be freed from rotating with the shaft and locks another to it. Gears are locked and unlocked by internal dog clutches which slide between the center of the gear and the shaft. The forward gears usually employ synchronizers; friction members which smoothly bring gear and shaft to the same speed before the toothed dog clutches are engaged.

Back-Up Light Switch

REMOVAL & INSTALLATION

Pickup and MPV

The switch is located on the upper left rear of the transmission case. To replace it, disconnect the wiring and unscrew the switch from the case. Don't lose the washer.

Navajo

♦ See Figure 1

1. Disconnect the negative battery cable.
2. Raise and support the vehicle safely.
3. Place the transmission in any position other than **R** or **N**.
4. Clean the area around the switch, then remove the switch.
5. Installation is the reverse of removal. Tighten the switch to 8–12 ft. lbs (11–16Nm).

Extension Housing Seal

REMOVAL & INSTALLATION

♦ See Figures 2 and 3

1. Disconnect the negative battery cable.
2. Raise and support the vehicle safely.
3. Place a suitable drain pan beneath the extension housing. Clean the area around the transfer extension housing seal.
4. Matchmark the driveshaft to the rear axle flange. Disconnect the driveshaft and pull it rearward from the unit.
5. Remove the extension housing seal using tool T74P–77248–A or equivalent, remove the extension housing seal.

To install:

6. To install, lubricate the inside diameter of the oil seal and install the seal into the extension housing using tool T74P–77052–A. Check to ensure that the oil seal drain hole faces downward. Install the remaining components in the reverse of removal. Make sure the marks on the driveshaft made during removal are in alignment.

Manual Transmission Assembly

REMOVAL & INSTALLATION

♦ See Figures 4, 5, 6 and 7

➡On Pickup and MPV models equipped with 4WD, the transmission and transfer case are removed as a unit.

1. Disconnect the negative battery cable.
2. Remove knobs from the transfer case shifter, if equipped, and the gear shifter.
3. Remove the shifter boot and floor console, if equipped.
4. If necessary, shift the transmission into **N**, then unbolt and remove the shift lever assembly. On Navajo models, the shift lever assembly must be removed from the control housing.
5. On Navajo, cover the opening in the control housing with a cloth to prevent dirt from falling into the unit.
6. Raise and safely support the vehicle.

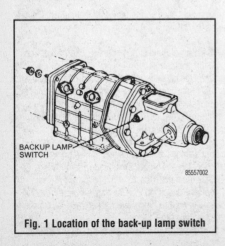

Fig. 1 Location of the back-up lamp switch

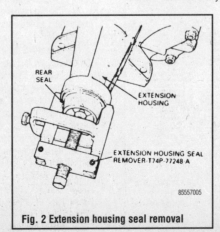

Fig. 2 Extension housing seal removal

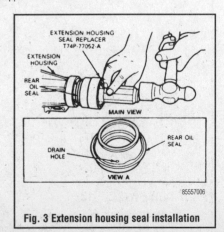

Fig. 3 Extension housing seal installation

DRIVE TRAIN 7-3

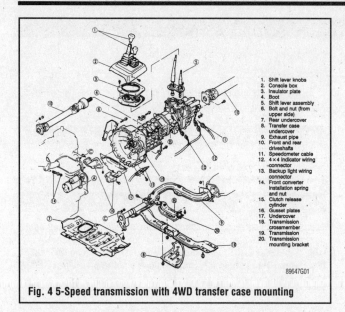

Fig. 4 5-Speed transmission with 4WD transfer case mounting

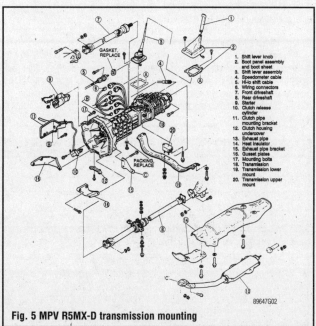

Fig. 5 MPV R5MX-D transmission mounting

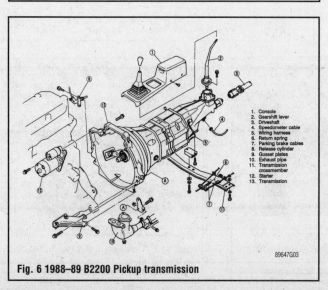

Fig. 6 1988–89 B2200 Pickup transmission

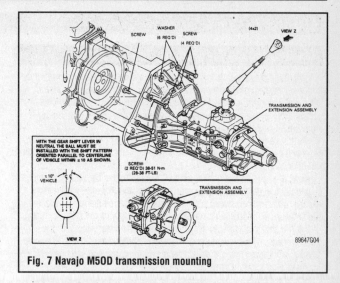

Fig. 7 Navajo M5OD transmission mounting

7. Drain the transmission and, if equipped, transfer case fluid.
8. If necessary, remove transmission and transfer case, if equipped, splash shields.
9. Matchmark and remove the driveshaft(s). On MPV, stuff a rag in the double offset joint to prevent damage to the boot from the driveshaft.
10. Disconnect the speedometer cable from the transmission/transfer case housing and, if equipped, the Hi-Lo shift cable.
11. Label and disconnect any and all wiring harnesses at the transmission or transmission/transaxle assembly that are required to remove the transmission unit from the vehicle.
12. Remove the front exhaust pipe and heat shield.
13. Remove the starter motor.
14. If necessary, disconnect the parking brake return spring and parking brake cables.
15. On MPV models with the 2.6L engine, unbolt the support bracket from the transmission, if equipped.
16. On Pickup and MPV models, remove the clutch slave cylinder and hydraulic line bracket. It is not necessary to disconnect the hydraulic line. Support the slave cylinder aside.
17. On Navajo, disconnect the clutch hydraulic line at the clutch housing. Plug the lines.
18. Place a wood block on a service jack and position the jack under the engine oil pan. Position a transmission jack, under the transmission.
19. Remove the transmission-to-engine gusset plates, if equipped, and clutch housing cover.
20. On Navajo models, remove the transfer case from the vehicle.
21. Remove the nuts and bolts attaching the transmission mount and damper to the crossmember.
22. Remove the nuts and bolts attaching the crossmember to the frame side rails and remove the crossmember.
23. Lower the transmission to gain access to the top bolts and remove the transmission-to-engine bolts.
24. Pull the transmission straight back on the jack until the mainshaft clears the clutch. Lower the transmission and pull it out from under the vehicle.

To install:
25. Check that the mating surfaces of the clutch housing, engine rear and dowel holes are free of burrs, dirt and paint.
26. Place the transmission on the transmission jack. Position the transmission under the vehicle, then raise it into position. Align the input shaft splines with the clutch disc splines and work the transmission forward into the locating dowels.
27. Install the transmission-to-engine retaining bolts and washers. Tighten the retaining bolts to specifications. Remove the transmission jack.
28. Install the starter motor. Tighten the attaching nuts.
29. Raise the engine and install the rear crossmember, insulator and damper and attaching nuts and bolts. Tighten and torque the bolts to specification.
30. On 4WD vehicles, install the transfer case.
31. On 2WD vehicles, insert the driveshaft into the transmission extension housing and install the center bearing attaching nuts, washers and lockwashers. Connect the driveshaft to the rear axle drive flange.
32. Install the remaining components in the reverse order of removal.

7-4 DRIVE TRAIN

CLUTCH

Understanding the Clutch

The purpose of the clutch is to disconnect and connect engine power at the transmission. A vehicle at rest requires a lot of engine torque to get all that weight moving. An internal combustion engine does not develop a high starting torque (unlike steam engines) so it must be allowed to operate without any load until it builds up enough torque to move the vehicle. To a point, torque increases with engine rpm. The clutch allows the engine to build up torque by physically disconnecting the engine from the transmission, relieving the engine of any load or resistance.

The transfer of engine power to the transmission (the load) must be smooth and gradual; if it weren't, drive line components would wear out or break quickly. This gradual power transfer is made possible by gradually releasing the clutch pedal. The clutch disc and pressure plate are the connecting link between the engine and transmission. When the clutch pedal is released, the disc and plate contact each other (the clutch is engaged) physically joining the engine and transmission. When the pedal is pushed in, the disc and plate separate (the clutch is disengaged) disconnecting the engine from the transmission.

Most clutch assemblies consists of the flywheel, the clutch disc, the clutch pressure plate, the throw out bearing and fork, the actuating linkage and the pedal. The flywheel and clutch pressure plate (driving members) are connected to the engine crankshaft and rotate with it. The clutch disc is located between the flywheel and pressure plate, and is splined to the transmission shaft. A driving member is one that is attached to the engine and transfers engine power to a driven member (clutch disc) on the transmission shaft. A driving member (pressure plate) rotates (drives) a driven member (clutch disc) on contact and, in so doing, turns the transmission shaft.

There is a circular diaphragm spring within the pressure plate cover (transmission side). In a relaxed state (when the clutch pedal is fully released) this spring is convex; that is, it is dished outward toward the transmission. Pushing in the clutch pedal actuates the attached linkage. Connected to the other end of this is the throw out fork, which hold the throw out bearing. When the clutch pedal is depressed, the clutch linkage pushes the fork and bearing forward to contact the diaphragm spring of the pressure plate. The outer edges of the spring are secured to the pressure plate and are pivoted on rings so that when the center of the spring is compressed by the throw out bearing, the outer edges bow outward and, by so doing, pull the pressure plate in the same direction — away from the clutch disc. This action separates the disc from the plate, disengaging the clutch and allowing the transmission to be shifted into another gear. A coil type clutch return spring attached to the clutch pedal arm permits full release of the pedal. Releasing the pedal pulls the throw out bearing away from the diaphragm spring resulting in a reversal of spring position. As bearing pressure is gradually released from the spring center, the outer edges of the spring bow outward, pushing the pressure plate into closer contact with the clutch disc. As the disc and plate move closer together, friction between the two increases and slippage is reduced until, when full spring pressure is applied (by fully releasing the pedal) the speed of the disc and plate are the same. This stops all slipping, creating a direct connection between the plate and disc which results in the transfer of power from the engine to the transmission. The clutch disc is now rotating with the pressure plate at engine speed and, because it is splined to the transmission shaft, the shaft now turns at the same engine speed. The clutch is operating properly if:

1. It will stall the engine when released with the vehicle held stationary.
2. The shift lever can be moved freely between 1st and reverse gears when the vehicle is stationary and the clutch disengaged.

Driven Disc and Pressure Plate

REMOVAL & INSTALLATION

✱✱ CAUTION

The clutch driven disc may contain asbestos, which has been determined to be a cancer causing agent. Never clean clutch surfaces with compressed air! Avoid inhaling any dust from any clutch surface! When cleaning clutch surfaces, use a commercially available brake cleaning fluid.

Pickup and MPV

▶ See Figures 8 thru 15

1. Remove the transmission.
2. Remove the four attaching and two pilot bolts holding the clutch cover to the flywheel. Loosen the bolts evenly and a turn or two at a time. If the clutch cover is to be reinstalled, mark the flywheel and clutch cover to show the location of the two pilot holes.
3. Remove the clutch disc.

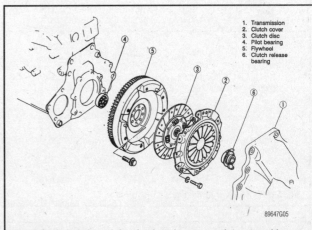

Fig. 8 Exploded view of the clutch and pressure plate assembly

Fig. 9 Loosen and remove the clutch and pressure plate bolts evenly, a little at a time . . .

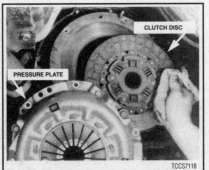

Fig. 10 . . . then carefully remove the pressure plate and clutch assembly from the flywheel

Fig. 11 Check the flywheel surface for flatness and, if necessary, remove it from the engine for truing

DRIVE TRAIN 7-5

Fig. 12 Be sure that the flywheel surface is clean before installing the clutch

Fig. 13 Typical clutch alignment tool, note how the splines match the transmission's input shaft

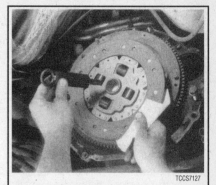
Fig. 14 Use the clutch alignment tool to align the clutch disc during assembly.

Fig. 15 Be sure to use a torque wrench to tighten all of the bolts

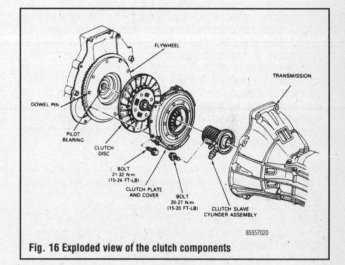
Fig. 16 Exploded view of the clutch components

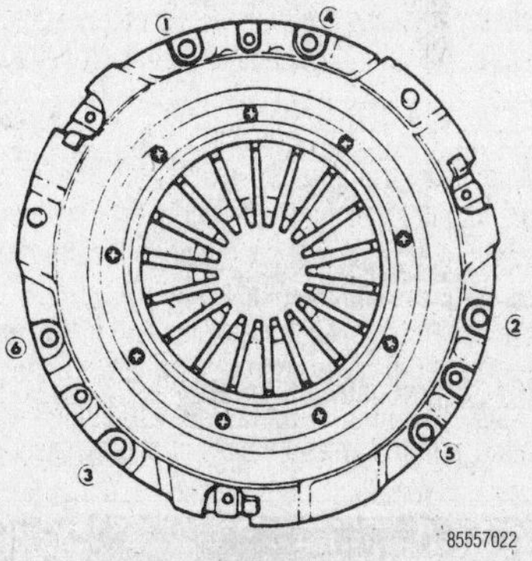

4. Install the clutch disc on the flywheel. Do not touch the facing or allow the facing to come in contact with grease or oil. The clutch disc can be aligned using a tool made for that purpose, or with an old mainshaft.
5. Install the clutch cover on the flywheel and install the four standard bolts and the two pilot bolts.
6. To avoid distorting the pressure plate, tighten the bolts evenly a few turns at a time until they are all tight.
7. Tighten the bolts to 13–20 ft.lbs. (18–27 Nm). using a crossing pattern.
8. Remove the aligning tool.
9. Apply a light film of lubricant to the release bearing, release lever contact area on the release bearing hub and to the input shaft bearing retainer.
10. Install the transmission.
11. Check the operation of the clutch and if necessary, adjust the pedal free-play and the release lever.

Navajo

♦ See Figures 9 thru 17

1. Disconnect the negative battery cable.
2. Disconnect the clutch hydraulic system master cylinder from the clutch pedal and remove.
3. Raise the vehicle and support it safely.
4. Remove the starter.
5. Disconnect the hydraulic coupling at the transmission.

➡ Clean the area around the hose and slave cylinder to prevent fluid contamination.

6. Remove the transmission from the vehicle.
7. Mark the assembled position of the pressure plate and cover the flywheel, to aid during re-assembly.
8. Loosen the pressure plate and cover attaching bolts evenly until the pressure plate springs are expanded, and remove the bolts.
9. Remove the pressure plate and cover assembly and the clutch disc from the flywheel. Remove the pilot bearing only for replacement.

Fig. 17 Pressure plate bolt torque sequence

To install:
10. Position the clutch disc on the flywheel so that the Clutch Alignment Shaft Tool T74P–7137–K or equivalent can enter the clutch pilot bearing and align the disc.
11. When reinstalling the original pressure plate and cover assembly, align the assembly and flywheel according to the marks made during the removal operations. Position the pressure plate and cover assembly on the flywheel, align the pressure plate and disc, and install the retaining bolts that fasten the assembly to the flywheel. Tighten the bolts to 15–25 ft.lbs. (21–35 Nm) in the proper sequence. Remove the clutch disc pilot tool.
12. Install the transmission into the vehicle.
13. Connect the coupling by pushing the male coupling into the slave cylinder.
14. Connect the hydraulic clutch master cylinder pushrod to the clutch pedal.

7-6 DRIVE TRAIN

ADJUSTMENTS

Free-Play/Pedal Height

EXCEPT NAVAJO

♦ See Figure 18

1. Measure the distance from the top of the clutch pedal pad to the carpet. The distance should be as follows:
   - B2200: 7.13–7.52 inches (181–191mm)
   - B2600: 7.52–7.91 inches (191–201mm)
   - MPV: 8.19–8.58 inches (208–218mm)
2. If the distance is not within specification, loosen the clutch switch locknut and turn the switch until the distance is correct. Tighten the lock nut.
3. Check the free-play by pressing the pedal by hand until clutch resistance is felt. The free-play should be 0.02–0.12 inch (0.6–3.0mm).
4. If the free-play is not within specification, loosen the locknut on the actuator rod and turn the actuator rod until the free-play is correct.
5. Check that the disengagement height from the upper surface of the pedal height to the carpet is correct when the pedal is fully depressed. The disengagement height should be as follows:
   - B2200: 2.60 inches (66mm)
   - B2600: 2.80 inches (71mm)
   - MPV: 1.38 inches (35mm)
6. Tighten the actuator rod locknut to 8.7–12.0 ft. lbs. (12–17 Nm).
7. Recheck the pedal height. Certain models are equipped with a clutch assist spring. The spring is mounted on the clutch linkage, and is adjustable. The correct length should be between 1.33–1.37 inches (33.8–34.8mm). Adjust by turning the locknut.

NAVAJO

The hydraulic clutch system provides automatic adjustment. No adjustment of clutch linkage or pedal position is required.

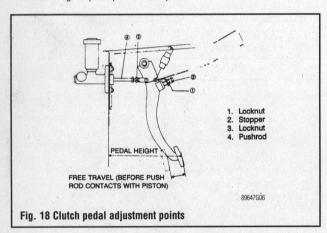

Fig. 18 Clutch pedal adjustment points

Clutch Master Cylinder

REMOVAL & INSTALLATION

Pickup and MPV

♦ See Figure 19

1. Disconnect and plug the fluid outlet line at the outlet fitting on the master cylinder one-way valve.
2. Remove the nuts and bolts attaching the master cylinder to the firewall.
3. Remove the cylinder straight out away from the firewall.
4. Start the pedal pushrod into the master cylinder and position the master cylinder on the firewall.
5. Install the attaching nuts and bolts. Tighten the nuts to 12–17 ft. lbs. (16–23 Nm).

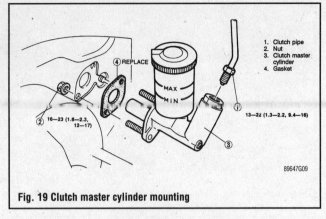

Fig. 19 Clutch master cylinder mounting

6. Connect the fluid outlet line to the master cylinder fitting.
7. Bleed the hydraulic system.
8. Check the clutch pedal free-travel and adjust as necessary.

Navajo

♦ See Figure 20

1. Disconnect the negative battery cable.
2. Disconnect the clutch master cylinder pushrod from the clutch pedal.
3. Remove the switch from the master cylinder assembly, if equipped.
4. Remove the screw retaining the fluid reservoir to the cowl access cover.
5. Disconnect the tube from the slave cylinder and plug both openings.
6. Remove the bolts retaining the clutch master cylinder to the dash panel and remove the clutch master cylinder assembly.
7. To install position the pushrod through the hole in the engine compartment. Make certain it is located on the correct side of the clutch pedal. Place the master cylinder assembly in position and install the retaining bolts. Tighten to 8–12 ft. lbs. (11–16 Nm).
8. Install the remaining components in the reverse of removal. Bleed the system.

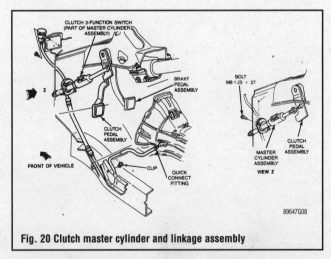

Fig. 20 Clutch master cylinder and linkage assembly

Clutch Slave Cylinder

REMOVAL & INSTALLATION

Pickup and MPV

♦ See Figures 21

1. Raise and support the front end on jackstands.
2. Back off the flare nut on the fluid pipe to free the slave cylinder hose.
3. Pull off the hose-to-bracket retaining clip and pull the hose from the bracket. Cap the pipe to prevent fluid loss.

DRIVE TRAIN 7-7

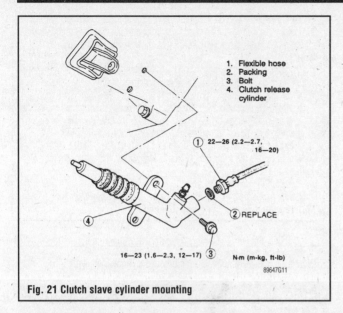

Fig. 21 Clutch slave cylinder mounting

4. Unbolt and remove the slave cylinder.
5. Installation is the reverse of removal. Tighten the bolt to 12–17 ft. lbs. (16–23 Nm).

Navajo

➡ Before performing any service that requires removal of the slave cylinder, the master cylinder and pushrod must be disconnected from the clutch pedal. If not disconnected, permanent damage to the master cylinder assembly will occur if the clutch pedal is depressed while the slave cylinder is disconnected.

1. Disconnect the negative battery cable.
2. Disconnect the coupling at the transmission, using the clutch coupling removal tool T88T–70522–A or equivalent. Slide the white plastic sleeve toward the slave cylinder while applying a slight tug on the tube.
3. Remove the transmission assembly.
4. Remove the slave cylinder-to-transmission retaining bolts.
5. Remove the slave cylinder from the transmission input shaft.

To install:

6. Fit the slave cylinder over the transmission input shaft with the bleed screws and coupling facing the left side of the transmission.
7. Install the slave cylinder retaining bolts. Torque to 13–19 ft. lbs. (18–26Nm).
8. Install the remaining components in the reverse of removal.

BLEEDING THE HYDRAULIC SYSTEM

Pickup and MPV

The clutch hydraulic system must be bled whenever the line has been disconnected or air has entered the system.

To bleed the system, remove the rubber cap from the bleeder valve and attach a rubber hose to the valve. Submerge the other end of the hose in a large jar of clean brake fluid. Open the bleeder valve. Depress the clutch pedal and allow it to return slowly. Continue this pumping action and watch the jar of brake fluid. When air bubbles stop appearing, close the bleeder valve and remove the tube.

During the bleeding process, the master cylinder must be kept at least ¾ full. After the bleeding operation is finished, install the cap on the bleeder valve and fill the master cylinder to the proper level. Always use fresh brake fluid, and above all, do not use the fluid that was in the jar for bleeding, since it contains air. Install the master cylinder reservoir cap.

Navajo

The following procedure is recommended for bleeding a hydraulic system installed on the vehicle. The largest portion of the filling is carried out by gravity. It is recommended that the original clutch tube with quick connect be replaced when servicing the hydraulic system because air can be trapped in the quick connect and prevent complete bleeding of the system. The replacement tube does not include a quick connect.

1. Clean the dirt and grease from the dust cap.
2. Remove the cap and diaphragm and fill the reservoir to the top with approved brake fluid C6AZ–19542–AA or BA, (ESA–M6C25–A) or equivalent.

➡ **To keep brake fluid from entering the clutch housing, route a suitable rubber tube of appropriate inside diameter from the bleed screw to a container.**

3. Loosen the bleed screw, located in the slave cylinder body, next to the inlet connection. Fluid will now begin to move from the master cylinder down the tube to the slave cylinder.

➡ **The reservoir must be kept full at all time during the bleeding operation, to ensure no additional air enters the system.**

4. Notice the bleed screw outlet. When the slave is full, a steady stream of fluid comes from the slave outlet. Tighten the bleed screw.
5. Depress the clutch pedal to the floor and hold for 1–2 seconds. Release the pedal as rapidly as possible. The pedal must be released completely. Pause for 1–2 seconds. Repeat 10 times.
6. Check the fluid level in the reservoir. The fluid should be level with the step when the diaphragm is removed.
7. Repeat Step 5 and 6 five times. Replace the reservoir diaphragm and cap.
8. Hold the pedal to the floor, crack open the bleed screw to allow any additional air to escape. Close the bleed screw, then release the pedal.
9. Check the fluid in the reservoir. The hydraulic system should now be fully bled and should release the clutch.
10. Check the vehicle by starting, pushing the clutch pedal to the floor and selecting reverse gear. There should be no grating of gears. If there is, and the hydraulic system still contains air, repeat the bleeding procedure from Step 5.

AUTOMATIC TRANSMISSION

Understanding the Automatic Transmission

The automatic transmission allows engine torque and power to be transmitted to the rear wheels within a narrow range of engine operating speeds. It will allow the engine to turn fast enough to produce plenty of power and torque at very low speeds, while keeping it at a sensible rpm at high vehicle speeds (and it does this job without driver assistance). The transmission uses a light fluid as the medium for the transmission of power. This fluid also works in the operation of various hydraulic control circuits and as a lubricant. Because the transmission fluid performs all of these functions, trouble within the unit can easily travel from one part to another.

Neutral Start Switch/Back-up Switch

REMOVAL & INSTALLATION

A4LD Transmission

▶ See Figure 22

The Park/Neutral Position (PNP) switch, mounted on the transmission, allows the vehicle to start only in **P** or **N**. The switch has a dual purpose, in that it is also the back-up lamp switch.

7-8 DRIVE TRAIN

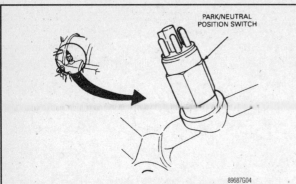

Fig. 22 View of the Park/Neutral Position (PNP) switch used on the A4LD transmission

1. Disconnect the negative battery cable.
2. Raise and support the vehicle safely.
3. Disconnect the harness connector from the neutral start switch.
4. Clean the area around the switch. Remove the switch and O-ring, using a thin wall socket (tool T74P-77247-A or equivalent).
5. Installation is the reverse of removal. Check the operation of the switch, with the parking brake engaged. The engine should only start in **N** or **P**. The back-up lamps should come ON only in **R**.

RA4A-EL and RA4AX-EL Transmissions

♦ See Figure 23

1. Disconnect the negative battery cable.
2. Raise and safely support the vehicle.
3. Disconnect the control linkage from the transmission manual shaft.
4. Disconnect the park/neutral switch harness connector, remove the switch mounting bolts, and remove the park/neutral switch.

To install:

5. Install the park/neutral switch over the manual shaft.
6. Install the switch mounting bolts but do not tighten.
7. Make sure the transmission manual shaft lever is positioned at the **L** position (fully forward). Turn the manual shaft lever fully rearward, then return it 2 notches (**N** position).
8. Insert a 0.157 inch (4.0mm) pin through the holes of the switch and the manual shaft lever.
9. Tighten the switch mounting bolts to 22–35 inch lbs. (2.5–3.9 Nm). Remove the pin.
10. Connect the electrical connector.
11. Reinstall the shift control linkage.
12. Safely lower the vehicle, connect the negative battery cable, and check for proper operation.

NA4A-HL Transmission

♦ See Figure 23

1. Disconnect the negative battery cable.
2. Raise and safely support the vehicle.

Fig. 23 View of the Park/Neutral Position (PNP) switch

3. Disconnect the control linkage from the transmission manual shaft.
4. Disconnect the wiring harness connector and remove the park/neutral switch mounting bolts. Remove the switch from the manual shaft lever.

To install:

5. Position the park/neutral switch over the manual shaft lever and install the mounting bolts. Do not fully tighten the mounting bolts.
6. Move the manual shaft lever to the **N** position.
7. Remove the screw on the switch body and move the switch so the screw hole is aligned with the small hole inside the switch. Check their alignment by inserting a 0.079 in. (2.0mm) diameter pin through the holes.
8. Tighten the switch mounting bolts to 43–61 inch lbs. (4.9–6.9 Nm).
9. Install the screw in the park/neutral switch body and reconnect the wiring harness connector.
10. Reinstall the control linkage.
11. Safely lower the vehicle, connect the negative battery cable, and check for proper operation.

ADJUSTMENTS

Refer to the removal and installation procedures for adjustment.

Extension Housing Seal

REMOVAL & INSTALLATION

♦ See Figure 24

1. Raise and support the vehicle safely.
2. Matchmark the driveshaft end yoke and rear axle companion flange to assure proper positioning during assembly. Remove the driveshaft.
3. Remove the oil seal from the extension housing, using seal remover T71P-7657-A or equivalent.

To install:

Before install the replacement seal, inspect the sealing surface of the universal joint yoke for scores. If scoring is found, replace the yoke.

4. Install the new seal, using seal installer T74P-77052-A or equivalent. Coat the inside diameter at the end of the rubber boot portion of the seal with long-life lubricant (C1AZ-19590-BA or equivalent).
5. Align the matchmarks and install the driveshaft.
6. Lower the vehicle.

Fig. 24 Examine the overall condition of the transmission extension housing seal and replace it if necessary

Automatic Transmission Assembly

REMOVAL & INSTALLATION

♦ See Figures 25 and 26

1. Disconnect the negative battery cable.
2. Raise the vehicle and support it safely.

DRIVE TRAIN 7-9

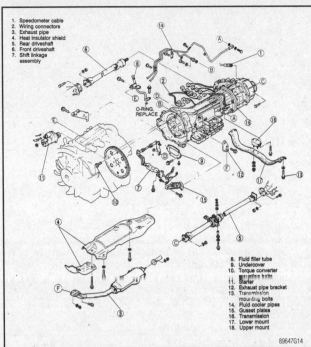

Fig. 25 MPV 4-wheel drive automatic transmission mounting—2-wheel drive similar

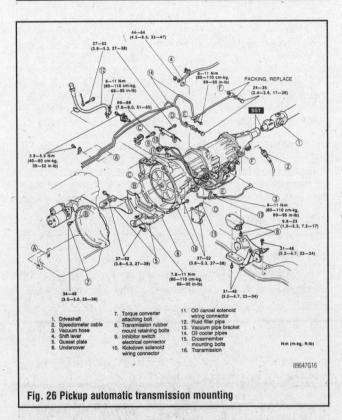

Fig. 26 Pickup automatic transmission mounting

3. Position a drain pan under the transmission pan and drain the transmission fluid.
4. Remove the converter access cover.
5. Remove the flywheel to converter attaching nuts. Use a socket and breaker bar on the crankshaft pulley attaching bolt. Rotate the pulley clockwise as viewed from the front to gain access to each of the nuts.

➡ On belt driven overhead cam engines, never rotate the pulley in a counterclockwise direction as viewed from the front.

6. Remove the speedometer cable and/or vehicle speed sensor from the transfer case (4WD) or extension housing (2WD).
7. On 2WD vehicles, scribe a mark indexing the driveshaft to the rear axle flange. Remove the driveshaft.
8. On 4WD vehicles, remove the transfer case.
9. Disconnect the shift rod or cable at the transmission manual lever and retainer bracket.
10. Disconnect the downshift cable from the downshift lever. Depress the tab on the retainer and remove the kickdown cable from the bracket.
11. Disconnect all of the transmission wire harness plugs.
12. Remove the starter mounting bolts and the ground cable. Remove the starter.
13. If equipped, remove the vacuum line from the transmission vacuum modulator.
14. Remove the filler tube from the transmission.
15. Position a transmission jack under the transmission and raise it slightly.
16. Remove the engine rear support to crossmember bolts.
17. Remove the crossmember to frame side support attaching nuts and bolts. Remove the crossmember.
18. Remove the converter housing to engine bolts.
19. Slightly lower the jack to gain access to the oil cooler lines. Disconnect the oil cooler lines at the transmission. Plug all openings to keep dirt and contamination out.
20. Move the transmission to the rear so it disengages from the dowel pins and the converter is disengaged from the flywheel. Lower the transmission from the vehicle.
21. If necessary, remove the torque converter from the transmission.

➡ If the transmission is to be removed for a period of time, support the engine with a safety stand and wood block.

To install:

✱✱ WARNING

Before installing an automatic transmission, always check that the torque converter is fully seated into the transmission. Typically, the converter has notches or tangs on the hub which must engage the transmission fluid pump. If they are not engaged in the pump, the transmission will not mate to the engine properly, as the converter will be holding it away. Severe damage to the pump, converter or transmission casing can occur if the transmission-to-engine bolts are tightened as if to force the transmission to mate to the engine.

Proper installation of the converter requires full engagement of the converter hub in the pump gear. To accomplish this, the converter must be pushed and at the same time rotated through what feels like 2 notches or bumps. When fully installed, rotation of the converter will usually result in a clicking noise heard, caused by the converter surface touching the housing to case bolts.

This should not be a concern, but an indication of proper converter installation since, when the converter is attached to the engine flywheel, it will be pulled slightly forward away from the bolt heads. Besides the clicking sound, the converter should rotate freely with no binding.

For reference, a properly installed converter will have a distance from the converter pilot nose from face to converter housing outer face of $13/32$–$7/16$ in. (10.5–14.5mm).

22. Install the converter on the transmission.
23. With the converter properly installed, position the transmission on the jack.
24. Rotate the converter so that the drive studs are in alignment with the holes in the flywheel.
25. Move the converter and transmission assembly forward into position, being careful not to damage the flywheel and converter pilot. The converter housing is piloted into position by the dowels in the rear of the engine block.

➡ During this move, to avoid damage, do not allow the transmission to get into a nose down position as this will cause the converter to move forward and disengage from the pump gear.

26. Install the converter housing to engine attaching bolts and tighten to specification. The 2 longer bolts are located at the dowel holes.
27. Remove the jack supporting the engine.
28. The rest of the installation procedure is the reverse of removal. Tighten all fasteners to the following specifications:

7-10 DRIVE TRAIN

Navajo Models
- Transmission-to-engine mounting bolts: 30–41 ft. lbs. (40–55 Nm)
- Torque converter-to-flywheel mounting bolts: 22–30 ft. lbs. (30–40 Nm)
- Transmission mount-to-transmission: 64–81 ft. lbs. (87–110 Nm)
- Transmission crossmember mounting bolts: 63–87 ft. lbs. (85–118 Nm)
- Transmission mount-to-crossmember: 60–82 ft. lbs. (82–111 Nm)

Pickup and MPV Models
- Transmission-to-engine mounting bolts: 28–38 ft. lbs. (38–51 Nm)
- Torque converter-to-flywheel mounting bolts: 27–39 ft. lbs. (37–53 Nm)
- Transmission mount-to-transmission: 32–44 ft. lbs. (44–60 Nm)
- Transmission crossmember mounting bolts: 32–44 ft. lbs. (44–60 Nm)
- Transmission mount-to-crossmember: 32–44 ft. lbs. (44–60 Nm)

29. Follow the procedures outlined in Section 1 when filling the transmission with fluid.

Adjustments

SHIFT LINKAGE

Pickups

L3N71B

1. Move the gearshift lever to the **P** range.
2. Loosen shift linkage locknuts.
3. Be sure to shift the transmission to the **P** range by moving the select lever of the transmission.
4. Push and hold the gearshift lever forward with a force of approximately 22 lbs.
5. Tighten linkage locknuts.

L4N71B

1. Move the gearshift lever to the **P** range.
2. Loosen the shift linkage locknuts.
3. Be sure to shift the transmission to the **P** range by moving the select lever of the transmission.
4. Using a feeler gauge, adjust the linkage until there is 0.039 in. (1mm) between the adjust lever and the locknut.
5. Remove the feeler gauge and tighten jam nut to 69–95 inch lbs. (7.8–11.0 Nm).
6. Move the gearshift lever to the **N**, **D** and **P** ranges and make sure there is clearance between the select lever bracket and the guide pin.
7. If there is no clearance, readjust the lever.

N4A-HL

1. Move the gearshift lever to the **P** range.
2. Loosen linkage locknuts so they are both at least 0.039 in. (1mm) away from the adjustment lever.
3. Shift the transmission to the **P** range by moving the manual shaft of the transmission.
4. With the link at 90 degrees to the lever, adjust the clearance using a feeler gauge to 0.039 in. (1mm) between the adjustment lever and the locknut.
5. Remove the feeler gauge and tighten the jam nut to 69–95 inch lbs. (8–11 Nm).
6. Measure the clearance between the guide plate and the guide pin in the **P** range. There should be 0.039 in. (1mm) clearance in front of the pin and 0.020 in. (0.5mm) clearance behind the pin.
7. Move the gearshift lever to the **N** and **D** ranges and check that the clearance between the guide plate and guide pin is the same in both ranges. If not, readjust.

R4AX-EL

1. Disconnect the negative battery cable to deactivate the shift-lock.
2. Remove the selector knob and console.
3. Loosen linkage locknuts and lock bolt.
4. Shift the transmission manual shaft to the **P** range.
5. Push and hold the selector lever forward with a force of approximately 22 lbs., then tighten the lock bolt to 67–95 inch lbs. (8–11 Nm).
6. Turn the locknut by hand until it just touches the spacer, then tighten the jam nut to 67–95 inch lbs. (8–11 Nm).

7. Check the lever so the clearance between the guide plate and the guide pin in the **P** range with the pushrod lightly depressed is as specified.
8. Move the selector lever to the **N** and **D** ranges and make sure there is the same clearance between the guide plate and guide pin. If not, readjust the lever.
9. Install the console. Clean and apply locking compound to the selector knob screw threads and tighten the screws to 13–26 inch lbs. (1.5–2.9 Nm).
10. Connect the negative battery cable.

MPV

▶ See Figure 27

1. Move the selector lever to the **P** range.
2. Remove the column covers.
3. Pull the selector lever rearward, toward the driver and insert a 0.197 in. (5mm) outer diameter pin into the gearshift rod assembly.
4. Remove the air intake tube.
5. Loosen the shift lever and the top lever mounting bolts.
6. Shift the transmission manual shaft to the **P** range position.
7. Adjust the clearance between the lower bracket and the shift lever bushing by sliding the shift lever assembly until there is no clearance.
8. Tighten the shift lever mounting bolts to 12–17 ft. lbs. (16–23 Nm).
9. Make sure the detent ball is positioned in the center of the **P** range detent. If not, loosen the linkage bolts and turn the bracket to adjust the position, then retighten the bolts to 61–87 inch lbs. (6.9–9.8 Nm).
10. Adjust the clearance between the lower bracket and the shift lever bushing by turning the top lever until there is no clearance. Tighten the top lever mounting bolt, then retighten the remaining bolt to 12–17 ft. lbs. (16–23 Nm).
11. Remove the pin from the gear shift rod assembly and install the column covers. Check selector lever operation.

Navajo

1. Raise and safely support the vehicle, as necessary. From inside the vehicle, place the column shift selector lever in the **OD** position. Hang an 8 lb. weight on the selector lever.
2. From below the vehicle, pull down the lock tab on the shift cable and remove the fitting from the transmission manual control lever ball stud.
3. Position the transmission manual control lever in the **OD** position by moving the lever all the way rearward and then moving it 3 detents forward.
4. Connect the cable end fitting to the transmission manual control lever. Push up on the lock tab to lock the cable in the correctly adjusted position.
5. Remove the 8 lb. weight from the column shift selector.
6. After adjustment, check for **P** engagement. Check the column shift selector lever in all detent positions with the engine running to ensure correct adjustment.

KICK-DOWN SWITCH

1987-89 L3N71B

1. Connect an ohmmeter across the terminals of the kick-down switch, depress the accelerator pedal fully and make sure that there is continuity. If not, proceed:

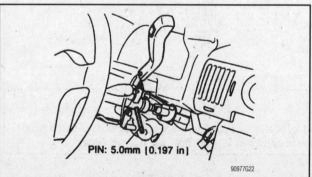

Fig. 27 Pull the selector lever rearward, toward the driver and insert a 0.197 in. (5mm) outer diameter pin into the gearshift rod assembly

DRIVE TRAIN 7-11

2. Loosen the switch locknut.
3. Depress the accelerator pedal fully and adjust the switch until continuity is observed.
4. Tighten the locknut.

L4N71B

1. Connect an ohmmeter across the 2 lower terminals of the kick-down switch. Make sure that there is continuity when the tip of the switch is depressed 6.0-6.5mm (0.236-0.256 in.). If not, proceed:
2. Loosen the switch locknut.
3. Depress the accelerator pedal fully and adjust the switch until continuity is observed when the tip of the switch is depressed 6.0-6.5mm (0.236-0.256 in.).
4. Tighten the locknut.

N4A-HL

♦ See Figures 28 and 29

1. Connect an ohmmeter across the 2 lower terminals of the kick-down switch. Make sure that there is continuity when the tip of the switch is depressed 6.0–6.5mm. If not, proceed:
2. Disconnect the wiring.
3. Loosen the switch locknut and back the switch out fully.
4. Depress the accelerator pedal fully and hold it.
5. With the pedal held down, turn the kick-down switch clockwise until a click is heard, then, turn it ¼ turn further.
6. Tighten the locknut.
7. Release the pedal and reconnect the wiring.
8. Apply 12 volts to the kick-down solenoid and observe that a click is heard. If not, replace the solenoid.

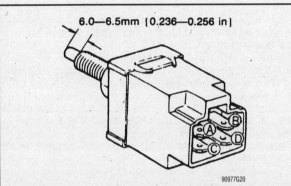

Fig. 28 Measure the continuity between terminals C and D with the switch depressed

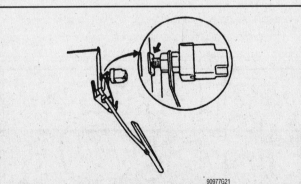

Fig. 29 With the pedal held down, turn the kick-down switch clockwise until a click is heard, then, turn it ¼ turn further

KICKDOWN CABLE

A4LD transmission

♦ See Figure 30

The kickdown cable is self-adjusting over a tolerance range of 1 in. (25mm). If the cable requires readjustment, reset the by depressing the semi-circular metal tab on the self-adjuster mechanism and pulling the cable forward (toward the front of the vehicle) to the "Zero" position setting. The cable will then automatically readjust to the proper length when kicked down.

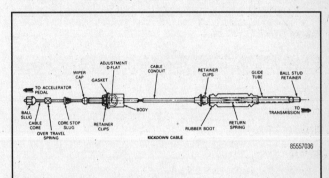

Fig. 30 View of the transmission kickdown cable assemble and its adjusting points

MANUAL LINKAGE

L3N71B

1. Place the shift lever in **P**.
2. Loosen the 3 locknuts on the linkage trunion.
3. Move the transmission lever to the **P** range.
4. Apply a forward force of about 20 lbs. to the shift handle and tighten the locknuts to 95 inch lbs.
5. Move the shift handle to **N**, **P**, and **D** ranges. Make sure that there is clearance between the select lever bracket and guide pin. If there is no clearance, adjust the lever by adjusting the locknuts.

L4N71B

1. Place the shift lever in **P**.
2. Remove the shifter boot.
3. Loosen the 2 locknuts on the cable at the base of the shifter.
4. Raise and support the truck on jackstands.
5. Move the transmission lever to the **P** range.
6. Using a feeler gauge, turn the locknuts until there is 1mm (0.039 in.) clearance between the forward locknut and the adjust lever.
7. Tighten the rear locknut.
8. Move the shift handle to **N**, **P**, and **D** ranges. Make sure that there is clearance between the select lever bracket and guide pin. If there is no clearance, adjust the lever by adjusting the locknuts.

A4LD

Before the linkage is adjusted, be sure the engine idle speed and anti-stall dashpot are properly adjusted.

1. Position the transmission selector control lever in **D** position and loosen the trunion bolt.

➡ Make sure that the shift lever detent pawl is held against the rearward Drive detent stop during the linkage adjustment procedure.

2. Position the transmission manual lever in the **D** position, by moving the bellcrank lever all the way rearward, then forward 3 detents.

7-12 DRIVE TRAIN

3. With the transmission selector lever/manual lever in the **D** position, apply light forward pressure to the shifter control tower arm while tightening the trunion bolt to 12–23 ft. lbs. Forward pressure on the shifter lower arm will ensure correct positioning within the **D** detent as noted in Step 1.

After adjustment, check for Park engagement. The control lever must move to the right when engaged in Park. Check the transmission control lever in all detent positions with the engine running to ensure correct detent transmission action. Readjust if necessary.

TRANSFER CASE

Rear Output Shaft Seal

REMOVAL & INSTALLATION

▶ See Figures 31 and 32

1. Raise and safely support the vehicle.
2. Remove the driveshaft after match-marking the flange and shaft for installation reference. On models with a double offset joint, push a rag into the joint to keep the shaft straight so boot damage will not occur.
3. Hold the companion flange with a suitable tool, and remove the mounting nut. Remove the flange. It may be necessary to use a puller, on some models, to remove the flange.
4. Pry out the old seal. Clean the mounting surface. Apply oil to the lips of the seal and install the seal using the appropriate driver. Install the flange and the driveshaft.

Front Output Shaft Seal

REMOVAL & INSTALLATION

▶ See Figure 33

1. Raise and safely support the vehicle.
2. Remove the skid plate, if equipped. Mark the driveshaft to flange for installation reference.

3. Drain the fluid from the transfer case if the support angle suggests an oil spill when the flange is removed from the transfer case. Remove the driveshaft.
4. Remove the retaining nut, washer, etc. Remove the companion flange.
5. Pry the seal from the housing. Clean the seal mounting surfaces.
6. Oil the seal lips. Install the seal using the proper driver. Install the flange and the driveshaft.

Transfer Case Assembly

REMOVAL & INSTALLATION

▶ See Figure 34

※※ CAUTION

The catalytic converter is located beside the transfer case. Be careful when working around the catalytic converter because of the extremely high temperatures generated by the converter.

1. Disconnect the negative battery cable.
2. Raise the vehicle and support it safely.
3. If so equipped, remove the skid plate from frame.
4. Drain the transfer case.
5. Remove the damper from the transfer case, if so equipped.
6. On electronic shift models, remove the wire connector from the feed wire harness at the rear of the transfer case. Be sure to squeeze the locking tabs, then pull the connectors apart.
7. Disconnect the front driveshaft from the axle input yoke.
8. If equipped, loosen the clamp retaining the front driveshaft boot to the transfer case, and pull the driveshaft and front boot assembly out of the transfer case front output shaft.
9. Disconnect the rear driveshaft from the transfer case output shaft yoke.
10. If equipped, disconnect the speedometer driven gear from the transfer case rear cover.
11. If equipped, disconnect the electrical wire harness plug from the Vehicle Speed Sensor (VSS).
12. Disconnect the vent hose from the mounting bracket.
13. On manual shift models, perform the following:
    a. Remove the shift lever retaining nut and remove the lever.
    b. Remove the bolts that retains the shifter to the extension housing. Note the size and location of the bolts to aid during installation. Remove the lever assembly and bushing.
14. If equipped, remove the heat shield from the transfer case.
15. Support the transfer case with a transmission jack.

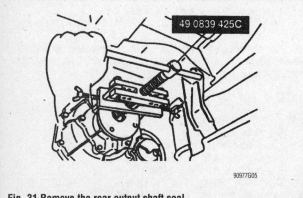

Fig. 31 Remove the rear output shaft seal

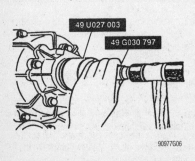

Fig. 32 Drive the new seal into the case housing using the appropriate seal installation tool

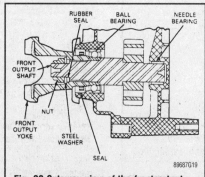

Fig. 33 Cutaway view of the front output shaft, seal and yoke assembly—rear components are similar

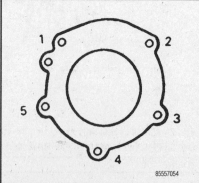

Fig. 34 Case-to-extension bolt torque sequence—Navajo and B Series Pick-up

DRIVE TRAIN 7-13

16. Remove the mounting bolts retaining the transfer case to the transmission.
17. Slide the transfer case rearward off the transmission output shaft and lower the transfer case from the vehicle. Remove the gasket or any old sealer from between the transfer case and the transmission.

To install:
18. Install the heat shield onto the transfer case, if equipped, and place a new gasket or silicone sealer between the transfer case and adapter.
19. Raise the transfer case with a suitable transmission jack or equivalent, raise it high enough so that the transmission output shaft aligns with the splined transfer case input shaft.
20. Slide the transfer case forward on to the transmission output shaft and onto the dowel pin. Install transfer case retaining bolts and torque them to specification. Tighten the mounting bolts to 35–46 ft. lbs. (47–63 Nm) on Navajo and 27–39 ft. lbs. (37–52 Nm) on Pickup and MPV models.
21. The remainder of the installation procedure is the reverse of removal. Check the fluid level and, if necessary, top off with the correct amount and type of transfer case fluid.

ADJUSTMENTS

Manual Shift Models

▶ See Figure 35

The following procedure should be used, if a partial or incomplete engagement of the transfer case shift lever detent is experienced or if the control assembly requires removal.
1. Disconnect the negative battery cable.
2. Raise the shift boot to expose the top surface of the cam plates.
3. Loosen the 1 large and 1 small bolt, approximately 1 turn. Move the transfer case shift lever to the **4L** position (lever down).
4. Move the cam plate rearward until the bottom chamfered corner of the neutral lug just contacts the forward right edge of the shift lever.

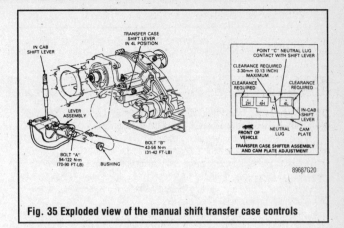

Fig. 35 Exploded view of the manual shift transfer case controls

5. Hold the cam plate in this position and torque the larger bolt first to 70–90 ft. lbs. (95–122 Nm) and torque the smaller bolt to 31–42 ft. lbs. (42–57 Nm).
6. Move the transfer case in cab shift lever to all shift positions to check for positive engagement. There should be a clearance between the shift lever and the cam plate in the **2H** front and **4H** rear (clearance not to exceed 3.3mm) and **4L** shift positions.
7. Install the shift boot assembly.
8. Reconnect the negative battery cable.

Except Manual Shift Models

Both of the electronic shift and the AWD model transfer cases do not require any linkage adjustments, nor are any possible.

DRIVELINE

Front Driveshaft and U-Joints

REMOVAL & INSTALLATION

▶ See Figures 36 thru 37

1. Raise and support the vehicle safely.

➡ **The driveshaft is a balanced unit. Before removing the drive shaft, matchmark the driveshaft in relationship to the end yoke so that it may be installed in its original position.**

2. Using a shop cloth or gloves, pull back on the dust slinger to remove the boot from the transfer case slip yoke.

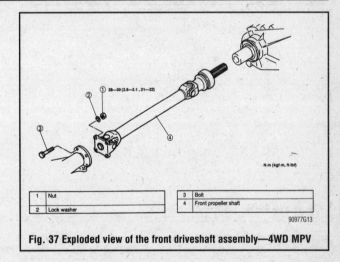

Fig. 37 Exploded view of the front driveshaft assembly—4WD MPV

3. Remove the bolts and straps that retains the driveshaft to the front driving axle yoke. Remove the U-joint assembly from the front driving axle yoke.
4. Slide the splined yoke assembly out of the transfer case and remove the driveshaft assembly.
5. Inspect the boot for rips or tears. Inspect the stud yoke splines for wear or damage. Replace any damage parts.
6. To install, first apply a light coating of Multi-purpose Long-Life lubricant C1AZ-19490-B or equivalent, to the yoke splines and the edge of the inner diameter of the rubber boot. Position the U-joint assembly in the front drive axle yoke in its original position. Install the retaining bolts and straps. Tighten the bolts to 10–15 ft. lbs. (14–20 Nm). Install the remaining components in the reverse of removal.

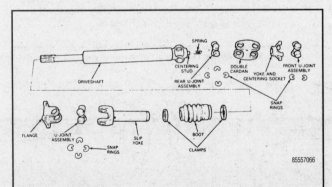

Fig. 36 Exploded view of the front driveshaft and double cardan type U-joint—4WD Navajo

7-14 DRIVE TRAIN

U-JOINT REPLACEMENT

Single Cardan Type U-Joint

▶ See Figures 38 and 39

1. Remove the driveshaft.
2. If the front yoke is to be disassembled, matchmark the driveshaft and sliding splined yoke (transmission yoke) so that driveline balance is preserved upon reassembly. Remove the snaprings which retain the bearing caps.
3. Select two sockets, one small enough to pass through the yoke holes for the bearing caps, the other large enough to receive the bearing cap.
4. Using a vise or a press, position the small and large sockets on either side of the U-joint. Press in on the smaller socket so that it presses the opposite bearing cap out of the yoke and into the larger socket. If the cap does not come all the way out, grasp it with a pair of pliers and work it out.
5. Reverse the position of the sockets so that the smaller socket presses on the cross. Press the other bearing cap out of the yoke.
6. Repeat the procedure on the other bearings.
7. To install, grease the bearing caps and needles thoroughly if they are not pregreased. Start a new bearing cap into one side of the yoke. Position the cross in the yoke.
8. Select two sockets small enough to pass through the yoke holes. Put the sockets against the cross and the cap, and press the bearing cap ¼ in. (6mm) below the surface of the yoke. If there is a sudden increase in the force needed to press the cap into place, or if the cross starts to bind, the bearings are cocked, They must be removed and restarted in the yoke. Failure to do so will greatly reduce the life of the bearing.
9. Install a new snapring.
10. Start a new bearing into the opposite side. Place a socket on it and press in until the opposite bearing contacts the snapring.
11. Install a new snapring. It may be necessary to grind the facing surface of the snapring slightly to permit easier installation.
12. Install the other bearings in the same manner.
13. Check the joint for free movement. If binding exists, smack the yoke ears with a brass or plastic faced hammer to seat the bearing needles. Do not strike the bearings, and support the shaft firmly. Do not install the driveshaft until free movement exists at all joints.

Double Cardan Type U-Joint

▶ See Figures 40 and 41

1. Place the driveshaft on a suitable workbench.
2. Matchmark the positions of the spiders, the center yoke and the centering socket yoke as related to the stud yoke which is welded to the front of the driveshaft tube.

➡ The spiders must be assembled with the bosses in their original position to provide proper clearance.

3. Remove the snaprings that secure the bearings in the front of the center yoke.
4. Position the U-joint tool, T74P-4635-C or equivalent, on the center yoke. Thread the tool clockwise until the bearing protrudes approximately ⅜ in. (10mm) out of the yoke.
5. Position the bearing in a vice and tap on the center yoke to free it from the bearing. Lift the 2 bearing cups from the spider.
6. Re-position the tool on the yoke and move the remaining bearing in the opposite direction so that it protrudes approximately ⅜ in. (10mm) out of the yoke.
7. Position the bearing in a vice. Tap on the center yoke to free it from the bearing. Remove the spider from the center yoke.
8. Pull the centering socket yoke off the center stud. Remove the rubber seal from the centering ball stud.
9. Remove the snaprings from the center yoke and from the driveshaft yoke.
10. Position the tool on the driveshaft yoke and press the bearing outward until the inside of the center yoke almost contacts the slinger ring at the front of the driveshaft yoke. Pressing beyond this point can distort the slinger ring interference point.
11. Clamp the exposed end of the bearing in a vice and drive on the center yoke with a soft-faced hammer to free it from the bearing.
12. Reposition the tool and press on the spider to remove the opposite bearing.
13. Remove the center yoke from the spider. Remove the spider form the driveshaft yoke.
14. Clean all serviceable parts in cleaning solvent. If using a repair kit, install all of the parts supplied in the kit.
15. Remove the clamps on the driveshaft boot seal. Discard the clamps.
16. Note the orientation of the slip yoke to the driveshaft tube for installation during assembly. Mark the position of the slip yoke to the driveshaft tube.
17. Carefully pull the slip yoke from the driveshaft. Be careful not to damage the boot seal.
18. Clean and inspect the spline area of the driveshaft.

To assemble:

19. Lubricate the driveshaft slip splines with Multi-purpose Long-Life lubricant C1AZ-19490-B or equivalent.
20. With the boot loosely installed on the driveshaft tube, install the slip yoke into the driveshaft splines in their original orientation.

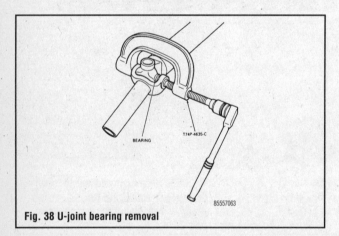

Fig. 38 U-joint bearing removal

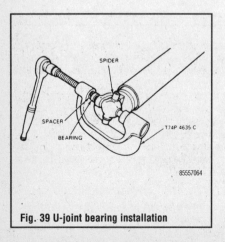

Fig. 39 U-joint bearing installation

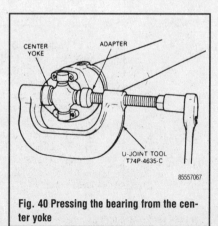

Fig. 40 Pressing the bearing from the center yoke

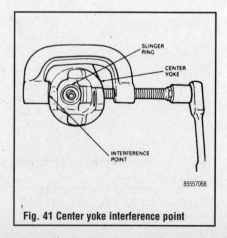

Fig. 41 Center yoke interference point

DRIVE TRAIN   7-15

21. Using new clamps, install the driveshaft boot in its original position.
22. To assemble the double cardan joint, position the spider in the driveshaft yoke. Make certain the spider bosses (or lubrication plugs on kits) will be in the same position as originally installed. Press in the bearing using the U-joint tool. Then, install the snaprings.
23. Pack the socket relief and the ball with Multi-purpose Long-Life lubricant C1AZ–19490–B or equivalent, then position the center yoke over the spider ends and press in the bearing. Install the snaprings.
24. Install a new seal on the centering ball stud. Position the centering socket yoke on the stud.
25. Place the front spider in the center yoke. Make certain the spider bosses (or lubrication plugs on kits) are properly positioned.
26. With the spider loosely positioned on the center stop, seat the first pair of bearings into the centering socket yoke. Then, press the second pair into the centering yoke. Install the snaprings.
27. Apply pressure on the centering socket yoke and install the remaining bearing cup.
28. If a kit was used, lubricate the U-joint through the grease fitting, using Multi-purpose Long-Life lubricant C1AZ–19490–B or equivalent.

Rear Driveshaft

REMOVAL & INSTALLATION

Navajo Models

▶ See Figure 42

1. Disconnect the negative battery cable.
2. Raise and support the vehicle safely.

➡ The driveshaft is a balanced unit. Before removing the driveshaft, matchmark the driveshaft yoke in relationship to the axle flange so that it may be installed in its original position.

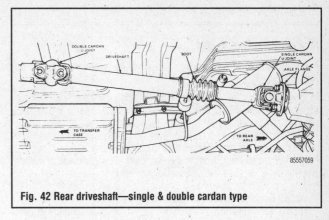

Fig. 42 Rear driveshaft—single & double cardan type

3. Remove the retaining bolts and disconnect the driveshaft from the axle companion flange.
4. Remove the retaining bolts that retains the driveshaft to the rear of the transfer case.
5. Remove the driveshaft.

To install:

6. Install the driveshaft into the rear of the transfer case. Make certain that the driveshaft is positioned with the slip yoke toward the front of the vehicle. Install the bolts and tighten to 41–55 ft. lbs. (55–74Nm).
7. Install the driveshaft so the index mark on the rear yoke is in line with the index mark on the axle companion flange.

Pickup and MPV Models

▶ See Figures 43 thru 52

1. Matchmark the rear U-joint with the rear companion flange. Remove the bolts attaching the driveshaft to the rear companion flange.
2. Remove the center support bearing bracket from the underbody.

Fig. 43 Stuff a rag into the CV U-joint to hold the driveshaft straight and prevent damage to the boot

Fig. 44 Before removing any mounting nuts, matchmark the position of the driveshaft to the differential for correct reassembly

Fig. 45 To separate the driveshaft from the differential housing, use two, 14 mm box wrenches and remove the four mounting nuts

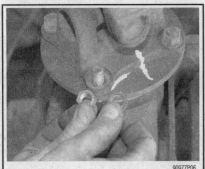

Fig. 46 Be careful not to lose the washer when removing the driveshaft-to-differential mounting nuts

Fig. 47 Using a 14mm socket, loosen the ground wire retaining bolt . . .

Fig. 48 . . . then remove the bolt, separating the ground wire from the driveshaft center bearing

7-16 DRIVE TRAIN

Fig. 49 Remove the driveshaft center bearing mounting bolts

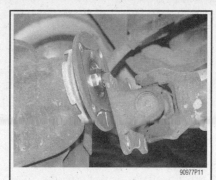
Fig. 50 After removing the center bearing mounting bolts, separate the driveshaft from the differntial housing . . .

Fig. 51 . . . then, while supporting the driveshaft assembly, slide the driveshaft straight out of the extension housing

Fig. 52 After removing the driveshaft, install a plug such as the one pictured, into the opening of the extension housing to prevent transmission fluid from dripping out

3. Pull the driveshaft rearward and out of the transmission. Plug the rear seal opening.
4. Installation is the reverse of removal. Make sure that you align the matchmarks. Torque the rear companion flange bolts to 39–47 ft. lbs.; the center bearing bracket nuts to 27–38 ft. lbs.

U-JOINT REPLACEMENT

For replacement procedures for the single cardan U-joint, refer to u-joint replacement under "Front Driveshaft and U-Joints" earlier in this section.

DRIVESHAFT BALANCING

▶ See Figures 53, 54 and 55

Unbalance

Propeller shaft vibration increases as the vehicle speed is increased. A vibration that occurs within a specific speed range is not usually caused by a propeller shaft being unbalanced. Defective universal joints, or an incorrect propeller shaft angle, are usually the cause of such a vibration.

If propeller shaft is suspected of being unbalanced, it can be verified with the following procedure.

➡Removing and re-indexing the propeller shaft 180° relative to the yoke may eliminate some vibrations.

1. Raise and safely support the vehicle securely on jackstands.
2. Clean all the foreign material from the propeller shaft and the universal joints.
3. Inspect the propeller shaft for missing balance weights, broken welds, and bent areas. If the propeller shaft is dented or bent, it must be replaced.
4. Inspect the universal joints to ensure that they are not worn, are properly installed, and are correctly aligned with the shaft.
5. Check the universal joint clamp bolt torque.
6. Remove the wheels and tires. Install the wheel lug nuts to retain the brake drums or rotors.
7. Mark and number the shaft six inches (15.24cm) from the yoke end at four positions 90° apart.
8. Run and accelerate the vehicle until vibration occurs. Note the intensity and speed the vibration occurred. Stop the engine.
9. Install a screw clamp at any position.
10. Start the engine and re-check for vibration. If there is little or no change in vibration, move the clamp to one of the other three positions. Repeat the vibration test.
11. If there is no difference in vibration at the other positions, the source of the vibration may not be propeller shaft.

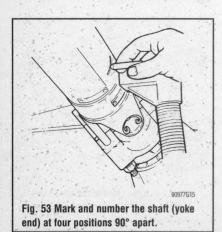
Fig. 53 Mark and number the shaft (yoke end) at four positions 90° apart.

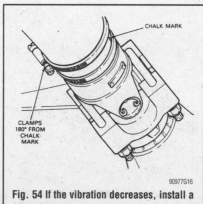
Fig. 54 If the vibration decreases, install a second clamp

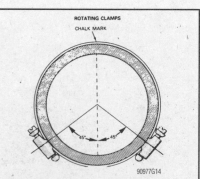
Fig. 55 If the second clamp causes additional vibration, rotate the clamps in opposite directions

DRIVE TRAIN 7-17

12. If the vibration decreased, install a second clamp and repeat the test.
13. If the additional clamp causes additional vibration, rotate the clamps (¼ inch above and below the mark). Repeat the vibration test.
14. Increase distance between the clamp screws and repeat the test until the amount of vibration is at the lowest level. Bend the slack end of the clamps so the screws will not loosen.
15. If the vibration remains unacceptable, apply the same steps to the front end of the propeller shaft.
16. Install the wheel and tires.
17. Lower the vehicle.

Run-out

♦ See Figure 56

1. Remove dirt, rust, paint, and undercoating from the propeller shaft surface where the dial indicator will contact the shaft.
2. The dial indicator must be installed perpendicular to the shaft surface.

➡ Measure front/rear run-out approximately 3 inches (76mm) from the weld seem at each end of the shaft tube for tube lengths over 30 inches (76.2cm).

3. Measure run-out at the center and ends of the shaft sufficiently far away from weld areas to ensure that the effects of the weld process will not enter into the measurements.
4. Replace the propeller shaft if the run-out exceeds 0.035 inch (0.89mm) on Navajo and 0.016 inch (0.40mm) on Pickup and MPV models.

Center Bearing

REPLACEMENT

♦ See Figures 57 thru 63

The center support bearing is a sealed unit which requires no periodic maintenance. The following procedure should be used if it becomes necessary to replace the bearing. You will need a pair of snapring pliers for this job.

1. Remove the driveshaft assembly.
2. To maintain driveline balance, matchmark the rear driveshaft, the center yoke and the front driveshaft so that they may be installed in their original positions.
3. Remove the center universal joint from the center yoke, leaving it attached to the rear driveshaft. See the following section for the correct procedure.
4. Remove the nut and washer securing the center yoke to the front driveshaft.
5. Slide the center yoke off the splines. The rear oil seal should slide off with it.
6. If the oil has remained on top of the snapring, remove and discard the seal. Remove the snapring from its groove. Remove the bearing.
7. Slide the center support and front oil seal from the front driveshaft. A puller tool may be necessary. Discard the seal.

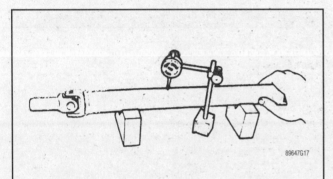

Fig. 56 Measure the front and rear shaft runout, replace if the runout is excessive

Fig. 57 Before separating the two halves of the rear driveshaft assembly, matchmark the positioning of the halves for correct reassembly

Fig. 58 When separating the front half from the rear half, use two, 14 mm box wrenches and remove the four mounting nuts

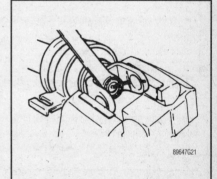

Fig. 59 Remove the nut and washer securing the center yoke to the front driveshaft

Fig. 60 Remove the snapring using snapring pliers

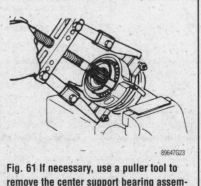

Fig. 61 If necessary, use a puller tool to remove the center support bearing assembly

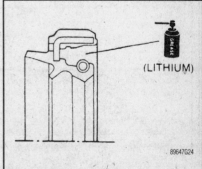

Fig. 62 Before installing the front seal, apply a coat of lithium grease to the seal lips

7-18 DRIVE TRAIN

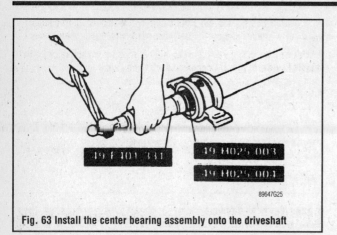

Fig. 63 Install the center bearing assembly onto the driveshaft

8. Install the new bearing into the center support. Secure it with the snapring.
9. Apply a coat of grease to the lips of the new oil seals, and install them into the center support on either side of the bearing.
10. Coat the splines of the front driveshaft with grease. Install the center support assembly and the center yoke onto the front driveshaft, being sure to match up the marks made during disassembly.
11. Install the washer and nut. Torque the nut to 116–130 ft. lbs.
12. Check that the center support assembly rotates smoothly around the driveshaft.
13. Align the mating marks on the center yoke and the rear driveshaft, and assemble the center universal joint.
14. Install the driveshaft. Be sure that the rear yoke and the axle flange are aligned properly.

FRONT DRIVE AXLE

The Navajo employs a Ford/Dana 35 integral carrier drive axle. Consequently, tool numbers and part numbers listed in the procedures for that axle will be Ford numbers.

Manual Locking Hubs

REMOVAL & INSTALLATION

B Series Pickup

♦ See Figure 64

1. Raise and safely support the vehicle. Remove the wheel and tire assembly.
2. Set the locking hub in the **FREE** position.
3. Remove the locking hub mounting bolts and remove the locking hub.
4. With the hub removed, install 2 bolts and nuts opposite each other to hold the hub together.
5. Check for smooth turning of the control handle.
6. Check for smooth rotation of the inner hub with the control lever in the **FREE** position.
7. Check for no rotation of the inner hub with the control lever in the **LOCK** position.

To install:

8. Coat the surface of the hub with sealant.
9. Place the control lever in the **FREE** position.
10. Install the hub on the vehicle and tighten the bolts to 22–25 ft. lbs. (29–34 Nm).
11. Install the wheel and tire assembly and check the operation of the hub. Lower the vehicle.

Navajo

♦ See Figure 65

1. Raise and support the vehicle safely.
2. Remove the lug nuts and remove the wheel and tire assembly.
3. Remove the retainer washers from the lug nut studs and remove the manual locking hub assembly. To remove the internal hub lock assembly from the outer body assembly, remove the outer lock ring seated in the hub body groove. The internal assembly, spring and clutch gear will now slide out of the hub body. Do not remove the screw from the plastic dial.
4. Rebuild the hub assembly in the reverse order of disassembly.
5. Adjust the wheel bearing if necessary. Install the manual locking hub assembly over the spindle and place the retainer washers on the lug nut studs.
6. Install the wheel and tire assembly and lower the vehicle.

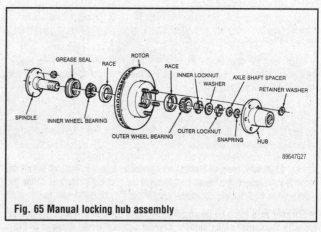

Fig. 65 Manual locking hub assembly

ADJUSTMENT

➡ This procedure applies to Navajo only.

1. Raise and safely support the vehicle. Remove the wheel and tire assembly.
2. Remove the retainer washers from the lug nut studs and remove the manual locking hub assembly from the spindle.
3. Remove the snapring from the end of the spindle shaft.
4. Remove the axle shaft spacer.
5. Remove the outer wheel bearing locknut from the spindle using locknut wrench 49 UN01 042 or equivalent. Make sure the tabs on the tool engage the slots in the locknut.

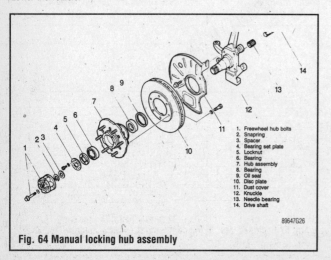

Fig. 64 Manual locking hub assembly

1. Freewheel hub bolts
2. Snapring
3. Spacer
4. Bearing set plate
5. Locknut
6. Bearing
7. Hub assembly
8. Bearing
9. Oil seal
10. Disc plate
11. Dust cover
12. Knuckle
13. Needle bearing
14. Drive shaft

DRIVE TRAIN 7-19

6. Remove the locknut washer from the spindle.
7. Loosen the inner wheel bearing locknut using the locknut wrench. Make sure the tabs on the tool engage the slots in the locknut and the slot in the tool is centered over the locknut pin.
8. Tighten the inner locknut to 35 ft. lbs. (47 Nm) to seat the bearings.
9. Spin the rotor and back off the inner locknut ¼ turn. Rethighten the inner locknut to 16 inch lbs. (1.8 Nm). Install the lockwasher on the spindle. It may be necessary to tighten the inner locknut slightly so the pin on the locknut aligns with the closest hole in the lockwasher.
10. Install the outer wheel bearing locknut using the locknut wrench. Tighten the locknut to 150 ft. lbs. (203 Nm).
11. Install the axle shaft spacer.
12. Clip the snapring onto the end of the spindle. Install the manual hub assembly over the spindle and install the retainer washers.
13. Install the wheel and tire assembly. Check the endplay of the wheel and tire assembly on the spindle. Final endplay should be 0–0.003 in. (0–0.08mm). The maximum torque to rotate the hub should be 25 inch lbs. (2.8 Nm).
14. Lower the vehicle.

Automatic Locking Hubs

REMOVAL & INSTALLATION

Navajo

♦ See Figure 66

1. Raise and support the vehicle safely. Remove the wheel lug nuts and remove the wheel and tire assembly.
2. Remove the retainer washers from the lug nut studs and remove the automatic locking hub assembly from the spindle.
3. Remove the snapring from the end of the spindle shaft.
4. Remove the axle shaft spacer.
5. Being careful not to damage the plastic moving cam or thrust spacers, pull the cam assembly off the wheel bearing adjusting nut. Remove the 2 plastic thrust spacers from the adjusting nut.
6. Using a magnet, remove the locking key. It may be necessary to rotate the adjusting nut slightly to relieve the pressure against the locking key, before the key can be removed.

➡ To prevent damage to the spindle threads, look into the spindle keyway under the adjusting nut and remove the separate locking key before removing the adjusting nut.

7. Loosen the wheel bearing adjusting nut from the spindle.
8. While rotating the hub and rotor assembly, tighten the wheel bearing adjusting nut to 35 ft. lbs. (47 Nm) to seat the bearings. Spin the rotor and back off the nut one-quarter turn.
9. Rethighten the adjusting nut to 16 inch lbs. (1.8 Nm) using a torque wrench. Align the closest hole in the wheel bearing adjusting nut with the center of the spindle keyway slot. Advance the nut to the next lug if required. Install the separate locking key in the spindle keyway under the adjusting nut.

➡ Extreme care must be taken when aligning the spindle nut adjustment lug with the center of the spindle keyway slot to prevent damage to the separate locking key.

10. Install the 2 thrust spacers. Push or press the cam assembly onto the locknut by lining up the key in the fixed cam with the spindle keyway.

➡ Extreme care must be taken when aligning the fixed cam key with the spindle keyway to prevent damage to the fixed cam.

11. Install the axle shaft spacer.
12. Clip the snapring onto the end of the spindle.
13. Install the automatic locking hub assembly over the spindle by lining up the 3 legs in the hub assembly with the 3 pockets in the cam assembly. Install the retainer washers.
14. Install the wheel and tire assembly. Check the endplay of the wheel and tire assembly on the spindle. Final endplay should be 0–0.003 in. (0–0.08mm). The maximum torque to rotate the hub should be 25 inch lbs. (2.8 Nm).
15. Lower the vehicle.

Freewheel Mechanism

➡ The Remote Freewheel Mechanism on 1989–93 4WD B Series pickup and the Automatic Freewheel Mechanism on 1990–93 4WD MPV are used in place of automatic locking hubs.

REMOVAL & INSTALLATION

♦ See Figures 67, 68 and 69

1. Disconnect the negative battery cable. Raise and safely support the vehicle. Remove the left front wheel and tire assembly.
2. Drain the fluid from the front differential.
3. Remove the left side halfshaft assembly.
4. Tag and disconnect the vacuum hoses and electrical connector from the control box assembly.
5. Remove and discard the snap pin at the control box assembly.
6. Remove the attaching bolts and remove the joint shaft assembly.
7. Remove the attaching bolts and remove the control box assembly.

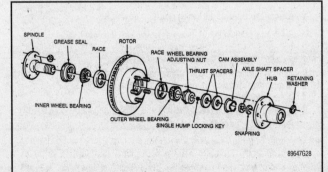

Fig. 66 Automatic locking hub assembly

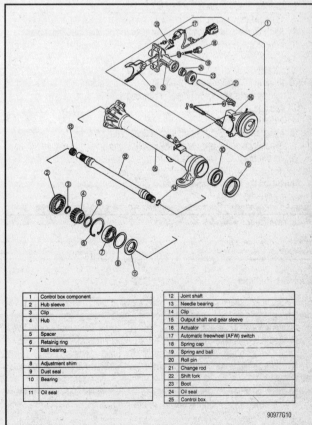

Fig. 67 Exploded view of the components to the automatic freewheel mechanism—MPV

7-20 DRIVE TRAIN

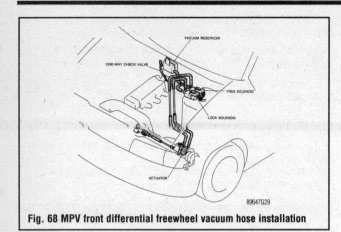

Fig. 68 MPV front differential freewheel vacuum hose installation

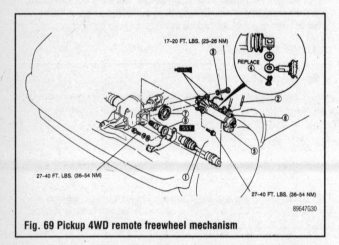

Fig. 69 Pickup 4WD remote freewheel mechanism

8. Remove the gear sleeve from the side of the differential, if necessary.
9. If necessary, remove the output shaft from the differential using a slide hammer.

To install:
10. If removed, install a new clip on the end of the output shaft and install in the differential. Install the gear sleeve, if removed.
11. Install the control box and tighten the attaching bolts to 17–20 ft. lbs. (23–26 Nm).
12. Install the joint shaft assembly and tighten the attaching bolts to 27–40 ft. lbs. (36–54 Nm). On MPV, install the attaching nut and tighten to 49–72 ft. lbs. (67–97 Nm).
13. Install a new snap pin at the control box assembly. The remaining assembly is the reverse of removal.

Spindle Bearings

REMOVAL & INSTALLATION

B Series Pickups

1. Raise and safely support the vehicle. Remove the wheel and tire assembly.
2. Remove the drive flange.
3. Remove the brake caliper. Support the caliper aside with rope or mechanics wire; do not let the caliper hang by the brake hose.
4. Remove the snapring and spacer. Remove the set bolts and bearing set plate.
5. Remove the bearing locknut using a suitable removal tool. Remove the hub and rotor without letting the washer and bearing fall.
6. Remove the dust cover.
7. Disconnect the tie rod end from the knuckle. Disconnect the stabilizer bar and the lower shock mount.

8. Support the lower control arm with a jack. Remove the lower ball joint nut and separate the knuckle from the lower arm using a suitable tool.
9. Remove the upper ball joint nut and separate the knuckle from the lower arm using a suitable tool.
10. Lower the lower control arm and remove the knuckle.
11. Inspect the knuckle, hub and bearings for wear and or damage. Replace components, as necessary.
12. Remove the oil seal and the bearing inner race from the knuckle. Using a suitable drift, remove the bearing outer race by tapping lightly with a hammer.
13. Using a slide hammer, remove the needle bearing from the knuckle.
14. Mark the position of the disc brake rotor on the hub, then remove the bolts and disassemble the rotor and hub.
15. Remove the oil seal and the bearing inner race from the hub. Using a suitable drift, remove the bearing outer race by tapping lightly with a hammer.

To install:
16. Press a new needle bearing into the knuckle using a suitable driver.
17. After installing the inner bearing into the knuckle, press in a new oil seal. Apply wheel bearing grease to the oil seal lip.
18. Press fit the outer side bearing outer race, then the inner side bearing outer race, into the hub using suitable drivers. Press in a new oil seal until it is flush with the hub end surface. Apply wheel bearing grease to the seal lip.
19. Align the matching marks of the hub and brake rotor and tighten the mounting bolts to 40–51 ft. lbs. (54–69 Nm).
20. Liberally apply high temperature wheel bearing grease to the inside of the hub. Install the outer bearing race and washer in the hub.
21. Insert the halfshaft into the knuckle and install the nut for the lower ball joint. Tighten the nut by hand.
22. Raise the lower control arm with the jack until the upper ball joint is connected to the knuckle. Install the nut and tighten by hand.
23. Tighten the upper ball joint nut to 22–38 ft. lbs. (29–51 Nm) and the lower ball joint nut to 87–116 ft. lbs. (118–157 Nm). Install new cotter pins.
24. Connect the tie rod end to the knuckle, tighten the nut to 23–43 ft. lbs. (44–59 Nm) and install a new cotter pin.
25. Install the dust cover to the knuckle and tighten to 14–19 ft. lbs. (19–26 Nm).
26. After loosely installing the lower shock absorber mount, install the stabilizer bar.
27. Install the hub and rotor assembly, then adjust the bearing preload as follows:
  a. Tighten the locknut, then turn the hub and rotor 2–3 times to seat the bearing.
  b. Loosen the locknut so they can be turned by hand.
  c. Attach a suitable pull scale to a wheel lug bolt and measure the frictional forces. The preload is the frictional force plus 1.3–2.6 lbs.
  d. Tighten the locknut until the preload is as specified.
  e. Install the bearing set plate using 2 bolts. Tighten the bolts to 43–61 inch lbs. (5–7 Nm).
  f. Coat the spacer with grease and install it. Install a new snapring.
28. Install the remaining components in the reverse of removal.

MPV

▶ See Figure 70

1. Raise and safely support the vehicle. Remove the wheel and tire assembly.
2. Remove and discard the locknut from the end of the halfshaft.
3. Remove the brake caliper and disc brake rotor. Support the caliper aside with rope or mechanics wire; do not let the caliper hang by the brake hose.
4. Remove the cotter pin and nut and, using a suitable tool, disconnect the tie rod end from the knuckle.
5. Remove the cotter pin and loosen the lower ball joint nut. Separate the lower arm from the knuckle using a suitable tool.
6. Remove the knuckle-to-strut bolts and nuts and remove the knuckle/hub assembly from the vehicle.
7. Pry out the inner oil seal from the knuckle.
8. Position the knuckle/hub assembly in a press and, using a suitable driver, press the hub from the knuckle.

➡ **If the inner bearing race remains on the hub, position the hub in a vise, secured by the flange. Move the race away from the hub using a hammer and chisel, then position the hub in a press and press the race off of the hub.**

Drive Train 7-21

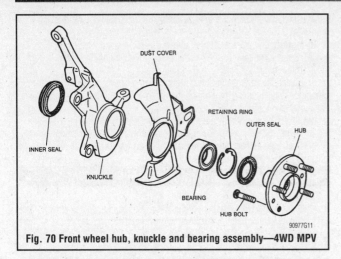

Fig. 70 Front wheel hub, knuckle and bearing assembly—4WD MPV

9. Pry out the outer oil seal from the knuckle.
10. Remove the retaining ring and position the knuckle in a press. Using a suitable driver, press the wheel bearing from the knuckle.
11. If necessary, mark the position of the dust shield on the knuckle and remove the dust shield, using a hammer and chisel. Do not reuse the dust cover, if removed.

To install:
12. If the dust cover was removed, mark the new cover in the same place as the old was marked during removal. Align the cover and knuckle marks and press the cover onto the knuckle.
13. Press a new wheel bearing into the knuckle, using a suitable driver. Install the retaining ring and a new outer seal. Apply grease to the seal lip.
14. Press the hub into the knuckle, using a suitable driver. Install a new inner seal and lubricate the seal lip with grease.
15. Install the remiaing components in the reverse of removal.

Navajo

1. Raise and safely support the vehicle. Remove the front wheel and tire assemblies.
2. Remove the disc brake caliper and wire it to the frame. Do not let the caliper hang by the brake hose.
3. Remove the hub locks, wheel bearings and locknuts.
4. Remove the hub, rotor and outer wheel bearing.
5. Remove the grease seal from the rotor with a seal removal tool. Remove the inner wheel bearing.
6. If the wheel bearings are to be replaced, remove the inner and outer bearing races with a suitable puller or a hammer and brass drift.
7. Remove the nuts retaining the spindle to the steering knuckle. Tap the spindle with a plastic hammer to jar the spindle from the knuckle. Remove the splash shield.
8. On the left side of the vehicle, remove the shaft and joint assembly by pulling the assembly out of the carrier. On the right side of the carrier, remove and discard the clamp from the shaft and joint assembly and the stub shaft. Pull the shaft and joint assembly from the splines of the stub shaft.
9. Place the spindle in a vise on the second step of the spindle. Wrap a shop towel around the spindle or use a brass-jawed vise to protect the spindle.
10. Remove the oil seal and needle bearing from the spindle with a slide hammer and seal remover TOOL–1175–AC or equivalent. If necessary, remove the slinger from the shaft by driving off with a hammer.
11. Remove the cotter pin from the tie rod nut and then remove the nut. Tap on the tie rod stud to free it from the steering arm.
12. Remove the upper ball joint snapring and remove the upper ball joint pinch bolt. Loosen the lower ball joint nut to the end of the stud.
13. Strike the inside of the knuckle near the upper and lower ball joints to break the knuckle loose from the ball joint studs.
14. Remove the camber adjuster sleeve. Note the position of the slot in the camber adjuster so it can be reinstalled in the same position during assembly.
15. Remove the lower ball joint nut. Place the knuckle in a vise and remove the snapring from the bottom ball joint socket, if equipped.
16. Assemble C-frame T74P–4635–C and ball joint remover T83T–3050–A or equivalents on the lower ball joint. Turn the forcing screw clockwise until the lower ball joint is removed from the steering knuckle.
17. Assemble the C-frame and ball joint remover on the upper ball joint and remove in the same manner.

➡ Always remove the lower ball joint first.

To install:
18. Clean the steering knuckle bore and insert the lower ball joint in the knuckle as straight as possible.
19. Assemble C-frame T74P–4635–C, ball joint installer T83T–3050–A and receiver cup T80T–3010–A3 or equivalents to install the lower ball joint. Turn the forcing screw clockwise until the lower ball joint is firmly seated. Install the snapring on the lower ball joint.

➡ The lower ball joint must always be installed first.

20. Assemble the C-frame, ball joint installer and receiver cup to install the upper ball joint. Turn the forcing screw clockwise until the ball joint is firmly seated.
21. Install the camber adjuster into the support arm, making sure the slot is in the original position.

➡ The torque sequence in Steps 22 and 23 must be followed exactly when securing the knuckle. Excessive knuckle turning effort may result in reduced steering returnability if this procedure is not followed.

22. Install a new nut on the bottom ball joint stud. Tighten the nut to 90 ft. lbs. (122 Nm) minimum, then tighten to align the next slot in the nut with the hole in the stud. Install a new cotter pin.
23. Install the snapring on the upper ball joint stud. Install the upper ball joint pinch bolt and tighten to 48–65 ft. lbs. (65–88 Nm).

➡ The camber adjuster will seat itself into the knuckle at a predetermined position during the tightening sequence. Do not attempt to adjust this position.

24. Clean all dirt and grease from the spindle bearing bore. The bearing bores must be free from nicks and burrs.
25. Place the bearing in the bore with the manufacturers identification facing outward. Drive the bearing into the bore using spindle bearing replacer T80T–4000–S and driver handle T80T–4000–W or equivalents.
26. Install the grease seal in the bearing bore with the lip side of the seal facing towards the tool. Drive the seal in the bore using the same tools as in Step 25. Coat the bearing seal lip with high-temperature lubricant.
27. If removed, press on a new shaft slinger.
28. On the right side of the carrier, install the rubber boot and new keystone clamps on the stub shaft slip yoke. Slide the right shaft and joint assembly into the slip yoke making sure the splines are fully engaged. Slide the boot over the assembly and crimp the keystone clamp using suitable pliers.

➡ The Dana model 35 axle does not have a blind spline, therefore pay special attention to make sure the yoke ears are in phase (in line) during assembly.

29. On the left side of the carrier, slide the shaft and joint assembly through the knuckle and engage the splines on the shaft in the carrier.
30. Install the splash shield and spindle onto the steering knuckle. Install and tighten the spindle nuts to 45 ft. lbs. (61 Nm).
31. If removed, drive the bearing races into the rotor using a suitable driver. Pack the inner and outer wheel bearings and the lip of a new seal with high-temperature wheel bearing grease.
32. Position the inner wheel bearing in the race and install the seal using a seal installer. Install the rotor on the spindle and install the outer wheel bearing in the race.
33. Install the wheel bearing, locknut, thrust bearing, snapring and locking hubs.
34. Install the caliper and the wheel and tire assemblies. Lower the vehicle.

7-22 DRIVE TRAIN

Halfshaft

REMOVAL & INSTALLATION

B Series Pickup

1. Raise and safely support the vehicle. Remove the wheel and tire assembly.
2. Remove the drive flange hub.
3. Remove the caliper, mounting support and knuckle arm. Support the caliper aside with rope or mechanics wire; do not let the caliper hang by the brake hose.
4. Disconnect the stabilizer bar and the tie rod end.
5. Remove the lower mount of the shock absorber.
6. Remove the snapring and spacer.
7. Support the lower control arm with a jack.
8. Disconnect the upper and lower ball joints and the knuckle.
9. Lower the lower control arm and remove the knuckle assembly.
10. Remove the splash shield.
11. Using a suitable prybar, pry out the halfshaft from the differential and remove the halfshaft from the vehicle. Be careful not to damage the dust cover or oil seal.

To install:

12. Install a new clip on the halfshaft. Coat the differential seal with clean transmission fluid.
13. Install the halfshaft in the differential, being careful not to damage the seal. After installation, attempt to pull the halfshaft outward to make sure it does not come out.
14. Install the knuckle and hub to the halfshaft and ball joints. Install the spacer and a new snapring.
15. Install the remaining components in the reverse of removal.

MPV

▶ See Figure 71

1. Raise and safely support the vehicle. Remove the wheel and tire assembly.
2. Remove and discard the halfshaft locknut.
3. Disconnect the tie rod end from the knuckle.
4. Remove the caliper and brake rotor from the knuckle. Support the caliper aside with rope or mechanics wire; do not let it hang by the brake hose.
5. Remove the nut and bolts and remove the lower ball joint. Remove the bolts and nuts and remove the knuckle/hub assembly from the strut.

➡ **If the halfshaft is stuck to the hub, install a used locknut so it is flush with the end of the shaft, then tap the nut with a soft mallet.**

6. Remove the splash shield.
7. Using a suitable prybar, pry out the halfshaft from the differential and remove the halfshaft from the vehicle. Be careful not to damage the dust cover or oil seal.

To install:

8. Install a new clip on the halfshaft. Coat the differential seal with clean transmission fluid.
9. Install the halfshaft in the differential, being careful not to damage the seal. After installation, attempt to pull the halfshaft outward to make sure it does not come out.
10. Install the knuckle/hub assembly to the strut and tighten the nuts to 69–86 ft. lbs. (93–117 Nm).
11. Install the remaining components in the reverse of removal.

Spindle, Shaft and Joint Assembly

REMOVAL & INSTALLATION

Navajo

▶ See Figure 72

1. Loosen the front wheel lug nuts.
2. Raise and support the vehicle safely. Remove the wheel and tire assembly.
3. Remove the disc brake calipers and support the caliper on the vehicle's frame rail.
4. Remove the hub locks and locknuts.
5. Remove the hub and rotor.
6. Remove the nuts retaining the spindle to the steering knuckle. Tap the spindle with a plastic or rawhide hammer to jar the spindle from the knuckle.
7. Remove the front disc brake rotor shield.

➡ **The left-hand axle shaft is engaged inside of the front carrier assembly. Depending on how the truck is sitting (especially if it is not level), some fluid may leak out of the front carrier assembly. A small drip pan should can be placed underneath the front carrier as a precautionary measure.**

8. Remove the left-hand side axle shaft by pulling the assembly out of the carrier and through the hole in the steering knuckle (spindle mount).
9. Remove the right-hand axle shaft by performing the following:
    a. Remove and discard the right front axle joint boot clamp from the outer axle assembly.
    b. Pull the right-hand axle shaft out of the axle joint boot and stub shaft and through the hole in the steering knuckle (spindle mount).
10. Inspect the seals on the outer axle shaft ends, and replace them if necessary. Replace the seals as follows:
    a. Remove the old seal from the axle shaft by driving them off with a hammer.
    b. Thoroughly clean the axle seal area of the shaft.
    c. Place the shaft in a press and install the new seal using the Spindle/Axle Seal Installer Tool T83T-3132-A, or equivalent.

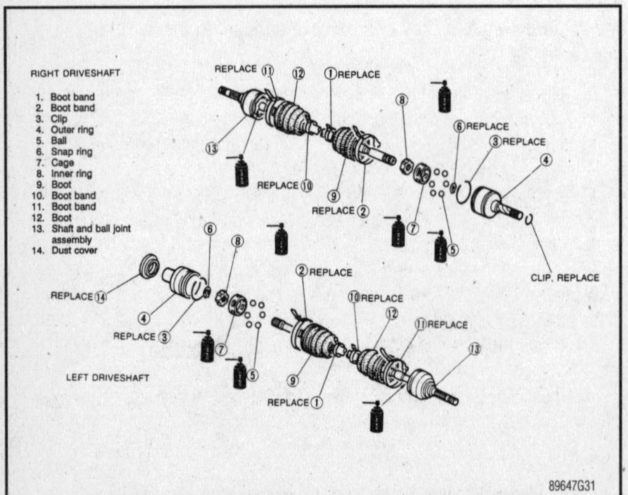

Fig. 71 Exploded view of axle shaft components—4WD MPV

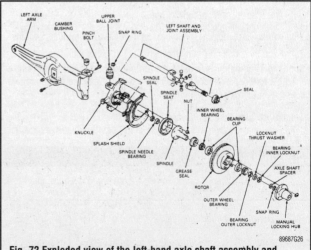

Fig. 72 Exploded view of the left-hand axle shaft assembly and related components

To install:

11. Install the right-hand axle shaft as follows:
   a. Ensure that the rubber boot is properly installed on the carrier stub shaft. Slide a new outer axle shaft boot clamp onto the rubber boot.

➡ The model 35 front axle does not use blind, or master, splines. Therefore, special attention should be made to ensure that the yoke ears are in line (in phase) during assembly.

   b. Slide the right axle shaft assemble through the hole in the steering knuckle, into the rubber boot and engage the splines of the stub shaft. Ensure the splines are fully engaged.
   c. Position the rubber boot and clamp onto the outer axle shaft and crimp the clamp securely on the rubber boot using Keystone Clamp Pliers T63P-9171-A.

12. Install the left-hand axle shaft by sliding it through the hole in the steering knuckle and engaging it into the carrier. Ensure that the shaft is fully seated into the carrier and engage to the splines inside.
13. Install the remaining components in the reverse of removal.

Pinion Seal

REMOVAL & INSTALLATION

➡ This service procedure disturbs the pinion bearing preload and this preload must be carefully reset when assembling.

1. Raise the vehicle and support it safely.
2. Remove the wheels and the brake drums.
3. Mark the driveshaft and the axle companion flange so the driveshaft can be reinstalled in the same position. Remove the driveshaft.
4. Using an inch pound torque wrench on the pinion nut, record the torque required to maintain rotation of the pinion through several revolutions.
5. While holding the companion flange with a suitable tool, remove the pinion nut. Mark the companion flange in relation to the pinion shaft so the flange can be reinstalled in the same position.
6. Using a suitable puller, remove the rear axle companion flange. Use a small prybar to remove the seal from the carrier.

To install:

7. Make sure the splines of the pinion shaft are free of burrs.
8. Apply grease to the lips of the pinion seal and install, using a seal installer.
9. Check the seal surface of the companion flange for scratches, nicks or a groove. Replace the companion flange, as necessary. Apply a small amount of lubricant to the splines. Align the mark on the flange with the mark on the pinion shaft and install the companion flange.

➡ The companion flange must never be hammered on or installed with power tools.

10. Install a new nut on the pinion shaft. Hold the companion flange with a suitable tool while tightening the nut.
11. Tighten the pinion nut, rotating the pinion occasionally to ensure proper bearing seating. Take frequent pinion bearing torque preload readings until the original recorded preload reading is obtained.

➡ Under no circumstances should the pinion nut be backed off to reduce preload. If reduced preload is required, a new collapsible pinion spacer and pinion nut must be installed.

12. Install the driveshaft and check the fluid level in the carrier. Lower the vehicle.

Differential Carrier

REMOVAL & INSTALLATION

B Series Pickup

➡ On 1989–92 vehicles, the differential is removed as a unit with the freewheel mechanism. After removal, the differential can then be separated from the freewheel mechanism, if necessary.

1. Raise and safely support the vehicle. Remove the wheel and tire assemblies.
2. Remove the splash shield and drain the differential fluid.
3. Remove the halfshafts.
4. Mark the position of the driveshaft on the axle flange and remove the driveshaft.
5. On 1989–92 vehicles, tag and disconnect the vacuum hoses and electrical connector from the freewheel mechanism control box.
6. Support the differential with a jack.
7. Remove the crossmember bolts adjacent to the lower control arm. Lower the differential/crossmembers assembly from the vehicle.
8. Remove the crossmembers from the differential, if necessary. On 1989–92 vehicles, remove the freewheel mechanism from the differential, if necessary.

To install:

9. If removed, install the freewheel mechanism.
10. If removed, install the differential to the crossmembers.
11. Raise the differential/crossmembers assembly into position. Install the crossmember mounting bolts and tighten to 69–85 ft. lbs. (93–116 Nm). Remove the jack.
12. Install the remaining components in the reverse order of their removal. Fill the differential with the proper type and quantity of fluid.

MPV

◆ See Figure 73

➡ The differential is removed as a unit with the freewheel mechanism. After removal, the differential can then be separated from the freewheel mechanism, if necessary.

1. Raise and safely support the vehicle. Remove the wheel and tire assemblies.
2. Remove the splash shield and drain the differential fluid.
3. Remove the halfshafts.
4. Mark the position of the driveshaft on the axle flange and remove the driveshaft.
5. Tag and disconnect the vacuum hoses and electrical connector from the freewheel mechanism control box.
6. Support the differential with a jack.
7. Remove the bolts/nuts attaching the differential/freewheel mechanism assembly in 3 places and lower the assembly from the vehicle.
8. If necessary, separate the freewheel mechanism from the differential.

To install:

9. If removed, install the freewheel mechanism.
10. Raise the differential/freewheel mechanism assembly into position and install the attaching bolts/nuts. Tighten to 49–72 ft. lbs. (67–97 Nm). Remove the jack.
11. Install the remaining components in the reverse order of their removal. Fill the differential with the proper type and quantity of fluid.

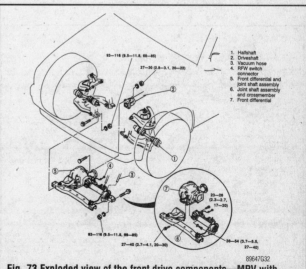

Fig. 73 Exploded view of the front drive components—MPV with 4WD

7-24 DRIVE TRAIN

REAR AXLE

Axle Shaft, Bearing and Seal

REMOVAL & INSTALLATION

Pickups

♦ See Figure 74

1. Raise and support the rear end on jackstands.
2. Remove the wheel and brake drum.
3. Remove the brake shoes.
4. Remove the parking brake cable retainer.
5. Disconnect and cap the brake lines at the wheel cylinders.
6. Remove the bolts securing the backing plate and bearing housing.
7. Slide the axle shaft from the axle housing. Be careful to avoid damaging the oil seal with the shaft.
8. If the seal in the axle housing is damaged in any way, it must be replaced. The seal can be removed using a slide hammer and adapter.
9. Remove two of the backing plate bolts, diagonally from each other.
10. Using a grinding wheel, grind down the bearing retaining collar in one spot, until about 5mm (0.197 in.) remains before you get to the axle shaft. Place a chisel at this point and break the collar. Be careful to avoid damaging the shaft.

✻✻ CAUTION

Wear some kind of protective goggles when grinding the collar and breaking the collar from the shaft!

11. Using a press or puller, remove the hub and bearing assembly from the shaft. Remove the spacer from the shaft.
12. Remove the bearing and seal from the hub.
13. Using a drift, tap the race from the hub.
14. Check all parts for wear or damage. If either race is to be replaced, both must be replaced. The race in the axle housing can be removed with a slide hammer and adapter. It's a good idea to replace the bearing and races as a set. It's also a good idea to replace the seals, regardless of what other service is being performed.

To install:

15. The outer race must be installed using an arbor press. The inner race can be driven into place in the axle housing.
16. Pack the hub with lithium based wheel bearing grease.
17. Tap a new oil seal into the axle housing until it is flush with the end of the housing. Coat the seal lip with wheel bearing grease.
18. Install a new spacer on the shaft with the larger flat surface up.
19. Install a new seal in the hub.
20. Thoroughly pack the bearing with clean, lithium based, wheel bearing grease. If one is available, use a grease gun adapter meant for packing bearings. These are available at all auto parts stores.
21. Place the bearing in the hub, and, using a press, press the hub and bearing assembly onto the shaft.
22. Press the new collar onto the shaft. The press pressure for the collar is critical. Press pressures should be 9,240-13,420 lb. (4,200-6100 kg).
23. Install one shaft in the housing being very careful to avoid damaging the inner seal.
24. If only on shaft was being serviced, the other must now be removed to check bearing play on the serviced axle. If both shafts were removed, leave the other one out for now.
25. Tighten the backing plate bolts on the one installed axle to 80 ft. lbs.
26. Mount a dial indicator on the backing plate, with the pointer resting on the axle shaft flange. Check the axial play. Standard bearing play should be 0.65-0.95mm (0.0256-0.0374 in.).
27. If play is not within specifications, shims are available for correcting it.
28. Install the other shaft and torque the backing plate bolts. Check the play as on the first shaft. Play should be 0.05-0.25mm (0.0019-0.0098 in.). If not, correct it with shims.
29. Install the brake drums and wheels. Bleed the brake system.

MPV

♦ See Figure 75

1. Raise and support the rear end on jackstands.
2. Remove the wheel and brake drum.
3. Remove the brake shoes.
4. Remove the parking brake cable retainer.
5. Disconnect and cap the brake lines at the wheel cylinders.
6. Remove the bolts securing the backing plate and bearing housing.
7. Slide the axle shaft from the axle housing. Be careful to avoid damaging the oil seal with the shaft.
8. If the seal in the axle housing is damaged in any way, it must be replaced. The seal can be removed using a slide hammer and adapter.
9. Remove two of the backing plate bolts, diagonally from each other.
10. Using a grinding wheel, grind down the bearing retaining collar in one spot, until about 5mm remains before you get to the axle shaft. Place a chisel at this point and break the collar. Be careful to avoid damaging the shaft.

✻✻ CAUTION

Wear some kind of protective goggles when grinding the collar and breaking the collar from the shaft!

11. Using a press or puller, remove the hub and bearing assembly from the shaft. Remove the spacer from the shaft.
12. Remove the bearing and seal from the hub.
13. Using a drift, tap the race from the hub.

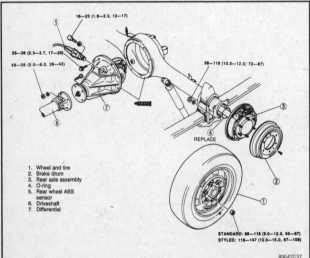

Fig. 74 Exploded view of the rear differential and axle shaft components

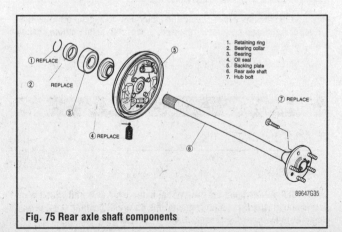

Fig. 75 Rear axle shaft components

DRIVE TRAIN 7-25

14. Check all parts for wear or damage. If either race is to be replaced, both must be replaced. The race in the axle housing can be removed with a slide hammer and adapter. It's a good idea to replace the bearing and races as a set. It's also a good idea to replace the seals, regardless of what other service is being performed.

To install:

15. The outer race must be installed using an arbor press. The inner race can be driven into place in the axle housing.
16. Pack the hub with lithium based wheel bearing grease.
17. Tap a new oil seal into the axle housing until it is flush with the end of the housing. Coat the seal lip with wheel bearing grease.
18. Install a new spacer on the shaft with the larger flat surface up.
19. Install a new seal in the hub.
20. Thoroughly pack the bearing with clean, lithium based, wheel bearing grease. If one is available, use a grease gun adapter meant for packing bearings. These are available at all auto parts stores.
21. Place the bearing in the hub, and, using a press, press the hub and bearing assembly onto the shaft.
22. Press the new collar onto the shaft. The press pressure for the collar is critical. Press pressures should be 9,240–13,420 lb. (4,200–6,100 kg).
23. Install one shaft in the housing being very careful to avoid damaging the inner seal.
24. If only on shaft was being serviced, the other must now be removed to check bearing play on the serviced axle. If both shafts were removed, leave the other one out for now.
25. Tighten the backing plate bolts on the one installed axle to 80 ft. lbs.
26. Mount a dial indicator on the backing plate, with the pointer resting on the axle shaft flange. Check the axial play. Standard bearing play should be 0.57mm.
27. If play is not within specifications, shims are available for correcting it. See the table below:
28. Install the other shaft and torque the backing plate bolts. Check the play as on the first shaft.
29. Install the brake drums and wheels. Bleed the brake system.

Navajo

♦ See Figures 76 thru 83

1. Disconnect the negative battery cable.
2. Raise and support the vehicle safely.
3. Remove the rear wheels and brake drums.
4. Drain the rear axle lubricant.
5. For all axles except 3.73:1 and 4.10:1 ratio.
   a. Remove the differential pinion shaft lock bolt and differential pinion shaft.

➡ The pinion gears may be left in place. Once the axle shafts are removed, reinstall the pinion shaft and lock bolt.

   b. Push the flanged end of the axle shafts toward the center of the vehicle and remove the C-lockwasher from the end of the axle shaft.
   c. Remove the axle shafts from the housing. If the seals and/or bearing are not being replaced, be careful not to damage the seals with the axle shaft splines upon removal.
6. For 3.73:1 and 4.10:1 ratio axles.
   a. Remove the pinion shaft lock bolt. Place a hand behind the differential case and push out the pinion shaft until the step contacts the ring gear.
   b. Remove the C-lockwasher from the axle shafts.
   c. Remove the axle shafts from the housing. If the seals and/or bearing are not being replaced, be careful not to damage the seals with the axle shaft splines upon removal.
7. Insert the wheel bearing and seal remover, T85L–1225–AH or equivalent, and a slide hammer into the axle bore and position it behind the bearing so the tanks on the tool engage the bearing outer race. Remove the bearing and seal as a unit.

To install:

8. If removed, lubricate the new bearing with rear axle lubricant and install the bearing into the housing bore. Use axle tube bearing replacer, T78P–1225–A or equivalent.
9. Apply Multi-Purpose Long-Life Lubricant, C1AZ–19590–B or equivalent, between the lips of the axle shaft seal.

Fig. 76 To remove the axle, first remove the rear axle cover and drain the fluid

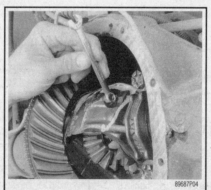

Fig. 77 Next, loosen the pinion shaft lock bolt . . .

Fig. 78 . . . and remove it from the axle carrier

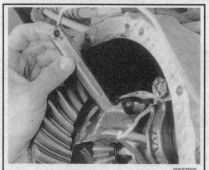

Fig. 79 Pull the pinion shaft out of the axle carrier. DO NOT rotate the axle with the pinion shaft removed!

Fig. 80 Push in on the axle flange (wheel side) and remove the axle C-lock (A) from the end of the axle (B)

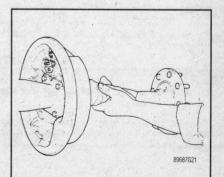

Fig. 81 Slide the axle out of the axle tube. Use care to not damage the bearing or seal

7-26 DRIVE TRAIN

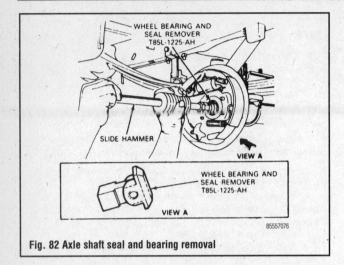

Fig. 82 Axle shaft seal and bearing removal

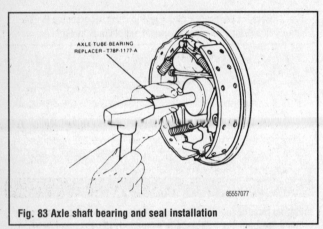

Fig. 83 Axle shaft bearing and seal installation

10. Install a new axle shaft seal using axle tube seal replacer T78P–1177–A or equivalent.

➡ To permit axle shaft installation on 3.73:1 and 4.10:1 ratio axles, make sure the differential pinion shaft contacts the ring gear before performing Step 11.

11. Carefully slide the axle shaft into the axle housing, making sure not to damage the oil seal. Start the splines into the side gear and push firmly until the button end of the axle shaft can be seen in the differential case.
12. Install the C-lockwasher on the end of the axle shaft splines, then pull the shaft outboard until the shaft splines engage the C-lockwasher seats in the counterbore of the differential side gear.
13. Position the differential pinion shaft through the case and pinion gears, aligning the hole in the shaft with the lock screw hole. Install the lock bolt and tighten to 15–22 ft. lbs. (21–29Nm).
14. Clean the gasket mounting surface on the rear axle housing and cover. Apply a continuous bead of Silicone Rubber Sealant ESE–M4G195–A or equivalent to the carrier casting face.
15. Install the cover and tighten the retaining bolts to 25–35 ft. lbs. (20–34Nm).

➡ The cover assembly must be installed within 15 minutes of application of the silicone sealant.

16. Add lubricant until it is ¼ in. (6mm) below the bottom of the filler hole in the running position. Install the filler plug and tighten to 15–30 ft. lbs. (20–41Nm).

Pinion Seal

REMOVAL & INSTALLATION

Pickups and MPV

▶ See Figures 84, 85 and 86

1. Raise and support the front end on jackstands.
2. Matchmark and remove the driveshaft.

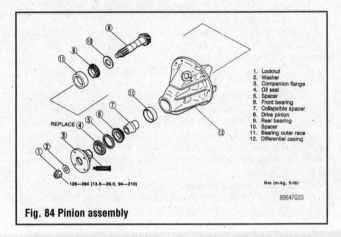

Fig. 84 Pinion assembly

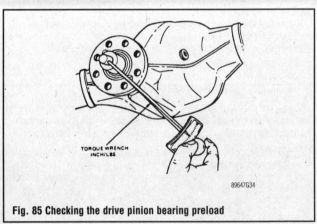

Fig. 85 Checking the drive pinion bearing preload

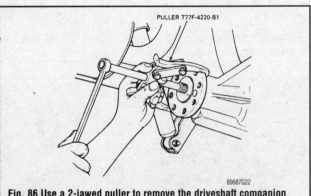

Fig. 86 Use a 2-jawed puller to remove the driveshaft companion flange

3. Remove the wheels and brake calipers.
4. Using an inch lbs. torque wrench on the companion flange nut, measure the rotational torque of the differential and note the reading.
5. Hold the companion flange from turning and remove the locknut.
6. Using a puller, remove the companion flange.
7. Using a center punch to deform the seal and pry it out of the bore.
8. Coat the outer edge of the new seal with sealer and drive it into place with a seal driver.
9. Coat the seal lip with clean gear oil.
10. Coat the companion flange with chassis lube and install it.
11. Install the nut and tighten it until the previously noted rotational torque is achieved. Torque on the nut should not exceed 130 ft. lbs.
12. Install the driveshaft.
13. Replace any lost gear oil.

Navajo

▶ **See Figures 86 and 87**

1. Disconnect the negative battery cable.
2. Raise and support the vehicle safely. Allow the axle to drop to rebound position for working clearance.
3. Remove the rear wheels and brake drums. No drag must be present on the axle.
4. Mark the companion flanges and U-joints for correct reinstallation position.
5. Remove the driveshaft.
6. Using an inch pound torque wrench and socket on the pinion yoke nut measure the amount of torque needed to maintain differential rotation through several clockwise revolutions. Record the measurement.
7. Use a suitable tool to hold the companion flange. Remove the pinion nut.
8. Place a drain pan under the differential. Clean the area around the seal and mark the yoke-to-pinion relation.
9. Use a 2-jawed puller to remove the companion flange.
10. Remove the seal with a small prybar.

To install:

11. Thoroughly clean the oil seal bore.

➡ If you are not absolutely certain of the proper seal installation depth, the proper seal driver must be used. If the seal is misaligned or damaged during installation, it must be removed and a new seal installed.

12. Drive the new seal into place with a seal driver such as T83T–4676–A. Coat the seal lip with clean, waterproof wheel bearing grease.
13. Coat the splines with a small amount of wheel bearing grease and install

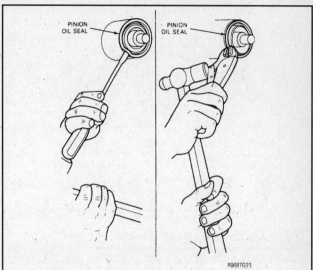

Fig. 87 Remove the seal by either prying it out, or grasping the lip edge with locking pliers and tapping it out of the housing

the yoke, aligning the matchmarks. Never hammer the yoke onto the pinion!

14. Install a new nut on the pinion.
15. Hold the yoke with a holding tool. Tighten the pinion nut, taking frequent turning torque readings until the original preload reading is attained. If the original preload reading, that you noted before disassembly, is lower than the specified reading of 8–14 inch lbs. for used bearings; 16–29 inch lbs. for new bearings, keep tightening the pinion nut until the specified reading is reached. If the original preload reading is higher than the specified values, torque the nut just until the original reading is reached.

✲✲ WARNING

Under no circumstances should the nut be backed off to reduce the preload reading! If the preload is exceeded, the yoke and bearing must be removed and a new collapsible spacer must be installed. The entire process of preload adjustment must be repeated.

16. Install the driveshaft using the matchmarks. Torque the nuts to 15 ft. lbs.
17. Lower the vehicle.
18. Reconnect the negative battery cable.

7-28 DRIVE TRAIN

TORQUE SPECIFICATIONS

Component	U.S.	Metric
Actuator rod locknut Pickup & MPV	8.7-12 ft. lbs.	12-16 Nm
Backup lamp switch bolt Navajo	18-26 ft. lbs.	25-35 Nm
Bearing cover bolts pickups	20 ft. lbs.	27 Nm
MPV	20 ft. lbs.	27 Nm
Bottom cover bolts MPV	95 inch lbs.	11 Nm
Cam plate bolts manual shift Borg-Wagner larger bolt	70-90 ft. lbs.	95-122 Nm
smaller bolt	31-42 ft. lbs.	42-57 Nm
Center bearing mount bolt	27-39 ft. lbs.	37-53 Nm
Center bearing support assembly nuts Navajo & 4WD pickups	116-130 ft. lbs.	158-177 Nm
Center bearing cover bolts Navajo	14-19 ft. lbs.	19-26 Nm
Clutch pressure plate pickups & MPV	13-20 ft. lbs.	18-27 Nm
Navajo	15-25 ft. lbs.	20-34 Nm
Clutch slave cylinder bolts B2200	12-17 ft. lbs.	16-23 Nm
Companion flange nuts 4WD pickups front	94-130 ft. lbs.	128-177 Nm
2WD & 4WD pickups Rear	39-47 ft. lbs.	53-64 Nm
Rear companion flange bolts MPV	39-47 ft. lbs.	53-64 Nm
Navajo rear differential	70-95 ft. lbs.	95-129 Nm
Navajo front differential	200 ft. lbs.	272 Nm
Counter lever fixing bolts	6-7 inch lbs.	0.7-0.8 Nm
Countershaft locknut	94-144 ft. lbs.	128-196 Nm
Crossmember-to-chassis bolts B2200	23-34 ft. lbs.	31-46 Nm
B2600	23-34 ft. lbs.	31-46 Nm
MPV	32-45 ft. lbs.	44-61 Nm
Navajo w/A4LD	20-30 ft. lbs.	27-41 Nm
Crossmember-to-mount bolts B2200	12-17 ft. lbs.	16-23 Nm
Crossmember-to-transmission B2600	23-34 ft. lbs.	31-46 Nm
MPV nuts	23-34 ft. lbs.	31-46 Nm
bolts/nuts	32-45 ft. lbs.	44-61 Nm
Damper-to-transfer case bolts Navajo	25-35 ft. lbs.	31-46 Nm
Detent ball bolts MPV 2 upper bolts	43 ft. lbs.	58 Nm
lower bolt	19 ft. lbs.	26 Nm
Differential carrier-to-housing fasteners pickups	12-17 ft. lbs.	16-23 Nm
MPV	12-17 ft. lbs.	16-23 Nm
Navajo	12-17 ft. lbs.	16-23 Nm
Differential carrier shear bolt/nut Navajo front differential	75-95 ft. lbs.	102-129 Nm
Differential housing flange bolts Navajo	40 ft. lbs.	54 Nm

TORQUE SPECIFICATIONS

Component	U.S.	Metric
Differential assembly mounting bolts Navajo front differential	40-50 ft. lbs.	54-68 Nm
Differential pinion shaft bolts Navajo	15-22 ft. lbs.	20-30 Nm
Differential side bearing caps 2WD B2200, 4WD B2600 front axle	27-38 ft. lbs.	37-52 Nm
2WD & 4WD B2600 rear axle	41-59 ft. lbs.	56-80 Nm
MPV	51-61 ft. lbs.	69-83 Nm
Navajo Rear	70-85 ft. lbs.	95-116 Nm
Front	47-67 ft. lbs.	64-91 Nm
Driveshaft mounting bolt MPV w/R4AX-E	36-43 ft. lbs.	49-58 Nm
Exhaust pipe-to-manifold nuts B2200	30-41 ft. lbs.	41-56 Nm
Exhaust pipe-to-transmission bracket bolt B2200	13-20 ft. lbs.	18-27 Nm
Extension housing bolts pickups w/2WD	34 ft. lbs.	46 Nm
MPV	35 ft. lbs.	47 Nm
Navajo	24-34 ft. lbs.	33-46 Nm
5th/reverse cam lockout plate bolts Navajo	6-7 ft. lbs.	8-10 Nm
5th/reverse shift rail fixing bolt Navajo	16-22 ft. lbs.	22-30 Nm
Front drive shaft-to-front output shaft yoke bolts	8-15 ft. lbs.	11-20 Nm
Front driveshaft-to-transfer case output shaft yoke bolt	12-16 ft. lbs.	16-22 Nm
Gusset plate bolts B2200	27-38 ft. lbs.	37-52 Nm
Height sensor connecting bolt MPV standard models	19 ft. lbs.	26 Nm
w/towing package	104 inch lbs.	12 Nm
Hub-to-axle bolts Navajo	25 ft. lbs.	34 Nm
Inner locknut-to-seat front axle bearings Navajo	35 ft. lbs.	48 Nm
Inner adjusting nut Navajo front axle bearings	35 ft. lbs.	48 Nm
Input shaft bearing cover bolts MPV	19 ft. lbs.	26 Nm
Lock plate bolts pickups	14-19 ft. lbs.	19-26 Nm
Mainshaft locknut pickups	174 ft. lbs.	237 Nm
MPV	174 ft. lbs.	237 Nm
Master cylinder bolts/nuts pickups	12-17 ft. lbs.	16-23 Nm
MPV	12-17 ft. lbs.	17-23 Nm
Navajo	8-12 ft. lbs.	11-17 Nm
Oil guide bolts MPV	95 inch lbs.	11 Nm
Oil passage Navajo	95 inch lbs.	11 Nm
Output shaft locknut MPV	6-7 ft. lbs.	8-10 Nm
Navajo	160-200 ft. lbs.	218-272 Nm

TORQUE SPECIFICATIONS

Component	U.S.	Metric
Pinion bearing preload		
Navajo rear axle		
new bearings	9-14 inch lbs.	12-19 Nm
old bearings	16-29 inch lbs.	22-39 Nm
Pinion locknut		
MPV	94-210 ft. lbs.	128-286 Nm
Navajo	160 ft. lbs.	218 Nm
Rear axle cover bolts		
Navajo	25-35 ft. lbs.	34-48 Nm
Rear driveshaft-to-rear output shaft yoke bolts		
Navajo	20-28 ft. lbs.	27-38 Nm
Rear driveshaft-to-transfer case output shaft yoke bolt		
Navajo	61-87 ft. lbs.	83-118 Nm
Rear side frame bolts		
Navajo	85 ft. lbs.	116 Nm
Reverse idler gear shaft base bolt		
MPV	122 inch lbs.	14 Nm
Reverse idler gear fixing bolt		
Navajo	58-86 ft. lbs.	79-117 Nm
Ring gear-to-differential assembly bolts		
pickups and MPV	51-61 ft. lbs.	69-83 Nm
Navajo rear axle	100-120 ft. lbs.	136-163 Nm
Navajo front differential	70-90 ft. lbs.	95-122 Nm
Shifter boot & insulator plate bolts		
B2600	25-37 ft. lbs.	34-50 Nm
Shift control case bolts		
pickups	22 ft. lbs.	30 Nm
Shift control lever-to-transfer case bolts		
MPV	22 ft. lbs.	30 Nm
Shift lever bolts		
B2200	69-95 inch lbs.	8-11 Nm
B2600	25-37 ft. lbs.	34-50 Nm
Navajo	20-27 ft. lbs.	27-34 Nm
Shift handle locknuts		
Navajo	95 inch lbs.	11 Nm
Skid plate bolts/nuts		
Navajo	15-20 ft. lbs.	20-27 Nm
Slave cylinder bolts		
pickups	12-17 ft. lbs.	16-23 Nm
MPV	12-17 ft. lbs.	16-23 Nm
Navajo	13-19 ft. lbs.	18-26 Nm
Speedometer drive gear bolt		
Navajo	20-25 ft. lbs.	27-34 Nm
pickups	69-95 inch lbs.	8-11 Nm
Starter motor nuts		
Navajo	15-20 ft. lbs.	20-27 Nm
Subframe bolts		
Navajo		
A	76 ft. lbs.	103 Nm
B	59 ft. lbs.	80 Nm
Top cover retaining bolts		
Navajo	12-16 ft. lbs.	16-22 Nm
Torque converter & drive plate bolts		
pickups	25-36 ft. lbs.	34-49 Nm
MPV w/R4AX-EL	27-40 ft. lbs.	37-54 Nm
Navajo w/A4LD	20-34 ft. lbs.	27-46 Nm
Transfer case bolts		
pickups	22 ft. lbs.	30 Nm
MPV	27-40 ft. lbs.	37-54 Nm
Navajo	26-43 ft. lbs.	35-58 Nm

TORQUE SPECIFICATIONS

Component	U.S.	Metric
Transfer case chain cover bolts		
pickups	14-19 ft. lbs.	19-26 Nm
Transfer case drain plug		
Navajo	14-22 ft. lbs.	19-30 Nm
Transfer case fill plug		
Navajo	14-22 ft. lbs.	19-30 Nm
Transmission cooler lines bolts		
MPV w/R4AX-EL	17-26 ft. lbs.	23-35 Nm
Transmission drain plug		
Navajo	29-43 ft. lbs.	39-58 Nm
Transmission-to-engine bolts		
B2200, B2600		
w/MT	51-65 ft. lbs.	69-88 Nm
2WD w/AT, exc. N4A-HL	23-34 ft. lbs.	31-46 Nm
4WD pickup w/R4AX-EL	27-38 ft. lbs.	37-52 Nm
N4A-HL	27-38 ft. lbs.	37-52 Nm
MPV w/3.0L	27-38 ft. lbs.	37-52 Nm
MPV w/2.6L	51-65 ft. lbs.	69-88 Nm
MPV w/N4A-HL	23-34 ft. lbs.	31-46 Nm
Navajo	28-38 ft. lbs.	38-52 Nm
Transmission-to-chassis bolts		
4WD pickup w/R4AX-EL	23-34 ft. lbs.	31-46 Nm
Transmission mount bolts		
2WD pickups w/N4A-HL	7.2-17 ft. lbs.	10-23 Nm
MPV w/R4AX-EL	27-38 ft. lbs.	37-52 Nm
Transmission mount bracket bolt		
2WD pickup w/N4A-HL	23-34 ft. lbs.	31-46 Nm
Transmission rear crossmember mount bolts		
MPV w/R4AX-EL	32-45 ft. lbs.	44-61 Nm
Navajo w/A4LD	60-80 ft. lbs.	82-109 Nm
Trunion bolt		
Navajo	12-23 ft. lbs.	16-31 Nm
U-joint assembly bolts		
pickups & Navajo	10-15 ft. lbs.	14-20 Nm
Double cardan U-joint bolts		
Navajo & 4WD pickups	12-16 ft. lbs.	16-22 Nm
U-joint end of driveshaft-to-rear axle		
Navajo & 4WD pickups	61-87 ft. lbs.	83-118 Nm
Vacuum diaphragm bolt		
Navajo A4LD transmission	80-106 inch lbs.	9-12 Nm

FRONT SUSPENSION 8-2
COIL SPRINGS 8-4
  REMOVAL & INSTALLATION 8-4
TORSION BAR 8-5
  REMOVAL & INSTALLATION 8-5
SHOCK ABSORBERS 8-7
  REMOVAL & INSTALLATION 8-7
  TESTING 8-8
STRUTS 8-8
  REMOVAL & INSTALLATION 8-8
  OVERHAUL 8-8
UPPER BALL JOINTS 8-10
  INSPECTION 8-10
  REMOVAL & INSTALLATION 8-10
LOWER BALL JOINTS 8-11
  INSPECTION 8-11
  REMOVAL & INSTALLATION 8-11
SWAY BAR 8-12
  REMOVAL & INSTALLATION 8-12
COMPRESSION RODS 8-13
  REMOVAL & INSTALLATION 8-13
RADIUS ARM 8-14
  REMOVAL & INSTALLATION 8-14
UPPER CONTROL ARM 8-15
  REMOVAL & INSTALLATION 8-15
  CONTROL ARM BUSHING REPLACEMENT 8-15
LOWER CONTROL ARM 8-15
  REMOVAL & INSTALLATION 8-15
  CONTROL ARM BUSHING REPLACEMENT 8-17
I-BEAM AXLE 8-17
  REMOVAL & INSTALLATION 8-17
KNUCKLE AND SPINDLE 8-18
  REMOVAL & INSTALLATION 8-18
FRONT WHEEL BEARINGS 8-20
  REMOVAL & INSTALLATION 8-20
  ADJUSTMENT 8-22
WHEEL ALIGNMENT 8-23
  CASTER 8-23
  CAMBER 8-23
  TOE 8-23
REAR SUSPENSION 8-24
COIL SPRINGS 8-25
  REMOVAL & INSTALLATION 8-25
LEAF SPRINGS 8-25
  REMOVAL & INSTALLATION 8-25
SHOCK ABSORBERS 8-26
  REMOVAL & INSTALLATION 8-26
CONTROL ARMS/LINKS 8-27
  REMOVAL & INSTALLATION 8-27
STABILIZER BAR 8-27
  REMOVAL & INSTALLATION 8-27
REAR WHEEL BEARINGS 8-28
  REMOVAL AND INSTALLATION 8-28
STEERING 8-29
STEERING WHEEL 8-29
  REMOVAL & INSTALLATION 8-29
COMBINATION SWITCH 8-30
  REMOVAL & INSTALLATION 8-30
IGNITION SWITCH 8-31
  REMOVAL & INSTALLATION 8-31
STEERING LINKAGE 8-31
  REMOVAL & INSTALLATION 8-31
MANUAL STEERING GEAR 8-33
  REMOVAL & INSTALLATION 8-33
POWER STEERING GEAR 8-34
  REMOVAL & INSTALLATION 8-34
POWER STEERING RACK AND PINION 8-35
  REMOVAL & INSTALLATION 8-35
POWER STEERING PUMP 8-36
  REMOVAL & INSTALLATION 8-36
  BLEEDING THE SYSTEM 8-38
COMPONENT LOCATIONS
FRONT SUSPENSION COMPONENT LOCATIONS 8-2
FRONT SUSPENSION COMPONENT LOCATIONS—MPV REAR WHEEL DRIVE SHOWN 8-3
REAR SUSPENSION COMPONENT LOCATIONS - MPV SHOWN 8-24
SPECIFICATIONS CHARTS
TORQUE SPECIFICATIONS 8-39

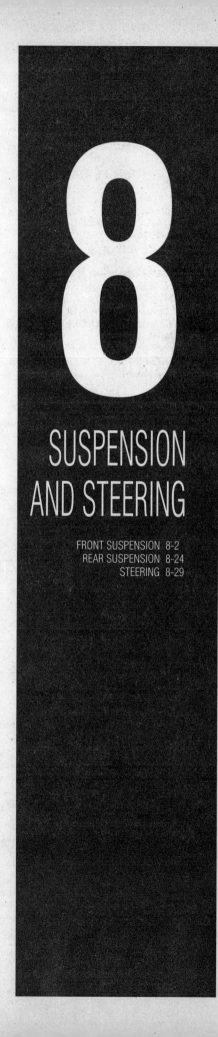

8
SUSPENSION AND STEERING

FRONT SUSPENSION 8-2
REAR SUSPENSION 8-24
STEERING 8-29

8-2 SUSPENSION AND STEERING

FRONT SUSPENSION

FRONT SUSPENSION COMPONENT LOCATIONS - 4WD MODELS

1. Lower ball joint & steering knuckle
2. Shock absorber
3. Radius arm
4. I-beam
5. Outer tie rod end
6. Adjusting sleeve
7. Drag link
8. Stabilizer bar
9. Stabilizer bar end link

SUSPENSION AND STEERING 8-3

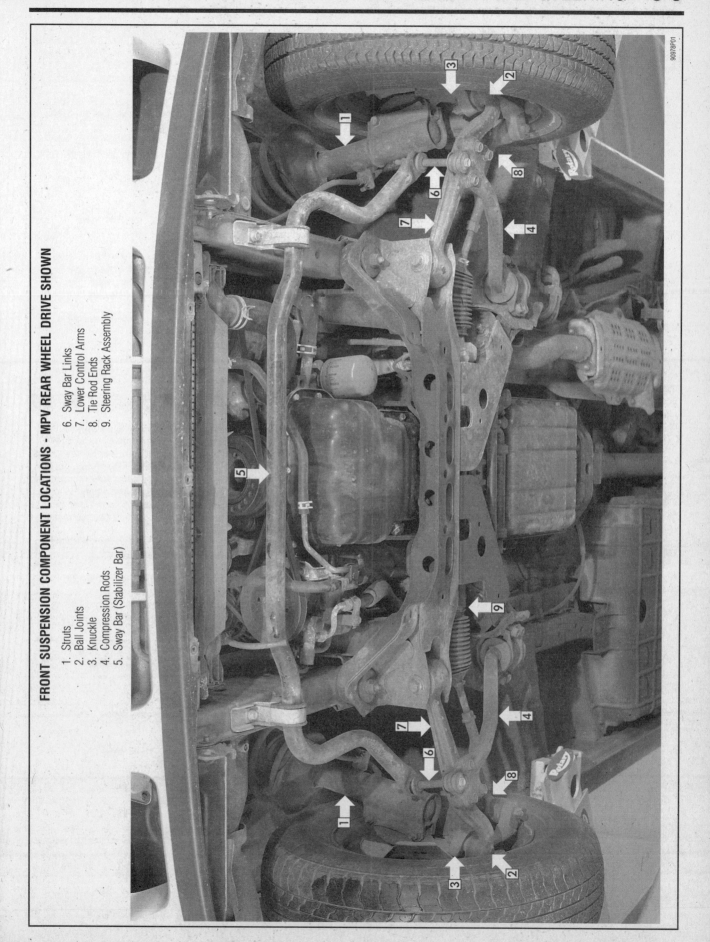

FRONT SUSPENSION COMPONENT LOCATIONS - MPV REAR WHEEL DRIVE SHOWN

1. Struts
2. Ball Joints
3. Knuckle
4. Compression Rods
5. Sway Bar (Stabilizer Bar)
6. Sway Bar Links
7. Lower Control Arms
8. Tie Rod Ends
9. Steering Rack Assembly

8-4 SUSPENSION AND STEERING

All Navajo models use a twin I-beam front suspension which utilizes coil springs to support the vehicle and radius rods to locate the I-beam.

All B Series Pick-up models use unequal length control arms (Short/Long Arm—SLA) which utilize torsion bars to support the vehicle.

All MPV models use a MacPherson-type strut front suspension to support the vehicle and compression rods to locate the lower control arms.

Coil Springs

REMOVAL & INSTALLATION

Navajo

REAR WHEEL DRIVE

♦ See Figure 1

1. Loosen the wheel lugs slightly. Raise the front of the vehicle and place jackstands under the frame and a jack under the axle. Remove the wheels.

✲✲✲ WARNING

The axle must not be permitted to hang by the brake hose. If the length of the brake hoses is not sufficient to provide adequate clearance for removal and installation of the spring, the disc brake caliper must be removed from the spindle. A Strut Spring Compressor tool may be used to compress the spring sufficiently, so that the caliper does not have to be removed. After removal, the caliper must be placed on the frame or otherwise supported to prevent suspending the caliper from the caliper hose. These precautions are absolutely necessary to prevent serious damage to the tube portion of the caliper hose assembly!

2. Remove the nut attaching the shock absorber to radius arm and slide the shock absorber off of the mounting stud. Remove the nut securing the lower retainer to spring seat. Remove the lower retainer.

3. Lower the axle as far as it will go without stretching the brake hose and tube assembly. The axle should now be unsupported without hanging by the brake hose. If not, then either remove the caliper or use a Strut Spring Compressor Tool. Remove the spring.

4. If there is a lot of slack in the brake hose assembly, a pry bar can be used to lift the spring over the bolt that passes through the lower spring seat.

5. Rotate the spring so the built-in retainer on the upper spring seat is cleared.

6. Remove the spring from the vehicle.

To install:

7. If removed, install the bolt in the axle arm and install the nut all the way down. Install the spring lower seat and lower insulator. Install the stabilizer bar mounting bracket and spring spacer.

8. With the axle in the lowest position, install the top of the spring in the upper seat. Rotate the spring into position.

9. Lift the lower end of the spring over the bolt.

10. Raise the axle slowly until the spring is seated in the lower spring upper seat. Install the lower retainer and nut. Tighten the nut to 70-100 ft. lbs. (95-136 Nm).

11. Install the shock absorber. Tighten the lower mounting nut to 39-53 ft. lbs. (53-72 Nm).

12. Install the front wheels. Remove the jack and jackstands and lower vehicle.

4-WHEEL DRIVE

♦ See Figure 2

1. Raise the vehicle and install jackstands under the frame. Position a jack beneath the spring under the axle. Raise the jack and compress the spring.

2. Remove the nut retaining the shock absorber to the radius arm. Slide the shock out from the stud.

3. Remove the nut that retains the spring to the axle and radius arm. Remove the retainer.

4. Slowly lower the axle until all spring tension is released and adequate clearance exists to remove the spring from its mounting.

5. Remove the spring by rotating the upper coil out of the tabs in the upper spring seat. Remove the spacer and the seat.

⊥⊥ WARNING

The axle must be supported on the jack throughout spring removal and installation, and must not be permitted to hang by the brake hose. If the length of the brake hose is not sufficient to provide adequate clearance for removal and installation of the spring, the disc brake caliper must be removed from the spindle. After removal, the caliper must be placed on the frame or otherwise supported to prevent suspending the caliper from the brake line hose. These precautions are absolutely necessary to prevent serious damage to the tube portion of the caliper hose assembly!

6. If required, remove the stud from the axle assembly.

To install:

7. If removed, install the stud on the axle and torque to 190–230 ft. lbs. Install the lower seat and spacer over the stud.

8. Place the spring in position and slowly raise the front axle. Ensure springs are positioned correctly in the upper spring seats.

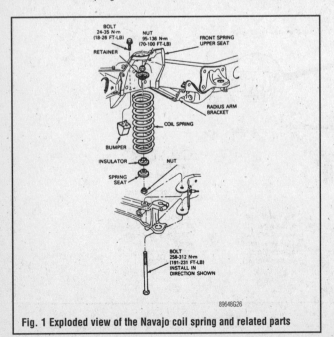

Fig. 1 Exploded view of the Navajo coil spring and related parts

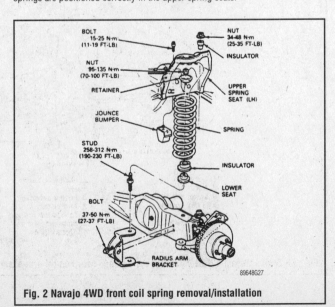

Fig. 2 Navajo 4WD front coil spring removal/installation

SUSPENSION AND STEERING   8-5

9. Position the spring lower retainer over the stud and lower seat and torque the attaching nut to 70–100 ft. lbs.

10. Position the shock absorber to the lower stud and install the attaching nut. Tighten the nut to 41–63 ft. lbs. Lower the vehicle.

Torsion Bar

REMOVAL & INSTALLATION

♦ See Figures 3 thru 11

1. Raise and safely support the vehicle. Remove the wheel and tire assembly.
2. Support the lower control arm with a jack.
3. Remove the cotter pin and nut from the lower ball joint stud. Separate the ball joint from the knuckle using tool 49 0727 575 or equivalent.
4. Remove the bolt, washer and nut attaching the shock absorber to the lower control arm.
5. Mark the position of the anchor bolt and swivel for reference during reassembly and remove the anchor bolt and swivel.
6. Mark the position of the torsion bar on the anchor arm and remove the anchor arm.
7. Mark the position of the torsion bar on the torque plate and remove the torsion bar. If removing the torsion bar from both sides of the vehicle, mark their positions as the torsion bars are not interchangeable.
8. Remove the attaching bolts and remove the torque plate.
9. Check the torsion bar for bending or for looseness between the serrations of the torsion bar and anchor arm and/or torque plate, replace as necessary.

To install:

10. Install the torque plate and tighten the attaching bolts to 55–69 ft. lbs. (75–93 Nm).

11. Coat the serrations of the torsion bar with grease and install in the torque plate, aligning the marks made during the removal procedure. If both torsion bars were removed, make sure the correct torsion bar is being installed.

12. Coat the serrations on the other end of the torsion bar with grease and install the anchor arm onto the torsion bar, aligning the marks made during the removal procedure.

13. Install the anchor bolt and swivel. Tighten the anchor bolt until the marks made during removal are aligned.

14. Connect the shock absorber to the lower control arm and loosely tighten the nut and bolt. Connect the lower ball joint to the knuckle; install the nut and tighten to 87–116 ft. lbs. (118–157 Nm). Install a new cotter pin.

15. Install the wheel and tire assembly and lower the vehicle. With the vehicle unladen, tighten the shock absorber-to-lower control arm bolt and nut to 41–59 ft. lbs. (55–80 Nm).

16. Check vehicle ride height as follows:
    a. Check the front and rear tire pressures and bring to specification.
    b. Measure the distance from the center of each front wheel to the fender brim. The difference must not be greater than 0.39 inch (10mm).
    c. If the difference is not as specified, turn the necessary torsion spring anchor bolt to adjust.
17. Check the front end alignment.

➡ **If, for some reason, you didn't matchmark the torsion bar anchor bolt, or the matchmarks were lost, or you're installing a new, unmarked torsion bar, here's a procedure to help you attain the correct ride height:**

   a. Install the anchor arm on the torsion bar so that there is 4.92 inches (125mm) between the lowest point on the arm and the crossmember directly above it.
   b. Tighten the anchor bolt until the anchor arm contacts the swivel. Then, tighten the bolt an additional 1.77 inches (45mm) travel.

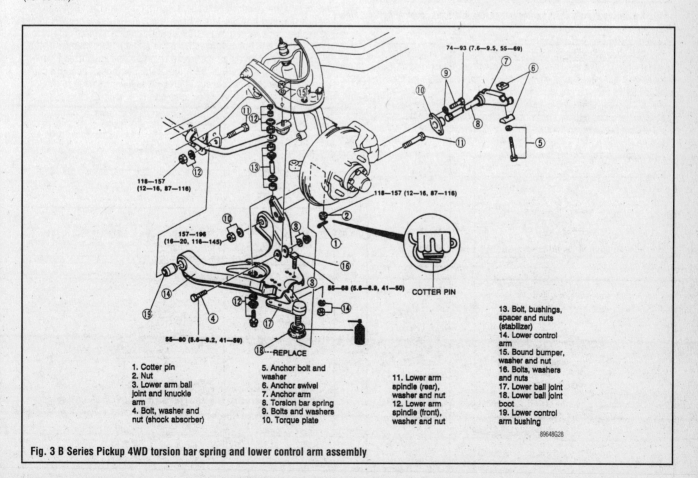

Fig. 3 B Series Pickup 4WD torsion bar spring and lower control arm assembly

8-6 SUSPENSION AND STEERING

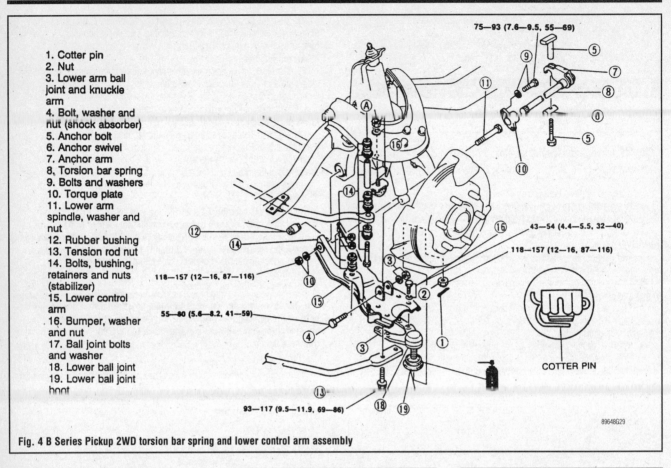

1. Cotter pin
2. Nut
3. Lower arm ball joint and knuckle arm
4. Bolt, washer and nut (shock absorber)
5. Anchor bolt
6. Anchor swivel
7. Anchor arm
8. Torsion bar spring
9. Bolts and washers
10. Torque plate
11. Lower arm spindle, washer and nut
12. Rubber bushing
13. Tension rod nut
14. Bolts, bushing, retainers and nuts (stabilizer)
15. Lower control arm
16. Bumper, washer and nut
17. Ball joint bolts and washer
18. Lower ball joint
19. Lower ball joint boot

Fig. 4 B Series Pickup 2WD torsion bar spring and lower control arm assembly

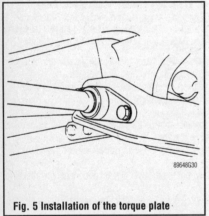

Fig. 5 Installation of the torque plate

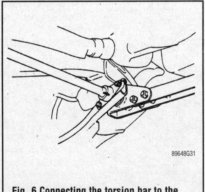

Fig. 6 Connecting the torsion bar to the torque plate

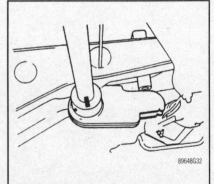

Fig. 7 Connecting the anchor arm to the torsion bar

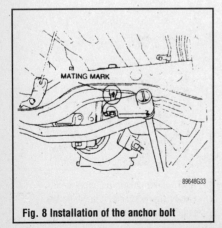

Fig. 8 Installation of the anchor bolt

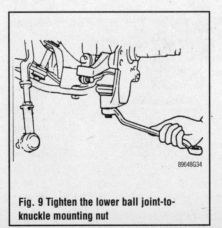

Fig. 9 Tighten the lower ball joint-to-knuckle mounting nut

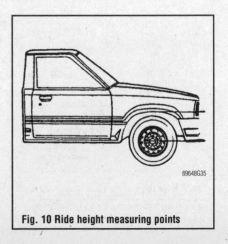

Fig. 10 Ride height measuring points

SUSPENSION AND STEERING 8-7

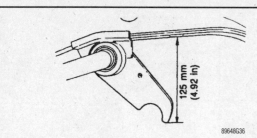

Fig. 11 Anchor arm installation when reference marks are not available

Shock Absorbers

REMOVAL & INSTALLATION

Pickup

▶ See Figure 12

1. Raise and safely support the vehicle. Remove the wheel and tire assembly.
2. Remove the upper shock absorber nuts, retainer and bushing.
3. Remove the lower shock absorber-to-lower control arm mounting bolt, nut and washer.
4. Slightly compress the shock absorber and remove it from the vehicle. Remove the remaining retainers and bushing from the upper shock absorber stud.

To install:

5. Install the shock absorber and install the mounting bolts, nuts, washers and bushings. Do not tighten at this time.
6. Install the wheel and tire assembly and lower the vehicle.
7. With the vehicle unladen, tighten the upper shock absorber mounting nuts until the stud protrudes 0.28 inch (7mm) above the upper nut. Tighten the lower mounting bolt and nut to 41–59 ft. lbs. (55–80 Nm).
8. Check the front end alignment.

Navajo

REAR WHEEL DRIVE

➡Low pressure gas shocks are charged with Nitrogen gas. Do not attempt to open, puncture or apply heat to them. Prior to installing a new shock absorber, hold it upright and extend it fully. Invert it and fully compress and extend it at least 3 times. This will bleed trapped air.

1. Raise the vehicle, as required to provide additional access and remove the bolt and nut attaching the shock absorber to the lower bracket on the radius arm.
2. Remove the nut, washer and insulator from the shock absorber at the frame bracket and remove the shock absorber.
3. Position the washer and insulator on the shock absorber rod and position the shock absorber to the frame bracket.
4. Position the insulator and washer on the shock absorber rod and install the attaching nut loosely.
5. Position the shock absorber to the lower bracket and install the attaching bolt and nut loosely.
6. Tighten the lower mounting nut to 39-53 ft. lbs. (53-72 Nm) and the upper to 25-35 ft. lbs. (34-48 Nm).

4-WHEEL DRIVE

▶ See Figures 13, 14, 15 and 16

➡Low pressure gas shocks are charged with Nitrogen gas. Do not attempt to open, puncture or apply heat to them. Prior to installing a new shock absorber, hold it upright and extend it fully. Invert it and fully compress and extend it at least 3 times. This will bleed trapped air.

1. Raise the vehicle, as required to provide additional access and remove the bolt and nut attaching the shock absorber to the radius arm.
2. Remove the nut, washer and insulator from the shock absorber at the frame bracket and remove the shock absorber.

To install:

3. Position the washer and insulator on the shock absorber rod and position the shock absorber to the frame bracket.
4. Position the insulator and washer on the shock absorber rod and install the attaching nut loosely.
5. Position the shock absorber to the lower bracket and install the attaching bolt and nut loosely.
6. Tighten the lower attaching bolts to 39–53 ft. lbs., and the upper attaching bolts to 25–35 ft. lbs.

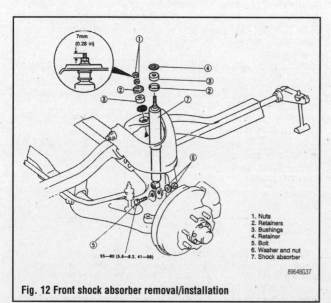

Fig. 12 Front shock absorber removal/installation

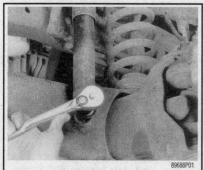

Fig. 13 To remove the front shock absorber, remove the lower radius arm shock retaining nut . . .

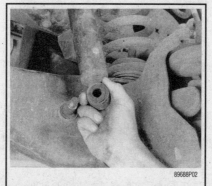

Fig. 14 . . . then pull the lower shock mount from the stud

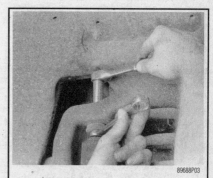

Fig. 15 Next, unbolt the upper shock mount using a second wrench on the mount stud to keep it from spinning . . .

8-8 SUSPENSION AND STEERING

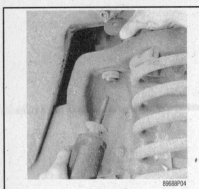
Fig. 16 . . . then pull the shock assembly from the upper spring mount

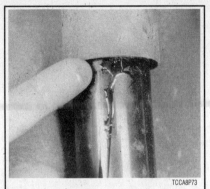
Fig. 17 When fluid is seeping out of the shock absorber, it's time to replace it

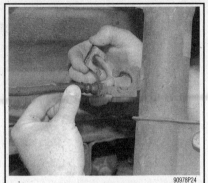
Fig. 18 Remove the retaining clip securing the rubber brake hose to the strut

TESTING

♦ See Figure 17

The purpose of the shock absorber is simply to limit the motion of the spring during compression and rebound cycles. If the vehicle is not equipped with these motion dampers, the up and down motion would multiply until the vehicle was alternately trying to leap off the ground and to pound itself into the pavement.

Contrary to popular rumor, the shocks do not affect the ride height of the vehicle. This is controlled by other suspension components such as springs and tires. Worn shock absorbers can affect handling; if the front of the vehicle is rising or falling excessively, the "footprint" of the tires changes on the pavement and steering is affected.

The simplest test of the shock absorber is simply push down on one corner of the unladen vehicle and release it. Observe the motion of the body as it is released. In most cases, it will come up beyond it original rest position, dip back below it and settle quickly to rest. This shows that the damper is controlling the spring action. Any tendency to excessive pitch (up-and-down) motion or failure to return to rest within 2-3 cycles is a sign of poor function within the shock absorber. Oil-filled shocks may have a light film of oil around the seal, resulting from normal breathing and air exchange. This should NOT be taken as a sign of failure, but any sign of thick or running oil definitely indicates failure. Gas filled shocks may also show some film at the shaft; if the gas has leaked out, the shock will have almost no resistance to motion.

While each shock absorber can be replaced individually, it is recommended that they be changed as a pair (both front or both rear) to maintain equal response on both sides of the vehicle. Chances are quite good that if one has failed, its mate is weak also.

Struts

Only MPV models utilize a MacPherson-type strut front suspension.

REMOVAL & INSTALLATION

♦ See Figures 18 thru 25

1. Raise and safely support the vehicle. Remove the wheel and tire assembly.
2. Support the lower control arm with a jack.
3. Remove the clip attaching the brake hose to the strut and disconnect the hose from the strut.
4. Remove the strut-to-knuckle attaching bolts and nuts.
5. Working in the engine compartment, remove the 4 attaching nuts from the strut tower and remove the strut assembly from the vehicle. If necessary, remove any components that can hinder the removal of the strut mounting nuts.
6. If the strut must be disassembled, or the coil spring removed, refer to the overhaul procedure later in this section.

To install:

7. Assemble the strut according to the overhaul procedure, if necessary.
8. Install the strut assembly in the strut tower, making sure the white mark on the upper mounting block is in the front-inside direction. Install the attaching nuts and tighten to 34–46 ft. lbs. (47–62 Nm).
9. Install the strut to the knuckle and tighten the attaching bolts and nuts to 69–86 ft. lbs. (93–117 Nm).
10. Position the brake hose on the strut and install the clip. Remove the jack from under the lower control arm.
11. Install the wheel and tire assembly and lower the vehicle. Check the front end alignment.

OVERHAUL

♦ See Figures 26 thru 35

1. Secure the strut assembly firmly in a bench vise. Remove the rubber cap from the upper mounting block. Loosen the upper attaching nut, but do not remove it.

Fig. 19 Using a socket and a backup wrench, remove the two strut-to-steering knuckle nuts and bolts

Fig. 20 Separate the lower portion of the strut from the steering knuckle

Fig. 21 After separating the strut from the knuckle, support the knuckle with a strong piece of wire to prevent possible stress damage to the rubber brake hose

SUSPENSION AND STEERING 8-9

Fig. 22 Remove the igniter, ignition coil and mounting bracket assembly from the top of the strut tower to access the upper strut mounting nuts

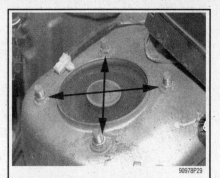
Fig. 23 Remove the four upper strut mounting nuts while supporting the strut assembly

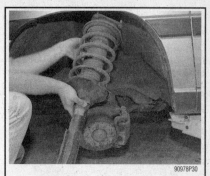
Fig. 24 Carefully remove the strut assembly from the vehicle through the fender well

Fig. 25 Note the marking to indicate how the strut assembly is installed into the vehicle (OUT w/arrow)

Fig. 26 Mount the complete strut assembly firmly in a bench vise, then remove the top plastic cap from the strut bearing plate

Fig. 27 Before attempting to compress the strut coil spring, loosen the top srut retaining nut a couple of turns, but do NOT remove

Fig. 28 Install a coil spring compressor tool as shown, then use a box wrench and socket to compress the coil . . .

Fig. 29 . . . to the compressed size as shown . . .

Fig. 30 . . . then remove the top retaining nut and two washers

Fig. 31 Pull the bearing plate straight off the top of the strut assembly. Note during installation that it may be necessary to pull up the strut rod through the bumper stop by installing the nut and using two screwdrivers to pry upward (arrows)

Fig. 32 Pull the spring seat/bumper stop/dust boot assembly up off of the strut rod

Fig. 33 Remove the coil spring with the compressor still installed

8-10 SUSPENSION AND STEERING

Fig. 34 Remove the rubber ring from the strut

Fig. 35 Strut assembly components

2. Install a suitable spring compressor and compress the coil spring.
3. Remove the upper attaching nut and slowly relieve the tension on the coil spring, using the spring compressor. When the spring is no longer under tension, remove the spring compressor.
4. Remove the upper mounting block, upper spring seat, spring seat, coil spring, bump stopper and ring rubber from the strut.
5. Apply a suitable rubber grease to the ring rubber and install it on the bump stopper. Install the bump stopper on the strut.
6. Attach the spring compressor to the coil spring and compress the spring.
7. Install the compressed spring on the strut and install the spring seat.
8. Install the upper spring seat. The flat of the strut rod must fit correctly into the upper spring seat.
9. Install the upper mounting block. Install and loosely tighten the upper attaching nut.
10. Remove the spring compressor. Make sure the spring is properly seated in the upper and lower spring seats.
11. Secure the upper spring seat in a vise and tighten the upper attaching nut to 47–59 ft. lbs. (64–80 Nm). Install the rubber cap on the upper mounting block.

Upper Ball Joints

INSPECTION

Pickup

1. Raise and safely support the vehicle. Remove the wheel and tire assembly.
2. Support the lower control arm with a jack.
3. Remove the clip attaching the brake hose to the upper control arm and disconnect the hose from the arm.
4. Remove the cotter pin and nut from the upper ball joint stud. Using tool 49 0727 575 or equivalent, separate the upper ball joint from the knuckle.
5. Install tool 49 0180 510B or equivalent to the ball joint stud and attach a suitable pull scale to the stud.
6. After rocking the ball joint stud back and forth 3–4 times, measure the pull scale reading while the ball joint stud is rotating. The pull scale reading should be 4.4–7.7 lbs.
7. If the pull scale reading is not as specified, replace the upper ball joint.

2WD Navajo

1. Raise the vehicle and position a jackstand under the I-beam axle beneath the coil spring. Check and adjust the front wheel bearings.
2. Have a helper grasp the lower edge of the tire and move the wheel assembly in and out.
3. While the wheel is being moved, observe the upper spindle arm and the upper part of the axle jaw.
4. One thirty-second inch (0.8mm) or greater movement between the upper part of the axle jaw and the upper spindle arm indicates that the upper ball joint must be replaced.

REMOVAL & INSTALLATION

Pickup

1. Raise and safely support the vehicle. Remove the wheel and tire assembly.
2. Support the lower control arm with a jack.
3. Remove the clip attaching the brake hose to the upper control arm and disconnect the hose from the arm.
4. Remove the cotter pin and nut from the upper ball joint stud. Using tool 49 0727 575 or equivalent, separate the upper ball joint from the knuckle.
5. Remove the upper ball joint-to-upper control arm attaching bolts and remove the upper ball joint.
6. Position the ball joint assembly on the upper control arm and secure the mounting hardware. Tighten the upper ball joint-to-upper control arm attaching bolts to 18–25 ft. lbs. (25–33 Nm) and the upper ball joint stud nut to 22–38 ft. lbs. (29–51 Nm). Install a new cotter pin.
7. Check the front end alignment.

Navajo With 2WD

♦ See Figures 36 and 37

1. Raise and safely support the vehicle. Remove the wheel and tire assembly.
2. Remove the brake caliper and support it aside with mechanics wire. Do not let the caliper hang by the brake hose.
3. Remove the dust cap, cotter pin, nut retainer, washer and outer bearing and remove the brake rotor from the spindle. Remove the brake dust shield.
4. Disconnect the steering linkage from the spindle and spindle arm by removing the cotter pin and nut. Remove the tie rod end from the spindle arm.
5. Remove the cotter pin and nut from the lower ball joint stud. Remove the axle clamp bolt from the axle.
6. Remove the camber adjuster from the upper ball joint stud and axle beam.
7. Strike the inside area of the axle to pop the lower ball joint loose from the axle beam. Remove the spindle and ball joint assembly from the axle.

➡ Do not use a pickle fork to separate the ball joint from the axle as this will damage the seal and ball joint socket.

8. Install the spindle assembly in a vise and remove the snapring from the lower ball joint. Remove the lower ball joint from the spindle using C-frame T74P–4635–C or equivalent and a suitable receiver cup to press the ball joint from the spindle.

➡ The lower ball joint must be removed first.

9. Repeat the procedure in Step 8 to remove the upper ball joint.

➡ Do not heat the ball joints or the spindle to aid in removal.

SUSPENSION AND STEERING   8-11

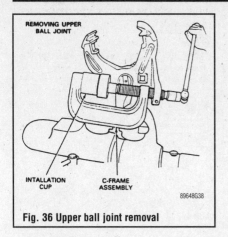

Fig. 36 Upper ball joint removal

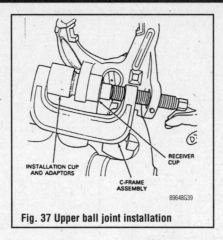

Fig. 37 Upper ball joint installation

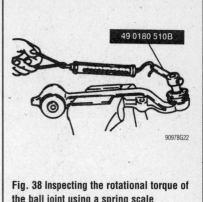

Fig. 38 Inspecting the rotational torque of the ball joint using a spring scale

To install:

10. Assemble the C-frame and receiver cup and press in the upper ball joint.
11. Repeat the procedure in Step 10 to install the lower ball joint.

➡ Do not heat the ball joints or axle to aid in installation.

12. Install the snapring onto the ball joint.
13. Place the spindle and ball joints into the axle. Install the camber adjuster in the upper spindle over the ball joint stud making sure it is properly aligned.
14. Tighten the lower ball joint stud nut to 104–146 ft. lbs. (141–198 Nm). Continue tightening the castellated nut until it lines up with the hole in the stud, then install the cotter pin.
15. Install the clamp bolt into the axle boss and tighten to 48–65 ft. lbs. (65–88 Nm).
16. Install the remaining components.

Lower Ball Joints

INSPECTION

Pickup

1. Raise and safely support the vehicle. Remove the wheel and tire assembly.
2. Support the lower control arm with a jack.
3. Remove the cotter pin and nut from the lower ball joint stud. Separate the ball joint from the knuckle using tool 49 0727 575 or equivalent.
4. Install tool 49 0180 510B or equivalent to the ball joint stud and attach a suitable pull scale to the stud.
5. After rocking the ball joint stud back and forth 3–4 times, measure the pull scale reading while the ball joint stud is rotating. The pull scale reading should be 4.4–7.7 lbs.
6. If the pull scale reading is not as specified, replace the lower ball joint.

MPV

▶ See Figure 38

Defective ball joints are determined by checking the rotational torque with a special preload attachment and spring scale.
1. Remove the lower control arm from the vehicle.
2. Inspect the control arm for damage and the boots for cracks. Replace them if necessary.
3. Inspect the ball joint for looseness and replace, if necessary.
4. Shake the ball joint stud at least 5 times.
5. Install tool 49 0180 510B or equivalent to the ball joint stud and attach a suitable pull scale to the stud.
6. After shaking the ball joint stud back and forth 5 times, measure the rotational torque while the ball joint stud is rotating. The rotational torque should be 18–30 inch lbs. (2–3 Nm). The pull scale reading should be 4.4–7.7 lbs.
7. If the reading is not as specified, replace the lower ball joint (4WD) or control arm assembly (rear wheel drive).

2WD Navajo

1. Raise the vehicle and position a jackstand under the I-beam axle beneath the coil spring. Check and adjust the front wheel bearings.
2. Have a helper grasp the upper edge of the tire and move the wheel assembly in and out.
3. While the wheel is being moved, observe the lower spindle arm and the lower part of the axle jaw.
4. One thirty-second inch (0.8mm) or greater movement between the lower part of the axle jaw and the lower spindle arm indicates that the lower ball joint must be replaced.

REMOVAL & INSTALLATION

Pickup

1. Raise and safely support the vehicle. Remove the wheel and tire assembly.
2. Support the lower control arm with a jack.
3. On 2WD vehicles, remove the bolts attaching the tension rod to the lower control arm.
4. Remove the cotter pin and nut from the lower ball joint stud. Separate the ball joint from the knuckle using tool 49 0727 575 or equivalent.
5. Remove the bolts/nuts attaching the lower ball joint to the lower control arm and remove the lower ball joint.
6. Position the ball joint assembly to the lower control arm and secure it with the mounting hardware. Tighten the lower ball joint-to-lower control arm bolts/nuts to 32–40 ft. lbs. (43–54 Nm) on 2WD vehicles or 41–50 ft. lbs. (55–68 Nm) on 4WD vehicles. Tighten the lower ball joint stud nut to 87–116 ft. lbs. (118–157 Nm) and install a new cotter pin.
7. Check the front end alignment.

MPV

▶ See Figure 39

On 2WD models, the lower ball joints are pressed into the lower control arm. The ball joints cannot be removed from the lower control arms and in the event of the defective ball joint, the lower control arm and ball joint must be replaced as an assembly. Defective ball joints are determined by checking the rotational torque with a special preload attachment and spring scale. Ripped ball joint dust boots can be replaced. Replacement of the ball joint dust boot is accomplished by removing the lower control arm from the vehicle and chiseling the off old boot. Coat the inside of the new dust boot with lithium grease and press it into the ball joint using the proper tool. Check the ball joint stud threads for damage and repair as necessary. Check the ball joint preload and install the lower control arm by reversing the removal procedure.
1. On 4WD models, raise and safely support the vehicle. Remove the wheel and tire assembly.
2. Disconnect the stabilizer bar from the lower control arm.
3. Remove the cotter pin and nut from the lower ball joint stud. Separate the ball joint from the knuckle using tool 49 0727 575 or equivalent.
4. Install tool 49 0180 510B or equivalent to the ball joint stud and attach a suitable pull scale to the stud.

8-12 SUSPENSION AND STEERING

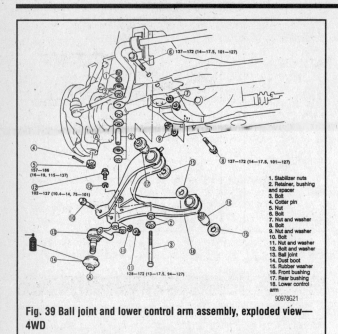

Fig. 39 Ball joint and lower control arm assembly, exploded view—4WD

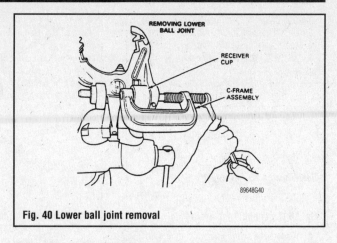

Fig. 40 Lower ball joint removal

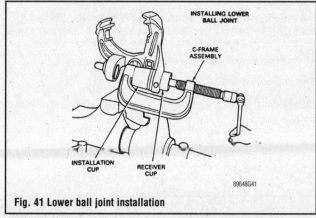

Fig. 41 Lower ball joint installation

5. After rocking the ball joint stud back and forth 3–4 times, measure the pull scale reading while the ball joint stud is rotating. The pull scale reading should be 4.4–7.7 lbs.
6. If the pull scale reading is not as specified, replace the lower ball joint. The ball joint assembly is bolted onto the lower control arm. Remove the two upper and one through bolt and remove the ball joint. Install in the reverse order.

2WD Navajo

♦ See Figures 40 and 41

1. Raise and safely support the vehicle. Remove the wheel and tire assembly.
2. Remove the brake caliper and support it aside with mechanics wire. Do not let the caliper hang by the brake hose.
3. Remove the dust cap, cotter pin, nut retainer, washer and outer bearing and remove the brake rotor from the spindle. Remove the brake dust shield.
4. Disconnect the steering linkage from the spindle and spindle arm by removing the cotter pin and nut. Remove the tie rod end from the spindle arm.
5. Remove the cotter pin and nut from the lower ball joint stud. Remove the axle clamp bolt from the axle.
6. Remove the camber adjuster from the upper ball joint stud and axle beam.
7. Strike the inside area of the axle to pop the lower ball joint loose from the axle beam. Remove the spindle and ball joint assembly from the axle.

➡Do not use a pickle fork to separate the ball joint from the axle as this will damage the seal and ball joint socket.

8. Install the spindle assembly in a vise and remove the snapring from the lower ball joint. Remove the lower ball joint from the spindle using C-frame T74P-4635-C or equivalent and a suitable receiver cup to press the ball joint from the spindle.

➡Do not heat the ball joint or the spindle to aid in removal.

To install:

9. Assemble the C-frame and receiver cup and press in the lower ball joint.

➡Do not heat the ball joint or axle to aid in installation.

10. Install the snapring onto the ball joint.
11. Place the spindle and ball joints into the axle. Install the camber adjuster in the upper spindle over the ball joint stud making sure it is properly aligned.
12. Tighten the lower ball joint stud nut to 104–146 ft. lbs. (141–198 Nm). Continue tightening the castellated nut until it lines up with the hole in the stud, then install the cotter pin.
13. Install the clamp bolt into the axle boss and tighten to 48–65 ft. lbs. (65–88 Nm).
14. Install the remaining components.

Sway Bar

REMOVAL & INSTALLATION

Pickup

♦ See Figure 42

1. Raise and support the front end on jackstands.

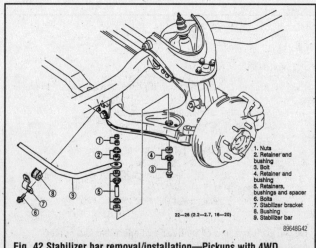

Fig. 42 Stabilizer bar removal/installation—Pickups with 4WD

SUSPENSION AND STEERING   8-13

2. Unbolt the stabilizer bar-to-frame clamps.
3. Unbolt the stabilizer bar from the lower control arms. Keep all the bushings, washers and spacers in order.
4. Check all parts for wear or damage and replace anything which looks suspicious.
5. Install the stabilizer bar to the lower control arms and mount the frame brackets. Tighten all fasteners lightly, then torque them to specifications with the wheels on the ground.
- Stabilizer bar-to-control arm nut: 34 ft. lbs.
- Stabilizer-to-frame clamp bolts: 19 ft. lbs.

MPV

♦ See Figures 43, 44 and 45

1. Raise and support the front end on jackstands.
2. Remove the splash shield.
3. Disconnect the end links at the compression rods.
4. Remove the clamp bolts. Lift out the stabilizer bar.
5. Inspect all parts for wear and/or damage. Replace as necessary.
6. Install the Stabilizer Bar and components in reverse order. The end link bushings have alignment marks on the bar. Torque the clamp bolts to 45 ft. lbs. Tighten the end link bolts until 13mm ± 1mm of thread is visible above the nut.

Navajo

REAR WHEEL DRIVE

♦ See Figure 46

1. As required, raise and support the vehicle safely.
2. Remove the nuts and washer and disconnect the stabilizer link assembly from the front I-beam axle.

3. Remove the mounting bolts and remove the stabilizer bar retainers from the stabilizer bar assembly.
4. Remove the stabilizer bar from the vehicle.

To install:

5. Place stabilizer bar in position on the frame mounting brackets.
6. Install retainers and tighten retainer bolt to 30–50 ft. lbs. If removed, install the stabilizer bar link assembly to the stabilizer bar. Install the nut and washer and tighten to 30–40 ft. lbs.
7. Position the stabilizer bar link in the I-beam mounting bracket. Install the bolt and tighten to 30–44 ft. lbs.

4-WHEEL DRIVE

1. As required, raise and support the vehicle safely. Remove the bolts and the retainers from the center and right hand end of the stabilizer bar.
2. Remove the nut, bolt and washer retaining the stabilizer bar to the stabilizer link.
3. Remove the stabilizer bar and bushings from the vehicle.
4. Installation is the reverse of the removal procedure. Tighten the retainer bolts to 35–50 ft. lbs. Tighten the stabilizer bar to link nut to 30–44 ft. lbs.

Compression Rods

All rear wheel drive MPV models are equipped with compression rods on each side of the front suspension.

REMOVAL & INSTALLATION

♦ See Figures 47 and 48

1. Raise and support the front end on jackstands.
2. Disconnect the stabilizer link at the compression rod.
3. Remove the compression rod-to-lower arm bolts.

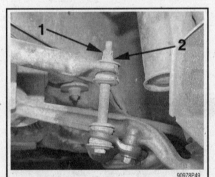

Fig. 43 Use a box wrench and a backup wrench, remove the locking nut (1) and then the retaining nut (2)

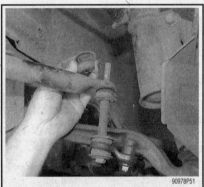

Fig. 44 Remove the top bushing and retainer from the top of the sway bar link

Fig. 45 Remove the two sway bar bushing retainer mounting bolts

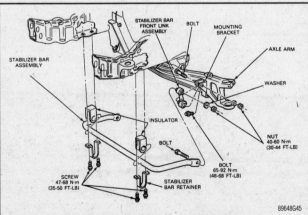

Fig. 46 Exploded view of the Navajo stabilizer bar and related parts—rear wheel drive shown

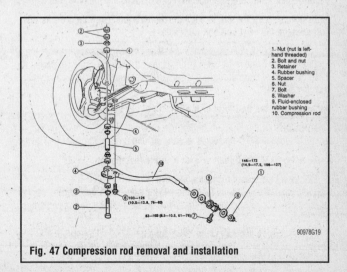

Fig. 47 Compression rod removal and installation

8-14 SUSPENSION AND STEERING

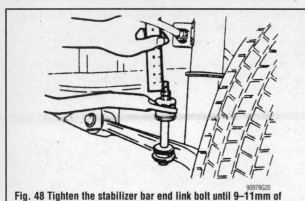

Fig. 48 Tighten the stabilizer bar end link bolt until 9–11mm of thread is visible above the nut

4. Remove the compression rod end nut and washer.

➡ The end nut on the left compression rod may be equipped with left handed threads.

5. Unbolt the compression rod fluid-filled bushing and remove the bushing and washers.
6. Check the bushing for signs of leakage. Replace it if leaking or damaged.
7. Install the compression rod and components in reverse order. Observe the following torques:
   - Fluid-filled bushing bolts: 61–76 ft. lbs. (83–103 Nm)
   - Compression rod end nut: 108–127 ft. lbs. (146–172 Nm)
   - Compression rod-to-lower arm: 76–93 ft. lbs. (103–126 Nm)
8. Tighten the stabilizer bar end link bolt until 9–11mm of thread is visible above the nut.

Radius Arm

REMOVAL & INSTALLATION

Navajo

REAR WHEEL DRIVE

♦ See Figure 49

1. Raise the front of the vehicle, place jackstands under the frame. Place a jack under the axle.

❋❋ WARNING

The axle must be supported on the jack throughout spring removal and installation, and must not be permitted to hang by the brake hose. If the length of the brake hose is not sufficient to provide adequate clearance for removal and installation of the spring, the disc brake caliper must be removed from the spindle. After removal, the caliper must be placed on the frame or otherwise supported to prevent suspending the caliper from the caliper hose. These precautions are absolutely necessary to prevent serious damage to the tube portion of the caliper hose assembly.

2. Disconnect the lower end of the shock absorber from the shock lower bracket (bolt and nut).
3. Remove the front spring. Loosen the axle pivot bolt.
4. Remove the spring lower seat from the radius arm, and then remove the bolt and nut that attaches the radius arm to the axle and front bracket.
5. Remove the nut, rear washer and insulator from the rear side of the radius arm rear bracket.
6. Remove the radius arm from the vehicle, and remove the inner insulator and retainer from the radius arm stud.

To install:

7. Position the front end of the radius arm to the axle. Install the attaching bolt from underneath, and install the nut finger tight.
8. Install the retainer and inner insulator on the radius arm stud and insert the stud through the radius arm rear bracket.
9. Install the rear washer, insulator and nut on the arm stud at the rear side of the arm rear bracket. Tighten the nut to 81–120 ft. lbs.
10. Tighten the nut on the radius arm-to-axle bolt to 160–220 ft. lbs.
11. Install the spring lower seat and spring insulator on the radius arm so that the hole in the seat goes over the arm-to-axle bolt.
12. Install the front spring.
13. On 1990–91 vehicles connect the lower end of the shock absorber to the stud on the radius arm with the retaining nut. Torque the nut to 40–63 ft. lbs.

4-WHEEL DRIVE

♦ See Figure 50

1. Raise the front of the vehicle, place jackstands under the frame. Place a jack under the axle.

❋❋ WARNING

The axle must be supported on the jack throughout spring removal and installation, and must not be permitted to hang by the brake hose. If the length of the brake hose is not sufficient to provide adequate clearance for removal and installation of the spring, the disc brake caliper must be removed from the spindle. After removal, the caliper must be placed on the frame or otherwise supported to prevent suspending the caliper from the caliper hose. These precau-

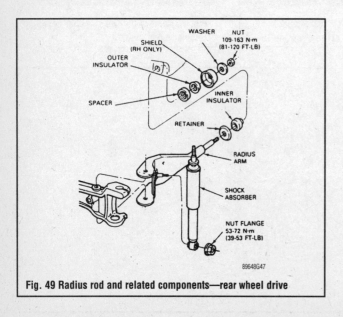

Fig. 49 Radius rod and related components—rear wheel drive

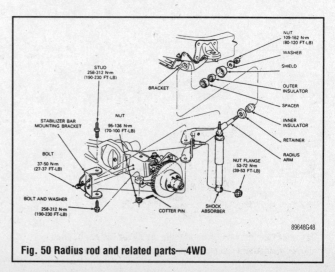

Fig. 50 Radius rod and related parts—4WD

SUSPENSION AND STEERING 8-15

tions are absolutely necessary to prevent serious damage to the tube portion of the caliper hose assembly.

2. Disconnect the lower end of the shock absorber from the lower stud. Remove the front spring from the vehicle.

3. Remove the spring lower seat and stud from the radius arm. Remove the bolts that attach the radius arm to the axle and front bracket.

4. Remove the nut, rear washer and insulator from the rear side of the radius arm rear bracket.

5. Remove the radius arm from the vehicle. Remove the inner insulator and retainer from the radius arm stud.

To install:

6. Position the front end of the radius arm from bracket to axle. Install the retaining bolts and stud in the bracket finger tight.

7. Install the retainer and inner insulator on the radius arm stud and insert the stud through the radius arm rear bracket.

8. Install the rear washer, insulator and nut on the arm stud at the rear side of the arm rear bracket. Tighten the nut to 80–120 ft. lbs.

9. Tighten the stud to 190–230 ft. lbs. Tighten the front bracket to axle bolts to 37–50 ft. lbs. and the lower bolt and washer to 190–230 ft. lbs.

10. Install the spring lower seat and spring insulator on the radius arm so that the hole in the seat goes over the arm to axle bolt. Tighten the axle pivot bolt to 120–150 ft. lbs.

11. Install the front spring. Connect the lower end of the shock absorber to the stud of the radius arm and torque the retaining nut to 39–53 ft. lbs.

Upper Control Arm

REMOVAL & INSTALLATION

Pickup

♦ See Figure 51

1. Loosen the wheel lugs nuts slightly. Raise and support the front end on jackstands placed under the frame.

2. Remove the wheels. Support the lower arm with a floor jack.

3. Remove the cotter pin and nut from the upper ball joint and separate the ball joint from the upper arm using a ball joint separator tool.

4. Remove the nuts and bolts that retain the upper arm shaft to the support bracket. Note the number and location of the shims under the nuts. These must be installed in their exact locations for proper wheel alignment. Check all parts for wear or damage. Replace any suspect parts.

5. Place the control arm assembly in position. Install the mounting bolts and nuts with the alignment shims in correct positions. Torque the upper arm shaft mounting bolts to: 2WD 60–68 ft. lbs. 4WD 69–85 ft. lbs. Tighten the ball joint nut to 30–37 ft. lbs. and install a new cotter pin.

CONTROL ARM BUSHING REPLACEMENT

♦ See Figure 51

The control arm bushings are threaded separately into the upper control arm. If the bushings require service, the front suspension upper control arm must be removed and the bushings unthreaded.

Lower Control Arm

Only the MPV and B Series Pick-up models have control arms.

REMOVAL & INSTALLATION

Pickup

1. Raise and safely support the vehicle. Remove the wheel and tire assembly.

2. Support the lower control arm with a jack.

3. Remove the cotter pin and nut from the lower ball joint stud. Separate the ball joint from the knuckle using tool 49 0727 575 or equivalent.

4. Remove the bolt, washer and nut attaching the shock absorber to the lower control arm.

5. Remove the torsion bar, anchor arm and torque plate assembly.

6. Remove the bolt(s) and nut(s) attaching the lower control arm to the frame.

7. On 2WD vehicles, remove the bolts attaching the tension rod to the lower control arm.

8. Remove the bolts, bushings, retainers, spacer and nuts connecting the stabilizer bar to the lower control arm.

9. Remove the lower control arm from the vehicle. Remove the lower ball joint, if necessary.

To install:

10. Position the lower control arm to the frame and install the attaching bolt(s) and nut(s), but do not tighten at this time.

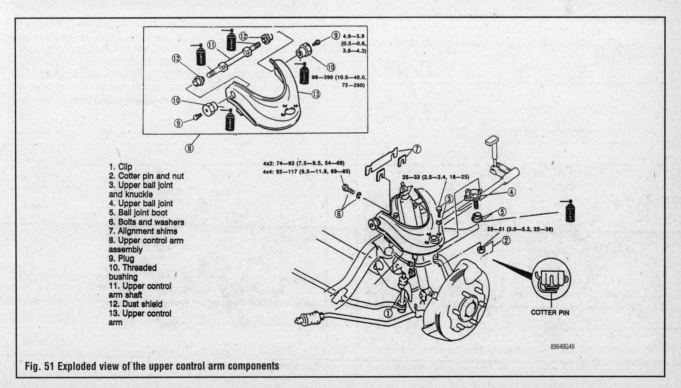

1. Clip
2. Cotter pin and nut
3. Upper ball joint and knuckle
4. Upper ball joint
5. Ball joint boot
6. Bolts and washers
7. Alignment shims
8. Upper control arm assembly
9. Plug
10. Threaded bushing
11. Upper control arm shaft
12. Dust shield
13. Upper control arm

Fig. 51 Exploded view of the upper control arm components

8-16 SUSPENSION AND STEERING

11. Install the torsion bar, anchor arm and torque plate assembly.
12. On 2WD vehicles, install the tension rod bolt and tighten to 69–86 ft. lbs. (93–117 Nm).
13. Attach the stabilizer bar to the control arm with the bolts, bushings, retainers, spacer and nuts. Tighten the nuts so 0.73 in. (18.5mm) of thread is exposed at the end of the bolt.
14. Install the shock absorber to the lower control arm and loosely tighten the mounting bolt and nut.
15. Install the wheel and tire assembly and lower the vehicle.
16. With the vehicle unladen, tighten the lower control arm-to-frame bolt and nut on 2WD vehicles and the front side lower control arm-to-frame bolt and nut on 4WD vehicles to 87–116 ft. lbs. (118–157 Nm). Tighten the rear side lower control arm-to-frame bolt and nut on 4WD vehicles to 116–145 ft. lbs. (157–196 Nm).
17. With the vehicle unladen, tighten the shock absorber-to-lower control arm bolt and nut to 41–59 ft. lbs. (55–80 Nm).
18. Check vehicle ride height as follows:
    a. Check the front and rear tire pressures and bring to specification.
    b. Measure the distance from the center of each front wheel to the fender brim. The difference must not be greater than 0.39 in. (10mm).
    c. If the difference is not as specified, turn the necessary torsion spring anchor bolt to adjust.
19. Check the front end alignment.

MPV

REAR WHEEL DRIVE

♦ See Figures 52, 53, 54 and 55

1. Raise and safely support the vehicle. Remove the wheel and tire assembly.
2. Remove the brake caliper and support it aside with mechanics wire, do not let it hang by the brake hose.
3. Remove the nuts, bolts, spacer, washers and bushings and remove the compression rod from the lower control arm and chassis and disconnect the stabilizer bar from the lower control arm.
4. Remove the cotter pin and nut and, using tool 49 0118 850C or equivalent, separate the tie rod end from the knuckle.
5. Remove the bolts and nuts and disconnect the strut from the knuckle.
6. Remove the cotter pin and nut from the lower ball joint stud. Using tool 49 0727 575 or equivalent, separate the lower ball joint from the knuckle.
7. Remove the mounting bolt and nut and remove the lower control arm from the vehicle.

To install:

8. Position the lower control arm to the chassis and install the bolt and nut, but do not tighten at this time.
9. Install the knuckle to the lower control arm. Tighten the lower ball joint stud nut to 87–116 ft. lbs. (118–157 Nm) and install a new cotter pin.

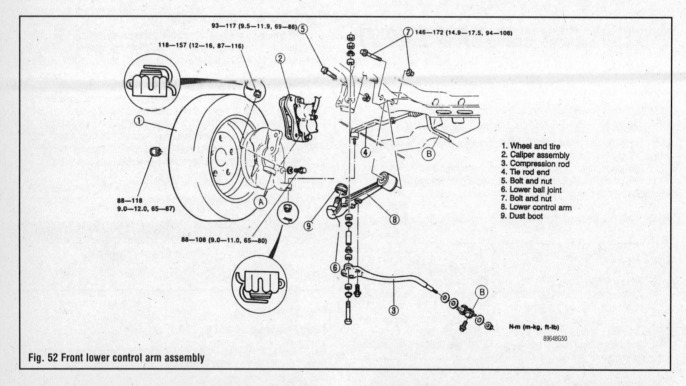

Fig. 52 Front lower control arm assembly

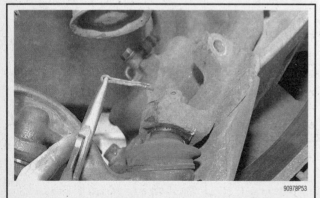

Fig. 53 Remove and discard the lower ball joint stud cotter pin

Fig. 54 Using a ball joint separator tool, separate the steering knuckle from the lower ball joint

SUSPENSION AND STEERING 8-17

Fig. 55 After separating the ball joint from the steering knuckle and sway bar link from the lower control arm, remove the 3 compression rod-to-control arm bolts (B), control arm bushing nut and through bolt (C) and 4 tailing arm-to-frame rail bolts (A)

To install:

5. Position the lower control arm to the chassis and install the bolts and nuts. Do not tighten at this time.
6. Connect the lower ball joint to the knuckle and tighten the ball joint stud nut to 115–137 ft. lbs. (157–186 Nm). Install a new cotter pin.
7. Install the bolt, retainers, bushings, spacer and nuts and connect the stabilizer bar to the lower control arm. Tighten the nuts so 0.24 in. (6mm) of thread is exposed at the end of the bolt.
8. Install the wheel and tire assembly and lower the vehicle. With the vehicle unladen, tighten the lower control arm-to-chassis nuts and bolts to 101–127 ft. lbs. (137–172 Nm).
9. Check the front end alignment.

CONTROL ARM BUSHING REPLACEMENT

Rear Wheel Drive MPV and All B Series Pick-up Models

The control arm bushings are not serviced separately. If the bushings require service, the front suspension upper control arm will have to be replaced.

MPV Models With 4WD

♦ See Figures 39, 56, 57 and 58

1. Remove the lower control arm.
2. Secure the lower control arm in a bench vise protected with brass mounting pads.
3. Cut the rim section of the front lower arm bushing.
4. Remove the front bushing from the control arm by using special bushing press tools 49-G033-102, 49-F027-009 and a press.
5. Using special press tools 49-U034-202, 49-G026-103 and a press, remove the rear lower control arm bushing.
6. Apply soapy water to the outside surface of the new front lower control arm bushing.
7. Using special tools 49-G033-102, 49-F027-009 and a press, install the new front bushing into the lower control arm.
8. Set a new rear control arm bushing into the lower control arm with the direction marks on the bushing aligned with marks on the control arm.
9. Using special tools 49-U034-202, 49-G026-103 and a press, install the new rear bushing into the lower control arm.

I-Beam Axle

Al Navajo models use an I-beam axle front suspension.

REMOVAL & INSTALLATION

Rear Wheel Drive

♦ See Figures 59, 60 and 61

1. Raise and safely support the vehicle. Remove the front wheel spindle. Remove the front spring. Remove the front stabilizer bar, if equipped.

10. Connect the strut to the knuckle and tighten the attaching bolts and nuts to 69–86 ft. lbs. (93–117 Nm).
11. Connect the tie rod end to the knuckle. Tighten the tie rod end stud nut to 43–58 ft. lbs. (59–78 Nm) and install a new cotter pin.
12. Install the compression rod to the lower control arm and chassis. Tighten the compression rod-to-lower control arm mounting bolts to 76–93 ft. lbs. (103–126 Nm) and the compression rod bushing-to-chassis bolts to 61–76 ft. lbs. (83–103 Nm). Install the compression rod nut but do not tighten at this time.

➡ The left-hand compression rod nut has left-hand thread.

13. Connect the stabilizer bar to the control arm with the bolt, washers, bushings, spacer and nuts. Tighten the nuts so 0.24 in. (6mm) of thread is exposed at the end of the bolt.
14. Install the caliper and the wheel and tire assembly. Lower the vehicle.
15. With the vehicle unladen, tighten the lower control arm-to-chassis bolt and nut to 94–108 ft. lbs. (146–172 Nm). Tighten the compression rod nut to 108–127 ft. lbs. (146–172 Nm).
16. Check the front end alignment.

4-WHEEL DRIVE

♦ See Figure 39

1. Raise and safely support the vehicle. Remove the wheel and tire assembly.
2. Remove the bolt, retainers, bushings, spacer and nuts and disconnect the stabilizer bar from the lower control arm.
3. Remove the cotter pin and nut from the lower ball joint stud. Separate the ball joint from the knuckle using tool 49 0727 575 or equivalent.
4. Remove the lower control arm-to-chassis nuts and bolts and remove the lower control arm.

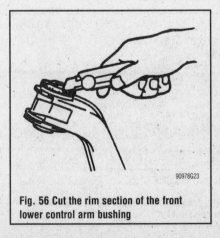

Fig. 56 Cut the rim section of the front lower control arm bushing

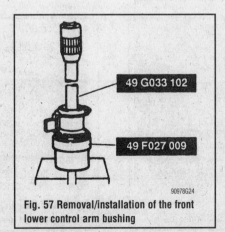

Fig. 57 Removal/installation of the front lower control arm bushing

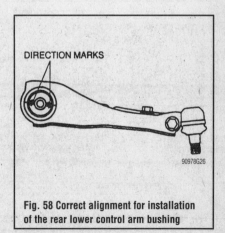

Fig. 58 Correct alignment for installation of the rear lower control arm bushing

8-18 SUSPENSION AND STEERING

2. Remove the spring lower seat from the radius arm, and then remove the bolt and nut that attaches the stabilizer bar bracket, and the radius arm to the (I-Beam) front axle.
3. Remove the axle-to-frame pivot bracket bolt and nut.

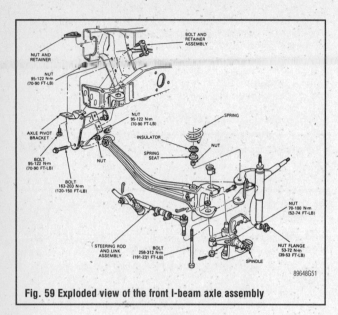

Fig. 59 Exploded view of the front I-beam axle assembly

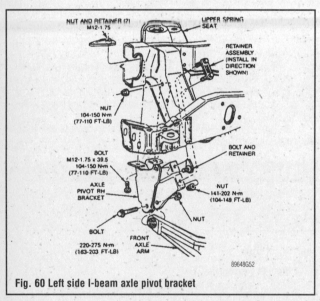

Fig. 60 Left side I-beam axle pivot bracket

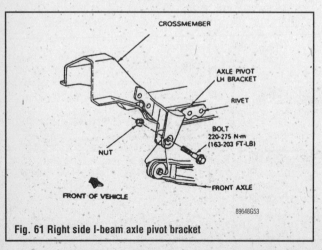

Fig. 61 Right side I-beam axle pivot bracket

To install:

4. Position the axle to the frame pivot bracket and install the bolt and nut finger tight.
5. Position the opposite end of the axle to the radius arm, install the attaching bolt from underneath through the bracket, the radius arm,, and the axle. Install the nut and tighten to 191–220 ft. lbs.
6. Install the spring lower seat on the radius arm so that the hole in the seat indexes over the arm-to-axle bolt.
7. Install the front spring.

➡ Lower the vehicle on its wheels or properly support the vehicle at the front springs before tightening the axle pivot bolt and nut.

8. Tighten the axle-to-frame pivot bracket bolt to 120–150 ft. lbs.
9. Install the front wheel spindle.

4-Wheel Drive

The I-beam axle is part of the front drive axle assembly. Refer to Section 7 for front drive axle housing removal and installation procedures.

Knuckle and Spindle

REMOVAL & INSTALLATION

Pickup 2WD Models

➡ See Section 7 for servicing 4WD models.

1. Raise and support the front end on jackstands.
2. Remove the wheels.
3. Remove brake calipers. Suspend the calipers out of the way with a wire. Don't disconnect the brake line.
4. Remove the brake rotor, hub and bearings.
5. Remove the tie rod-to-knuckle nut, and, using a ball joint separator, remove the tie rod end from the knuckle.
6. Support the lower arm with a floor jack.
7. Remove the cotter pin and nut from the lower ball joint, and, using a ball joint separator, disconnect the lower ball joint from the knuckle.
8. Remove the cotter pin and nut from the upper ball joint, and, using a ball joint separator, disconnect the upper ball joint from the knuckle.
9. Pull the knuckle and spindle assembly from the control arms.
10. The knuckle arm may now be removed.
11. Clean and inspect all parts for wear or damage. Replace parts as necessary.
12. Secure the knuckle in a vise and install the knuckle arm, if removed. Torque the bolts to 70–74 ft. lbs.
13. Install the knuckle assembly on the ball joints and secure the mounting nuts. Install the tie rod to knuckle arm, and install the remaining components. Observe the following torques.

- Upper ball joint-to-knuckle: 35–38 ft. lbs.
- Lower ball joint-to-knuckle: 116 ft. lbs.
- Tie rod end-to-knuckle: 22–29 ft. lbs.

MPV Models

REAR WHEEL DRIVE

1. Raise and safely support the vehicle.
2. Remove the wheel assembly.
3. Remove the wheel hub/bearing assembly, if necessary.
4. Remove the brake caliper.
5. Remove the disc plate.
6. Disconnect the tie-rod end from the knuckle/spindle.
7. Disconnect the lower arm.
8. Remove the knuckle/spindle assembly.

To install:

9. Install the knuckle/spindle assembly. Torque the strut mounting nut to 69–86 ft. lbs. (94–116 Nm).
10. Install the lower arm, ball joint to the knuckle/spindle assembly. Torque the ball joint nut to 87–115 ft. lbs. (118–156 Nm).
11. Connect the tie-rod end, torque the nut to 44–57 ft. lbs. (59–78 Nm).
12. Install wheel hub/bearing assembly, if removed.

SUSPENSION AND STEERING 8-19

13. Install the disc plate.
14. Install the brake caliper, torque the mounting bolts to 66–79 ft. lbs. (89–107 Nm).
15. Install the locknut, torque to 131–173 ft. lbs. (117–235 Nm).
16. Install the hub dust cap.
17. Install the wheel assembly.
18. Lower the vehicle.

4-WHEEL DRIVE

1. Raise and safely support the vehicle.
2. Remove the wheel assembly.
3. Remove the locknut.
4. Remove the brake caliper.
5. Remove the disc plate retaining screw(s).
6. Disconnect the tie-rod end from the knuckle.
7. Disconnect the lower ball joint.
8. Remove the disc plate.
9. Remove the ball joint mounting nuts and bolts.
10. Remove the knuckle, wheel hub and dustplate as an assembly.
11. Remove the wheel bearings from the hub assembly, if needed.

To install:

12. Install wheel bearings to the hub assembly, if removed.
13. Install the knuckle assembly. Torque the strut mounting nut to 69–86 ft. lbs. (94–116 Nm).
14. Install the ball joint mounting nuts and bolts. Torque the upper mounting bolts to 76–101 ft. lbs. (102–137 Nm). Torque the through-bolt nut to 95–106 ft. lbs. (128–171 Nm).
15. Replace the disc plate.
16. Install ball joint to the knuckle assembly. Torque the ball joint nut to 116–137 ft. lbs. (157–186 Nm).
17. Connect the tie-rod end, torque the nut to 44–57 ft. lbs. (59–78 Nm).
18. Install the brake caliper, torque the mounting bolts to 66–79 ft. lbs. (89–107 Nm).
19. Install the disc plate retaining nut.
20. Install the locknut, torque to 174–231 ft. lbs. (236–313 Nm).
21. Install the wheel assembly.
22. Lower the vehicle.

Navajo

REAR WHEEL DRIVE

♦ See Figure 62

1. Raise the front of the vehicle and install jackstands.
2. Remove the wheel and tire assembly.
3. Remove the caliper assembly from the rotor and hold it out of the way with wire.

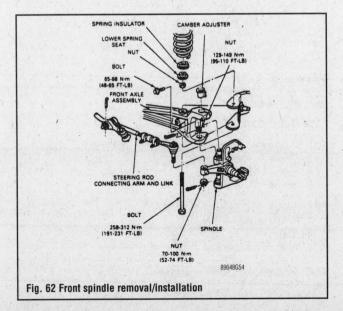

Fig. 62 Front spindle removal/installation

4. Remove the dust cap, cotter pin, nut, nut retainer, washer, and outer bearing, and remove the rotor from the spindle.
5. Remove inner bearing cone and seal. Discard the seal.
6. Remove brake dust shield.
7. Disconnect the steering linkage from the spindle and spindle arm by removing the cotter pin and nut.
8. With Tie Rod removal tool 3290–D or equivalent remove the tie rod end from the spindle arm.
9. Remove the cotter pin and the castellated nut from the lower ball joint stud.
10. Remove the axle clamp bolt from the axle. Remove the camber adjuster from the upper ball joint stud and axle beam.
11. Strike the area inside the top of the axle to pop the lower ball joint loose from the axle beam.

✱✱ WARNING

Do not use a ball joint fork to separate the ball joint from the spindle, as this will damage the seal and the ball joint socket!

12. Remove the spindle and the ball joint assembly from the axle.

To install:

➡ A 3 step sequence for tightening ball joint stud nuts must be followed to avoid excessive turning effort of spindle about axle.

13. Prior to assembly of the spindle, make sure the upper and lower ball joints seals are in place.
14. Place the spindle and the ball joint assembly into the axle.
15. Install the camber adjuster in the upper over the upper ball joint. If camber adjustment is necessary, special adapters must be installed.
16. Tighten the lower ball joint stud to 95–110 ft. lbs. Continue tightening the castellated nut until it lines up with the hole in the ball joint stud. Install the cotter pin. Install the dust shield.
17. Pack the inner and outer bearing cones with high temperature wheel bearing grease. Use a bearing packer. If a bearing packer is unavailable, pack the bearing cone by hand working the grease through the cage behind the rollers.
18. Install the inner bearing cone and seal. Install the hub and rotor on the spindle.
19. Install the outer bearing cone, washer, and nut. Adjust bearing endplay and install the cotter pin and dust cap.
20. Install the caliper.
21. Connect the steering linkage to the spindle. Tighten the nut to 52–74 ft. lbs. and advance the nut as required for installation of the cotter pin.
22. Install the wheel and tire assembly. Lower the vehicle. Check, and if necessary, adjust the toe setting.

4-WHEEL DRIVE

1. Raise the vehicle and support on jackstands.
2. Remove the wheel and tire assembly.
3. Remove the caliper.
4. Remove hub locks, wheel bearings, and locknuts.
5. Remove the hub and rotor. Remove the outer wheel bearing cone.
6. Remove the grease seal from the rotor with seal remover tool 1175–AC and slide hammer 750T-100–A or equivalent. Discard seal and replace with a new one upon assembly.
7. Remove the inner wheel bearing.
8. Remove the inner and outer bearing cups from the rotor with a bearing cup puller.
9. Remove the nuts retaining the spindle to the steering knuckle. Tap the spindle with a plastic or rawhide hammer to jar the spindle from the knuckle. Remove the splash shield.
10. On the left side of the vehicle remove the shaft and joint assembly by pulling the assembly out of the carrier.
11. On the right side of the carrier, remove and discard the keystone clamp from the shaft and joint assembly and the stub shaft. Slide the rubber boot onto the stub shaft and pull the shaft and joint assembly from the splines of the stub shaft.
12. Place the spindle in a vise on the second step of the spindle. Wrap a shop towel around the spindle or use a brass-jawed vise to protect the spindle.
13. Remove the oil seal and needle bearing from the spindle with slide hammer T50T-100–A and seal remover tool 1175–A-C or equivalent.

SUSPENSION AND STEERING

14. If required, remove the seal from the shaft, by driving off with a hammer.
15. If the tie rod has not been removed, then remove cotter pin from the tie rod nut and then remove nut. Tap on the tie rod stud to free it from the steering arm.
16. Remove the upper ball joint cotter pin and nut. Loosen the lower ball joint nut to the end of the stud.
17. Strike the inside of the spindle near the upper and lower ball joints to break the spindle loose from the ball joint studs.
18. Remove the camber adjuster sleeve. If required, use pitman arm puller, T64P–3590–F or equivalent to remove the adjuster out of the spindle. Remove the lower ball joint nut.
19. Place knuckle in vise and remove snapring from bottom ball joint socket if so equipped.
20. Assemble the C-frame, T74P–4635–C, forcing screw, D79T–3010–AE and ball joint remover T83T–3050–A or equivalent on the lower ball joint.
21. Turn forcing screw clockwise until the lower ball joint is removed from the steering knuckle.
22. Repeat Steps 20 and 21 for the upper ball joint.

➡ **Always remove lower ball joint first.**

To install:

23. Clean the steering knuckle bore and insert lower ball joint in knuckle as straight as possible. The lower ball joint doesn't have a cotter pin hole in the stud.
24. Assemble the C-frame, T74P–4635–C, forcing screw, D790T–3010–AE, ball joint installer, T83T–3050–A and receiver cup T80T–3010–A3 or equivalent tools, to install the lower ball joint.
25. Turn the forcing screw clockwise until the lower ball joint is firmly seated. Install the snapring on the lower ball joint.

➡ **If the ball joint cannot be installed to the proper depth, realignment of the receiver cup and ball joint installer will be necessary.**

26. Repeat Steps 24 and 25 for the upper ball joint.
27. Install the camber adjuster into the support arm. Position the slot in its original position.

✴✴ CAUTION

The following torque sequence must be followed exactly when securing the spindle. Excessive spindle turning effort may result in reduced steering returnability if this procedure is not followed.

28. Install a new nut on the bottom of the ball joint stud and torque to 90 ft. lbs. (minimum). Tighten to align the nut to the next slot in the nut with the hole in the ball joint stud. Install a new cotter pin.
29. Install the snapring on the upper ball joint stud. Install the upper ball joint pinch bolt and torque the nut to 48–65 ft. lbs.

➡ **The camber adjuster will seat itself into the knuckle at a predetermined position during the tightening sequence. Do not attempt to adjust this position.**

30. Clean all dirt and grease from the spindle bearing bore. Bearing bores must be free from nicks and burrs.
31. Place the bearing in the fore with the manufacturer's identification facing outward. Drive the bearing into the bore using spindle replacer, T80T–4000S and driver handle T80T–4000–W or equivalent.
32. Install the grease seal in the bearing bore with the lip side of the seal facing towards the tool. Drive the seal in the bore with spindle bearing replacer, T83T–3123–A and driver handle T80–4000–W or equivalent. Coat the bearing seal lip with Lubriplate®.
33. If removed, install a new shaft seal. Place the shaft in a press, and install the seal with spindle/axle seal installer, T83T–3132–A, or equivalent.
34. On the right side of the carrier, install the rubber boot and new keystone clamps on the stub slip yoke.

➡ **This axle does not have a blind spline. Therefore, special attention should be made to assure that the yoke ears are in line during assembly.**

35. Slide the boot over the assembly and crimp the keystone clamp using keystone clamp pliers, T63P–9171–A or equivalent.
36. On the left side of the carrier slide the shaft and joint assembly through the knuckle and engage the splines on the shaft in the carrier.
37. Install the splash shield and spindle onto the steering knuckle. Install and tighten the spindle nuts to 40–50 ft. lbs.
38. Drive the bearing cups into the rotor using bearing cup replacer T73T–4222–B and driver handle, T80T–4000–W or equivalent.
39. Pack the inner and outer wheel bearings and the lip of the oil seal with Multi–Purpose Long-Life Lubricant, C1AZ–19590–B or equivalent.
40. Place the inner wheel bearing in the inner cup. Drive the grease seal into the bore with hub seal replacer, T80T–4000–T and driver handle, T80T–4000–W or equivalent. Coat the bearing seal lip with multipurpose long life lubricant, C1AZ–19590–B or equivalent.
41. Install the rotor on the spindle. Install the outer wheel bearing into cup.

➡ **Verify that the grease seal lip totally encircles the spindle.**

42. Install the wheel bearing, locknut, thrust bearing, snapring, and locking hubs.

Front Wheel Bearings

REMOVAL & INSTALLATION

Pickup With Rear Wheel Drive

➡ **See Section 7 for servicing 4WD models.**

1. Raise and safely support the vehicle. Remove the wheel and tire assembly.
2. Remove the brake caliper and support it with mechanics wire. Do not let the caliper hang by the brake hose.
3. Remove the grease cap, cotter pin, retainer, adjusting nut and washer. Discard the cotter pin.
4. Remove the outer bearing and pull the hub and rotor off the spindle. Remove the grease seal using a seal removal tool. Discard the grease seal.
5. Remove the inner bearing from the hub. Remove all traces of old lubricant from the bearings, hub and spindle with solvent and dry thoroughly.
6. Inspect the bearings and bearing races for scratches, pits or cracks. If the bearings and/or races are worn or damaged, remove the races with a brass drift.

To install:

7. If the bearing races were removed, install new races in the hub with suitable installation tools. Make sure the races are properly seated.
8. Using a bearing packer, pack the bearings with high-temperature wheel bearing grease. If a packer is not available, work as much grease as possible between the rollers and cages by hand.
9. Place a small amount of grease within the hub and grease the races. Install the inner bearing. Install a new wheel seal using a seal installer. Apply grease to the lips of the seal.
10. Install the hub and rotor assembly on the spindle. Install the outer bearing, washer and adjusting nut. Adjust the bearings.
11. Install the retainer, a new cotter pin and the grease cap.
12. Install the caliper and the wheel and tire assembly. Lower the vehicle.
13. Before driving the vehicle, pump the brake pedal several times to restore normal brake travel.

MPV Models

REAR WHEEL DRIVE

◆ See Figures 63 thru 70

1. Raise and safely support the vehicle.
2. Remove the wheel assembly.
3. Remove the hub dust cap.
4. Remove the locknut.
5. Remove the brake caliper.
6. Remove the disc plate.
7. Remove the hub assembly.

To install:

8. Install the hub assembly.
9. Install the disc plate.
10. Install the brake caliper, torque the mounting bolts to 66–79 ft. lbs. (89–107 Nm).

SUSPENSION AND STEERING 8-21

Fig. 63 Remove the dust cap from the wheel hub, using pliers, if necessary

Fig. 64 Installing the dust cap onto the wheel hub using a hammer and a 2 3/8 inch socket

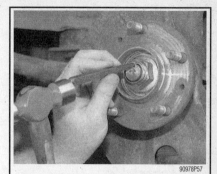
Fig. 65 If the wheel hub retaining nut was staked previously, open it up using a hammer and a punch or drift

Fig. 66 Using a 34mm deep well socket, remove the wheel hub retaining nut

Fig. 67 Remove the wheel hub/bearing assembly by sliding straight off of the spindle

Fig. 68 Be sure to clean the spindle area of any debris using a towel

Fig. 69 After installing the wheel hub/bearing assembly and retaining nut, use a torque wrench to tighten the nut

Fig. 70 Using a hammer and a punch or drift, stake the hub/bearing retaining nut onto the spindle

11. Install the locknut, torque to 131–173 ft. lbs. (117–235 Nm).
12. Using a hammer and a punch or drift, stake the hub/bearing retaining nut onto the spindle.
13. Install the hub dust cap.
14. Install the wheel assembly.
15. Lower the vehicle.

4-WHEEL DRIVE

▶ See Figure 71

1. Raise and safely support the vehicle.
2. Remove the wheel assembly.
3. Remove the locknut.
4. Remove the brake caliper.
5. Remove the disc plate retaining screw(s).
6. Disconnect the tie-rod end from the knuckle.
7. Disconnect the lower ball joint.
8. Remove the disc plate.
9. Remove the ball joint mounting nuts and bolts.
10. Remove the knuckle, wheel hub and dustplate as an assembly.
11. Remove the wheel hub/bearing assembly from the knuckle.

To install:

12. Install wheel hub/bearing assembly to the knuckle.
13. Install the knuckle assembly. Torque the strut mounting nut to 69–86 ft. lbs. (94–116 Nm).
14. Install the ball joint mounting nuts and bolts. Torque the upper mounting bolts to 76–101 ft. lbs. (102–137 Nm). Torque the through-bolt nut to 95–106 ft. lbs. (128–171 Nm).
15. Replace the disc plate.
16. Install ball joint to the knuckle assembly. Torque the ball joint nut to 116–137 ft. lbs. (157–186 Nm).
17. Connect the tie-rod end, torque the nut to 44–57 ft. lbs. (59–78 Nm).

8-22 SUSPENSION AND STEERING

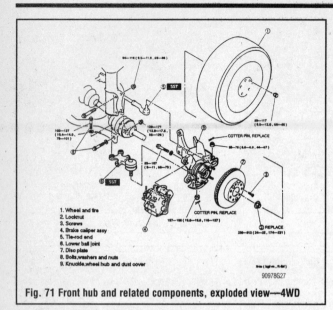

Fig. 71 Front hub and related components, exploded view—4WD

1. Wheel and tire
2. Locknut
3. Screws
4. Brake caliper assy
5. Tie-rod end
6. Lower ball joint
7. Disc plate
8. Bolts, washers and nuts
9. Knuckle, wheel hub and dust cover

18. Install the brake caliper, torque the mounting bolts to 66–79 ft. lbs. (89–107 Nm).
19. Install the disc plate retaining nut.
20. Install the locknut, torque to 174–231 ft. lbs. (236–313 Nm).
21. Install the wheel assembly.
22. Lower the vehicle.

Navajo With Rear Wheel Drive

➡ See Section 7 for servicing 4WD models.

1. Raise and support the vehicle safely. Remove the tire and wheel assembly from the hub and rotor.
2. Remove the caliper from its mounting and position it to the side with mechanics wire in order to prevent damage to the brake line hose.
3. Remove the grease cap from the hub. Remove the cotter pin, retainer, adjusting nut and flatwasher from the spindle.
4. Remove the outer bearing cone and roller assembly from the hub. Remove the hub and rotor from the spindle.
5. Using seal removal tool 1175–AC or equivalent remove and discard the grease seal. Remove the inner bearing cone and roller assembly from the hub.
6. Clean the inner and outer bearing assemblies in solvent. Inspect the bearings and the cones for wear and damage. Replace defective parts, as required.
7. If the cups are worn or damaged, remove them with front hub remover tool T81P–1104–C and tool T77F–1102–A or equivalent.
8. Wipe the old grease from the spindle. Check the spindle for excessive wear or damage. Replace defective parts, as required.

To install:

9. If the inner and outer cups were removed, use bearing driver handle tool T80–4000–W or equivalent and replace the cups. Be sure to seat the cups properly in the hub.
10. Use a bearing packer tool and properly repack the wheel bearings with the proper grade and type grease. If a bearing packer is not available work as much of the grease as possible between the rollers and cages. Also, grease the cone surfaces.
11. Position the inner bearing cone and roller assembly in the inner cup. A light film of grease should be included between the lips of the new grease retainer (seal).
12. Install the retainer using the proper installer tool. Be sure that the retainer is properly seated.
13. Install the hub and rotor assembly onto the spindle. Keep the hub centered on the spindle to prevent damage to the spindle and the retainer.
14. Install the outer bearing cone and roller assembly and flatwasher on the spindle. Install the adjusting nut. Adjust the wheel bearings.
15. Install the retainer, a new cotter pin and the grease cap. Install the caliper.
16. Lower the vehicle and tighten the lug nuts to 100 ft. lbs. Before driving the vehicle pump the brake pedal several times to restore normal brake pedal travel.

✱✱ CAUTION

Tighten the wheel lug nuts to specification after about 500 miles of driving. Failure to do this could result in the wheel coming off while the vehicle is in motion possibly causing loss of vehicle control or collision.

ADJUSTMENT

Pickup With Rear Wheel Drive

1. Raise and safely support the vehicle. Remove the wheel and tire assembly.
2. Remove the brake caliper and suspend it aside with rope or mechanics wire; do not let the caliper hang by the brake hose.
3. Remove the dust cap and cotter pin.
4. Tighten the locknut to 14–22 ft. lbs. (20–29 Nm) and turn the hub and rotor 2–3 times to seat the bearings.
5. Loosen the locknut until it can be turned by hand.
6. Attach a suitable pull scale to a wheel lug bolt and measure the frictional force.
7. Tighten the locknut until the pull scale reading, the initial turning torque, reaches the frictional force plus 1.3–2.4 lbs. Insert the retainer and secure with a new cotter pin.
8. Install the dust cap and the caliper. Install the wheel and tire assembly and lower the vehicle.
9. Before driving the vehicle, pump the brake pedal several times to restore normal brake travel.

Navajo With Rear Wheel Drive

◆ See Figure 72

1. Raise and support the vehicle safely. Remove the wheel cover. Remove the grease cap from the hub.
2. Wipe the excess grease from the end of the spindle. Remove the cotter pin and retainer. Discard the cotter pin.
3. Loosen the adjusting nut 3 turns.

✱✱ CAUTION

Obtain running clearance between the disc brake rotor surface and shoe linings by rocking the entire wheel assembly in and out several times in order to push the caliper and brake pads away from the rotor. An alternate method to obtain proper running clearance is to tap lightly on the caliper housing. Be sure not to tap on any other area that may damage the disc brake rotor or the brake lining sur-

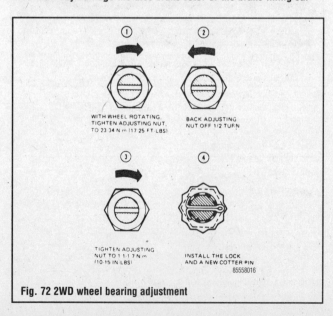

Fig. 72 2WD wheel bearing adjustment

SUSPENSION AND STEERING 8-23

faces. Do not pry on the phenolic caliper piston. The running clearance must be maintained throughout the adjustment procedure. If proper clearance cannot be maintained, the caliper must be removed from its mounting.

4. While rotating the wheel assembly, tighten the adjusting nut to 17–25 ft. lbs. in order to seat the bearings. Loosen the adjusting nut a half turn. Tighten the adjusting nut 18–20 inch lbs.
5. Place the retainer on the adjusting nut. The castellations on the retainer must be in alignment with the cotter pin holes in the spindle. Once this is accomplished install a new cotter pin and bend the ends to insure its being locked in place.
6. Check for proper wheel rotation. If correct, install the grease cap and wheel cover. If rotation is noisy or rough recheck your work and correct as required.
7. Lower the vehicle and tighten the lug nuts to 100 ft. lbs., if the wheel was removed. Before driving the vehicle pump the brake pedal several times to restore normal brake pedal travel.

✲✲ CAUTION

If the wheel was removed, tighten the wheel lug nuts to specification after about 500 miles of driving. Failure to do this could result in the wheel coming off while the vehicle is in motion possibly causing loss of vehicle control or collision.

MPV Models

The wheel hub/bearing assembly used on both rear- and 4-wheel drive models is a non-adjustable component. No adjustments can be made, nor are any possible. However, the wheel bearing can be inspected as follows:
1. Raise and support the vehicle safely. Remove the tire and wheel assembly.
2. Remove and properly support the caliper assembly.
3. Position a dial indicator gauge against the dust cap. Push and pull the disc brake rotor or brake drum in and out in the axial direction and measure the end-play of the wheel bearing.
4. End-play should not exceed 0.002 in. (0.05mm).
5. If end-play is excessive, check the hub nut torque or replace the bearing.

Wheel Alignment

If the tires are worn unevenly, if the vehicle is not stable on the highway or if the handling seems uneven in spirited driving, the wheel alignment should be checked. If an alignment problem is suspected, first check for improper tire inflation and other possible causes. These can be worn suspension or steering components, accident damage or even unmatched tires. If any worn or damaged components are found, they must be replaced before the wheels can be properly aligned. Wheel alignment requires very expensive equipment and involves minute adjustments which must be accurate; it should only be performed by a trained technician. Take your vehicle to a properly equipped shop.

Following is a description of the alignment angles which are adjustable on most vehicles and how they affect vehicle handling. Although these angles can apply to both the front and rear wheels, usually only the front suspension is adjustable.

CASTER

▶ See Figure 73

Looking at a vehicle from the side, caster angle describes the steering axis rather than a wheel angle. The steering knuckle is attached to a control arm or strut at the top and a control arm at the bottom. The wheel pivots around the line between these points to steer the vehicle. When the upper point is tilted back, this is described as positive caster. Having a positive caster tends to make the wheels self-centering, increasing directional stability. Excessive positive caster makes the wheels hard to steer, while an uneven caster will cause a pull to one side. Overloading the vehicle or sagging rear springs will affect caster, as will raising the rear of the vehicle. If the rear of the vehicle is lower than normal, the caster becomes more positive.

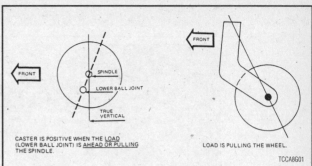

Fig. 73 Caster affects straight-line stability. Caster wheels used on shopping carts, for example, employ positive caster

CAMBER

▶ See Figure 74

Looking from the front of the vehicle, camber is the inward or outward tilt of the top of wheels. When the tops of the wheels are tilted in, this is negative camber; if they are tilted out, it is positive. In a turn, a slight amount of negative camber helps maximize contact of the tire with the road. However, too much negative camber compromises straight-line stability, increases bump steer and torque steer.

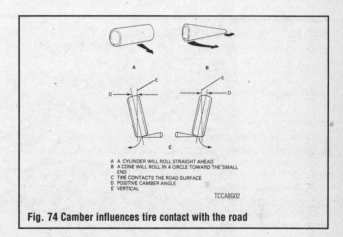

Fig. 74 Camber influences tire contact with the road

TOE

▶ See Figure 75

Looking down at the wheels from above the vehicle, toe angle is the distance between the front of the wheels, relative to the distance between the back of the wheels. If the wheels are closer at the front, they are said to be toed-in or to have negative toe. A small amount of negative toe enhances directional stability and provides a smoother ride on the highway.

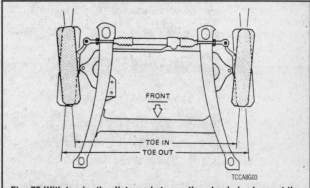

Fig. 75 With toe-in, the distance between the wheels is closer at the front than at the rear

8-24 SUSPENSION AND STEERING

REAR SUSPENSION

REAR SUSPENSION COMPONENT LOCATIONS - MPV SHOWN

1. Coil Springs
2. Upper Control Arms
3. Lower Control Arms
4. Stabilizer Bar (Sway Bar)
5. Sway Bar Links

SUSPENSION AND STEERING 8-25

Coil Springs

REMOVAL & INSTALLATION

MPV

♦ See Figures 76, 77, 78 and 79

1. Raise and safely support the vehicle. Remove the splash shield.
2. Remove the stabilizer bar.
3. Remove the nut and disconnect the height sensor from the rear axle.
4. Remove the bolt attaching the parking brake cable bracket.
5. Support the rear axle housing with a jack. Raise the jack slightly to take the load off the shock absorbers.
6. Remove the attaching bolts and nuts and disconnect the shock absorbers from the lower axle housing.
7. Slowly lower the axle housing until the spring tension is relieved. Remove the coil springs.
8. Remove the spring seats and bump stopper, if equipped.

To install:

9. Install the upper and lower spring seats and the bump stopper, if removed.
10. Install the coil springs, making sure the larger diameter coil is toward the axle housing.
11. Raise the axle housing enough to connect the shock absorbers. Install the attaching bolts and nuts and tighten to 56–76 ft. lbs. (76–103 Nm). Remove the jack.
12. Install the bolt attaching the parking brake cable bracket and the nut attaching the height sensor.
13. Install the stabilizer bar. Tighten the link bolt nut until 0.28 in. (7mm) of thread is exposed at the top of the link bolt. Do not tighten the stabilizer bar bushing bracket bolts at this time.
14. Lower the vehicle. With the vehicle unladen, tighten the stabilizer bar bushing bracket bolts to 23–38 ft. lbs. (34–51 Nm).
15. Install the splash shield.

Leaf Springs

REMOVAL & INSTALLATION

Pickups

1. Raise and support the rear of the truck on jackstands under the frame.

✵✵ CAUTION

The rear leaf springs are under considerable tension. Be very careful when removing and installing them, they can exert enough force to cause serious injuries.

2. Place a floor jack under the rear axle to take up its weight.
3. Disconnect the lower end of the shock absorbers.
4. Remove the spring U bolts and plate.
5. Remove the spring front bolt.
6. Remove the rear shackle nuts and the shackle.
7. Lift the spring from the truck.
8. Place the leaf spring in position and install the rear shackle and front mount bolt. Install the spring U bolts. Secure all fasteners snugly and lower the truck. When the truck is on its wheels, torque the nuts and bolts. Observe the following torques:
   - Spring rear shackle nuts: 58 ft. lbs.
   - 2-wheel drive U-bolt nuts: 58 ft. lbs.
   - 4-wheel drive U-bolt nuts: 101 ft. lbs.
   - Front spring pin nut: 72 ft. lbs.
   - Shock absorber: 58 ft. lbs.

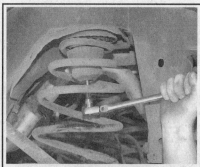

Fig. 76 Using a 14mm deep well socket, loosen the rubber coil spring bump stopper mounting bolt

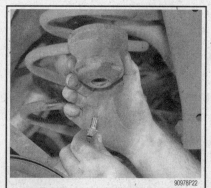

Fig. 77 Remove the bump stopper and mounting bolt

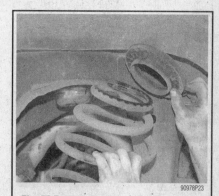

Fig. 78 Remove the rear coil spring and upper rubber mount from the vehicle

Fig. 79 Remove the lower coil spring mount from the top of the differential housing

Navajo

♦ See Figure 80

1. Raise the vehicle and install jackstands under the frame. The vehicle must be supported in such a way that the rear axle hangs free with the tires still touching the ground.
2. Remove the nuts from the spring U-bolts and drive the U-bolts from the U-bolt plate.
3. Remove the spring to bracket nut and bolt at the front of the spring.
4. Remove the shackle upper and lower nuts and bolts at the rear of the spring.
5. Remove the spring and shackle assembly from the rear shackle bracket.

To install:

6. Position the spring in the shackle. Install the upper shackle spring bolt and nut with the bolt head facing outward.
7. Position the front end of the spring in the bracket and install the bolt and nut.
8. Position the shackle in the rear bracket and install the nut and bolt.

8-26 SUSPENSION AND STEERING

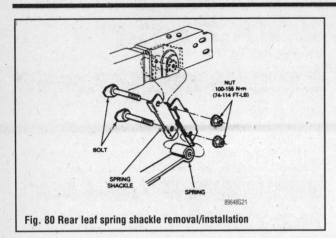

Fig. 80 Rear leaf spring shackle removal/installation

9. Position the spring on top of the axle with the spring tie bolt centered in the hole provided in the seat.
10. Lower the vehicle to the floor. Torque the spring U-bolt nuts to 65–75 ft. lbs. Torque the front spring bolt to 75–115 ft. lbs. Torque the rear shackle nuts and bolts to 75–115 ft. lbs.

Shock Absorbers

REMOVAL & INSTALLATION

Pickups

1. Raise and support the rear end on jackstands.
2. Remove the wheels.
3. Unbolt the shock absorber at each end and remove it.
4. Place the shock absorber in position and secure the mounting hardware. Torque each bolt to 58 ft. lbs. (78 Nm).

MPV

▶ See Figures 81 thru 88

1. Raise and support the rear end on jackstands.
2. Disconnect the shock absorbers at the lower, then upper, end. Remove them.
3. Place the shock absorber in position and secure the mounting hardware snugly. Lower the van to the ground and torque the bolts to 76 ft. lbs. (103 Nm).

Navajo

1. Raise the vehicle and position jackstands under the axle or wheel, in order to take the load off of the shock absorber.
2. Remove the shock absorber lower retaining nut and bolt. Swing the lower end free of the mounting bracket on the axle housing.

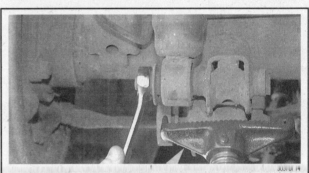

Fig. 81 Support the rear axle assembly by placing a jack stand underneath the lower shock absorber mounting bracket, then using a 21mm box wrench, loosen the mounting nut

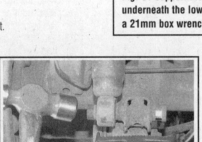

Fig. 82 Remove the mounting nut, washer and retaining plate from the bottom of the shock absorber

Fig. 83 Place the nut onto the end of the retaining bolt to prevent the bolt from being damaged by the hammer blows and if necessary, hammer the mounting bolt out of the shock and axle mounting bracket

Fig. 84 Pull out the lower shock absorber-to-rear axle retaining bolt

Fig. 85 The top rear shock absorber mounting nut is accessible through a hole in the inner fenderwell splash shield

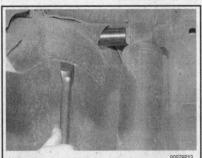

Fig. 86 Loosen the upper shock absorber mounting bolt using a 21mm socket placed through the access hole in the inner fender shield

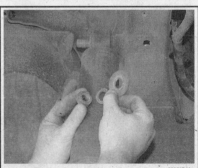

Fig. 87 Remove the mounting nut, washer and retaining plate from the top of the shock absorber

SUSPENSION AND STEERING   8-27

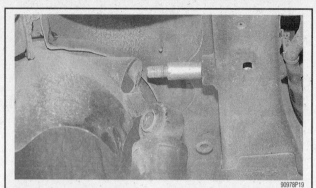

Fig. 88 Separate the top of the rear shock absorber from the top mounting stud

3. Remove the retaining nut(s) from the upper shock absorber mounting
4. Remove the shock absorber from the vehicle.
5. Position and secure the shock absorber with the mounting hardware. Torque the lower shock absorber retaining bolt to 39–53 ft. lbs.
6. Torque the upper shock absorber retaining nuts to 15–21 ft. lbs.

Control Arms/Links

REMOVAL & INSTALLATION

MPV

♦ See Figure 89

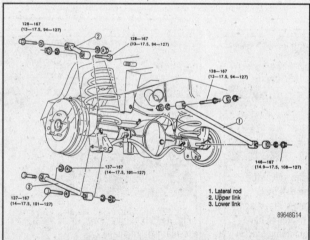

Fig. 89 Exploded view of MPV 5 link rear suspension components

LATERAL ROD

♦ See Figure 90

1. Raise and safely support the vehicle.
2. Support the axle housing with a jack.
3. Remove the lateral rod-to-chassis stud bolt and nut and the lateral rod-to-axle housing nut.
4. Remove the lateral rod.
5. Place the lateral rod in position and secure it with the mounting hardware. Make sure the lateral rod is installed with the identification mark toward the body.
6. Tighten the lateral rod-to-axle housing nut to 108–127 ft. lbs. (146–167 Nm). Tighten the lateral rod-to-chassis stud bolt and nuts to 94–127 ft. lbs. (128–167 Nm).

UPPER CONTROL ARMS

♦ See Figure 91

1. Raise and safely support the vehicle.
2. Support the axle housing with a jack.
3. Remove the upper control arm-to-chassis bolt and nut and the upper control arm-to-axle housing bolt and nut.
4. Remove the upper control arm.
5. Place the control arm in position and secure it with the mounting bolts. Tighten the upper control arm attaching bolts and nuts to 94–127 ft. lbs. (128–167 Nm).

LOWER CONTROL ARMS

♦ See Figure 92

1. Raise and safely support the vehicle.
2. Support the axle housing with a jack.
3. Remove the lower control arm-to-chassis bolt and nut and the lower control arm-to-axle housing bolt and nut.
4. Remove the lower control arm.
5. Position the lower control arm and install the mounting bolts. Tighten the upper control arm attaching bolts and nuts to 101–127 ft. lbs. (137–167 Nm).

Stabilizer Bar

REMOVAL & INSTALLATION

Navajo

♦ See Figure 93

1. As required, raise and support the vehicle.
2. Remove the nuts, bolts and washers and disconnect the stabilizer bar from the links.
3. Remove the U-bolts and nuts from the mounting bracket and retainers. Remove the mounting brackets, retainers and stabilizer bars.

Fig. 90 Using a backup wrench, remove the nuts and bolts securing the lateral rod to the vehicle underbody and rear axle

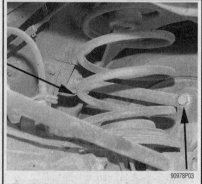

Fig. 91 Remove the nuts and bolts to the upper trailing arm

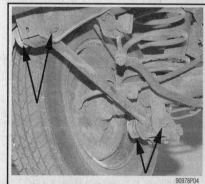

Fig. 92 Remove the nuts and bolts to the lower trailing arm

8-28 SUSPENSION AND STEERING

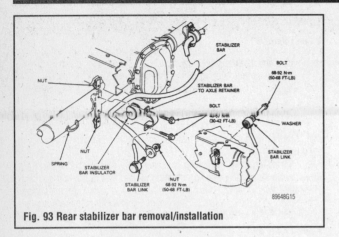

Fig. 93 Rear stabilizer bar removal/installation

To install:

4. Position the U-bolts and mounting brackets on the axle with the brackets having the **UP** marking in the proper position.
5. Install the stabilizer bar and retainers on the mounting brackets with the retainers having the **UP** marking in the proper position.
6. Connect the stabilizer bar to the rear links. Install the nuts, bolts, and washers and tighten.
7. Tighten the mounting bracket U-bolt nuts to 30–42 ft. lbs.

MPV Models

♦ See Figures 94 thru 100

1. Raise and safely support the vehicle.
2. Remove the stabilizer bar links from both sides of the stabilizer bar.
3. Remove the stabilizer bar bracket and bushing from the axle and remove the stabilizer bar.

To install:

4. Install the bushing on the bar and install the bar finger-tight on the axle housing.
5. Install the stabilizer bar links to the frame brackets.
6. Lower the vehicle to the floor and tighten the stabilizer bar to axle bolts to 26–37 ft. lbs. (35–50 Nm).

Rear Wheel Bearings

REMOVAL AND INSTALLATION

For replacement of the rear wheel bearing, please refer to the Axle Shaft removal and installation procedure, located in Section 7.

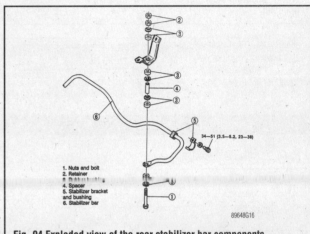

Fig. 94 Exploded view of the rear stabilizer bar components

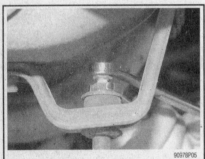

Fig. 95 Using a 12mm box wrench and a 12mm open end wrench as a backup wrench, loosen then remove the sway bar link locking nut

Fig. 96 Holding the sway bar link bolt with a 12mm box wrench, remove the retaining nut

Fig. 97 After removing the sway bar link retaining nut, remove the top bushing washer and rubber bushing

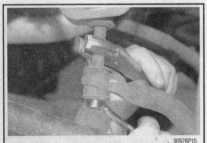

Fig. 98 The sway bar link may be corroded to the metal sleeve. If this is so, secure the sleeve in a pair of locking pliers while using a box wrench to loosen the corrosion bond between the two components

Fig. 99 Spray the sway bar link with a rust penetrating lubricant to aid in the removal of the sway bar link from the sleeve

Fig. 100 Remove the sway bar bushing retainer mounting bolt

SUSPENSION AND STEERING

STEERING

Steering Wheel

REMOVAL & INSTALLATION

♦ See Figures 101 thru 108

1. Disconnect the negative battery cable.
2. Remove the steering wheel pad from the steering wheel. On some models, pull the horn pad straight up from the steering wheel. On other models, remove the mounting screws from behind the steering wheel. Pull the pad back and disconnect the horn switch and, if equipped, cruise control wires. Remove the steering wheel pad.
3. Remove the steering wheel attaching bolt or nut. Check to see if the steering wheel and steering shaft have alignment marks or flats. If there are no steering wheel-to-steering column shaft alignment marks or flats, matchmark the steering wheel and column shaft so they can be reassembled in the same position.
4. Using a suitable puller, remove the steering wheel from the steering column shaft.

→Do not hammer on the steering wheel or steering shaft or use a knock-off type steering wheel puller, as either will damage the steering column.

To install:

5. Install the steering wheel on the steering column shaft, aligning the marks or flats on the steering wheel with the marks or flats on the steering shaft.
6. On all except Navajo, install the steering wheel attaching nut and tighten to 35 ft. lbs. (48 Nm). On Navajo, install the steering wheel attaching bolt and tighten to 23–33 ft. lbs. (31–45 Nm).

Fig. 101 Remove the screws that retain the steering wheel center hub/horn pad cover on the steering wheel

Fig. 102 Lift the center hub/horn pad cover up off of the steering wheel . . .

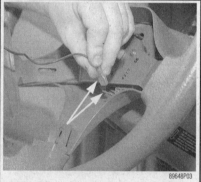

Fig. 103 . . . then disengage the steering wheel horn wiring connector

Fig. 104 Remove the 2 screws that retain the steering wheel bottom cover to the steering shaft

Fig. 105 Remove the steering wheel-to-column shaft lock bolt

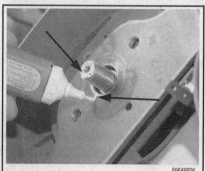

Fig. 106 Matchmark the position of the steering wheel to the column shaft for proper installation

Fig. 107 Use a steering wheel puller tool to remove the steering wheel from the column shaft

Fig. 108 Remove the steering wheel from the vehicle

8-30 SUSPENSION AND STEERING

7. Connect the horn switch and, if equipped, cruise control wires and install the steering wheel pad.
8. Connect the negative battery and check the steering column for proper operation.

Combination Switch

The combination turn signal, windshield wiper, and headlight switch (except Navajo) is mounted on the steering column, and must be replaced as an assembly.

REMOVAL & INSTALLATION

Pickups

1. Disconnect the negative battery cable.
2. Remove the steering wheel.
3. Remove the "Lights-Hazard" Indicator and the steering column shroud.
4. Unplug the electrical multiple connectors at the base of the steering column.
5. Pull the headlight knob from its shaft.
6. Remove the snapring, which retains the switch, from the steering shaft. Pull the turn indicator canceling cam from the shaft.
7. Remove the single retaining bolt near the bottom of the switch. Remove the complete switch from the column.
8. Place the switch in position and secure it. Install the turn indicator cam and snapring. Install the remaining components. Check the operation of the switch before installing the steering wheel.

MPV

▶ See Figure 109

1. Disconnect the battery. Remove the horn cover cap.
2. Remove the steering wheel attaching nut, and pull off the wheel with a puller.
3. Remove the attaching screws, and remove the right and left steering column covers.
4. Disconnect the connector for the combination switch or, if the ignition switch is being replaced, disconnect connectors for both that and the combination switch.
5. Remove the retaining ring from the steering column.
6. Remove the combination retaining screw, and remove the switch.

To install:

7. Install the combination switch and secure with the retaining screw.
8. Install the steering column retaining ring (if so equipped).
9. Connect all combination or ignition switch connectors at this time.
10. Install the steering column covers with retaining screws.
11. Install the steering wheel and horn cap. Connect the negative battery cable and check all the functions of the combination switch for proper operation.

Navajo

▶ See Figure 110

1. Disconnect the negative battery cable. Remove the steering wheel.
2. On vehicles equipped with tilt wheel, remove the tilt lever.
3. On vehicles equipped with tilt wheel, remove the steering column collar by pressing on the collar from the top and bottom while removing the collar.
4. Remove the instrument panel trim cover retaining screws. Remove the trim cover.
5. Remove the 2 screws from the bottom of the steering column shroud. Remove the bottom half of the shroud by pulling the shroud down and toward the rear of the vehicle.
6. If the vehicle is equipped with automatic transmission, move the shift lever as required to aid in removal of the shroud. Lift the top half of the shroud from the column.
7. If the vehicle is equipped with automatic transmission, disconnect the selector indicator actuation cable by removing the screw from the column casting and the plastic plug at the end of the cable.

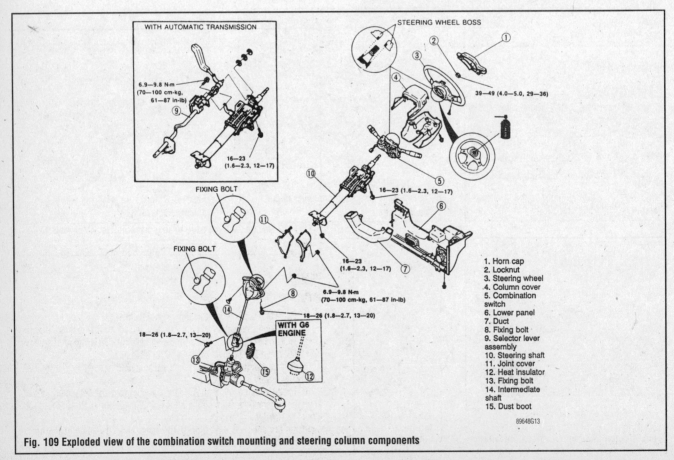

Fig. 109 Exploded view of the combination switch mounting and steering column components

SUSPENSION AND STEERING   8-31

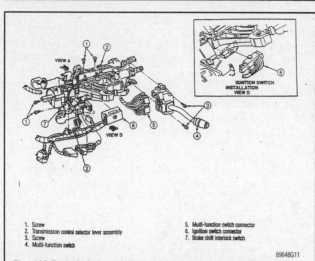

Fig. 110 Exploded view of the combination switch mounting and column related wire harness connectors

1. Screw
2. Transmission control selector lever assembly
3. Screw
4. Multi-function switch
5. Multi-function switch connector
6. Ignition switch connector
7. Brake shift interlock switch

8. To remove the plastic plug from the shift lever socket casting push on the nose of the plug until the head clears the casting and pull the plug from the casting.
9. Remove the plastic clip that retains the combination switch wiring to the steering column bracket.
10. Remove the 2 self taping screws that retain the combination switch to the steering column casting. Disengage the switch from the casting.
11. Disconnect the 3 electrical connectors, using caution not to damage the locking tabs. Be sure not to damage the PNDRL cable.
12. Installation is the reverse of the removal procedure. Torque the combination switch retaining screws to 18–27 inch lbs. (2–3 Nm).

Ignition Switch

REMOVAL & INSTALLATION

Pickups and MPV

♦ See Figure 111

1. Disconnect the battery ground cable.
2. Remove the steering column covers.
3. Disconnect the wiring harness connector at the switch.
4. Remove the attaching screw and lift out the switch.
5. Installation is the reverse of removal.

Navajo

1. Disconnect the negative battery cable.
2. Remove the steering wheel.

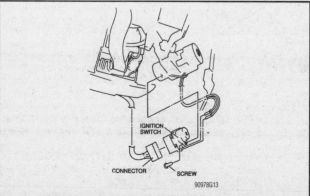

Fig. 111 Ignition switch removal/installation—MPV shown

3. As necessary, remove all under dash panels in order to gain access to the ignition switch.
4. As necessary, lower the steering column to gain working clearance.
5. Disconnect the ignition switch electrical connectors.
6. Remove the ignition switch retaining screws from the studs. Disengage the ignition switch from switch rod. Remove the switch from the vehicle.

To install:
7. Position the lock cylinder in the **LOCK** position.
8. To set the switch, position a wire in the opening in the outer surface of the switch through its positions until the wire drops down into the slot.

➡The slot is in the bottom of the switch where the rod must be inserted to allow full movement through the switch positions.

9. Position the ignition switch on the column studs and over the actuating rod. Torque the retaining nuts to 3.3–5.3 ft. lbs. (5–7 Nm).
10. Remove the wire from the slot in the housing. Continue the installation in the reverse order of the removal procedure.

Steering Linkage

REMOVAL & INSTALLATION

♦ See Figures 112 and 113

Idler Arm

PICKUPS

1. Raise and support the front end on jackstands.
2. Remove the idler arm-to-center link nut and cotter pin. Disconnect the center link from the idler arm using a ball joint separator.
3. Unbolt and remove the idler arm.
4. Install and secure the idler arm. Connect the center link. Always install new cotter pins. Torque the center link nut to 43 ft. lbs. (58 Nm); the frame mounting bolts to 69 ft. lbs. (94 Nm).

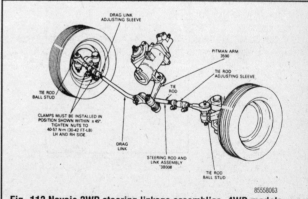

Fig. 112 Navajo 2WD steering linkage assemblies, 4WD models similar

Pitman Arm

PICKUPS

1. Raise and support the front end on jackstands.
2. Remove the cotter pin and nut attaching the center link to the pitman arm.
3. Disconnect the center link from the pitman arm with a ball joint separator.
4. Matchmark the pitman arm and sector shaft.
5. Remove the pitman arm-to-sector shaft nut and remove the pitman arm. It may be necessary to use a puller.
6. Install and attach the pitman arm and components. Make sure you align the matchmarks. Tighten the pitman arm-to-sector shaft nut to 130 ft. lbs. (177 Nm); the pitman arm-to-center link nut to 43 ft. lbs. (58 Nm). If the cotter pin does not align, tighten the nut to make it line up, never loosen it!

8-32 SUSPENSION AND STEERING

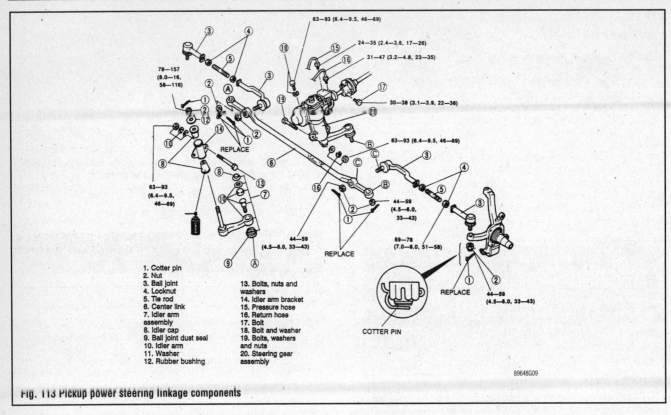

1. Cotter pin
2. Nut
3. Ball joint
4. Locknut
5. Tie rod
6. Center link
7. Idler arm assembly
8. Idler cap
9. Ball joint dust seal
10. Idler arm
11. Washer
12. Rubber bushing
13. Bolts, nuts and washers
14. Idler arm bracket
15. Pressure hose
16. Return hose
17. Bolt
18. Bolt and washer
19. Bolts, washers and nuts
20. Steering gear assembly

Fig. 113 Pickup power steering linkage components

NAVAJO

1. As required, raise and safely support the vehicle using jackstands.
2. Remove the cotter pin and nut from the drag link ball stud at the pitman arm.
3. Remove the drag link ball stud from the pitman arm using pitman arm removal tool T64P-3590-F or equivalent.
4. Remove the pitman arm retaining nut and washer. Remove the pitman arm from the steering gear sector shaft using tool T64P-3590-F or equivalent.
5. Install and secure the pitman arm and components. Torque the pitman arm attaching washer and nut to 170-230 ft. lbs. (231-313 Nm). Torque the drag link ball stud nut to 50-70 ft. lbs. (68-95 Nm) and install a new cotter pin.
6. Check and adjust front end alignment, as required.

Center (Drag) Link

PICKUPS

1. Raise and support the front end on jackstands.
2. Disconnect the center link at the tie rods, pitman arm and idler arm.
3. Install and secure the center link and components. Tighten all of the nuts to 43 ft. lbs. (58 Nm).

NAVAJO

1. Raise and support the vehicle using jackstands. Be sure that the front wheels are in the straight ahead position.
2. Remove the nuts and cotter pins from the ball stud at the pitman arm and steering tie rod. Remove the ball studs from the linkage using pitman arm removal tool T64P-3590-F or equivalent.
3. Loosen the bolts on the drag link adjusting sleeve. Be sure to count and record the number of turns it takes to remove the drag link.

To install:

4. Install the drag link in the same number of turns it took to remove it. Tighten the adjusting sleeve nuts to 30-42 ft. lbs. (41-57 Nm). Be sure that the adjusting sleeve clamps are pointed down approximately 45°.
5. Position the drag link ball stud in the pitman arm. Position the steering tie rod ball stud in the drag link. With the vehicle wheels in the straight ahead position install and torque the nuts to 50-75 ft. lbs. (68-102 Nm). Install a new cotter pin.
6. Check and adjust front end alignment, as required.

Tie Rod

NAVAJO

1. Raise and support the vehicle using jackstands. Be sure that the front wheels are in the straight ahead position.
2. Remove the nut and cotter pin from the ball stud on the drag link. Remove the ball stud from the drag link using pitman arm removal tool T64P-3590-F or equivalent.
3. Loosen the bolts on the tie rod adjusting sleeve. Be sure to count and record the number of turns it takes to remove the tie rod from the tie rod adjusting sleeve. Remove the tie rod from the vehicle.

To install:

4. Install the tie rod in the tie rod sleeve in the same number of turns it took to remove it. Torque the tie rod adjusting sleeve nuts to 30-42 ft. lbs. (41-57 Nm).
5. Be sure that the adjusting sleeve clamps are pointed down approximately 45°. Tighten the tie rod ball stud to drag link retaining bolt to 50-75 ft. lbs. (68-102 Nm). Install a new cotter pin.
6. Check and adjust front end alignment, as required.

Tie Rod Ends

PICKUPS

1. Loosen the tie rod jam nuts.
2. Remove and discard the cotter pin from the ball socket end, and remove the nut.
3. Use a ball joint puller to loosen the ball socket stud from the center link. Remove the stud from the kingpin steering arm in the same way.
4. Unscrew the tie rod end from the threaded sleeve, counting the number of threads until it's off. The threads may be left or right hand threads. Tighten the jam nuts to 58 ft. lbs. (79 Nm).
5. To install, lightly coat the threads with grease, and turn the new end in as many turns as were required to remove it. This will give the approximate correct toe-in.
6. Install the ball socket studs into center link and kingpin steering arm. Tighten the nuts to 43 ft. lbs. (58 Nm). Install a new cotter pin. You may tighten the nut to fit the cotter pin, but don't loosen it.
7. Check and adjust the toe-in, and tighten the tie rod clamps or jam nuts.

SUSPENSION AND STEERING 8-33

MPV

▶ See Figures 114 thru 120

1. Raise and support the front end on jackstands.
2. Remove the wheels.
3. Matchmark the tie rod end and tie rod and loosen the locknut.
4. Loosen the tie rod end ball stud nut and separate the tie rod end from the knuckle arm with a separator tool.
5. Unscrew the tie rod end, counting the number of turns until it's off, for installation purposes.

To install:

6. Install the tie rod end onto the tie rod the same number of turns that it took to remove it. Tighten the locknut to 58 ft. lbs. (79 Nm).
7. Install the ball stud into the steering knuckle. Tighten the ball stud nut to 58 ft. lbs. (79 Nm) and install the cotter pin.
8. Lower the vehicle and install a new cotter pin. Always advance the nut to align the cotter pin hole. Never back it off. Check the front alignment.

NAVAJO

1. Raise and support the vehicle using jackstands. Be sure that the front wheels are in the straight ahead position.
2. Remove the nut and cotter pin from the ball stud on the drag link. Remove the ball stud from the drag link using pitman arm removal tool T64P–3590–F or equivalent.
3. Loosen the bolts on the tie rod adjusting sleeve. Be sure to count and record the number of turns it takes to remove the sleeve from the ball stud.

To install:

4. Install the adjusting sleeve on the tie rod ball stud in the same number of turns it took to remove it. Loosely assemble the ball stud in the spindle arm.
5. Torque the retaining nuts to 30–42 ft. lbs. (41–57 Nm). Be sure that the adjusting sleeve clamps are pointed down approximately 45°.
6. With the vehicle wheels in the straight ahead position, install and torque the nut to 50–75 ft. lbs. (68–102 Nm). Install a new cotter pin.
7. Check and adjust front end alignment, as required.

Manual Steering Gear

REMOVAL & INSTALLATION

▶ See Figure 121

1. Raise and support the front end on jackstands.
2. Remove the pinch bolt securing the wormshaft to the steering shaft coupling.
3. Remove the cotter pin and nut securing the pitman arm to the center link and separate the pitman arm from the link with a ball joint tool.
4. Unbolt the steering gear from the frame.
5. If the pitman arm is to be removed from the sector shaft, first matchmark their positions, relative to each other.
6. Install the steering gear and connect the center link. Observe the following torques:

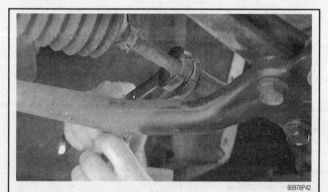

Fig. 114 Using a backup wrench, loosen the jam nut from the tie rod

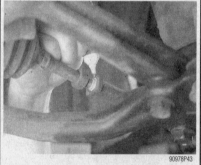

Fig. 115 Matchmark the location of the tie rod end on the tie rod. This will help during installation

Fig. 116 Remove and discard the cotter pin from tie rod ball stud

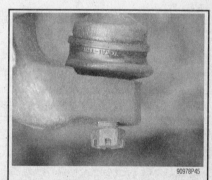

Fig. 117 Install the castle nut onto the end of the ball stud to prevent the stud from "mushrooming" during removal

Fig. 118 Using a puller tool with the castle nut on the end of the ball stud, separate the tie rod end ball joint from the steering knuckle

Fig. 119 Remove the castle nut and pull the tie rod end ball stud straight out of the steering knuckle

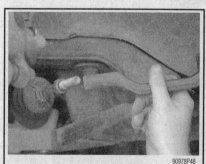

Fig. 120 Count the number of turns required to remove the tie rod end from the tie rod. This will help during the installation process

8-34 SUSPENSION AND STEERING

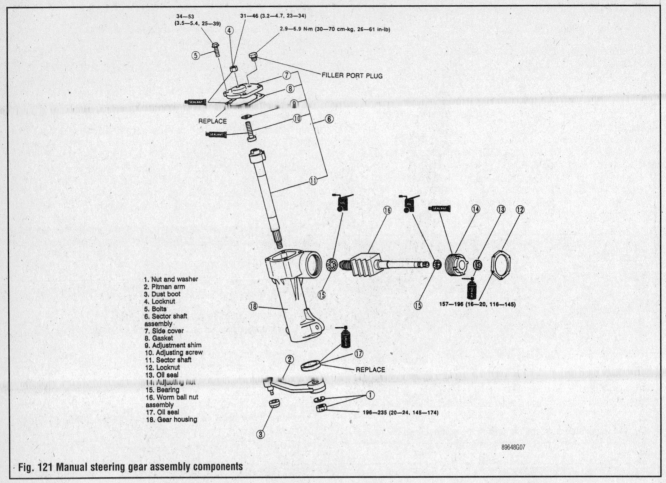

Fig. 121 Manual steering gear assembly components

- Steering gear-to-frame: 69 ft. lbs. (94 Nm)
- Wormshaft-to-steering shaft yoke: 38 ft. lbs. (52 Nm)
- Pitman arm-to-sector shaft: 130 ft. lbs. (177 Nm)
- Pitman arm-to-center link: 43 ft. lbs. (58 Nm)

Power Steering Gear

REMOVAL & INSTALLATION

Pickups

▶ See Figure 122

1. Raise and support the front end on jackstands.
2. Disconnect the pressure and return lines at the gear box. Have a drain pan underneath to catch the fluid.
3. Remove the pinch bolt securing the wormshaft to the steering shaft coupling.
4. Remove the cotter pin and nut securing the pitman arm to the center link and separate the pitman arm from the link with a ball joint tool.
5. Unbolt the steering gear from the frame.
6. If the pitman arm is to be removed from the sector shaft, first matchmark their positions, relative to each other.
7. Place the steering gear in position and install the mounting hardware. Connect the components. Observe the following torques:
- Pressure line: 26 ft. lbs. (35 Nm)
- Return line: 35 ft. lbs. (48 Nm)
- Steering gear-to-frame: 69 ft. lbs. (94 Nm)
- Wormshaft-to-steering shaft yoke: 38 ft. lbs. (52 Nm)
- Pitman arm-to-sector shaft: 130 ft. lbs. (177 Nm)
- Pitman arm-to-center link: 43 ft. lbs. (58 Nm)

Navajo

▶ See Figure 123

1. Disconnect the pressure and return lines from the steering gear. Plug the lines and the ports in the gear to prevent entry of dirt.
2. Remove the upper and lower steering gear shaft U-joint shield from the flex coupling. Remove the bolts that secure the flex coupling to the steering gear and to the column steering shaft assembly.
3. Raise the vehicle and remove the pitman arm attaching nut and washer.
4. Remove the pitman arm from the sector shaft using tool T64P–3590–F. Remove the tool from the pitman arm. Do not damage the seals.
5. Support the steering gear, and remove the steering gear attaching bolts.
6. Work the steering gear free of the flex coupling. Remove the steering gear from the vehicle.

To install:

7. Install the lower U-joint shield onto the steering gear lugs. Slide the upper U-joint shield into place on the steering shaft assembly.
8. Slide the flex coupling into place on the steering shaft assembly. Turn the steering wheel so that the spokes are in the horizontal position. Center the steering gear input shaft.
9. Slide the steering gear input shaft into the flex coupling and into place on the frame side rail. Install the attaching bolts and tighten to 50–62 ft. lbs. (60–84 Nm). Tighten the flex coupling bolt 26–34 ft. lbs. (35–46 Nm).
10. Be sure the wheels are in the straight ahead position, then install the pitman arm on the sector shaft. Install the pitman arm attaching washer and nut. Tighten nut to 170–230 ft. lbs. (231–313 Nm).
11. Connect and tighten the pressure and the return lines to the steering gear.
12. Disconnect the coil wire. Fill the reservoir. Turn on the ignition and turn the steering wheel from left to right to distribute the fluid.
13. Recheck fluid level and add fluid, if necessary. Connect the coil wire, start the engine and turn the steering wheel from side to side. Inspect for fluid leaks.

SUSPENSION AND STEERING 8-35

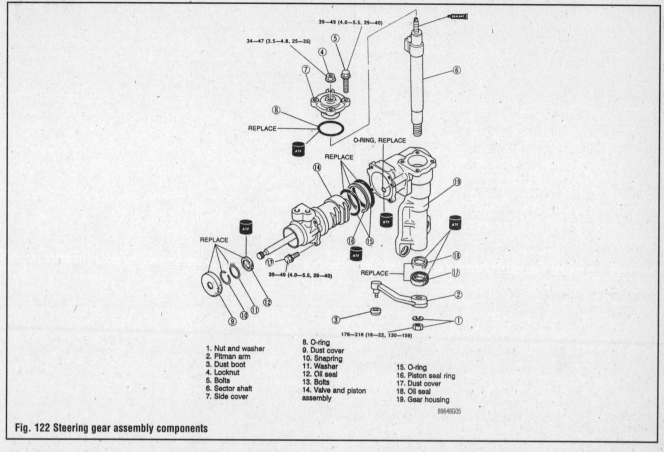

1. Nut and washer
2. Pitman arm
3. Dust boot
4. Locknut
5. Bolts
6. Sector shaft
7. Side cover
8. O-ring
9. Dust cover
10. Snapring
11. Washer
12. Oil seal
13. Bolts
14. Valve and piston assembly
15. O-ring
16. Piston seal ring
17. Dust cover
18. Oil seal
19. Gear housing

Fig. 122 Steering gear assembly components

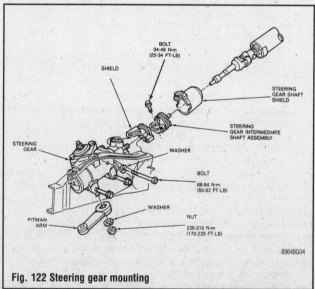

Fig. 122 Steering gear mounting

Power Steering Rack and Pinion

REMOVAL & INSTALLATION

MPV—2WD

♦ See Figure 124

1. Place the front wheels in the straight ahead position. Raise and safely support the vehicle.
2. Remove the wheel and tire assemblies. Remove the splash shield.
3. Remove the cotter pins and nuts from both tie rod end studs. Use separator tool 49 0727 575 or equivalent, to separate the tie rod ends from the knuckles.
4. Remove the pinch bolt from the intermediate shaft-to-pinion shaft coupling.
5. Disconnect and plug the pressure line from the rack and pinion assembly. Loosen the clamp and disconnect the return line from the rack and pinion assembly. Plug the line.
6. If equipped with automatic transmission, remove the change counter assembly to remove the protector plate mounting bolt.
7. Remove the steering bracket mounting bolts and remove the rack and pinion assembly and brackets.
8. If necessary, remove the brackets.

To install:

9. If removed, install the brackets and tighten the mounting bolts, in sequence, to 54–69 ft. lbs. (74–93 Nm).
10. Install the rack and pinion assembly and brackets in the vehicle. Tighten the bracket-to-chassis bolts to 46–69 ft. lbs. (63–93 Nm).
11. If equipped with automatic transmission, install the change counter assembly.
12. Connect the return line and tighten the clamp. Connect the pressure line and tighten the nut to 23–35 ft. lbs. (31–47 Nm).
13. Install the pinch bolt in the intermediate shaft-to-pinion shaft coupling and tighten to 13–20 ft. lbs. (18–26 Nm).
14. Position the tie rod end studs in the knuckles and install the nuts. Tighten the nuts to 43–58 ft. lbs. (59–78 Nm) and install new cotter pins.
15. Install the splash shield and the wheel and tire assemblies. Lower the vehicle and bleed the power steering system.

MPV—4WD

1. Place the front wheels in the straight ahead position. Raise and safely support the vehicle.
2. Remove the wheel and tire assemblies. Remove the splash shield.
3. Remove the cotter pins and nuts from both tie rod end studs. Use separator tool 49 0727 575 or equivalent, to separate the tie rod ends from the knuckles.

8-36 SUSPENSION AND STEERING

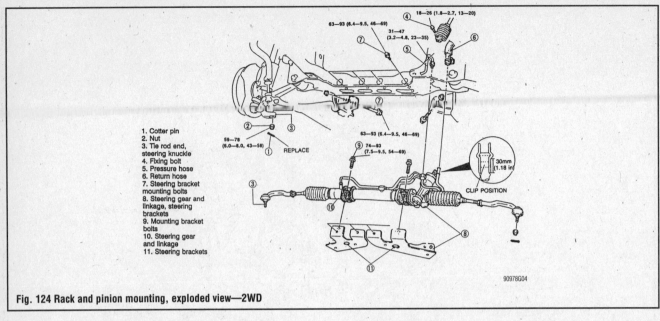

Fig. 124 Rack and pinion mounting, exploded view—2WD

1. Cotter pin
2. Nut
3. Tie rod end, steering knuckle
4. Fixing bolt
5. Pressure hose
6. Return hose
7. Steering bracket mounting bolts
8. Steering gear and linkage, steering brackets
9. Mounting bracket bolts
10. Steering gear and linkage
11. Steering brackets

4. Disconnect and plug the pressure and return hoses at the pressure and return lines.
5. Remove the pressure and return lines from the rack and pinion assembly.
6. Remove the pinch bolt from the intermediate shaft-to-pinion shaft coupling.
7. Working inside the vehicle, remove the lower panel and column cover from under the steering column. Remove the steering column mounting bolts and nuts and pull the column and intermediate shaft rearward to separate the intermediate shaft from the pinion shaft.
8. Mark the position of the front driveshaft on the axle flange and remove the front driveshaft.
9. Remove the rack and pinion assembly mounting bracket bolts and the front differential/joint shaft assembly mounting bolts.
10. Slide the differential/joint shaft assembly rearward. Slide the rack and pinion assembly rearward and turn it 90 degrees, then remove it from the left side of the vehicle.

To install:

11. Install the rack and pinion assembly from the left side of the vehicle, turn it 90 degrees and move it forward into position. Install the mounting bolts and tighten, in sequence, to 54–69 ft. lbs. (74–93 Nm).
12. Move the differential/joint shaft assembly forward, install the mounting bolts and tighten to 49–72 ft. lbs. (67–97 Nm).
13. Install the driveshaft, aligning the marks made during removal.
14. Working inside the vehicle, move the steering column and intermediate shaft forward to engage the intermediate shaft with the pinion shaft. Install and tighten the steering column nuts and bolts to 12–17 ft. lbs. (16–23 Nm). Install the lower panel and column cover.
15. Install the pinch bolt in the intermediate shaft-to-pinion shaft coupling and tighten to 13–20 ft. lbs. (18–26 Nm).
16. Install the pressure and return lines on the rack and pinion assembly. Connect the pressure and return hoses to the lines.
17. Position the tie rod end studs in the knuckles and install the nuts. Tighten the nuts to 43–58 ft. lbs. (59–78 Nm) and install new cotter pins.
18. Install the splash shield and the wheel and tire assemblies. Lower the vehicle and bleed the power steering system.

Power Steering Pump

REMOVAL & INSTALLATION

B2200

♦ See Figure 125

1. Raise and support the front end on jackstands.
2. Loosen the idler pulley bolt and remove the drive belt.

3. Remove the pump pulley nut.
4. Using a puller, remove the pulley from the pump.
5. Disconnect the return hose from the pump. Have a drain pan ready to catch the fluid.
6. Remove the pressure hose bracket bolt and unscrew the pressure hose from the pump. Always use a back-up wrench.
7. Support the pump and remove the front and rear pump-to-bracket bolts. Remove the pump.
8. Place the pump in position and install the mounting bolts. Connect the pressure and return lines. Install the drive pulley and belt. Adjust the drive belt to the proper tension. Fill and bleed the system. Observe the following torques:
 - Pump pulley: 43 ft. lbs. (58 Nm)
 - Pressure line connection: 35 ft. lbs. (48 Nm)
 - Pressure line bracket: 17 ft. lbs. (23 Nm)
 - Pump-to-bracket: 34 ft. lbs. (46 Nm)

B2600

1. Raise and support the front end on jackstands.
2. Remove the pump adjusting bolt and remove the drive belt.
3. Remove the lower pump-to-bracket through-bolt and spacer.
4. Rotate the pump to get to the hoses.
5. Disconnect the return hose from the pump. Have a drain pan ready to catch the fluid.
6. Remove the pressure hose from the pump. Always use a back-up wrench.
7. Support the pump and remove the upper pump-to-bracket bolt. Remove the pump.
8. Place the pump in position and install the upper mounting bolt loosely. Connect the pressure and return hoses. Install the lower mounting and adjusting bolts. Tighten the mounting bolts enough so that the pump can still be move for belt adjustment. Install the drive belt and adjust to proper tension. Tighten the mounting bolts. Fill and bleed the system. Observe the following torques:
 - Pressure line connection: 35 ft. lbs. (48 Nm)
 - Lower pump-to-bracket: 34 ft. lbs. (46 Nm)
 - Upper pump-to-bracket: 20 ft. lbs. (27 Nm)
 - Adjusting bolt: 34 ft. lbs. (46 Nm)

MPV w/6-Cylinder Engine

♦ See Figure 126

1. Loosen the idler pulley locknut.
2. Loosen the pump adjusting bolt.
3. Remove the drive belt.
4. Remove the pump pulley nut.
5. Using a puller, remove the pulley.
6. Disconnect the pressure switch wiring connector.

SUSPENSION AND STEERING 8-37

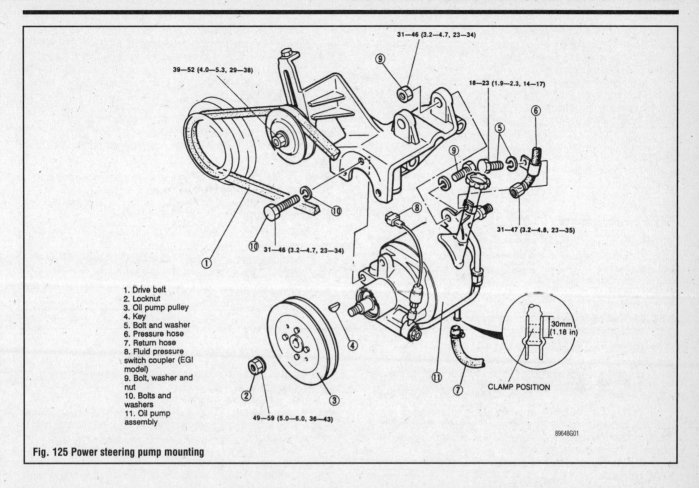

Fig. 125 Power steering pump mounting

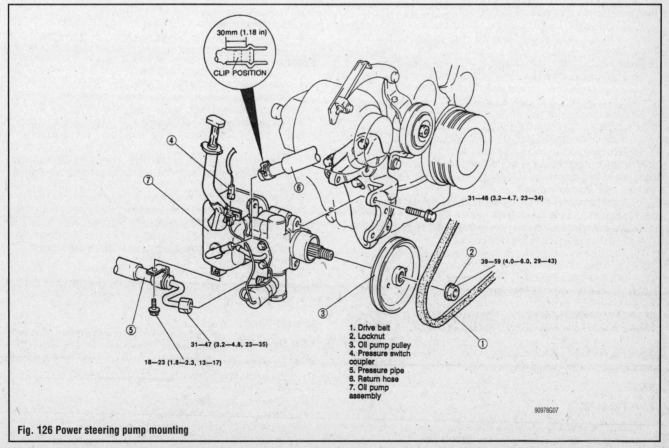

Fig. 126 Power steering pump mounting

8-38 SUSPENSION AND STEERING

7. Place a drain pan under the pump.
8. Matchmark the pressure line connection and disconnect it.
9. Remove the pressure line bracket bolt.
10. Disconnect the return line.
11. Remove the pump-to-bracket bolts and lift out the pump.
12. Place the pump in position and install the mounting bolts. Connect the return and pressure. Install the drive pulley and belt. Adjust the drive belt to the proper tension. Fill and bleed the system. Observe the following torques:
- mounting bolts to 34 ft. lbs. (46 Nm)
- pressure line connection to 35 ft. lbs. (48 Nm)
- pressure line bracket bolt: 17 ft. lbs. (23 Nm)
- pulley bolt: 43 ft. lbs. (58 Nm)
- adjusting bolt: 35 ft. lbs. (48 Nm)
- idler pulley locknut: 38 ft. lbs. (52 Nm)

MPV w/4-Cylinder Engine

1. Place a drain pan under the pump.
2. Loosen the idler pulley locknut and the pump adjusting bolt and remove the pump drive belt.
3. Remove the pump pulley nut.
4. Remove the pulley.
5. Unplug the pressure switch coupler.
6. Matchmark the high pressure line coupling and disconnect the line.
7. Remove the return hose.
8. Remove the pump-to-bracket bolts and lift out the pump.
9. Place the pump in position and install and tighten the mounting bolts. Connect the return and pressure hoses. Install the drive pulley. Install and adjust the drive belt. Fill and bleed the system. Tighten the high pressure coupling to 23–35 ft. lbs. (31–47 Nm), aligning the high pressure coupling matchmarks. Tighten the pump mounting bolts to 23–34 ft. lbs. (31–46 Nm). Tighten the pulley nut to 29–43 ft. lbs. (39–58 Nm). Tighten the idler pulley locknut to 27–38 ft. lbs. (37–52 Nm).

Navajo

1. Disconnect the negative battery cable.
2. Remove some power steering fluid from the reservoir by disconnecting the fluid return line hose at the reservoir. Drain the fluid into a container and discard it.
3. Remove the pressure hose from the pump. If equipped, disconnect the power steering pump pressure switch.
4. Slacken belt tension by lifting the tensioner pulley in a clockwise direction. Remove the drive belt from under the tensioner pulley and slowly lower the pulley to its stop.
5. Remove the drive belt from the pulley. If necessary, remove the oil dipstick tube.
6. If equipped, remove the power steering pump bracket support brace.

7. Install power steering pump pulley removal tool T69L–10300–B or equivalent. Hold the pump and rotate the tool counterclockwise to remove the pulley. Do not apply in and out pressure to the pump shaft, as internal pump damage will occur.
8. Remove the power steering retaining bolts. Remove the power steering pump from the vehicle.

To install:
9. Position the pump on the bracket. Install and tighten the retaining bolts.
10. Install the pulley removal tool and install the power steering pump pulley to the power steering pump.

➥**Fore and aft location of the pulley on the power steering pump shaft is critical. Incorrect belt alignment may cause belt squeal or chirp. Be sure that the pull off groove on the pulley is facing front and flush with the end of the shaft ± 0.010 inch (0.25mm).**

11. Continue the installation in the reverse order of the removal procedure. Adjust the belt tension to specification.
12. While lifting the tensioner pulley in a clockwise direction, slide the belt under the tensioner pulley and lower the pulley to the belt.

BLEEDING THE SYSTEM

Pickups and MPV

1. Raise and support the front end on jackstands.
2. Check the fluid level and fill it, if necessary.
3. Start the engine and let it idle. Turn the steering wheel lock-to-lock, several times. Recheck the fluid level.
4. Lower the truck to the ground.
5. With the engine idling, turn the wheel lock-to-lock several times again. If noise is heard in the fluid lines, air is present.
6. Put the wheels in the straight ahead position and shut off the engine.
7. Check the fluid level. If it is higher than when you last checked it, air is in the system. Repeat step 5. Keep repeating step 5 until no air is present.

Navajo

1. Disconnect the coil wire.
2. Crank the engine and continue adding fluid until the level stabilizes.
3. Continue to crank the engine and rotate the steering wheel about 30° to either side of center.
4. Check the fluid level and add as required.
5. Connect the coil wire and start the engine. Allow it to run for several minutes.
6. Rotate the steering wheel from stop to stop.
7. Shut of the engine and check the fluid level. Add fluid as necessary.

SUSPENSION AND STEERING 8-39

TORQUE SPECIFICATIONS
WHEELS AND SUSPENSION

Component	U.S.	Metric
Axle pivot bolt		
4WD Navajo	120-150 ft. lbs.	163-204 Nm
Axle-to-radius arm nut		
2WD Navajo	120-150 ft. lbs.	163-204 Nm
Axle-to-frame pivot bracket bolt		
2WD Navajo	120-150 ft. lbs.	163-204 Nm
Ball joint stud nut		
Lower, 2WD pickups	116 ft. lbs.	158 Nm
Lower, 2WD MPV	116 ft. lbs.	158 Nm
4WD pickups	70 ft. lbs.	95 Nm
4WD pickups	115 ft. lbs.	156 Nm
2WD Navajo	85-110 ft. lbs.	116-149 Nm
Ball joint-to-arm bolts		
2WD pickups		
old	15-20 ft. lbs.	21-27 Nm
New	70 ft. lbs.	95 Nm
4WD pickups	15-20 ft. lbs.	21-27 Nm
Cap nut		
2WD MPV	36-43 ft. lbs.	49-58 Nm
Clamp bolts		
4WD pickups	19 ft. lbs.	26 Nm
2WD MPV	45 ft. lbs.	61 Nm
Compression rod end nuts		
2WD MPV	127 ft. lbs.	173 Nm
Compression rod-to-lower arm		
2WD MPV	93 ft. lbs.	126 Nm
Fluid-filled bushing bolts		
2WD MPV	76 ft. lbs.	103 Nm
Front bracket-to-axle bolts		
4WD Navajo	37-50 ft. lbs.	50-68 Nm
Height sensor link		
MPV	104 inch lbs.	12 Nm
Height sensor bolts		
MPV	20 ft. lbs.	27 Nm
Insulator-to-arm stud nut		
4WD Navajo	80-120 ft. lbs.	109-163 Nm
Knuckle arm bolts		
2WD pickups	70-74 ft. lbs.	95-100 Nm
Lateral rod bolts		
MPV	127 ft. lbs.	173 Nm
Lower arm-to-frame nut		
2WD pickups	115 ft. lbs.	156 Nm
4WD pickups	115 ft. lbs.	156 Nm
2WD MPV	108 ft. lbs.	147 Nm
Lower arm end bolts		
2WD pickups	85 ft. lbs.	116 Nm
4WD pickups	85 ft. lbs.	116 Nm
Lower link bolts		
MPV	127 ft. lbs.	173 Nm
Lower strut bolts		
2WD MPV	86 ft. lbs.	117 Nm
Lug nuts		
Pickups		
non-styled wheels	65-87 ft. lbs.	89-118 Nm
styled wheels	87-108 ft. lbs.	119-146 Nm
MPV	65-87 ft. lbs.	89-118 Nm
Navajo	100 ft. lbs.	136 Nm

TORQUE SPECIFICATIONS
WHEELS AND SUSPENSION

Component	U.S.	Metric
Mounting block		
Lower Bolts		
2WD MPV	85 ft. lbs.	116 Nm
Upper Nuts		
2WD MPV	25 ft. lbs.	34 Nm
Parking brake cable clamp		
MPV	19 ft. lbs.	26 Nm
Radius arm-to-axle bolt/nut		
2WD Navajo	160-220 ft. lbs.	218-299 Nm
Radius rod bushing end nut	90 ft. lbs.	122 Nm
Rebound spring bracket		
MPV	19 ft. lbs.	26 Nm
Shock absorber		
Pickups	55-59 ft. lbs.	75-80 Nm
Navajo		
lower	40-63 ft. lbs.	55-85 Nm
upper	25-35 ft. lbs.	34-47 Nm
MPV	76 ft. lbs.	103 Nm
Springs		
Front Shackle Bolt		
Navajo	75-115 ft. lbs.	102-156 Nm
Front Spring Pin Nut		
Pickups	72 ft. lbs.	98 Nm
Lever Retainer Nut		
4WD Navajo	35-105 ft. lbs.	95-136 Nm
Rear Shackle Nuts		
Pickups	58 ft. lbs.	79 Nm
Navajo	75-115 ft. lbs.	102-156 Nm
U-Bolt Nuts		
2-Wheel Drive Pickups	65-75 ft. lbs.	89-102 Nm
Navajo	58 ft. lbs.	79 Nm
4-wheel Drive Pickups	101 ft. lbs.	137 Nm
Stabilizer bar bolt		
2WD pickups	19 ft. lbs.	26 Nm
4WD pickups	19 ft. lbs.	26 Nm
Stabilizer bar-to-control arm nut		
2WD pickups	34 ft. lbs.	46 Nm
Stabilizer bar-to-frame clamp bolts		
2WD pickups	16 ft. lbs.	22 Nm
Stabilizer bar clamp		
MPV	38 ft. lbs.	52 Nm
Stabilizer bar link-to-stabilizer bar		
2WD Navajo	30-40 ft. lbs.	41-54 Nm
Stabilizer bar link-to-mounting bracket		
2WD Navajo	30-44 ft. lbs.	41-59 Nm
Stabilizer bar-to-link nut		
4WD Navajo	30-44 ft. lbs.	41-59 Nm
Steering linkage-to-spindle nut		
2WD Navajo	52-74 ft. lbs.	71-100 Nm
Tie rod end-to-knuckle bolt		
2WD pickups	22-29 ft. lbs.	30-39 Nm
Tie rod end nut		
2WD MPV	58 ft. lbs.	79 Nm
Torque plate bolts		
2WD pickups	68 ft. lbs.	92 Nm
4WD pickups	68 ft. lbs.	92 Nm

SUSPENSION AND STEERING

TORQUE SPECIFICATIONS

WHEELS AND SUSPENSION

Component	U.S.	Metric
Upper control arm shaft mounting bolts 2WD pickups	60-68 ft. lbs.	82-92 Nm
Upper ball joint-to-knuckle bolt 2WD pickups	35-38 ft. lbs.	48-52 Nm
Upper arm shaft mounting bolts 4WD pickups	60-68 ft. lbs.	82-92 Nm
Upper ball joint pinch bolt nut 4WD Navajo	48-65 ft. lbs.	65-88 Nm
Upper link bolts MPV	127 ft. lbs.	173 Nm

STEERING

Component	U.S.	Metric
Ball socket stud nuts Pickups	45 ft. lbs.	61 Nm
Ball stud nut MPV	58 ft. lbs.	79 Nm
Navajo	30-42 ft. lbs.	41-57 Nm
Center link nut MPV	43 ft. lbs.	58 Nm
Combination switch retaining screws	18-27 in. lbs.	2-3 Nm
Drag link Navajo	50-70 ft. lbs.	68-95 Nm
Pickups	43 ft. lbs.	58 Nm
Ignition switch nuts Navajo	3.3-5.3 ft. lbs.	5-7 Nm
Pitman arm nut Navajo	170-230 ft. lbs.	231-313 Nm
Pickups	130 ft. lbs.	177 Nm
Pitman arm-to-center link nut Pickups	43 ft. lbs.	58 Nm
Power steering pump pulley B2200	43 ft. lbs.	58 Nm
MPV	29-43 ft. lbs.	40-58 Nm
Power steering pump pressure line connection B2200	35 ft. lbs.	48 Nm
B2600	35 ft. lbs.	48 Nm
MPV	35 ft. lbs.	48 Nm
Power steering pump pressure line bracket B2200	17 ft. lbs.	23 Nm
MPV	17 ft. lbs.	23 Nm
Power steering return Line Pickups	35 ft. lbs.	48 Nm
Power steering pump mounting bolts B2200	34 ft. lbs.	46 Nm
B2600	34 ft. lbs.	46 Nm
MPV	34 ft. lbs.	46 Nm

TORQUE SPECIFICATIONS

Component	U.S.	Metric
Power steering pump adjusting bolt	34 ft. lbs.	46 Nm
Power steering pump pulley bolt MPV	43 ft. lbs.	58 Nm
Power steering pump belt idler pulley locknut MPV	38 ft. lbs.	52 Nm
Steering column bracket Pickups	16 ft. lbs.	22 Nm
Steering gear input shaft bolts Navajo	50-62 ft. lbs.	68-84 Nm
Steering gear-to-frame mounting bolts Pickups	69 ft. lbs.	94 Nm
MPV	69 ft. lbs.	94 Nm
Steering gear-to-lower bracket clamp bolts MPV	69 ft. lbs.	94 Nm
Steering shaft coupling bolt Pickups	18 ft. lbs.	24 Nm
Navajo	26-34 ft. lbs.	35-46 Nm
Steering wheel attaching nut Pickups & MPV	35 ft. lbs.	48 Nm
Steering wheel lock bolt	23-33 ft. lbs.	31-45 Nm
Tie rod ball stud nuts Navajo	50-75 ft. lbs.	68-102 Nm
MPV	58 ft. lbs.	79 Nm
Tie rod adjusting sleeve nuts Navajo	30-42 ft. lbs.	41-57 Nm
Tie rod end jam nuts Pickups	58 ft. lbs.	79 Nm
MPV	58 ft. lbs.	79 Nm
Wormshaft locknut Pickups w/manual steering	140 ft. lbs.	191 Nm
Wormshaft-to-steering shaft yoke Pickups w/manual steering	38 ft. lbs.	52 Nm

BRAKE OPERATING SYSTEM 9-2
BASIC OPERATING PRINCIPLES 9-2
  DISC BRAKES 9-2
  DRUM BRAKES 9-2
BRAKE LIGHT SWITCH 9-3
  REMOVAL & INSTALLATION 9-3
MASTER CYLINDER 9-3
  REMOVAL & INSTALLATION 9-3
POWER BOOSTER 9-3
  REMOVAL & INSTALLATION 9-3
  PUSHROD CLEARANCE 9-5
PRESSURE DIFFERENTIAL VALVE 9-5
  REMOVAL & INSTALLATION 9-5
  CENTRALIZING THE PRESSURE
    DIFFERENTIAL VALVE 9-5
LOAD SENSING G-VALVE 9-5
  REMOVAL & INSTALLATION 9-5
PROPORTIONING VALVE 9-5
  REMOVAL & INSTALLATION 9-5
BRAKE HOSES AND LINES 9-5
  REMOVAL & INSTALLATION 9-6
BRAKE BLEEDING 9-6
  BLEEDING SEQUENCE 9-6
  MANUAL BLEEDING 9-7
  FLUSHING HYDRAULIC BRAKE
    SYSTEMS 9-7
DISC BRAKES 9-7
BRAKE PADS 9-7
  INSPECTION 9-7
  REMOVAL & INSTALLATION 9-7
CALIPER 9-9
  REMOVAL & INSTALLATION 9-9
  OVERHAUL 9-10
BRAKE ROTOR 9-11
  REMOVAL & INSTALLATION 9-11
  INSPECTION 9-11
REAR DRUM BRAKES 9-12
BRAKE DRUMS 9-12
  REMOVAL & INSTALLATION 9-12
  INSPECTION 9-12
BRAKE SHOES 9-12
  INSPECTION 9-12
  REMOVAL & INSTALLATION 9-12
  ADJUSTMENTS 9-16
WHEEL CYLINDER 9-17
  REMOVAL & INSTALLATION 9-17
  OVERHAUL 9-17
PARKING BRAKE 9-18
CABLES 9-18
  REMOVAL & INSTALLATION 9-18
  ADJUSTMENT 9-19
**REAR WHEEL ANTI-LOCK BRAKE
  SYSTEM 9-20**
DESCRIPTION AND OPERATION 9-20
SYSTEM SELF TEST 9-20
  PRECAUTIONS 9-23
ELECTRONIC CONTROL UNIT 9-23
  REMOVAL & INSTALLATION 9-23
ELECTRO-HYDRAULIC VALVE 9-24
  REMOVAL & INSTALLATION 9-24
SPEED SENSOR 9-24
  TESTING 9-24
  REMOVAL & INSTALLATION 9-24
**4-WHEEL ANTI-LOCK BRAKE
  SYSTEM (4WABS) 9-25**
SYSTEM SELF-TEST 9-25
HYDRAULIC CONTROL UNIT (HCU) 9-25
  PUMP TESTING 9-25
  REMOVAL & INSTALLATION 9-25
ELECTRONIC CONTROL UNIT 9-25
  REMOVAL & INSTALLATION 9-25
FRONT WHEEL SPEED SENSOR 9-25
  TESTING 9-25
  REMOVAL & INSTALLATION 9-25
REAR SPEED SENSOR 9-26
  TESTING 9-26
  REMOVAL & INSTALLATION 9-26
SPECIFICATIONS CHART
  BRAKE SPECIFICATIONS 9-26

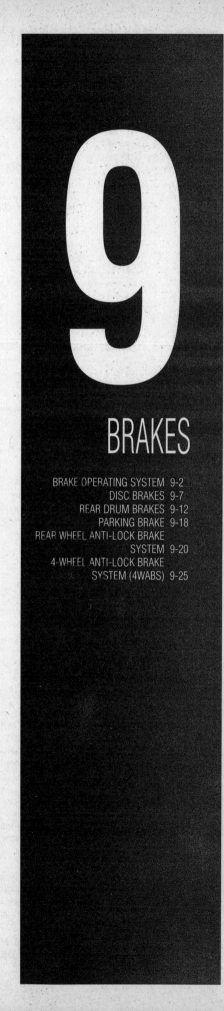

9
BRAKES

BRAKE OPERATING SYSTEM 9-2
DISC BRAKES 9-7
REAR DRUM BRAKES 9-12
PARKING BRAKE 9-18
REAR WHEEL ANTI-LOCK BRAKE
  SYSTEM 9-20
4-WHEEL ANTI-LOCK BRAKE
  SYSTEM (4WABS) 9-25

9-2 BRAKES

BRAKE OPERATING SYSTEM

Basic Operating Principles

Hydraulic systems are used to actuate the brakes of all modern automobiles. The system transports the power required to force the frictional surfaces of the braking system together from the pedal to the individual brake units at each wheel. A hydraulic system is used for two reasons.

First, fluid under pressure can be carried to all parts of an automobile by small pipes and flexible hoses without taking up a significant amount of room or posing routing problems.

Second, a great mechanical advantage can be given to the brake pedal end of the system, and the foot pressure required to actuate the brakes can be reduced by making the surface area of the master cylinder pistons smaller than that of any of the pistons in the wheel cylinders or calipers.

The master cylinder consists of a fluid reservoir along with a double cylinder and piston assembly. Double type master cylinders are designed to separate the front and rear braking systems hydraulically in case of a leak. The master cylinder coverts mechanical motion from the pedal into hydraulic pressure within the lines. This pressure is translated back into mechanical motion at the wheels by either the wheel cylinder (drum brakes) or the caliper (disc brakes).

Steel lines carry the brake fluid to a point on the vehicle's frame near each of the vehicle's wheels. The fluid is then carried to the calipers and wheel cylinders by flexible tubes in order to allow for suspension and steering movements.

In drum brake systems, each wheel cylinder contains two pistons, one at either end, which push outward in opposite directions and force the brake shoe into contact with the drum.

In disc brake systems, the cylinders are part of the calipers. At least one cylinder in each caliper is used to force the brake pads against the disc.

All pistons employ some type of seal, usually made of rubber, to minimize fluid leakage. A rubber dust boot seals the outer end of the cylinder against dust and dirt. The boot fits around the outer end of the piston on disc brake calipers, and around the brake actuating rod on wheel cylinders.

The hydraulic system operates as follows: When at rest, the entire system, from the piston(s) in the master cylinder to those in the wheel cylinders or calipers, is full of brake fluid. Upon application of the brake pedal, fluid trapped in front of the master cylinder piston(s) is forced through the lines to the wheel cylinders. Here, it forces the pistons outward, in the case of drum brakes, and inward toward the disc, in the case of disc brakes. The motion of the pistons is opposed by return springs mounted outside the cylinders in drum brakes, and by spring seals, in disc brakes.

Upon release of the brake pedal, a spring located inside the master cylinder immediately returns the master cylinder pistons to the normal position. The pistons contain check valves and the master cylinder has compensating ports drilled in it. These are uncovered as the pistons reach their normal position. The piston check valves allow fluid to flow toward the wheel cylinders or calipers as the pistons withdraw. Then, as the return springs force the brake pads or shoes into the released position, the excess fluid reservoir through the compensating ports. It is during the time the pedal is in the released position that any fluid that has leaked out of the system will be replaced through the compensating ports.

Dual circuit master cylinders employ two pistons, located one behind the other, in the same cylinder. The primary piston is actuated directly by mechanical linkage from the brake pedal through the power booster. The secondary piston is actuated by fluid trapped between the two pistons. If a leak develops in front of the secondary piston, it moves forward until it bottoms against the front of the master cylinder, and the fluid trapped between the pistons will operate the rear brakes. If the rear brakes develop a leak, the primary piston will move forward until direct contact with the secondary piston takes place, and it will force the secondary piston to actuate the front brakes. In either case, the brake pedal moves farther when the brakes are applied, and less braking power is available.

All dual circuit systems use a switch to warn the driver when only half of the brake system is operational. This switch is usually located in a valve body which is mounted on the firewall or the frame below the master cylinder. A hydraulic piston receives pressure from both circuits, each circuit's pressure being applied to one end of the piston. When the pressures are in balance, the piston remains stationary. When one circuit has a leak, however, the greater pressure in that circuit during application of the brakes will push the piston to one side, closing the switch and activating the brake warning light.

In disc brake systems, this valve body also contains a metering valve and, in some cases, a proportioning valve. The metering valve keeps pressure from traveling to the disc brakes on the front wheels until the brake shoes on the rear wheels have contacted the drums, ensuring that the front brakes will never be used alone. The proportioning valve controls the pressure to the rear brakes to lessen the chance of rear wheel lock-up during very hard braking.

Warning lights may be tested by depressing the brake pedal and holding it while opening one of the wheel cylinder bleeder screws. If this does not cause the light to go on, substitute a new lamp, make continuity checks, and, finally, replace the switch as necessary.

The hydraulic system may be checked for leaks by applying pressure to the pedal gradually and steadily. If the pedal sinks very slowly to the floor, the system has a leak. This is not to be confused with a springy or spongy feel due to the compression of air within the lines. If the system leaks, there will be a gradual change in the position of the pedal with a constant pressure.

Check for leaks along all lines and at wheel cylinders. If no external leaks are apparent, the problem is inside the master cylinder.

DISC BRAKES

Instead of the traditional expanding brakes that press outward against a circular drum, disc brake systems utilize a disc (rotor) with brake pads positioned on either side of it. An easily-seen analogy is the hand brake arrangement on a bicycle. The pads squeeze onto the rim of the bike wheel, slowing its motion. Automobile disc brakes use the identical principle but apply the braking effort to a separate disc instead of the wheel.

The disc (rotor) is a casting, usually equipped with cooling fins between the two braking surfaces. This enables air to circulate between the braking surfaces making them less sensitive to heat buildup and more resistant to fade. Dirt and water do not drastically affect braking action since contaminants are thrown off by the centrifugal action of the rotor or scraped off the by the pads. Also, the equal clamping action of the two brake pads tends to ensure uniform, straight line stops. Disc brakes are inherently self-adjusting. There are three general types of disc brake:

1. A fixed caliper.
2. A floating caliper.
3. A sliding caliper.

The fixed caliper design uses two pistons mounted on either side of the rotor (in each side of the caliper). The caliper is mounted rigidly and does not move.

The sliding and floating designs are quite similar. In fact, these two types are often lumped together. In both designs, the pad on the inside of the rotor is moved into contact with the rotor by hydraulic force. The caliper, which is not held in a fixed position, moves slightly, bringing the outside pad into contact with the rotor. There are various methods of attaching floating calipers. Some pivot at the bottom or top, and some slide on mounting bolts. In any event, the end result is the same.

DRUM BRAKES

Drum brakes employ two brake shoes mounted on a stationary backing plate. These shoes are positioned inside a circular drum which rotates with the wheel assembly. The shoes are held in place by springs. This allows them to slide toward the drums (when they are applied) while keeping the linings and drums in alignment. The shoes are actuated by a wheel cylinder which is mounted at the top of the backing plate. When the brakes are applied, hydraulic pressure forces the wheel cylinder's actuating links outward. Since these links bear directly against the top of the brake shoes, the tops of the shoes are then forced against the inner side of the drum. This action forces the bottoms of the two shoes to contact the brake drum by rotating the entire assembly slightly (known as servo action). When pressure within the wheel cylinder is relaxed, return springs pull the shoes back away from the drum.

Most modern drum brakes are designed to self-adjust themselves during application when the vehicle is moving in reverse. This motion causes both shoes to rotate very slightly with the drum, rocking an adjusting lever, thereby causing rotation of the adjusting screw. Some drum brake systems are designed to self-adjust during application whenever the brakes are applied. This on-board adjustment system reduces the need for maintenance adjustments and keeps both the brake function and pedal feel satisfactory.

Brakes 9-3

Brake Light Switch

REMOVAL & INSTALLATION

Pickups and MPV

The switch is located at the top of the brake pedal.
1. Disconnect the wiring from the switch.
2. Loosen the locknut and adjusting nut and unscrew the switch from the bracket.
3. Install the new switch. Adjust the brake pedal. Tighten the locknut.

Navajo

♦ See Figure 1

1. Lift the locking tab on the switch connector and disconnect the wiring.
2. Remove the hairpin retainer, slide the stoplamp switch, pushrod and nylon washer off of the pedal. Remove the washer, then the switch by sliding it up or down.

➡On some vehicles equipped with speed control, the spacer washer is replaced by the dump valve adapter washer.

3. To install the switch, position it so that the U-shaped side is nearest the pedal and directly over/under the pin.
4. Slide the switch up or down, trapping the master cylinder pushrod and bushing between the switch side plates.
5. Push the switch and pushrod assembly firmly towards the brake pedal arm. Assemble the outside white plastic washer to the pin and install the hairpin retainer.

➡Don't substitute any other type of retainer. Use only the specified hairpin retainer.

6. Assemble the connector on the switch.
7. Check stoplamp operation.

➡Make sure that the stoplamp switch wiring has sufficient travel during a full pedal stroke.

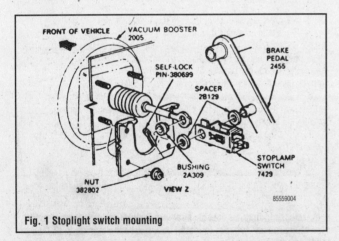

Fig. 1 Stoplight switch mounting

Master Cylinder

REMOVAL & INSTALLATION

♦ See Figures 2 and 3

➡Be careful not to spill brake fluid on the painted surfaces, it will harm the painted finish. If the reservoir is removed for any reason, install new mounting grommets. If the fluid has been emptied from the cylinder it is probably easier to bench bleed the cylinder prior to installation. See bleeding. Before installation the clearance between the power

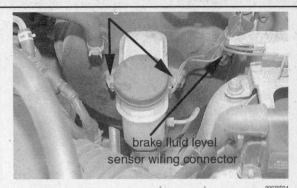

Fig. 2 Location of the master cylinder mounting nuts and fluid level sensor connector—MPV

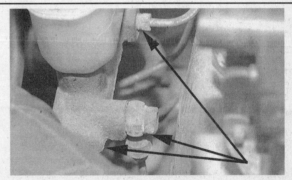

Fig. 3 Disconnect the brake lines from the master cylinder using a flare nut wrench and a box wrench—MPV

booster pushrod and the piston of the master cylinder should be check. Refer to the following booster procedures.

1. Clean all dirt and grease from the master cylinder and lines. Disconnect and cap the brake lines from the master cylinder.
2. Disconnect the fluid level sensor coupling, on models equipped.
3. Unbolt and remove the master cylinder from the firewall or power booster. Service as required. Replace the cylinder to booster mounting gasket.
4. Place the master cylinder in position and loosely install the mounting nuts. Connect the hydraulic lines to the master cylinder, but do not tighten fully at this time. Tighten the mounting nuts to specification. Tighten the hydraulic lines.
5. Fill and bleed the system. Check system operation.

Power Booster

REMOVAL & INSTALLATION

Pickups

♦ See Figure 4

1. Remove the master cylinder and proportioning bypass valve bracket (if equipped).
2. Disconnect the vacuum line at the booster.
3. Disconnect the pushrod at the brake pedal.
4. Unbolt and remove the power booster from the firewall (under the dash).
5. Installation is the reverse of removal. Check the clearance between the master cylinder piston and the power booster pushrod. Clearance should be 0. If not, adjust it at the pushrod. Torque the mounting nuts to 17 ft. lbs.

9-4 BRAKES

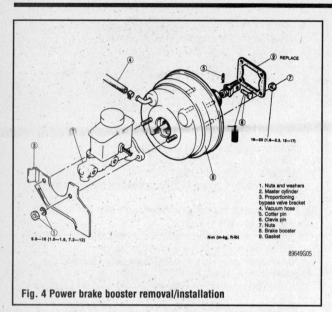

Fig. 4 Power brake booster removal/installation

MPV

6-CYLINDER ENGINE

▶ See Figure 5

1. Remove the wiper arms.
2. Remove the drive link nuts from the top of the cowl.
3. Working under the hood, disconnect the battery ground cable.
4. Remove the wiper motor and linkage mounting bolts and lift out the assembly.
5. Remove the master cylinder.
6. Disconnect the pushrod at the pedal.
7. Disconnect the vacuum line at the booster.
8. Unbolt and remove the power booster from the firewall.

To install:

9. Check the clearance between the master cylinder piston and the power booster pushrod. Clearance should be 0, but the piston should not depress the pushrod. Adjust the clearance at the pushrod.
10. Position a new mounting gasket, coated with sealant, on the firewall.
11. Position the vacuum unit on the firewall and install the nuts. Torque the mounting nuts to 19 ft. lbs.
12. Connect the pushrod at the pedal.

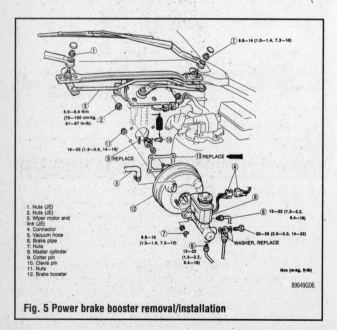

Fig. 5 Power brake booster removal/installation

13. Connect the vacuum line.
14. Install the master cylinder.
15. Install the wiper motor and linkage. When installing the wiper arms, make sure that the at-rest position gives a gap of 30mm between the blade tips and the lower windshield molding.
16. Bleed the brakes.

4-CYLINDER ENGINE

1. Disconnect the fluid sensor line.
2. Remove the master cylinder.
3. Disconnect the pushrod at the pedal.
4. Disconnect the vacuum line at the booster.
5. Unbolt and remove the power booster from the firewall.

To install:

6. Check the clearance between the master cylinder piston and the power booster pushrod. Clearance should be 0, but the piston should not depress the pushrod. Adjust the clearance at the pushrod.
7. Position a new mounting gasket, coated with sealant, on the firewall.
8. Position the vacuum unit on the firewall and install the nuts. Torque the mounting nuts to 19 ft. lbs.
9. Connect the pushrod at the pedal.
10. Connect the vacuum line.
11. Install the master cylinder.
12. Bleed the brakes.

Navajo

➡ Make sure that the booster rubber reaction disc is properly installed if the master cylinder push rod is removed or accidentally pulled out. A dislodged disc may cause excessive pedal travel and extreme operation sensitivity. The disc is black compared to the silver colored valve plunger that will be exposed after the push rod and front seal is removed. The booster unit is serviced as an assembly and must be replaced if the reaction disc cannot be properly installed and aligned, or if it cannot be located within the unit itself.

1. Disconnect the stop lamp switch wiring to prevent running the battery down.
2. Support the master cylinder from the underside with a prop.
3. Remove the master cylinder-to-booster retaining nuts.
4. Loosen the clamp that secures the manifold vacuum hose to the booster check valve, and remove the hose. Remove the booster check valve.
5. Pull the master cylinder off the booster and leave it supported by the prop, far enough away to allow removal of the booster assembly.
6. From inside the cab on vehicles equipped with push rod mounted stop lamp switch, remove the retaining pin and slide the stop lamp switch, push rod, spacers and bushing off the brake pedal arm.
7. From the engine compartment remove the bolts that attach the booster to the dash panel.

To install:

8. Mount the booster assembly on the engine side of the dash panel by sliding the bracket mounting bolts and valve operating rod in through the holes in the dash panel.

➡ Make certain that the booster push rod is positioned on the correct side of the master cylinder to install onto the push pin prior to tightening the booster assembly to the dash.

9. From inside the cab, install the booster mounting bracket-to-dash panel retaining nuts.
10. Position the master cylinder on the booster assembly, install the retaining nuts, and remove the prop from underneath the master cylinder.
11. Install the booster check valve. Connect the manifold vacuum hose to the booster check valve and secure with the clamp.
12. From inside the cab on vehicles equipped with push rod mounted stop lamp switch, install the bushing and position the switch on the end of the push rod. Then install the switch and rod on the pedal arm, along with spacers on each side, and secure with the retaining pin.
13. Connect the stop lamp switch wiring.
14. Start the engine and check brake operation.

BRAKES 9-5

PUSHROD CLEARANCE

♦ See Figure 6

1. On all except Navajo, adjust the power brake booster pushrod clearance as follows: Install clearance adjusting tool 49 F043 001 or equivalent on the rear of the master cylinder. Turn the adjusting bolt until it bottoms in the pushrod hole in the piston. Apply 19.7 in.Hg vacuum to the power brake booster with a vacuum pump. Invert the clearance adjusting tool and place it on the power brake booster. Turn the pushrod locknut until there is no clearance between the tool and the pushrod.

2. On the Navajo, measure the distance between the outer end of the booster pushrod and the front face of the booster assembly. The distance should be 0.995 inch (25.3mm). Turn the pushrod adjusting screw in or out until the distance is as specified.

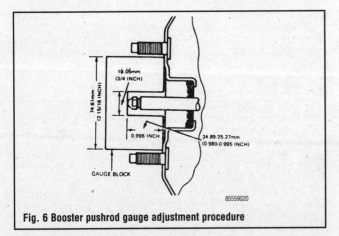

Fig. 6 Booster pushrod gauge adjustment procedure

Pressure Differential Valve

→This applies to 2-wheel drive pickups only.

REMOVAL & INSTALLATION

1. Disconnect the brake warning light switch connector, at the switch.
2. Disconnect the brake lines at the valve, and plug the lines.
3. Unbolt and remove the valve.
4. Installation is the reverse of removal. Bleed the system.

CENTRALIZING THE PRESSURE DIFFERENTIAL VALVE

After the brake system has been opened for repairs, or bled, the brake light may remain on. The pressure differential valve must be centered to make the light go off.
1. Turn the ignition switch ON, but don't start the engine.
2. Make sure that the master cylinder reservoirs are filled.
3. Slowly depress the brake pedal. The valve should center itself and the light go off. If not, bleed the brakes again and repeat the above procedure.

Load Sensing G-Valve

REMOVAL & INSTALLATION

4-Wheel Drive Pickups

1. Raise and support the front end on jackstands.
2. Disconnect and cap the brake lines.
3. Remove the valve mounting bolts.
4. Install the new valve, tighten the bolts, connect the brake lines and bleed the system.
5. Adjust the inclination angle of the valve:
   a. Place the truck on level ground.
   b. Make sure that there is no cargo in the truck and no people.
   c. Fill the tires to the recommended inflation pressure.
   d. Attach angle gauge 49 U043 003, or equivalent, to the valve. The correct angle should be 7° ± 1½°. If not:
   e. Loosen the mounting bolts and move the valve until the angle is correct.
   f. Tighten the bolts.

MPV

1. Raise and support the front end on jackstands.
2. Disconnect and cap the brake lines.
3. Remove the valve mounting bolts.
4. Install the new valve, tighten the bolts, connect the brake lines and bleed the system.
5. Adjust the inclination angle of the valve:
   a. Place the van on level ground.
   b. Make sure that there is no cargo in the van and no people.
   c. Fill the tires to the recommended inflation pressure.
   d. Attach angle gauge 49 U043 003, or equivalent, to the valve. The correct angle should be 8½° ± 1°. If not:
   e. Loosen the mounting bolts and move the valve until the angle is correct.
   f. Tighten the bolts.

Proportioning Valve

REMOVAL & INSTALLATION

MPV

♦ See Figure 7

1. Raise and support the front end on jackstands.
2. Disconnect and cap the brake lines at the valve.
3. Remove the attaching bolts.
4. Installation is the reverse of removal.
5. Bleed the system.

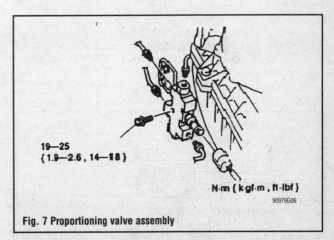

Fig. 7 Proportioning valve assembly

Brake Hoses and Lines

Metal lines and rubber brake hoses should be checked frequently for leaks and external damage. Metal lines are particularly prone to crushing and kinking under the vehicle. Any such deformation can restrict the proper flow of fluid and therefore impair braking at the wheels. Rubber hoses should be checked for cracking or scraping; such damage can create a weak spot in the hose and it could fail under pressure.

Any time the lines are removed or disconnected, extreme cleanliness must be observed. Clean all joints and connections before disassembly (use a stiff bristle brush and clean brake fluid); be sure to plug the lines and ports as soon as they are opened. New lines and hoses should be flushed clean with brake fluid before installation to remove any contamination.

9-6 BRAKES

REMOVAL & INSTALLATION

♦ See Figures 8, 9, 10 and 11

1. Disconnect the negative battery cable.
2. Raise and safely support the vehicle on jackstands.
3. Remove any wheel and tire assemblies necessary for access to the particular line you are removing.
4. Thoroughly clean the surrounding area at the joints to be disconnected.
5. Place a suitable catch pan under the joint to be disconnected.
6. Using two wrenches (one to hold the joint and one to turn the fitting), disconnect the hose or line to be replaced.
7. Disconnect the other end of the line or hose, moving the drain pan if necessary. Always use a back-up wrench to avoid damaging the fitting.
8. Disconnect any retaining clips or brackets holding the line and remove the line from the vehicle.

→If the brake system is to remain open for more time than it takes to swap lines, tape or plug each remaining clip and port to keep contaminants out and fluid in.

To install:

9. Install the new line or hose, starting with the end farthest from the master cylinder. Connect the other end, then confirm that both fittings are correctly threaded and turn smoothly using finger pressure. Make sure the new line will not rub against any other part. Brake lines must be at least 1/2 in. (13mm) from the steering column and other moving parts. Any protective shielding or insulators must be reinstalled in the original location.

✱✱ WARNING

Make sure the hose is NOT kinked or touching any part of the frame or suspension after installation. These conditions may cause the hose to fail prematurely.

10. Using two wrenches as before, tighten each fitting.
11. Install any retaining clips or brackets on the lines.
12. If removed, install the wheel and tire assemblies, then carefully lower the vehicle to the ground.
13. Refill the brake master cylinder reservoir with clean, fresh brake fluid, meeting DOT 3 specifications. Properly bleed the brake system.
14. Connect the negative battery cable.

Brake Bleeding

♦ See Figures 12 and 13

The hydraulic brake system must be free of air to operate properly. Air can enter the system when hydraulic parts are disconnected for servicing or replacement, or when the fluid level in the master cylinder reservoirs is very low. Air in the system will give the brake pedal a spongy feeling upon application.

The quickest and easiest of the two ways for system bleeding is the pressure method, but special equipment is needed to externally pressurize the hydraulic system. The other, more commonly used method of brake bleeding is done manually.

BLEEDING SEQUENCE

1. Master cylinder. If the cylinder is not equipped with bleeder screws, open the brake line(s) to the wheels slightly while pressure is applied to the brake pedal. Be sure to tighten the line before the brake pedal is released. The procedure for bench bleeding the master cylinder is covered below.
2. Pressure Differential Valve: If equipped with a bleeder screw.
3. Front/Back Split Systems: Start with the wheel farthest away from the master cylinder, usually the right rear wheel. Bleed the other rear wheel then the right front and left front.

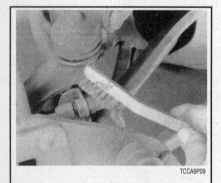
Fig. 8 Use a brush to clean the fittings of any debris

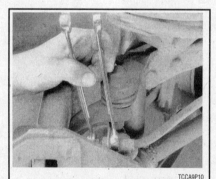
Fig. 9 Use two wrenches to loosen the fitting. If available, use flare nut type wrenches

Fig. 10 Any gaskets/crush washers should be replaced with new ones during installation

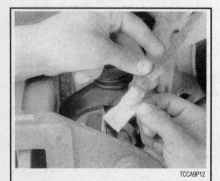
Fig. 11 Tape or plug the line to prevent contamination

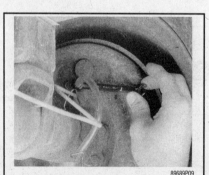
Fig. 12 Bleed the rear brakes first, ensuring that no air bubbles remain visible moving through the tubing

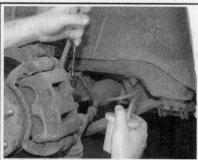
Fig. 13 Bleed the caliper until you can see clean, air bubble free brake fluid moving through the tube

BRAKES 9-7

✳✳ CAUTION

Do not allow brake fluid to spill on the truck's finish, it will remove the paint. Flush the area with water.

➡ If the vehicle is equipped with anti-lock brakes, the electro-hydraulic valve must also be bled. It is not necessary to energize the valve to bleed it.

MANUAL BLEEDING

1. Clean the bleed screw at each wheel.
2. Start with the wheel farthest from the master cylinder (right rear), or the left rear if the right side wheel cylinder is not equipped with a bleeder.
3. Attach a small rubber hose to the bleed screw and place the end in a container of clear brake fluid.
4. Fill the master cylinder with brake fluid. (Check often during bleeding). Have an assistant slowly pump up the brake pedal and hold pressure.
5. Open the bleed screw about one-quarter turn, press the brake pedal to the floor, close the bleed screw and slowly release the pedal. Continue until no more air bubbles are forced from the cylinder on application of the brake pedal.
6. Repeat procedure on remaining wheel cylinders and calipers, still working from cylinder/caliper farthest from the master cylinder.

Master cylinders equipped with bleed screws may be bled independently. When bleeding the Bendix-type dual master cylinder it is necessary to solidly cap one reservoir section while bleeding the other to prevent pressure loss through the cap vent hole.

✳✳ CAUTION

The bleeder valves must be closed at the end of each stroke, and before the brake pedal is released, to insure that no air can enter the system. It is also important that the brake pedal be returned to the full up position so the piston in the master cylinder moves back enough to clear the bypass outlets.

FLUSHING HYDRAULIC BRAKE SYSTEMS

Hydraulic brake systems must be totally flushed if the fluid becomes contaminated with water, dirt or other corrosive chemicals. To flush, simply bleed the entire system until all fluid has been replaced with the correct type of new fluid.

DISC BRAKES

✳✳ CAUTION

Brake pads may contain asbestos, which has been determined to be a cancer causing agent. Never clean the brake surfaces with compressed air! Avoid inhaling any dust from any brake surface! When cleaning brake surfaces, use a commercially available brake cleaning fluid.

Brake Pads

INSPECTION

▶ See Figure 14

The brake pads can be visually inspected after the front wheels have been removed. A cut out in the top of the caliper is provided. Check the lining thickness of both the outboard and inboard pads. If the lining is worn within 1/8 inch (3mm) of its metal backing (check local inspection requirements) replace all of the brake pads.

REMOVAL & INSTALLATION

Pickups

1. Raise and support the front end on jackstands.
2. Remove the wheels.
3. Remove the lower lock pin bolt from the caliper.
4. Rotate the caliper upward and remove the brake pads and shims.
5. Remove the master cylinder reservoir cap and remove about half of the fluid from the reservoir.
6. Using a large C-clamp, depress the caliper piston until it bottoms in its bore.
7. Install the shims and new pads.
8. Reposition the caliper and install the lock pin bolt. Torque the bolt to 30 ft. lbs. (41 Nm).
9. Install the wheels, lower the truck, refill the master cylinder and depress the brake pedal a few times to restore pressure. Bleed the system if required.

MPV

▶ See Figures 15 thru 21

1. Raise and support the front end on jackstands.
2. Remove the wheels.
3. Remove the lower lock pin bolt from the caliper.
4. Rotate the caliper upward and remove the brake pads and shims.
5. Remove the master cylinder reservoir cap and remove about half of the fluid from the reservoir.

Fig. 14 If inspecting the thickness of the brake pads, the pads are visible through the two openings in the caliper unit

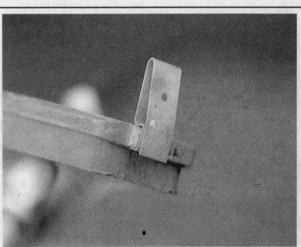

Fig. 15 The bottom end of the inner brake pad is equipped with a pad thickness sensor, which makes a squealing noise to alert you that the pads must be changed

9-8 BRAKES

Fig. 16 When removing the disc brake pads, you only have to remove the bottom caliper slide bolts

Fig. 17 After removing the bottom sliding bolt, swing up and support the caliper with a strong piece of wire. Then remove the inner and outer disc brake pads

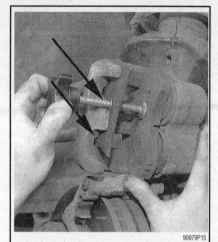
Fig. 18 Using a brake caliper piston and compressor tool and an old brake pad, push the pistons into the bore

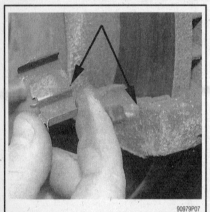
Fig. 19 If necessary, remove and clean, or replace, the four anti-rattle clips located in the caliper on the ends of each brake pad

Fig. 20 Before installing the pads and caliper, clean the anti-rattle clips with a brush

Fig. 21 Before installing the sliding caliper bolts, clean them and then lubricate them with brake grease

6. Using a large C-clamp and suitable piece of wood, depress the caliper pistons until they bottom in their bores.
7. Install the shims and new pads.
8. Reposition the caliper and install the lock pin bolt. Torque the bolt to 69 ft. lbs. (94 Nm).
9. Install the wheels, lower the vehicle, refill the master cylinder and depress the brake pedal a few times to restore pressure. Bleed the system if required.

Navajo

▶ See Figures 22, 23, 24, 25 and 26

1. To avoid fluid overflow when the caliper piston is pressed into the cylinder bore, siphon part of the brake fluid out of the master cylinder reservoir. Discard the removed fluid.
2. Raise the vehicle and install jackstands. Remove a front wheel and tire assembly.
3. Place an 8 inch (203mm) C-clamp on the caliper and tighten the clamp to move the piston into the cylinder bore about 1/8 (3mm). Avoid clamp contact with the outer pad spring clip. Place the screw end of the clamp below the spring clip. Avoid pad displacement beyond locking tab engagement. Remove the clamp.

➡Do not use a screwdriver or similar tool to pry piston away from the rotor.

4. Clean excess dirt from the areas around the pin tabs.
5. Using a 1/4 inch drive socket, 3/8 inch deep and a light hammer, tap the upper caliper pin towards the outboard side until the pin tabs pass the spindle face.
6. Place one end of a 1/6 inch (11mm) diameter punch against the end of the caliper pin and tap the pin out of the caliper slide groove.
7. Repeat Steps 5 and 6 to remove the lower pin.
8. Remove the caliper and hang in out of the way with mechanic's wire. Do not allow the caliper to hang by the brake hose.
9. Compress the anti-rattle clip and remove the inner brake pad from the caliper. Press each ear of the outer pad away from the caliper and slide the torque buttons out of the retaining notches. Remove the brake pad.

To install:

10. Bottom the caliper piston using the C-clamp. Place a piece of wood or the worn out brake pad on the piston before tightening the clamp. Do not attempt to bottom the piston with the outer pad installed. Place a new anti-rattle clip on the lower end of the inner pad. Be sure the tabs on the clip are positioned properly and the clip is fully seated.
11. Position the inner pads and anti-rattle clip in the abutment with the anti-rattle clip tab against the pad abutment and the loop-type spring away from the rotor. Compress the anti-rattle clip and slide the upper end of the pad in position.
12. Install the outer pad, making sure the torque buttons on the pad spring clip are seated solidly in the matching holes in the caliper.

BRAKES 9-9

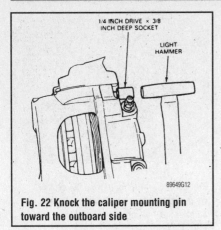
Fig. 22 Knock the caliper mounting pin toward the outboard side

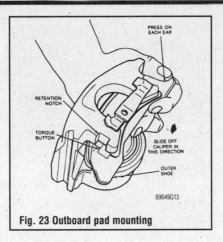
Fig. 23 Outboard pad mounting

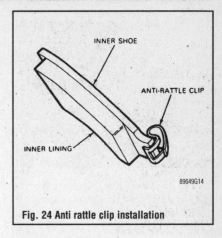
Fig. 24 Anti rattle clip installation

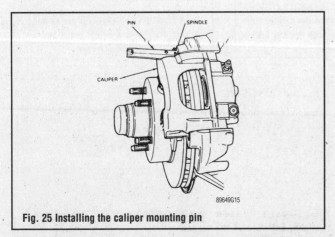
Fig. 25 Installing the caliper mounting pin

Fig. 26 Tap the mounting pin on the outboard end until the retention tabs on the side of the pin contact the spindle face

13. Install the caliper, using new pins, make sure the mounting surfaces are free of dirt and lubricate the caliper grooves with Disc Brake Caliper Grease. From the outboard side, position the upper pin between the caliper and spindle grooves. The pin must be positioned so the tabs will be installed against the spindle outer face. Tap the pin, on the outboard end with a soft hammer until the retention tabs on the sides of the pin contact the spindle face. Repeat the procedure for the lower pin.

※※ WARNING

Never reuse caliper pins. Always install new pins whenever a caliper is removed.

14. Pump the brake pedal several times to position the pads to the rotor. Fill the master cylinder.
15. Bleed the brake system if required.
16. Install the wheel and tire assembly. Torque the lug nuts to 85–115 ft. lbs. (115–156 Nm).
17. Remove the jackstands and lower the vehicle. Check the brake fluid level and fill as necessary. Check the brakes for proper operation.

Caliper

REMOVAL & INSTALLATION

Pickups and MPV

▶ See Figures 27, 28, 29 and 30

1. Raise and safely support the vehicle. Remove the wheel and tire assembly.
2. Remove the banjo bolt and disconnect the brake hose from the caliper. Plug the hose to prevent fluid leakage.
3. On pickups, remove the caliper mounting bolt and pivot the caliper about the mounting pin and off of the brake rotor. Remove the caliper from the pin.
4. On MPV, remove the caliper mounting bolts and remove the caliper.
5. To install, place the caliper in position and secure the mounting bolts following the reverse order of removal. Lubricate the caliper mounting bolts or bolt and pin prior to installation.
6. Tighten the caliper mounting bolt(s) to 23–30 ft. lbs. (31–41 Nm) on pickups; 61–69 ft. lbs. (83–93 Nm) on MPV. Bleed the brake system.

Navajo

▶ See Figure 31

1. Siphon part of the brake fluid out of the master cylinder to avoid overflow when the caliper piston is pressed into the caliper bore.
2. Raise the vehicle and support it safely. Remove the wheel and tire assembly.
3. Position an 8 inch (203mm) C-clamp on the caliper and tighten the clamp to move the caliper piston into the bore approximately 1/8 inch (3mm). Avoid clamp contact with the outer shoe spring clip. Remove the clamp.

➡ **Do not pry the piston away from the rotor.**

4. Clean excess dirt from the pin tab area.
5. Using a 1/4 inch drive socket, 3/8 inch deep and a light hammer, tap the upper caliper pin towards the outboard side until the pin tabs pass the spindle face.
6. Place one end of a 1/6 inch (11mm) diameter punch against the end of the caliper pin and tap the pin out of the caliper slide groove.
7. Repeat the procedure to remove the lower caliper pin.
8. Disconnect and plug the brake hose at the caliper. Remove the caliper from the rotor.

To install:

9. Make sure the caliper mounting surfaces are free of dirt. Lubricate the caliper grooves with disc brake caliper grease and install the caliper.

9-10 BRAKES

Fig. 27 Clean any dirt or debris from the hose and banjo bolt connection using a wire brush

Fig. 28 Remove the banjo bolt from the caliper connection

Fig. 29 Be careful not to lose the crush washer that is located between the caliper and the hose fitting

Fig. 30 Inspect the condition of the caliper sliding bolt rubber boots and replace if necessary

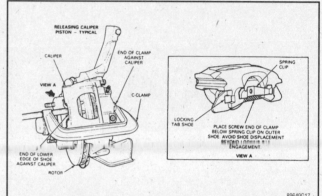
Fig. 31 Use a C-clamp to push the caliper piston back into the bore slightly before removing the caliper

10. From the caliper outboard side, position the pin between the caliper and spindle grooves. The pin must be positioned so the tabs will be installed against the spindle outer face.
11. Tap the pin on the outboard end with a hammer until the retention tabs on the sides of the pin contact the spindle face.
12. Repeat the procedure to install the lower pin.

➡ During installation, do not allow the tabs of the caliper pin to be tapped too far into the spindle groove. If this happens, it will be necessary to tap the other end of the caliper pin until the tabs snap in place. The tabs on each end of the pin must be free to catch on the spindle face.

13. Connect the brake hose to the caliper. Bleed the brake system.
14. Install the wheel and tire assembly and lower the vehicle. Check the brake fluid level and check the brakes for proper operation.

OVERHAUL

♦ See Figures 32 thru 39

➡ Some vehicles may be equipped dual piston calipers. The procedure to overhaul the caliper is essentially the same with the exception of multiple pistons, O-rings and dust boots.

1. Remove the caliper from the vehicle and place on a clean workbench.

✷✷ CAUTION

NEVER place your fingers in front of the pistons in an attempt to catch or protect the pistons when applying compressed air. This could result in personal injury!

➡ Depending upon the vehicle, there are two different ways to remove the piston from the caliper. Refer to the brake pad replacement procedure to make sure you have the correct procedure for your vehicle.

2. The first method is as follows:
   a. Stuff a shop towel or a block of wood into the caliper to catch the piston.
   b. Remove the caliper piston using compressed air applied into the caliper inlet hole. Inspect the piston for scoring, nicks, corrosion and/or worn or damaged chrome plating. The piston must be replaced if any of these conditions are found.
3. For the second method, you must rotate the piston to retract it from the caliper.
4. If equipped, remove the anti-rattle clip.
5. Use a prytool to remove the caliper boot, being careful not to scratch the housing bore.
6. Remove the piston seals from the groove in the caliper bore.
7. Carefully loosen the brake bleeder valve cap and valve from the caliper housing.
8. Inspect the caliper bores, pistons and mounting threads for scoring or excessive wear.
9. Use crocus cloth to polish out light corrosion from the piston and bore.
10. Clean all parts with denatured alcohol and dry with compressed air.

To assemble:

11. Lubricate and install the bleeder valve and cap.
12. Install the new seals into the caliper bore grooves, making sure they are not twisted.
13. Lubricate the piston bore.
14. Install the pistons and boots into the bores of the calipers and push to the bottom of the bores.
15. Use a suitable driving tool to seat the boots in the housing.
16. Install the caliper in the vehicle.
17. Install the wheel and tire assembly, then carefully lower the vehicle.
18. Properly bleed the brake system.

BRAKES 9-11

Fig. 32 For some types of calipers, use compressed air to drive the piston out of the caliper, but make sure to keep your fingers clear

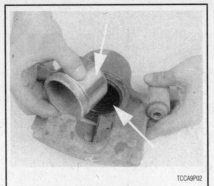

Fig. 33 Withdraw the piston from the caliper bore

Fig. 34 On some vehicles, you must remove the anti-rattle clip

Fig. 35 Use a prytool to carefully pry around the edge of the boot . . .

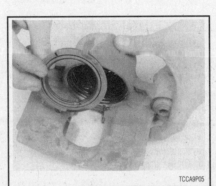

Fig. 36 . . . then remove the boot from the caliper housing, taking care not to score or damage the bore

Fig. 37 Use extreme caution when removing the piston seal; DO NOT scratch the caliper bore

Fig. 38 Use the proper size driving tool and a mallet to properly seal the boots in the caliper housing

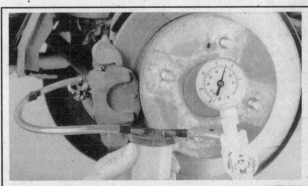

Fig. 39 There are tools, such as this Mighty-Vac, available to assist in proper brake system bleeding

Brake Rotor

REMOVAL & INSTALLATION

1. Raise and safely support the vehicle. Remove the wheel and tire assembly.
2. Remove the caliper and support it aside with mechanic's wire; do not let the caliper hang by the brake hose. On all except Navajo, remove the disc brake pads and mounting support.
3. On 2WD Pickups and Navajo, remove the dust cap, cotter pin, nut, washer and outer bearing and remove the rotor from the spindle.
4. On 4WD Pickups, remove the locking hub or drive flange, snapring and spacer, set bolts and bearing set plate. Remove the bearing locknut using a suitable puller and remove the hub and rotor assembly, being careful not to let the washer and bearing fall.
5. On 4WD Navajo, remove the locking hub and remove the brake rotor.
6. On MPV models, remove the attaching screw and remove the rotor.
7. Inspect the rotor for scoring, wear and runout. Machine or replace as necessary.
8. If rotor replacement is necessary on pickup models, remove the attaching bolts and separate the rotor from the hub.
9. Install the rotor. On B Series pickup, tighten the rotor-to-hub bolts to 40–51 ft. lbs. (54–69 Nm). Adjust the wheel bearings.

INSPECTION

▶ See Figure 40

Brake roughness is a shudder, vibration or pedal pulsation occurring during braking operation. It may be caused by a foreign material build-up on the rotor braking surfaces or by excessive rotor variation or distortion.

9-12 BRAKES

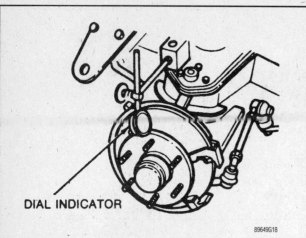

Fig. 40 Use a dial indicator to check rotor runout

If there is a foreign material build-up or contamination found on the rotor or lining surfaces, hand sand both the linings and rotor braking surfaces. Brake squeal or squeak is a higher frequency vibration which can result from foreign material build-up or glazing of the lining and rotor surfaces. Hand sanding the rotor and linings may correct the condition.

Always replace the linings if they are excessively worn. Check the caliper piston for roll back release. A sticking caliper piston can cause rapid lining wear and possible rotor damage. Check the sliding motion of the caliper on the mounting pinrails. Service as necessary.

Rotors should be replaced for structural damage to the hub casting, for excessive rotor wear when the minimum rotor thickness is below spec, and for heavy scoring or excessive runout.

Lathe turn the rotor to remove minor rotor scoring or scratching that cannot be removed by hand sanding. Deep scoring usually occurs when the lining is worn down to the point where the rivets or metal pad backing contacts the rotor.

Minor rotor runout may be corrected by lathe turning. To check rotor runout, first eliminate wheel bearing end play by correctly adjusting the wheel bearing. Clamp a dial indicator to the steering knuckle assembly so that the stylus contacts the rotor about one inch from the outer edge. Rotate the rotor and take an indicator reading. If the reading exceeds 0.008–0.010 inch total lateral runout within a six-inch radius, replace or resurface the rotor.

REAR DRUM BRAKES

✱✱ CAUTION

Brake shoes may contain asbestos, which has been determined to be a cancer causing agent. Never clean the brake surfaces with compressed air! Avoid inhaling any dust from any brake surface! When cleaning brake surfaces, use a commercially available brake cleaning fluid.

Brake Drums

REMOVAL & INSTALLATION

Pickups and MPV

▶ See Figures 41, 42 and 43

1. Raise and support the rear end on jackstands.
2. Remove the wheels.
3. Remove the drum attaching screws and insert them in the threaded holes in the drum. Turn the screws inward, evenly, to force the rum off the hub.
4. Thoroughly inspect the drum. Discard a cracked drum. If the drum is suspected of being out of round, or shows signs of wear or has a ridged or rough surface, have it turned on a lathe at a machine shop. The maximum oversize is stamped into the drum.
5. Installation is the reverse of removal. Make sure that the holes are aligned for the attaching screws. Tighten the screws evenly to install the drum.

Navajo

1. Raise the vehicle so that the wheel to be worked on is clear of the floor and install jackstands under the vehicle.
2. Remove the hub cap and the wheel/tire assembly. Remove the 3 retaining nuts and remove the brake drum. It may be necessary to back off the brake shoe adjustment in order to remove the brake drum. This is because the drum might be grooved or worn from being in service for an extended period of time.
3. Before installing a new brake drum, be sure and remove any protective coating with carburetor degreaser.
4. Install the brake drum in the reverse order of removal and adjusts the brakes.

INSPECTION

▶ See Figure 44

After the brake drum has been removed from the vehicle, it should be inspected for runout, severe scoring cracks, and the proper inside diameter.

Minor scores on a brake drum can be removed with fine emery cloth, provided that all grit is removed from the drum before it is installed on the vehicle.

A badly scored, rough, or out-of-round (runout) drum can be ground or turned on a brake drum lathe. Do not remove any more material from the drum than is necessary to provide a smooth surface for the brake shoe to contact. The maximum diameter of the braking surface is shown on the inside of each brake drum. Brake drums that exceed the maximum braking surface diameter shown on the brake drum, either through wear or refinishing, must be replaced. This is because after the outside wall of the brake drum reaches a certain thickness (thinner than the original thickness) the drum loses its ability to dissipate the heat created by the friction between the brake drum and the brake shoes, when the brakes are applied. Also the brake drum will have more tendency to warp and/or crack.

The maximum braking surface diameter specification, which is shown on each drum, allows for a 0.060 inch (1.5mm) machining cut over the original nominal drum diameter plus 0.030 inch (0.76mm) additional wear before reaching the diameter where the drum must be discarded. Use a brake drum micrometer to measure the inside diameter of the brake drums.

Brake Shoes

INSPECTION

After wheel and brake drum removal, visually inspect the brake shoe lining for wear. Replace the brake shoe set if the lining is worn to 1/32 inch (0.8mm) above the rivet head or the shoes metal webbing (check local inspection requirements).

REMOVAL & INSTALLATION

▶ See Figure 41

2-Wheel Drive Pickups

1. Raise and support the rear end on jackstands.
2. Remove the drums.
3. Remove the retracting springs.
4. Remove the holddown springs and guide pins by turning the collars 90° with a pliers, or spring tool, releasing the springs.
5. Remove the parking brake link and disconnect the parking brake cable from the lever.
6. Remove the adjusting pawl and spring.
7. Remove the shoes, noting in which place the shoe with the longer lining is installed.
8. Inspect the shoes for cracks, heat checking or contamination by oil or grease. Minimum lining thickness is 1.00mm (0.039 in.). If heat checking or

BRAKES 9-13

REAR DRUM BRAKE COMPONENTS

1. Secondary shoe
2. Adjusting screw assembly
3. Primary shoe
4. Adjuster spring
5. Adjuster lever
6. Hold-down pin
7. Hold-down spring
8. Hold-down assembly
9. Adjuster cable guide
10. Parking brake lever
11. Parking brake link
12. Link spring
13. Primary shoe return spring
14. Anchor pin plate
15. Secondary shoe return spring

9-14 BRAKES

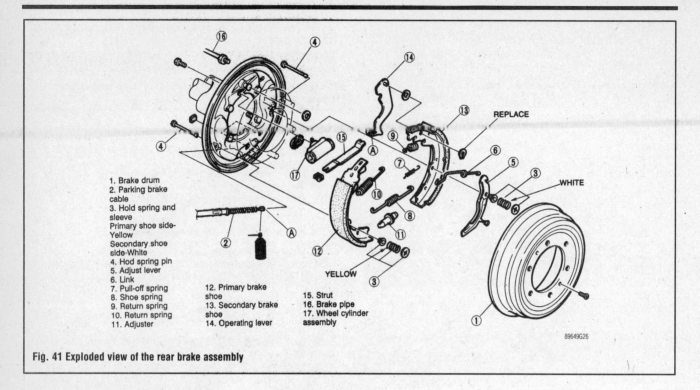

Fig. 41 Exploded view of the rear brake assembly

1. Brake drum
2. Parking brake cable
3. Hold spring and sleeve
   Primary shoe side-Yellow
   Secondary shoe side-White
4. Hod spring pin
5. Adjust lever
6. Link
7. Pull-off spring
8. Shoe spring
9. Return spring
10. Return spring
11. Adjuster
12. Primary brake shoe
13. Secondary brake shoe
14. Operating lever
15. Strut
16. Brake pipe
17. Wheel cylinder assembly

Fig. 42 Loosen the rear drum mounting screw. If the screw will not rotate very easily, an impact driver can be utilized first

Fig. 43 Once the mounting screw is removed, pull the drum right off

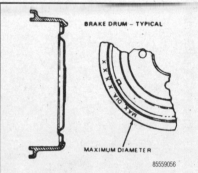

Fig. 44 Drum brake service limits. The maximum inside diameter is cast into the drum

discoloration is noted, the wheel cylinders are probably at fault and will have to be rebuilt or replaced.

➡ **Never replace the shoes on one side of the truck, only! Always replace shoes on both sides!**

9. Clean the backing plate with an approved cleaning fluid.
10. Lubricated the threads of the starwheel with lithium based or silicone based grease. Apply a small dab of lithium or silicone based grease to the pads on which the brake shoes ride.

To install:
11. Transfer the parking brake lever to the new shoe.
12. Installation is the reverse of the removal procedure.
13. Adjust the brake shoes.

➡ **The adjustment should be the same on both wheels.**

14. Adjust the parking brake.
15. Operate the brake pedal a few times. If the brakes feel at all spongy, bleed the system.

4-Wheel Drive Pickups

1. Raise and support the rear end on jackstands.
2. Remove the wheels.
3. Remove the brake drum.
4. Disconnect the parking brake cable.
5. Remove the holddown spring assemblies.
6. Remove the self-adjusting lever.
7. Remove the lever link.
8. Remove the pull-off spring.
9. Remove the lower shoe-to-shoe spring.
10. Remove the 2 upper return springs.
11. Remove the star-wheel adjuster.
12. Remove the brake shoes and strut.
13. Remove the parking brake cable lever.
14. Inspect the shoes for cracks, heat checking or contamination by oil or grease. Minimum lining thickness is 1.00mm (0.039 in.). If heat checking or discoloration is noted, the wheel cylinders are probably at fault and will have to be rebuilt or replaced.

✱✱ CAUTION

Never replace the shoes on one side of the truck, only! Always replace shoes on both sides!

15. Clean the backing plate with an approved cleaning fluid.
16. Lubricated the threads of the starwheel with lithium based or silicone based grease. Apply a small dab of lithium or silicone based grease to the pads on which the brake shoes ride.

BRAKES 9-15

To install:

17. Transfer the parking brake lever to the new shoe.
18. Installation is the reverse of the removal procedure.
19. Adjust the brake shoes.

➡ **The adjustment should be the same on both wheels.**

20. Adjust the parking brake.
21. Operate the brake pedal a few times. If the brakes feel at all spongy, bleed the system.

MPV

♦ See Figures 45, 46 and 47

1. Raise and support the rear end on jackstands.
2. Remove the wheels.
3. Remove the brake drum.
4. Disconnect the parking brake cable.
5. Remove the holddown spring assemblies.
6. Remove the self-adjusting lever.
7. Remove the lever link.
8. Remove the pull-off spring.
9. Remove the lower shoe-to-shoe spring.
10. Remove the 2 upper return springs.
11. Remove the star-wheel adjuster.
12. Remove the brake shoes and strut.
13. Remove the parking brake cable lever.
14. Inspect the shoes for cracks, heat checking or contamination by oil or grease. Minimum lining thickness is 1.00mm. If heat checking or discoloration is noted, the wheel cylinders are probably at fault and will have to be rebuilt or replaced.

※※ **CAUTION**

Never replace the shoes on one side of the truck, only! Always replace shoes on both sides!

15. Clean the backing plate with an approved cleaning fluid.
16. Lubricated the threads of the starwheel with lithium based or silicone based grease. Apply a small dab of lithium or silicone based grease to the pads on which the brake shoes ride.

To install:

17. Transfer the parking brake lever to the new shoe.
18. Installation is the reverse of the removal procedure.
19. Adjust the brake shoes.

➡ **The adjustment should be the same on both wheels.**

20. Adjust the parking brake.
21. Operate the brake pedal a few times. If the brakes feel at all spongy, bleed the system.

Navajo

♦ See Figure 48

1. Raise and safely support the vehicle. Remove the wheel and tire assembly and the brake drum.
2. Pull backward on the adjusting lever cable to disengage the adjusting lever from the adjusting screw. Move the outboard side of the adjusting screw upward and back off the pivot nut as far as it will go.
3. Pull the adjusting lever, cable and automatic adjuster spring down and toward the rear to unhook the pivot hook from the large hole in the secondary shoe web. Do not pry the pivot hook from the hole.
4. Remove the automatic adjuster spring and adjusting lever.
5. Remove the secondary shoe-to-anchor spring using a suitable brake spring removal/installation tool. Using the tool, remove the primary shoe-to-anchor spring and unhook the cable anchor. Remove the anchor pin plate, if equipped.
6. Remove the cable guide from the secondary shoe.
7. Remove the shoe hold-down springs, shoes, adjusting screw, pivot nut and socket. Note the color and position of each hold-down spring so they can be reassembled in the same position.
8. Remove the parking brake link and spring. Disconnect the parking brake cable from the parking brake lever.
9. Remove the secondary brake shoe. Remove the retainer clip and spring washer and remove the parking brake lever.

To install:

10. Clean the backing plate ledge pads and sand lightly. Apply a light coating of high temperature lithium grease to the points where the brake shoes touch the backing plate.
11. Install the parking brake lever on the secondary shoe and secure with the spring washer and retaining clip.
12. Position the brake shoes on the backing plate and install the hold-down spring pins, springs and cups. Install the parking brake link, spring and washer. Connect the parking brake cable to the parking brake lever.

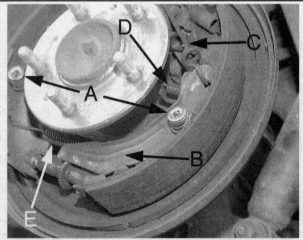

Fig. 45 Remove the following components: (A) hold-down springs, (B) self-adjusting lever, (C) lever link, (D) pull-off spring, (E) lower shoe-to-shoe spring

Fig. 46 Remove the 2 upper return springs

Fig. 47 Remove the (A) star-wheel adjuster, then remove the (B) brake shoes and strut

Fig. 48 Apply a small dab of lube on the backing plate platforms

9-16 BRAKES

13. Install the anchor pin plate, if equipped, and place the cable anchor over the anchor pin with the crimped side toward the backing plate.
14. Install the primary shoe-to-anchor spring using the brake spring removal/installation tool.
15. Install the cable guide on the secondary shoe with the flanged hole fitted into the hole in the secondary shoe. Thread the cable around the cable guide groove.

➥ Make sure the cable is positioned in the groove and not between the guide and shoe web.

16. Install the secondary shoe-to-anchor (long) spring.

➥ Make sure the cable end is not cocked or binding on the anchor pin when installed. All parts should be flat on the anchor pin.

17. Apply high temperature lithium grease to the threads and the socket end of the adjusting screw. Turn the adjusting screw into the adjusting pivot nut to the end of the threads and then loosen, ½ turn.
18. Place the adjusting socket on the screw and install the assembly between the shoe ends with the adjusting screw nearest the secondary shoe.

➥ Be sure to install the adjusting screw on the same side of the vehicle from which it came. To prevent incorrect installation, the socket end of each adjusting screw is stamped with R or L, to indicate installation on the right or left side of the vehicle. The adjusting pivot nuts have lines machined around the body of the nut, 2 lines indicating the right side nut and 1 line indicating the left side nut.

19. Hook the cable hook into the hole in the adjusting lever from the outboard plate side. The adjusting levers are also stamped with an R or L to indicate right or left side installation.
20. Place the hooked end of the adjuster spring in the large hole in the primary shoe web and connect the loop end of the spring to the adjuster lever hole.
21. Pull the adjuster lever, cable and automatic adjuster spring down toward the rear to engage the pivot hook in the large hole in the secondary shoe web.
22. After installation, check the action of the adjuster by pulling the section of the cable between the cable guide and the adjusting lever toward the secondary shoe web far enough to lift the lever past a tooth on the adjusting screw wheel. The lever should snap into position behind the next tooth and releasing the cable should cause the adjuster spring to return the lever to its original position. This return action will turn the adjusting screw 1 tooth.
23. If pulling the cable does not produce the action described in Step 22 or if lever action is sluggish instead of positive and sharp, check the position of the lever on the adjusting screw toothed wheel. With the brake in a vertical position, anchor at the top, the lever should contact the adjusting wheel 1 tooth above the center line of the adjusting screw. If the contact point is below the center line, the lever will not lock on the adjusting screw wheel teeth and the screw will not turn as the lever is actuated by the cable.
24. To find the cause of the condition described in Step 23, proceed as follows:
    a. Check the cable and fittings. The cable should completely fill or extend slightly beyond the crimped section of the fittings. If this does not happen, the cable assembly may be damaged and should be replaced.
    b. Check the cable guide for damage. The cable groove should be parallel to the shoe web and the body of the guide should lie flat against the web. Replace the guide if it shows damage.
    c. Check the pivot hook on the lever. The hook surfaces should be square with the body on the lever for proper pivoting. Repair the hook or replace the lever if the hook shows damage.
    d. Be sure the adjusting screw socket is properly seated in the notch in the shoe web.
25. Adjust the brake shoes using either a brake adjustment gauge or manually with the drums installed.
26. If using a brake adjustment gauge, proceed as follows:
    a. Measure the inside diameter of the brake drum with the gauge.
    b. Reverse the tool and adjust the brake shoes until they touch the gauge. The gauge contact points on the shoes must be parallel to the vehicle with the center line through the center of the axle.
    c. Install the drum and wheel and tire assembly. Lower the vehicle.
    d. Apply the brakes sharply several times while driving the vehicle in reverse. Check brake operation by making several stops while driving forward.
27. If manually adjusting the brakes, proceed as follows:
    a. Install the brake drum and wheel and tire assembly.
    b. Remove the cover from the adjusting hole at the bottom of the backing plate and turn the adjusting screw, using a suitable brake adjusting tool, to expand the brake shoes until they drag against the brake drum.
    c. When the shoes are against the drum, insert a narrow prybar through the brake adjusting hole and disengage the adjusting lever from the adjusting screw. While holding the adjusting lever away from the adjusting screw, loosen the adjusting screw with the brake adjusting tool, until the drum rotates freely without drag.
    d. Install the adjusting hole cover and lower the vehicle.
    e. Apply the brakes. If the pedal travels more than halfway to the floor, there is too much clearance between the brake shoes and drums. Repeat the adjustment procedure.

ADJUSTMENTS

◆ See Figures 49, 50 and 51

The drum brakes are self-adjusting and require a manual adjustment only after the brake shoes have been replaced.

➥ Disc brakes are not adjustable.

To adjust the rear brakes with drums installed, follow the procedure given below:

1. Raise the vehicle and support it with safety stands.
2. Remove the rubber plug from the adjusting slot on the backing plate.
3. Turn the adjusting screw using a Brake Shoe Adjustment Tool or equivalent inside the hole to expand the brake shoes until they drag against the brake drum and lock the drum.
4. Insert a small screwdriver or piece of firm wire (coat hanger wire) into the adjusting slot and push the automatic adjusting lever out and free of the starwheel on the adjusting screw and hold it there.
5. Engage the topmost tooth possible on the starwheel with the brake adjusting spoon. Move the end of the adjusting spoon upward to move the adjusting

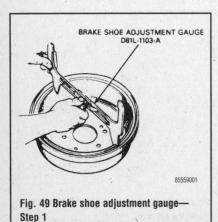

Fig. 49 Brake shoe adjustment gauge—Step 1

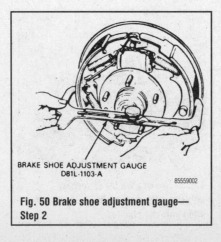

Fig. 50 Brake shoe adjustment gauge—Step 2

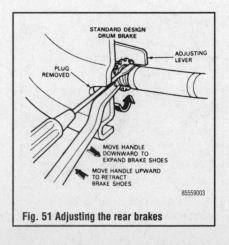

Fig. 51 Adjusting the rear brakes

BRAKES 9-17

screw starwheel downward and contract the adjusting screw. Back off the adjusting screw starwheel until the wheel spins FREELY with a minimum of drag about 10 to 12 notches. Keep track of the number of turns that the starwheel is backed off, or the number of strokes taken with the brake adjusting spoon.

6. Repeat this operation for the other side. When backing off the brakes on the other side, the starwheel adjuster must be backed off the same number of turns to prevent side-to-side brake pull.

7. When all drum brakes are adjusted, remove the safety stands and lower the vehicle and make several stops while backing the vehicle, to equalize the brakes at all of the wheels.

8. Road test the vehicle. PERFORM THE ROAD TEST ONLY WHEN THE BRAKES WILL APPLY AND THE VEHICLE CAN BE STOPPED SAFELY!

Wheel Cylinder

REMOVAL & INSTALLATION

♦ See Figures 52 and 53

1. Raise and support the rear end on jackstands.
2. Remove the brake drum and shoes.
3. Disconnect and plug the brake line(s) at the wheel cylinder.
4. Remove the attaching nuts from behind the backing plate and remove the wheel cylinder.
5. Installation is the reverse of removal.

OVERHAUL

♦ See Figures 54 thru 63

Wheel cylinder overhaul kits may be available, but often at little or no savings over a reconditioned wheel cylinder. It often makes sense with these components to substitute a new or reconditioned part instead of attempting an overhaul.

If no replacement is available, or you would prefer to overhaul your wheel cylinders, the following procedure may be used. When rebuilding and installing wheel cylinders, avoid getting any contaminants into the system. Always use clean, new, high quality brake fluid. If dirty or improper fluid has been used, it will be necessary to drain the entire system, flush the system with proper brake fluid, replace all rubber components, then refill and bleed the system.

1. Remove the wheel cylinder from the vehicle and place on a clean workbench.
2. First remove and discard the old rubber boots, then withdraw the pistons. Piston cylinders are equipped with seals and a spring assembly, all located behind the pistons in the cylinder bore.
3. Remove the remaining inner components, seals and spring assembly. Compressed air may be useful in removing these components. If no compressed air is available, be VERY careful not to score the wheel cylinder bore when removing parts from it. Discard all components for which replacements were supplied in the rebuild kit.
4. Wash the cylinder and metal parts in denatured alcohol or clean brake fluid.

※※ WARNING

Never use a mineral-based solvent such as gasoline, kerosene or paint thinner for cleaning purposes. These solvents will swell rubber components and quickly deteriorate them.

5. Allow the parts to air dry or use compressed air. Do not use rags for cleaning, since lint will remain in the cylinder bore.
6. Inspect the piston and replace it if it shows scratches.
7. Lubricate the cylinder bore and seals using clean brake fluid.
8. Position the spring assembly.
9. Install the inner seals, then the pistons.
10. Insert the new boots into the counterbores by hand. Do not lubricate the boots.
11. Install the wheel cylinder.

Fig. 52 Remove the brake shoes and the brake line, then remove the bolts and tilt the cylinder inwards . . .

Fig. 53 . . . then lift it up and off of the backing plate

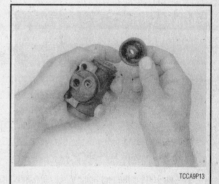

Fig. 54 Remove the outer boots from the wheel cylinder

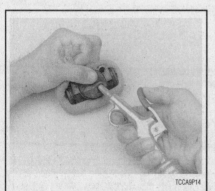

Fig. 55 Compressed air can be used to remove the pistons and seals

Fig. 56 Remove the pistons, cup seals and spring from the cylinder

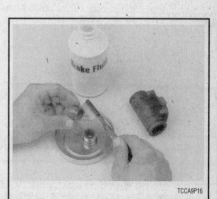

Fig. 57 Use brake fluid and a soft brush to clean the pistons . . .

9-18 BRAKES

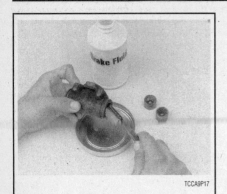

Fig. 58 . . . and the bore of the wheel cylinder

Fig. 59 Once cleaned and inspected, the wheel cylinder is ready for assembly

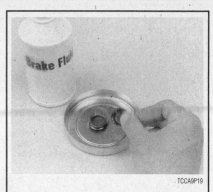

Fig. 60 Lubricate the cup seals with brake fluid

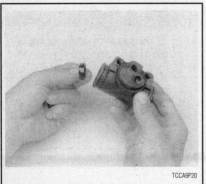

Fig. 61 Install the spring, then the cup seals in the bore

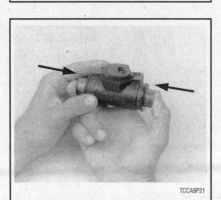

Fig. 62 Lightly lubricate the pistons, then install them

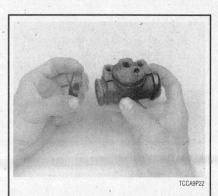

Fig. 63 The boots can now be installed over the wheel cylinder ends

PARKING BRAKE

Cables

REMOVAL & INSTALLATION

♦ See Figures 64 and 65

Front

PICKUPS AND MPV

1. Make sure the parking brake is fully released. Remove the parking brake lever adjusting nut from the forward end of the front cable.
2. Remove the seat(s) and roll back the front format, as required. On MPV, remove the cable cover. Raise and safely support the vehicle, as necessary.
3. Disengage the rear cables from the equalizer and remove the spring. Disconnect the front cable from the equalizer.
4. Remove the bolts from the cable retaining straps and remove the cable.
5. Install in reverse order. Adjust the parking brake.

NAVAJO

1. Raise and safely support the vehicle.
2. Back off the equalizer nut and remove the cable end of the intermediate cable from the tension limiter.
3. Remove the intermediate cable from the bracket and disconnect the intermediate cable from the front cable.
4. Lower the vehicle. Remove the forward ball end of the parking brake cable from the control assembly clevis.
5. Remove the cable from the control assembly.
6. Using a cord attached to the control lever end of the cable, remove the cable from the vehicle pulling it up into the passenger compartment.

To install:

7. Install in reverse order. Adjust the parking brake.

Rear

PICKUPS AND MPV

♦ See Figures 64 and 65

1. Make sure the parking brake is fully released. Loosen the parking brake lever adjusting nut.
2. Remove the seat(s) and roll back the front format, as required. On MPV, remove the cable cover.
3. Raise and safely support the vehicle. Disconnect the rear cable from the equalizer.
4. Remove the rear wheel and tire assembly, brake drum and brake shoes. Disconnect the cable from the backing plate.
5. Remove the bolts from the cable retaining straps and disconnect the spring from the cable. Remove the cable.
6. Install in reverse order. Adjust the parking brake.

NAVAJO

♦ See Figure 66

1. Release the parking brake control.
2. Raise and safely support the vehicle. Remove the wheel and tire assembly, brake drum and brake shoes.
3. Remove the locknut on the threaded rod at the equalizer. Disconnect the rear parking brake cable from the equalizer.
4. Compress the prongs that retain the cable housing to the frame bracket or crossmember and pull out the cable and housing.
5. Working on the wheel side of the backing plate, compress the prongs on the cable retainer so they can pass through the hole in the brake backing plate.
6. Lift the cable out of the slot in the parking brake lever, attached to the secondary brake shoe, and remove the cable through the brake backing plate hole.

BRAKES 9-19

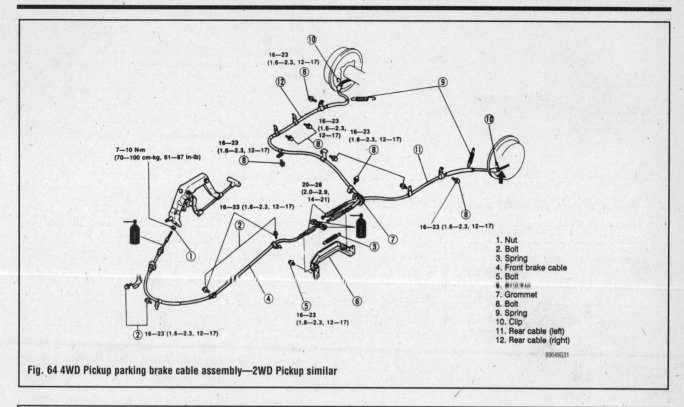

Fig. 64 4WD Pickup parking brake cable assembly—2WD Pickup similar

1. Nut
2. Bolt
3. Spring
4. Front brake cable
5. Bolt
6.
7. Grommet
8. Bolt
9. Spring
10. Clip
11. Rear cable (left)
12. Rear cable (right)

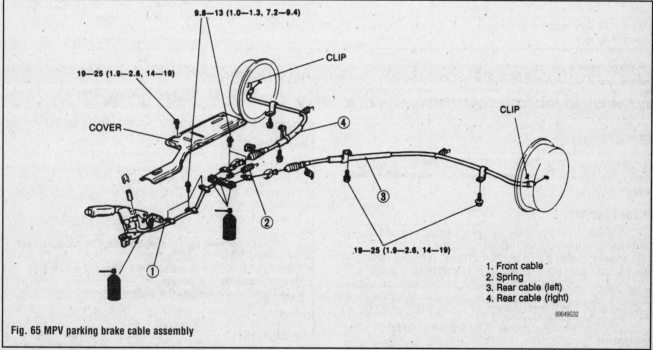

Fig. 65 MPV parking brake cable assembly

1. Front cable
2. Spring
3. Rear cable (left)
4. Rear cable (right)

To install:

7. Route the cable through the hole in the backing plate. Insert the cable anchor behind the slot in the parking brake lever. Make sure the cable is securely engaged in the parking brake lever so the cable return spring is holding the cable in the parking brake lever.
8. Push the retainer through the hole in the backing plate so the retainer prongs engage the backing plate.
9. Properly route the cable and insert the front of the cable through the frame bracket or crossmember until the prongs expand. Connect the rear cables to the equalizer.
10. Rotate the equalizer 90° and recouple the threaded rod to the equalizer.
11. Install the brake shoes, brake drum and wheel and tire assembly. Adjust the rear brakes.
12. Adjust the parking brake tension using the initial adjustment or the field adjustment procedure, as necessary.
13. Apply and release the parking brake control several times. Rotate both wheels to make sure the parking brakes are applied and released and not dragging.

ADJUSTMENT

Pickups and MPV

♦ See Figures 64, 65 and 67

1. Make sure the rear brake shoes are properly adjusted.
2. Start the engine and depress the brake pedal several times while the vehicle is moving in reverse.

9-20 BRAKES

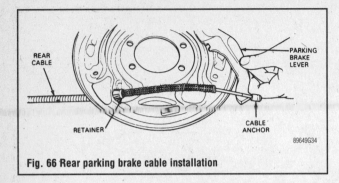

Fig. 66 Rear parking brake cable installation

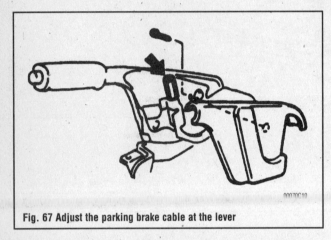

Fig. 67 Adjust the parking brake cable at the lever

3. Stop the engine.
4. On MPV, remove the screw and remove the parking brake lever cover. Remove the adjusting nut clip.

5. On B Series pickup, loosen the locknut at the end of the front cable, near the parking brake lever.
6. Turn the adjusting nut until the parking brake is fully applied when the lever is pulled 7–12 notches on B Series pickup or 5–7 notches on MPV.
7. Tighten the locknut on B Series pickup.
8. Install the adjusting nut clip and the parking brake lever cover on MPV.

Navajo

➡ Adjust the drum brakes before adjusting the parking brake. The brake drums must be cold for correct adjustment.

INITIAL ADJUSTMENT

Use this procedure when a new tension limiter is installed.
1. Apply the parking brake pedal to the fully engaged position.
2. Raise and safely support the vehicle, as necessary. Hold the threaded rod end of the right brake cable to keep it from spinning and thread the equalizer nut 2½ inches (63.5mm) up the rod.
3. Check to make sure the cinch strap has slipped and there are less than 1⅜ inches (35mm) remaining.
4. Release the parking brake and check for proper operation.

FIELD ADJUSTMENT

Use this procedure to correct a slack system if a new tension limiter is not installed.
1. Apply the parking brake pedal to the fully engaged position.
2. Raise and safely support the vehicle, as necessary. Grip the threaded rod to keep it from spinning and tighten the equalizer nut 6 full turns past its original position on the threaded rod.
3. Attach a suitable cable tension gauge in front of the equalizer assembly on the front cable and measure the cable tension. The cable tension should be 400–600 lbs. (181.5–272 kg) with the parking brake pedal in the last detent position. If tension is low, repeat Steps 2 and 3.
4. Release parking brake and check for rear wheel drag. There should be no brake drag.

REAR WHEEL ANTI-LOCK BRAKE SYSTEM

A Rear Wheel Anti-Lock Brake System is provided on Pickups, MPV and 1991–92 Navajo models.

Description and Operation

▶ See Figures 68 and 69

The rear wheel anti-lock system continually monitors rear wheel speed with a sensor mounted on the rear axle. When the teeth on an excitor ring, mounted on the differential ring gear, pass the sensor pole piece, an AC voltage is induced in the sensor circuit with a frequency proportional to the average rear wheel speed. In the event of an impending lockup condition during braking, the anti-lock system modulates hydraulic pressure to the rear brakes inhibiting rear wheel lockup.

When the brake pedal is applied, a control module senses the drop in rear wheel speed. If the rate of deceleration is too great, indicating that wheel lockup is going to occur, the module activates the electro-hydraulic valve causing the isolation valve to close. With the isolation closed, the rear wheel cylinders are isolated from the master cylinder and the rear brake pressure cannot increase. If the rate of deceleration is still too great, the module will energize the dump solenoid with a series of rapid pulses to bleed off rear cylinder fluid into an accumulator built into the electro-hydraulic valve. This will reduce the rear wheel cylinder pressure and allow the rear wheels to spin back to the vehicle speed. Continuing under module control, the dump and isolation solenoids will be pulsed in a manner that will keep the rear wheels rotating while still maintaining high levels of deceleration during braking.

At the end of the stop, when the operator releases the brake pedal, the isolation valve de-energizes and any fluid in the accumulator is returned to the master cylinder. Normal brake operation is resumed.

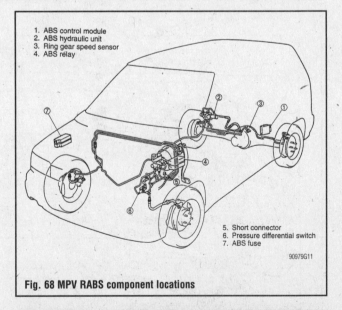

1. ABS control module
2. ABS hydraulic unit
3. Ring gear speed sensor
4. ABS relay
5. Short connector
6. Pressure differential switch
7. ABS fuse

Fig. 68 MPV RABS component locations

System Self Test

▶ See Figures 70 thru 76

The control module performs system tests and self-tests during start up and normal operation. The valve, sensor, and fluid level circuits are monitored for proper operation. If a fault is found, the anti-lock system will be deactivated and

Brakes 9-21

RABS Component Location

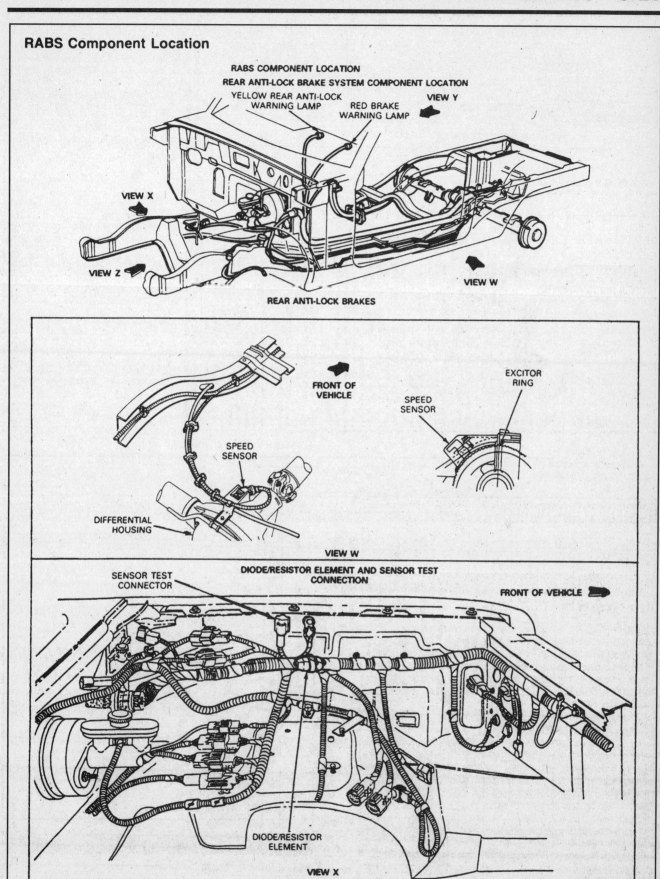

Fig. 69 Navajo RABS component locations

9-22 BRAKES

FLASHOUT CODES CHART	
CONDITION	
No Flashout Code	
Yellow REAR ANTILOCK Light Flashes 1 Time This Code Should Not Occur	
Yellow REAR ANTILOCK Light Flashes 2 Times Open Isolate Circuit	
Yellow REAR ANTILOCK Light Flashes 3 Times Open Dump Circuit	
Yellow REAR ANTILOCK Light Flashes 4 Times Red Brake Warning Light Illuminated RABS Valve Switch Closed or Open Dump Valve	
Yellow REAR ANTILOCK Light Flashes 5 Times System Dumps Too Many Times in 2WD Condition Occurs While Making Normal or Hard Stops. Rear Brake May Lock	
Yellow REAR ANTILOCK Light Flashes 6 Times (Sensor Signal Rapidly Cuts In and Out) Condition Only Occurs While Driving	
Yellow REAR ANTILOCK Light Flashes 7 Times No Isolate Valve Self Test	
Yellow REAR ANTILOCK Light Flashes 8 Times No Dump Valve Self Test	
Yellow REAR ANTILOCK Light Flashes 9 Times High Sensor Resistance	
Yellow REAR ANTILOCK Light Flashes 10 Times Low Sensor Resistance	
Yellow REAR ANTILOCK Light Flashes 11 Times Stop Lamp Switch Circuit Defective. Condition Indicated Only When Driving Above 5 mph	
Yellow REAR ANTILOCK Light Flashes 12 Times Low Brake Fluid Level Detected During Antilock Stop	
Yellow REAR ANTILOCK Light Flashes 13 Times Speed Processor Check	
Yellow REAR ANTILOCK Light Flashes 14 Times Program Check	
Yellow REAR ANTILOCK Light Flashes 15 Times Memory Failure	
Yellow REAR ANTILOCK Light Flashes 16 Times or More 16 or More Flashes Should Not Occur	

CAUTION: WHEN CHECKING RESISTANCE IN THE RABS SYSTEM, ALWAYS DISCONNECT THE BATTERY. IMPROPER RESISTANCE READINGS MAY OCCUR WITH THE VEHICLE BATTERY CONNECTED.

Fig. 71 RABS trouble code index—Navajo models

Number of flashing	Failure location	Failure condition
1	—	(1 flash should not occur)
2	Hydraulic unit	Open in isolation solenoid circuit
3	Hydraulic unit	Open in dump solenoid circuit
4		Solenoid valve switch closed
5	—	System dumps too many times in 4x2 (4x2 and 4x4 vehicles) (condition occurs while making normal or hard stops. Rear brake may lock.)
6	Speed sensor	(Speed sensor signal rapidly cuts in and out) condition only occurs while driving
7	Hydraulic unit	Shorted ground circuit (Isolation solenoid)
8		Shorted ground circuit (Dump solenoid)
9	Speed sensor	High speed sensor resistance
10		Low speed sensor resistance
11	Stoplight switch	Stoplight switch circuit defective. (Condition indicated only when driving above 56 km/h [35 mph])
12	—	(12 flashes should not occur)
13	Control unit	Control unit speed circuit phase lock loop failure detected during self-test
14		Control unit program check sum failure detected during self-test
15		Control unit RAM failure detected during self-test
16	—	(16 or more flashes should not occur)

Fig. 70 RABS trouble code index—Pickup and MPV models

BRAKES 9-23

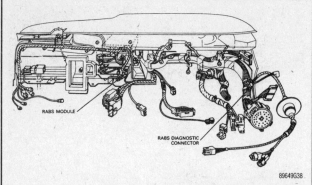

Fig. 72 Location of the RABS diagnostic connector(s)—Navajo models

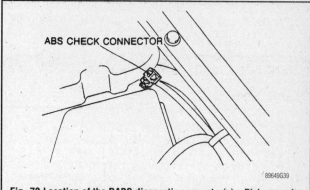

Fig. 73 Location of the RABS diagnostic connector(s)—Pickup models

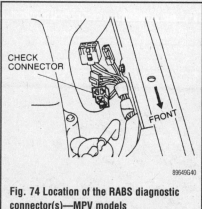

Fig. 74 Location of the RABS diagnostic connector(s)—MPV models

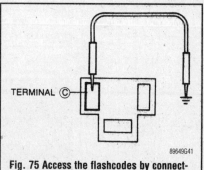

Fig. 75 Access the flashcodes by connecting a jumper wire between terminal C and ground of the diagnostic connector—Pickup and MPV models

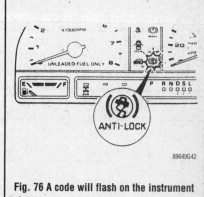

Fig. 76 A code will flash on the instrument cluster

the warning light will be illuminated. Most faults will cause the warning light to stay on until the ignition is turned off. While the light is illuminated a diagnostic flashout code may be obtained. However, there are certain faults (those associated with the fluid level switch or loss of power to the module), which cause the system to be deactivated and the warning light to be illuminated, but will not provide a diagnostic flashout code.

In most cases, the code will be lost if the vehicle is shut off. In other cases, the code may reappear when the vehicle is restarted, or the vehicle may have to be driven to reproduce the problem. If the problem was associated with an intermittent condition, it may be difficult to reproduce. Whenever possible, the code should be read before the vehicle is shut off.

✲✲ CAUTION

Place blocks behind the rear wheels and in front of the front wheels to prevent the vehicle from moving while the flashout code is being taken.

If the red brake light is also on, due to a grounding of the fluid level circuit (perhaps low brake fluid), no flashout code will be flashed and the anti-lock warning lamp will remain on steadily. If there is more than one system fault, only the first recognized code may be obtained.

A flashout code may be obtained only when the anti-lock warning light is on. No code will be flashed when the system is OK.

To check the anti-lock warning light for normal operation, insert the key in the ignition lock and turn it to the ON or START positions. The light should perform a self-check, glowing for about two seconds.

To obtain the flashout code, keep the ignition key in the **ON** position.

✲✲ CAUTION

Place blocks behind the rear wheels and in front of the front wheels to prevent the vehicle from moving while the flashout code is being taken.

1. Locate the diagnostic connector. On pickups and MPVs, the blue three pin connector is located in the left of the engine compartment. See the illustration for Navajo connector, which is located on the main wire bundle inside of cab under the dash, slightly rearward driver side.
2. Attach a jumper wire to the terminal C. On pickups, the yellow wire; on MPVs, the gray/white wire.
3. On Navajo models, attach a jumper wire to the connector with the black/orange wire.
4. Momentarily ground the jumper to the chassis. When the ground is made and broken, the anti-lock warning lamp should begin to flash.

➡ A flashing pattern consists of short flashes and ends with a long flash. Count the short flashes and include the long flash in the count. A same flashing pattern repeats until the ignition switch is turned OFF. Count the flash sequence several times to verify the number of flashes.

PRECAUTIONS

Use caution when disassembling any hydraulic components as the system will contain residual pressure. Cover the area around the component to be removed with a shop cloth to catch any brake fluid spray. Do not allow brake fluid to come in contact with painted surfaces.

Electronic Control Unit

REMOVAL & INSTALLATION

1. Disconnect the negative battery cable.
2. On B Series pickup, remove the driver's seat. On MPV, remove the inside trim panel from the left rear of the vehicle. The control unit is located under the dash on Navajo.
3. Disconnect the electrical connector from the control unit.

9-24 BRAKES

4. Remove the attaching screws and remove the control unit.
5. Installation is the reverse of the removal procedure. Check the system for proper operation.

Electro-Hydraulic Valve

REMOVAL & INSTALLATION

1. Disconnect the negative battery cable.
2. Disconnect and plug the 2 brake lines connected to the valve.
3. Disconnect the wiring harness from the valve harness.
4. Remove the screw(s) retaining the valve and remove the valve.
5. Installation is the reverse of the removal procedure. Tighten the valve retaining screw(s) to 14–19 ft. lbs. (19–25 Nm) on Pickup/MPV and 11–14 ft. lbs. (15–20 Nm) on Navajo. Bleed the brake system.

Speed Sensor

TESTING

▶ See Figure 77

1. Remove the RABS sensor from the axle housing.
2. Connect a Digital Volt/Ohm Meter (DVOM) across the two sensor terminals and record the reading.

Fig. 78 To remove the rear wheel sensor, first disconnect the electrical harness plug . . .

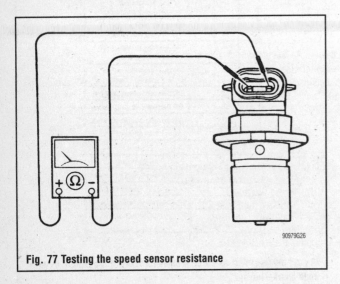
Fig. 77 Testing the speed sensor resistance

3. The reading should be between 0.9 and 2.5k Ohms.
4. If not, replace the sensor.

REMOVAL & INSTALLATION

▶ See Figures 78, 79 and 80

1. Thoroughly clean the axle housing around the sensor.
2. Disconnect the electrical harness plug from the sensor.
3. Remove the sensor hold-down bolt.
4. Remove the sensor by pulling it straight out of the axle housing.
5. Ensure that the axle surface is and that no dirt can enter the housing.
6. If a new sensor is being installed, lubricate the O-ring with clean engine oil. If the old sensor is being installed, clean it thoroughly and install a new O-ring coated with clean engine oil. Carefully push the sensor into the housing aligning the mounting flange hole with the threaded hole in the housing.
7. Tighten the hold-down bolt to 25–30 ft. lbs. (34–40 Nm) for the Navajo and 12–16 ft. lbs. (16–22 Nm) for the Pickup/MPV.

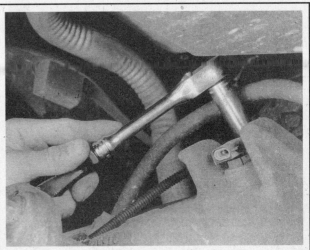
Fig. 79 . . . then remove the sensor-to-axle housing hold-down bolt

Fig. 80 Pull the sensor up and out of the axle housing

BRAKES 9-25

4-WHEEL ANTI-LOCK BRAKE SYSTEM (4WABS)

This system is used on the 1993 Navajo.
The 4WABS system consists of the following components:
a. The anti-lock Hydraulic Control Unit (HCU)
b. 4WABS electronic control module
c. The front wheel speed sensors—located on the steering knuckles
1. The rear speed sensor—located on the rear axle housing.
2. The front speed sensor tone rings
3. The rear speed sensor tone rings
4. The G-switch
5. The ABS main system and pump motor relays

System Self-Test

▶ See Figure 81

The 4WABS module performs system tests and self-tests during startup and normal operation. The valve, sensor and fluid level circuits are monitored for proper operation. If a fault is found, the 4WABS will be deactivated and the amber ANTI LOCK light will be lit until the ignition is turned OFF. When the light is lit, the Diagnostic Trouble Code (DTC) may be obtained. Under normal operation, the light will stay on for about 2 seconds while the ignition switch is in the ON position and will go out shortly after.

The Diagnostic Trouble Codes (DTC) are an alphanumeric code and a scan tool, such as the Special Service Tool NGS (New Generation Star) Tester 49-T088-0A0, is required to retrieve the codes. The DTC code chart has been included.

Hydraulic Control Unit (HCU)

PUMP TESTING

1. Disconnect the HCU pump motor electrical plug.
2. Connect a Digital Volt/Ohm Meter (DVOM) across the two pump terminals and record the reading.
3. The reading should be between 52–68 Ohms
4. If not, replace the pump motor.

REMOVAL & INSTALLATION

1. Disconnect the battery ground cable.
2. Unplug the 8-pin connector from the unit, and the 4-pin connector from the pump motor.
3. Disconnect the 5 inlet and outlet tubes from the unit. Immediately plug the ports.
4. Remove the 3 unit attaching nuts and lift out the unit.
5. Installation is the reverse of removal. Torque the mounting nuts to 12–18 ft. lbs. and the tube fittings to 10–18 ft. lbs.

➡ After reconnecting the battery, it may take 10 miles or more of driving for the Powertrain Control Module to relearn its driveability codes.

6. Bleed the brakes.

Electronic Control Unit

REMOVAL & INSTALLATION

1. Disconnect the battery ground cable.
2. Unplug the wiring from the ECU.
3. Remove the mounting bolts, slide the ECU off its bracket.
4. Installation is the reverse of removal. Torque the mounting screw to 5–6 ft. lbs. and the connector bolt to 4–5 ft. lbs.

➡ After reconnecting the battery, it may take 10 miles or more of driving for the Powertrain Control Module to relearn its driveability codes.

Service Code	Component
11	ECU Failure
16	System OK
17	Reference Voltage
22	Front Left Inlet Valve
23	Front Left Outlet Valve
24	Front Right Inlet Valve
25	Front Right Outlet Valve
26	Rear Axle Inlet Valve
27	Rear Axle Outlet Valve
31	FL Sensor Electrical Failure
32	FR Sensor Electrical Failure
33	RA Sensor Electrical Failure
35	FL Sensor Erratic Output
36	FR Sensor Erratic Output
37	RA Sensor Erratic Output
41	FL Sensor Mismatched Output
42	FR Sensor Mismatched Output
43	RA Sensor Mismatched Output
—	Front Left Valve Pair Function Check
—	Front Right Valve Pair Function Check
—	Rear Axle Valve Pair Function Check
55	FL Sensor Output Dropout
56	FR Sensor Output Dropout
57	RA Sensor Output Dropout
63	Pump Motor
65	G Switch
67	Pump Motor
75	FL Sensor Erratic Output
76	FR Sensor Erratic Output
77	RA Sensor Erratic Output
No Code Obtained	No ECU Initialization

Fig. 81 4-Wheel Anti-lock Brake System (4WABS) diagnostic trouble code chart—Navajo

Front Wheel Speed Sensor

TESTING

1. Disconnect the speed sensor wire harness plug from the sensor or sensor pigtail.
2. Connect a Digital Volt/Ohm Meter (DVOM) across the two sensor terminals and record the reading.
3. The reading should be within 0.270–0.330k ohms.

REMOVAL & INSTALLATION

▶ See Figure 82

1. Inside the engine compartment, disconnect the sensor from the harness.
2. Unclip the sensor cable from the brake hose clips.
3. Remove the retaining bolt from the spindle and slide the sensor from its hole.

9-26 BRAKES

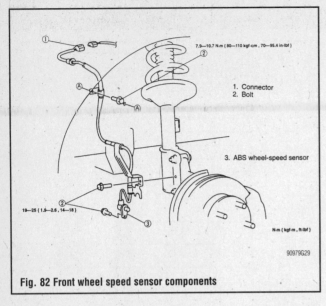

Fig. 82 Front wheel speed sensor components

4. Installation is the reverse of removal. Tighten the retaining bolt to 40–60 inch lbs. (5–7 Nm).

Rear Speed Sensor

TESTING

1. Remove the rear wheel speed sensor from the axle housing.
2. Connect a Digital Volt/Ohm Meter (DVOM) across the two sensor terminals and record the reading.
3. The reading should be between 0.9 and 2.5k Ohms.
4. If not, replace the sensor.

REMOVAL & INSTALLATION

1. Disconnect the wiring from the harness.
2. Remove the sensor hold-down bolt and remove the sensor from the axle.

To install:

3. Thoroughly clean the mounting surfaces. Make sure no dirt falls into the axle. Clean the magnetized sensor pole piece. Metal particles can cause sensor problems. Replace the O-ring.
4. Coat the new O-ring with clean engine oil.
5. Position the new sensor on the axle. It should slide into place easily. Correct installation will allow a gap of 0.005–0.045 inch.
6. Torque the hold-down bolt to 25–30 ft. lbs. (34–41 Nm).
7. Connect the wiring.

BRAKE SPECIFICATIONS
All measurements in inches unless noted

Year	Model	Master Cylinder Bore	Brake Disc Original Thickness	Brake Disc Minimum Thickness	Brake Disc Maximum Runout	Brake Drum Diameter Original Inside Diameter	Max. Wear Limit	Maximum Machine Diameter	Minimum Lining Thickness Pad	Minimum Lining Thickness Shoe
1987	B2200	0.875	0.790	0.710	0.006	10.24	10.30	NA	0.118	0.040
	B2600	0.875	④	0.790	0.006	10.24	10.30	NA	0.118	0.040
1988	B2200	0.875	0.790	0.710	0.006	10.24	10.30	NA	0.118	0.040
	B2600	0.875	④	0.790	0.006	10.24	10.30	NA	0.118	0.040
1989	B2200	0.875	0.790	0.710	0.006	10.24	10.30	NA	0.118	0.040
	B2600i	0.875	④	0.790	0.006	10.24	10.30	NA	0.118	0.040
	MPV	0.940	0.940	0.870	0.004	10.24	10.30	NA	0.080	0.040
1990	B2200	0.875	0.790	0.710	0.006	10.24	10.30	NA	0.118	0.040
	B2600i	0.875	④	①	0.006	10.24	10.30	NA	0.118	0.040
	MPV	0.940	0.940	0.870	0.004	10.24	10.30	NA	0.080	0.040
1991	B2200	0.875	0.790	0.710	0.006	10.24	10.30	NA	0.118	0.040
	B2600i	0.875	④	①	0.006	10.24	10.30	NA	0.118	0.040
	MPV	0.940	③	②	0.004	10.24	10.30	NA	0.080	0.040
	Navajo	0.937	0.850	0.810	0.003	10.00	10.09	10.06	0.120	0.030
1992	B2200	0.875	0.790	0.710	0.006	10.24	10.30	NA	0.118	0.040
	B2600i	0.875	④	①	0.006	10.24	10.30	NA	0.118	0.040
	MPV	0.940	③	②	0.004	10.24	10.30	NA	0.080	0.040
	Navajo	0.937	0.850	0.810	0.003	10.00	10.09	10.06	0.120	0.030
1993	B2200	0.875	0.790	0.710	0.006	10.24	10.30	NA	0.118	0.040
	B2600i	0.875	④	①	0.006	10.24	10.30	NA	0.118	0.040
	MPV	0.940	③	②	0.004	10.24	10.30	NA	0.080	0.040
	Navajo	0.937	0.850	0.810	0.003	10.00	10.09	10.06	0.120	0.030

NA Not Available
① 2WD-0.710 inch
  4WD-0.790 inch
② 2WD-1.10 inch
  4WD-1.02 inch
③ 2WD-1.18 inch
  4WD-1.10 inch
④ 2WD-0.790 inch
  4WD-0.870 inch

EXTERIOR 10-2
DOORS 10-2
    ADJUSTMENT 10-2
HOOD 10-2
    ALIGNMENT 10-2
TAILGATE 10-3
    REMOVAL & INSTALLATION 10-3
LIFTGATE 10-3
    REMOVAL & INSTALLATION 10-3
    ALIGNMENT 10-5
GRILLE 10-5
    REMOVAL & INSTALLATION 10-5
OUTSIDE MIRRORS 10-5
    REMOVAL & INSTALLATION 10-5
ANTENNA 10-6
    REPLACEMENT 10-6
FENDERS 10-6
    REMOVAL & INSTALLATION 10-6
CHASSIS AND CAB MOUNT
  BUSHINGS 10-6
    REMOVAL & INSTALLATION 10-6
POWER SUNROOF 10-8
    REMOVAL & INSTALLATION 10-8
INTERIOR 10-9
INSTRUMENT PANEL AND PAD 10-9
    REMOVAL & INSTALLATION 10-9
FLOOR CONSOLE 10-11
    REMOVAL & INSTALLATION 10-11
DOOR PANELS 10-11
    REMOVAL & INSTALLATION 10-11
DOOR AND LIFTGATE LOCKS 10-13
    REMOVAL & INSTALLATION 10-13
DOOR GLASS AND REGULATOR 10-14
    REMOVAL & INSTALLATION 10-14
ELECTRIC WINDOW MOTOR 10-15
    REMOVAL & INSTALLATION 10-15
WINDSHIELD AND FIXED GLASS 10-15
    REMOVAL & INSTALLATION 10-15
    WINDSHIELD CHIP REPAIR 10-15
INSIDE REAR VIEW MIRROR 10-16
SEATS 10-16
    REMOVAL & INSTALLATION 10-16
POWER SEAT MOTOR 10-18
    REMOVAL & INSTALLATION 10-18
SPECIFICATIONS CHART
  TORQUE SPECIFICATIONS 10-18

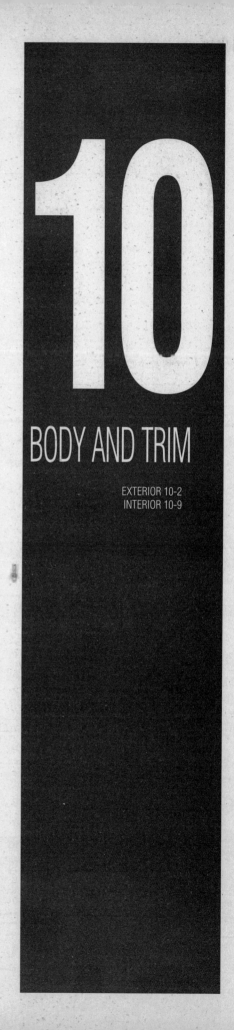

10

BODY AND TRIM

EXTERIOR 10-2
INTERIOR 10-9

10-2 BODY AND TRIM

EXTERIOR

Doors

ADJUSTMENT

▶ See Figures 1 and 2

When checking door alignment, look carefully at each seam between the door and body. The gap should be constant and even all the way around the door. Pay particular attention to the door seams at the corner farthest from the hinges, this is the area where errors will be most evident. Additionally, the door should pull against the weatherstrip when latched to seal out wind and water. The contact should be even all the way around and the stripping should be about half compressed.

The position of the door can be adjusted in three dimensions: fore and aft, up and down, in and out. The primary adjusting points are the hinge-to-body bolts. Apply tape to the fender (or door pillar and quarter panel) and door edges to protect the paint. Two layers of common masking tape works well. Loosen the bolts just enough to allow the hinge to move. With the help of an assistant, position the door and tighten the bolts. Inspect the door seams carefully and repeat the adjustment until correctly aligned.

The in-out adjustment (how far the door "sticks out" from the body) is adjusted by loosening the hinge-to-door bolts. Again, loosen the bolts, move the door into place, tighten the bolts. The dimension affects both the amount of crush on the weatherstrip and the amount of "bite" on the striker.

Further adjustment for closed position and smoothness of latching is made at the latch plate or striker. This piece is located at the rear edge of the door and is attached to the body work; it is the piece the door latch engages when the door is closed.

1. To adjust the striker: Loosen the large cross-pointed screws mounting the striker to the body. The striker mounting screws are usually very tight; an impact screwdriver is a handy tool to have for this job. Make sure you are using the proper size bit. On models equipped with a screw in post striker, loosen the post slightly counterclockwise.

2. With bolts, or post just loose enough to allow the striker to move, hold the outer door handle in the released position and close the door. The striker should move into the correct location to match the door latch. Open the door and tighten the bolts or post. The striker may be aligned towards or away from the center of the vehicle, thereby tightening or loosening the door fit.

The striker can be moved up and down to compensate for door position, but if the door is mounted correctly at the hinges this should not be necessary. Do not attempt to correct height variations (sag) by adjusting the striker.

3. Additionally, some models may use one or more spacers or shims behind the striker. These shims may be removed or added in combination to adjust the reach of the striker.

4. After tightening the striker bolts or post, open and close the door several times. Observe the motion of the door as it engages the striker; it should continue straight-in motion and deflect up or down as it hits the striker.

5. Check the feel of the latch during opening and closing. It must be smooth and linear, without any trace of grinding or binding during engagement or release.

It may be necessary to repeat the striker adjustment several times (and possibly re-adjust the hinges) before correct door to body match is produced.

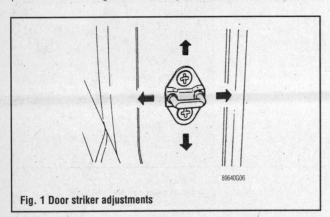

Fig. 1 Door striker adjustments

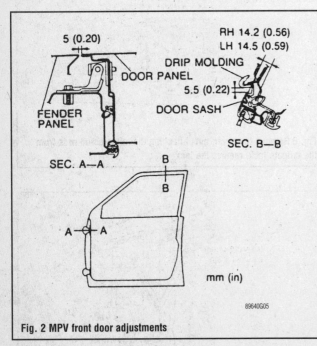

Fig. 2 MPV front door adjustments

Hood

ALIGNMENT

Pickups and MPV

▶ See Figures 3, 4 and 5

On self-supporting hoods, alignment can be adjusted front-to-rear or side-to-side by loosening the hood-to-hinge or hinge-to-body bolts. The front edge of the hood can be adjusted for closing height by adding or deleting shims under the hinges. The rear edge of the hood can be adjusted for closing height by raising or lowering the rear hood bumpers.

On hoods supported with a prop, alignment is accomplished by loosening the lock plate bolts and moving the lock plate up or down; side-to-side.

Navajo

▶ See Figures 3 and 4

1. Open the hood and match mark the hinge and latch positions.
2. Loosen the hinge-to-hood bolts just enough to allow movement of the hood.
3. Move the hood as required to obtain the proper fit and alignment between the hood and all adjoining body panels. When satisfactorily aligned, tighten the mounting bolts. To raise or lower the hood, loosen the hinge to body attaching bolts and raise or lower the hood as required.

BODY AND TRIM 10-3

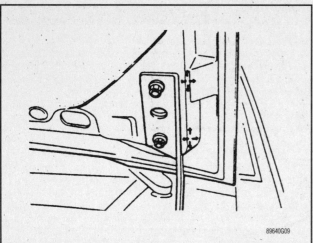

Fig. 3 Loosen these bolts to adjust the hood front and rear, side to side alignment

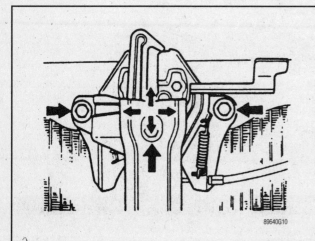

Fig. 4 Adjust the hood lock assembly after the hood has been aligned

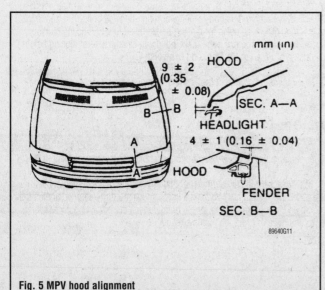

Fig. 5 MPV hood alignment

4. Loosen the latch attaching bolts.
5. Move the latch from side-to-side to align the latch with the striker. Tighten the latch bolts.
6. Lubricate the latch and hinges and check the hood fit several times.

Tailgate

REMOVAL & INSTALLATION

▶ See Figures 6 and 7

1. Open and support the tailgate.
2. Remove the hinge pins by removing the cotter pins and washers, then drive the hinge pins out.
3. Lift off the tailgate.
4. Position the tailgate and install the mounting bolts.

Liftgate

REMOVAL & INSTALLATION

MPV

▶ See Figure 8

1. Disconnect the battery ground cable.
2. Open the tailgate.
3. Disconnect the wiring connector.
4. Disconnect the washer hose.
5. Remove the strut-to-tailgate bolts.
6. Match mark the hinge position.
7. Support the tailgate and remove the hinge bolts. Lift off the tailgate.

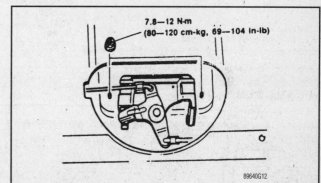

Fig. 6 Remove the cover, nuts attaching the tailgate and rods from the tailgate lock, remove the lock

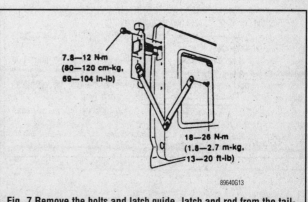

Fig. 7 Remove the bolts and latch guide, latch and rod from the tailgate

10-4 BODY AND TRIM

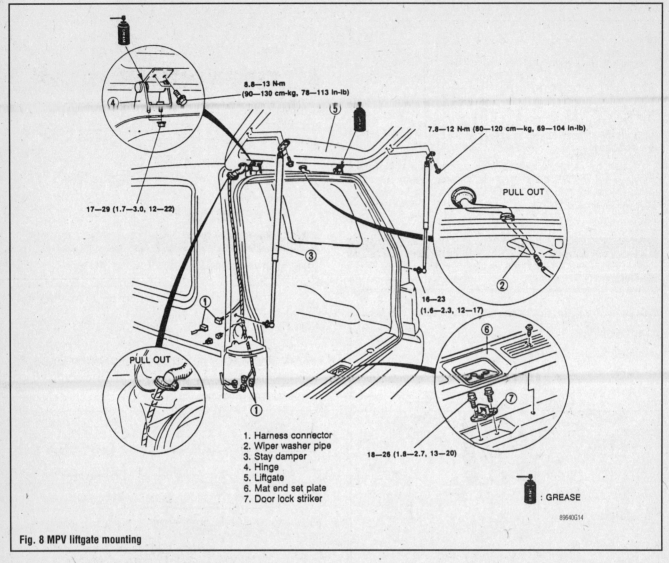

Fig. 8 MPV liftgate mounting

8. To install, position the liftgate and install the mounting bolts. Torque the hinge-to-tailgate bolts to 10 ft. lbs. (14 Nm); the strut-to-tailgate bolts to 104 inch lbs. (12 Nm).

Navajo

▶ See Figure 9

➡ The liftgate glass should not be open while the liftgate is open. Make sure the window is closed before opening the liftgate.

1. Open the liftgate door.
2. Remove the upper rear center garnish molding.
3. Support the door in the open position and disconnect the liftgate gas cylinder assist rod assemblies.
4. Carefully move the headliner out of position and remove the hinge-to-header panel attaching nuts.
5. Remove the hinge-to-liftgate attaching bolts and remove the complete assembly.
6. Installation is the reverse of removal. Adjust the liftgate hinge as necessary.

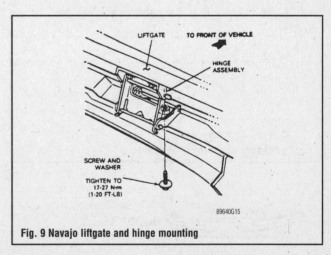

Fig. 9 Navajo liftgate and hinge mounting

BODY AND TRIM 10-5

ALIGNMENT

MPV

Both the striker and hinge bolt holes are made so that the striker and hinges can be moved to permit proper engagement and fit. Tighten the striker bolts to 20 ft. lbs. (27 Nm); the hinge bolts to 10 ft. lbs. (14 |Nm).

Navajo

➡ The liftgate glass should not be open while the liftgate is open. Make sure the window is closed before opening the liftgate.

The liftgate can be adjusted slightly in or out and side to side by loosening the hinge-to-header nut or bolt. Some up and down adjustment can be accomplished by loosening the hinge bolts on the liftgate and moving the gate up or down. The liftgate should be adjusted for even and parallel fit with adjoining panels.

Grille

REMOVAL & INSTALLATION

Pickups

♦ See Figure 10

1. Open the hood.
2. Remove the combination lamps from the grille.
3. Remove the grille retaining screws and lift off the grille.
4. Install in reverse order.

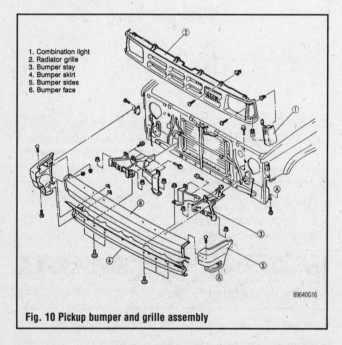

Fig. 10 Pickup bumper and grille assembly

MPV

♦ See Figure 11

The grille is in 3 pieces: 1 center piece and 2 side pieces. To remove any or all of these, remove the screws and disengage the retaining clips. To disengage the retaining clips, depress the tabs of the clip with a small screwdriver.
To install the clips, simply press them into place.

Navajo

1. Remove the plastic retainers attaching the grille to the air deflector.
2. Remove the screws attaching the grille to the radiator cross support.

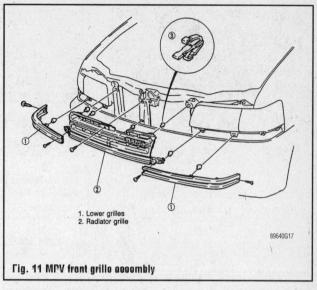

Fig. 11 MPV front grille assembly

Depress the spring tabs at the lower outboard openings and detach the grille from the headlamp housings.
3. Remove the grille.
4. Install in reverse order.

Outside Mirrors

REMOVAL & INSTALLATION

Pickup and MPV

♦ See Figures 12, 13, 14 and 15

The mirrors can be removed from the door without disassembling the door liner or other components except on Navajo models. The mirrors may be manual, manual remote, or electric remote. If the mirror glass is broken, replacements may be available through your dealer or a glass shop. If the housing is cracked or damaged, the entire mirror must be replaced. To remove the mirror:
1. If the mirror is manual remote, check to see if the adjusting handle is retained by a hidden screw, usually under the endcap on the lever. Or if it is retained by a threaded trim nut. If so, remove the screw, nut or other remote assembly retainer.

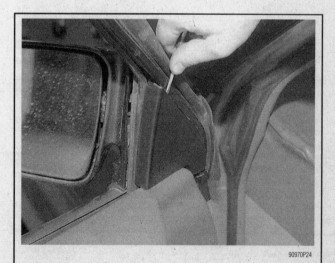

Fig. 12 Using a small prying tool, pry off the interior sideview mirror trim panel . . .

10-6 BODY AND TRIM

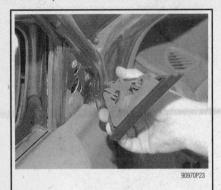

Fig. 13 . . . then pull away the trim panel from the door

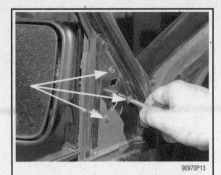

Fig. 14 Remove the 3 sideview mirror mounting screws using a phillips head screwdriver . . .

Fig. 15 . . . then pull the sideview mirror away and disengage the power mirror

2. Remove the plastic delta cover from the door window corner where the mirror is mounted. It can be removed with a blunt plastic or wooden tool. Don't use a screwdriver, the plastic will be marred.
3. Depending on the model and style mirror, there may be concealed plugs or other minor parts under the delta cover. If electrical connectors are present, disconnect them.
4. Installation is the reverse of removal.

Navajo

1. The door panel must first be removed to gain access to the mounting nuts. Disconnect the harness connector if equipped with power mirrors.
2. Remove the mounting screws or nuts and lift off the mirror. Remove and discard the gasket.
3. When installing, make sure the gasket is properly positioned before tightening the screws.

Antenna

REPLACEMENT

Depending on the year and model, it may be necessary to remove the glove box and or the instrument panel to gain working room.
1. If the antenna cable plugs directly into the radio, pull it straight out of the set. Otherwise, disconnect the antenna lead-in cable from the cable assembly in-line connector above the glove box.
2. Working under the instrument panel, disengage the cable from its retainers.
3. Outside, unsnap the cap from the antenna base. Or, on some models, unscrew the mounting tower nut.
4. Remove the retaining screws, or tower nut and lift off the antenna, pulling the cable with it, carefully.
5. Remove and discard the gasket.
6. Installation is the reverse of removal.

Fenders

REMOVAL & INSTALLATION

▶ See Figure 16

1. Clean all of the dirt from the fender mounting screws, bolts and nuts.
2. Remove the headlamp door, headlamp assembly, sidemarker lamp, or parking lamp assembly, depending on model.
3. Remove the bolt(s) attaching the rear of the front fender to the windshield cowl. This bolt is usually accessed from inside the vehicle with the door opened.
4. Remove the top bolts that mount the fender to the inter-body.
5. Loosen the wheel lugs slightly. Raise and safely support the vehicle. Remove the wheel and tire assembly.
6. Remove the bolts attaching the fender brace to the body. Remove the bolts attaching the fender to the radiator support, and remove the bolts attaching the fender to the fender apron (inner splash shield).

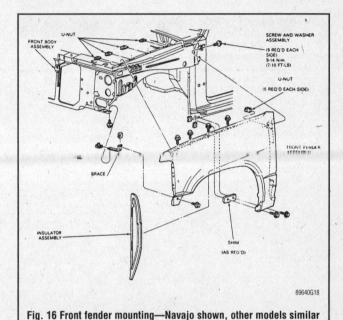

Fig. 16 Front fender mounting—Navajo shown, other models similar

7. Remove the bolts mounting the fender to the lower rocker sill. Check for any other mounting bolts, remove them. Remove the fender.
8. To install, place the fender in position and install the mounting bolts loosely. Align the lower edge of the fender to the rocker sill and secure the bolts. Align the upper edge of the fender to the cowl and tighten the bolts. Install the remaining components.

Chassis and Cab Mount Bushings

REMOVAL & INSTALLATION

▶ See Figures 17 and 18

➡ This applies to Navajo and Pickup models only.

Use the accompanying illustrations as a guide
1. Remove interior trim as required.
2. Back out the body mount bolt four or five turns.
3. If necessary, have an assistant hold the lower retainer with a wrench.
4. Strike the head of the bolt with a hammer to drive out the lower retainer.
5. Remove the bolt and lower retainer.
6. Use a jack and a block of wood and jack the body up enough to remove the old mount from between the frame and body.

➡ Some of the other body mounts may need to be loosened in order to lift the body enough to remove the mount.

7. Installation is the reverse of the removal procedure.

BODY AND TRIM 10-7

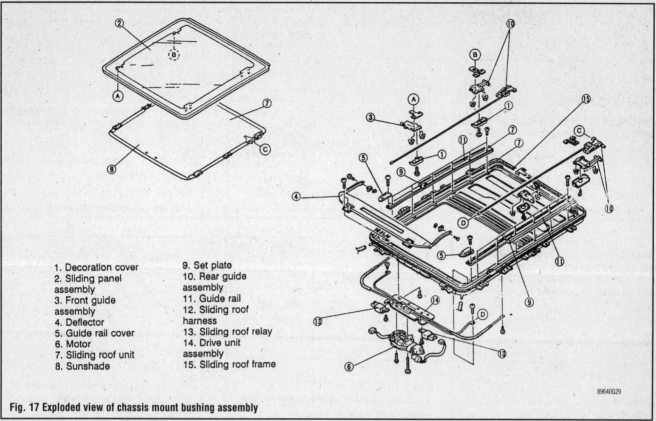

Fig. 17 Exploded view of chassis mount bushing assembly

1. Decoration cover
2. Sliding panel assembly
3. Front guide assembly
4. Deflector
5. Guide rail cover
6. Motor
7. Sliding roof unit
8. Sunshade
9. Set plate
10. Rear guide assembly
11. Guide rail
12. Sliding roof harness
13. Sliding roof relay
14. Drive unit assembly
15. Sliding roof frame

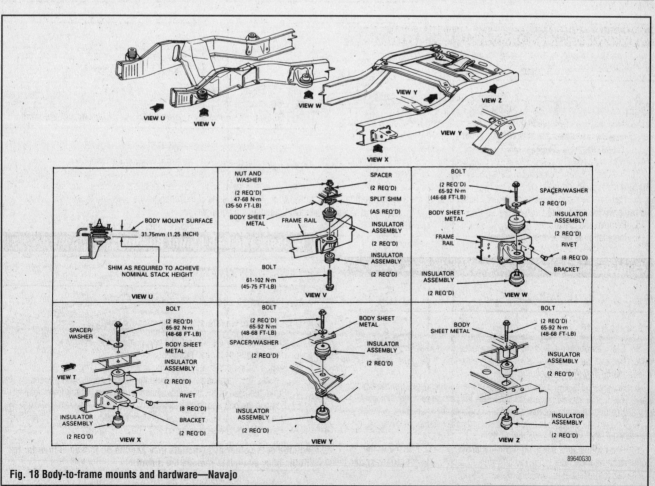

Fig. 18 Body-to-frame mounts and hardware—Navajo

10-8 BODY AND TRIM

Power Sunroof

REMOVAL & INSTALLATION

MPV

▶ See Figures 19 thru 28

➥Service to the sunroof harness, sunroof relay, drive unit assembly, sunroof frame, sunshade and guide rail require headliner removal.

1. Make sure the sunroof is fully closed, if possible. Disconnect the negative battery cable.
2. Remove the headliner.
3. Slide the sunshade all the way to the rear. Fully close the sliding panel. Remove the decoration cover mounting screws from the right and left decoration covers and remove the covers.
4. Remove the retaining nuts from the sliding panel and bracket. Remove the sliding panel by pushing it upward from inside the vehicle. Take care to remove the shims between the sliding panel and brackets before removing the sliding panel.
5. Remove the front guide assembly.
6. Remove the air deflector. Remove the E-ring at the rear of the deflector link, and remove the pin. Remove the screws and the deflector. Take care not to damage the deflector link or connector.
7. Remove the guide rail cover.
8. Remove the sliding roof unit. Remove the mounting nuts securing the drive unit and sliding roof unit to the body. With an assistant to help you, remove the unit.

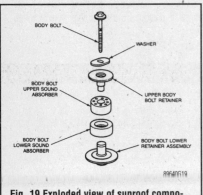

Fig. 19 Exploded view of sunroof components

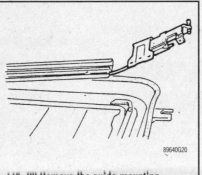

Fig. 20 Remove the guide mounting screws, lift the guide rail up and pull the rear guide assembly backward

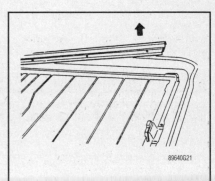

Fig. 21 Remove the guide rail from the sliding roof frame, lifting up the rear end of the guide rail

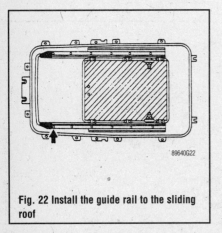

Fig. 22 Install the guide rail to the sliding roof

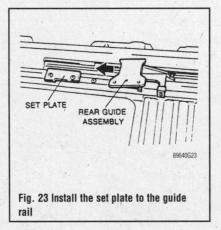

Fig. 23 Install the set plate to the guide rail

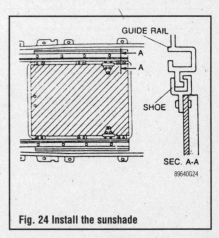

Fig. 24 Install the sunshade

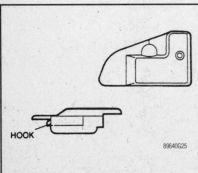

Fig. 25 Insert the guide rail cover hook into the sliding roof frame and secure with mounting screw

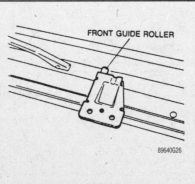

Fig. 26 Install the front guide roller assembly, roller faces forward

Fig. 27 Insert the sliding panel

BODY AND TRIM 10-9

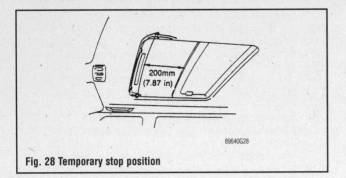

Fig. 28 Temporary stop position

To install:

11. Place the roof frame, drive unit, relay and harness in position.
12. Put the guide rail on the sliding roof frame after applying sealant to the underside of the guide rail.
13. Set the guide rail to the sliding roof frame and install bolts to the second screw holes.
14. Install the set plate to the guide rail. Lift up the rear end of the rail and insert the rear guide assembly until it reaches the set plate. Be sure to assemble the bracket and drive cable before insertion.
15. Lift up on the guide rail and install the sunshade sliding shoes into the guide rail. Secure the guide rail to the sliding roof frame. Secure the sliding roof unit to the body roof.
16. Install the guide rail covers. Insert the front guide assembly into the guide rail and set it to press the deflector link (sunroof in fully closed position). The front guide rollers face forward.
17. Install the sliding panel, rear end first. Insert the sliding panel bolts into the bracket holes of the front and rear guide assemblies, tighten the nuts. Be sure to install the shims in the same position from which they were removed. Install the direction covers.
18. Install the headliner.

9. Remove the sunshade from the guide rail. Remove the set plate. Remove the guide rail mounting screws. Lift up the rear end of the guide rail and pull out the rear guide assembly. Remove the guide rail from the sliding roof frame, lifting up the rear end of the guide rail.
10. Remove the sliding roof harness, sliding roof relay, drive unit assembly and the sliding roof frame.

INTERIOR

Instrument Panel and Pad

REMOVAL & INSTALLATION

Pickups

♦ See Figures 29, 30 and 31

1. Disconnect the negative battery cable.
2. Remove the steering wheel. Column upper and lower covers and the combination switch.
3. Remove the instrument cluster meter hood. Remove the instrument cluster assembly.
4. Remove the switch knob and the left side panel (next to the cluster mounting.
5. Remove the trim covers over the center panel mounting screws. Remove the center cover.
6. Remove the glove box lid and the glove box.
7. Remove the shift knob and boot. Remove the console box (models equipped).
8. Remove the radio assembly. Remove the small covers from each side of the instrument panel. Between the panel and each side body door frame.

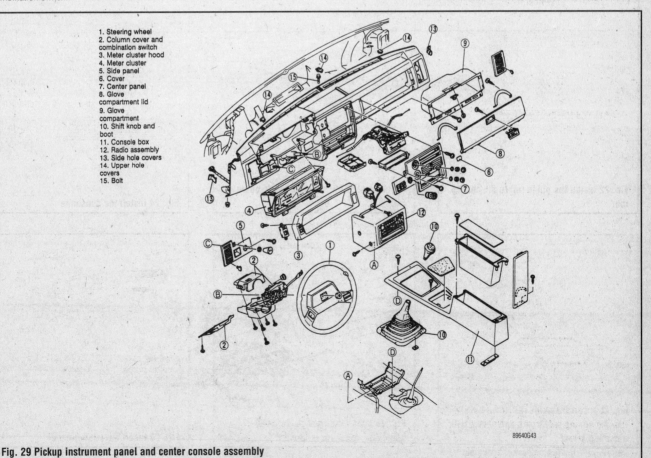

Fig. 29 Pickup instrument panel and center console assembly

10-10 BODY AND TRIM

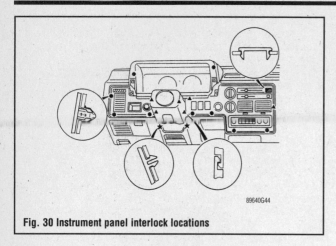

Fig. 30 Instrument panel interlock locations

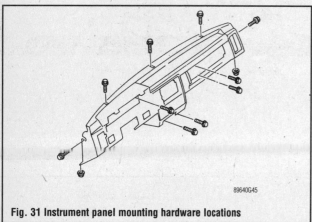

Fig. 31 Instrument panel mounting hardware locations

9. Remove the instrument panel mounting screw covers and the mounting screw. Remove the panel.
10. To install, place the panel in position and secure it with the mounting screws. Install the remaining components in the reverse of removal.

MPV

▶ See Figure 32

1. Disconnect the negative battery cable.
2. Remove the hood release knob, steering wheel and column cover.
3. Remove the combination switch.
4. Remove the instrument cluster assembly and meter set. Remove the side cover.
5. Remove the right side undercover pad (vehicles equipped). Remove the right and left side lower panel assemblies.
6. Remove the left side duct, ashtray and audio panel assembly.
7. Remove the audio unit. Remove the lower center panel.
8. Remove the switch knobs and the upper switch panel. Remove the temperature control, blower control and airflow mode control.
9. Remove the upper garnish and the dash panel.
10. Position and secure the dash panel. Install the remaining components in the reverse of removal.

Navajo

▶ See Figure 33

1. Disconnect the negative battery cable.
2. Remove the ashtray and retainer assembly. Remove the instrument cluster finish panel by unsnapping the seven snap-in retainers. Remove the cluster retaining screws. Pull the cluster forward and disconnect the speedometer and electrical connectors. Remove the cluster.
3. Remove the hood opening cable from the bottom of the steering column cover. Remove the two screws retaining the bottom of the steering column cover and remove the cover by disengaging the two retainers at the top corners of the cover.

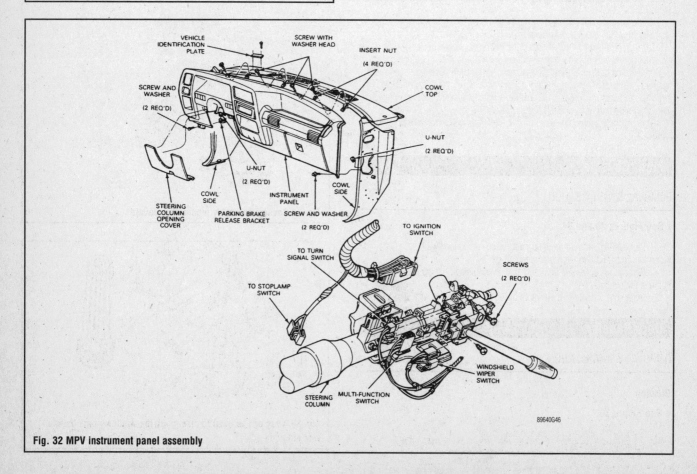

Fig. 32 MPV instrument panel assembly

BODY AND TRIM 10-11

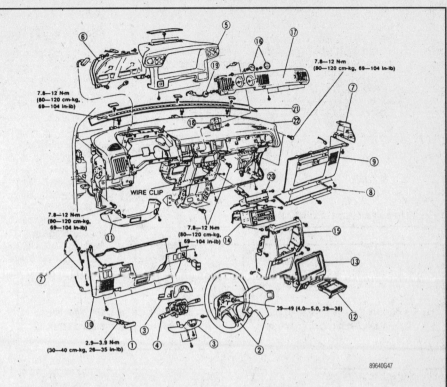

Fig. 33 Navajo instrument panel mounting

    4. Remove the screw that attaches the instrument panel to the brake/clutch support.
    5. Remove the steering column upper and lower shrouds.
    6. Disconnect the wiring harness connectors to the steering column switches.
    7. Remove the right and left cowl side trim covers. Remove the two bolts retaining the right and left instrument panel lower corners to the cowl side.
    8. Remove the front body inside pillar moldings. Remove the right lower insulator, if equipped. Remove the four screws retaining the top of the panel.
    9. Support the panel assembly and reach through the openings and underneath the instrument panel to disconnect the panel wiring, vacuum lines, radio antenna and heater/AC controls. (Label for reinstallation identification). Remove the panel mounting bolts and the panel.
    10. Install and secure the panel. Install the components in the reverse of removal.

Floor Console

REMOVAL & INSTALLATION

▶ See Figures 29 and 34

    1. Remove the arm rest mounting access covers. Unscrew the arm rest mounting bolts and remove the armrest.
    2. Remove the screws in the utility tray assembly and panel console. Remove the console.
    3. Install the console in the reverse order.

Door Panels

REMOVAL & INSTALLATION

Pickups

▶ See Figure 35

    1. Remove the attaching screw and door handle.
    2. Remove the armrest.

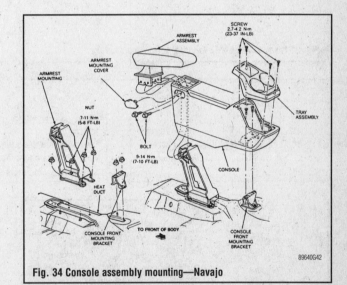

Fig. 34 Console assembly mounting—Navajo

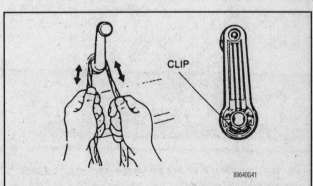

Fig. 35 A rag can be used to disengage the window crank handle retaining clip

10-12 BODY AND TRIM

3. Push in on the door panel, slightly, at the window handle and pry off the snapring retaining the handle to the regulator stem. Remove the handle.
4. Carefully slip a thin prying instrument behind the door panel and slide it along until you hit one of the retaining clips. Pry as close as possible to the clip to snap the clip from the door. Be very careful to avoid tearing the clip from the panel.
5. Pry out each clip, in turn, and lift off the door panel.
6. Installation is the reverse of removal. When snapping the clips into place, make sure that they are squarely over the holes to avoid bending them.

MPV

▶ See Figures 35 thru 43

1. Remove the attaching screw and door handle and plate.
2. Remove the plug in the top of the armrest and remove the panel support bracket screw underneath.
3. On models with power windows, remove the power window switch.
4. Push in on the door panel, slightly, at the window handle and pry off the snapring retaining the handle to the regulator stem. Remove the handle.

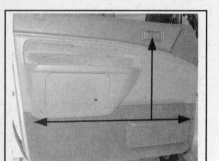

Fig. 36 Remove the interior door panel mounting screws before disengaging the retaining clips

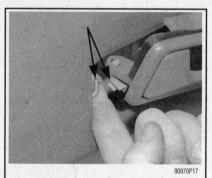

Fig. 37 Disengage the inside door handle release rod from the handle and retainer clip

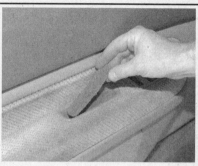

Fig. 38 Remove the felt pad at the bottom of the door panel pull cup

Fig. 39 Then, using a phillips head screwdriver, remove the screw at the bottom of the pull cup

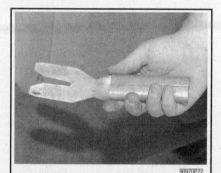

Fig. 40 It is a good idea to have a door trim panel removal tool when performing this kind of procedure

Fig. 41 Place the door panel trim removal tool between ther door and the trim panel and pry the clips away froim the door

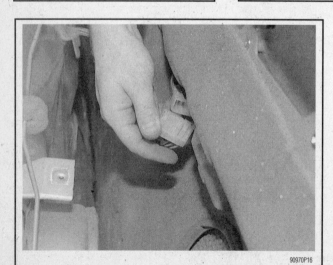

Fig. 42 If equipped with electric windows, disconnect the switches froom behind the door panel

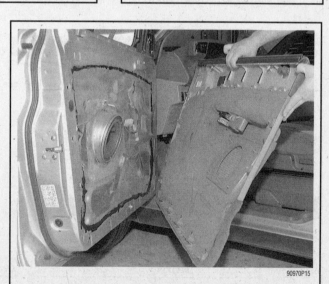

Fig. 43 Carefully pull the interior door panel away from the door

BODY AND TRIM 10-13

5. Carefully slip a thin prying instrument, like a wood spatula, behind the door panel and slide it along until you hit one of the retaining clips. Pry as close as possible to the clip to snap the clip from the door. Be very careful to avoid tearing the clip from the panel.

6. Pry out each clip, in turn, and lift off the door panel.

7. Installation is the reverse of removal. When snapping the clips into place, make sure that they are squarely over the holes to avoid bending them.

Navajo

♦ See Figure 44

1. Open the window. Remove the 2 screws retaining the trim panel located above the door handle.
2. Remove the rim cup behind the door handle using a small prying tool. Retention nibs will flex for ease of removal.
3. If equipped with power accessories, use the notch at the lower end of the plate and pry the plate off. Remove the plate from the trim panel and pull the wiring harness from behind the panel. Disconnect the harness from the switches.
4. Using a flat wood spatula, insert it carefully behind the panel and slide it along to find the push pins. When you encounter a pin, pry the pin outward. Do this until all the pins are out. NEVER PULL ON THE PANEL TO REMOVE THE PINS!
5. Lift slightly to disengage the panel from the flange at the top of the door.
6. Disconnect the door courtesy lamp and remove the panel completely. Replace any damaged or bent attaching clips.
7. Installation is the reverse of removal.

Door and Liftgate Locks

REMOVAL & INSTALLATION

➡ A key code is stamped on the lock cylinder to aid in replacing lost keys.

Pickups and MPV

♦ See Figures 45 and 46

1. Remove the door trim panel.
2. Pull the weathersheet, gently, away from the door lock access holes.
3. Using a screwdriver, push the lock cylinder retaining clip upward, noting the position of the lock cylinder.
4. Remove the lock cylinder from the door.
5. Install the lock cylinder in reverse of removal. It's a good idea to open the window before checking the lock operation, just in case it doesn't work properly.

Navajo

♦ See Figure 46

1. Remove the trim panel and watershield.
2. Disconnect the actuating rod from the lock control link clip.
3. Slide the retainer away from the lock cylinder.
4. Remove the cylinder from the door.

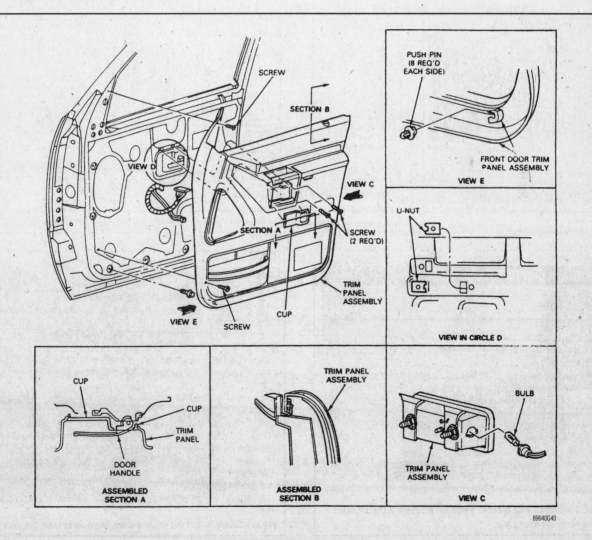

Fig. 44 Navajo door panel mounting

10-14 BODY AND TRIM

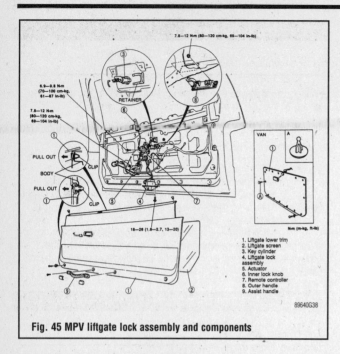

Fig. 45 MPV liftgate lock assembly and components

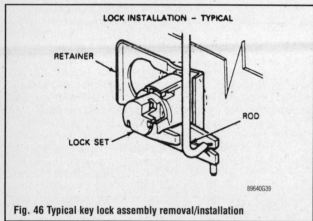

Fig. 46 Typical key lock assembly removal/installation

5. Use a new gasket when installing to ensure a watertight fit. Lubricate the cylinder with suitable oil recommended for this application.

Door Glass and Regulator

REMOVAL & INSTALLATION

Pickups

1. Remove the door panel.
2. Remove the weatherscreening carefully, so that it can be reused.
3. Position the glass so that the mounting screws can be accessed through one of the holes in the door frame. Remove the door glass mounting screws.
4. Remove the inner and outer weatherstripping around the frame.
5. Remove the glass guide mounting bolt.
6. Pull the glass up and out of the door.
7. Remove the mounting bolts and pull the regulator assembly from the access hole.
8. Install in the reverse order. Adjust the door glass so that it closes properly.

MPV

♦ See Figure 47

1. Remove the door trim panel.
2. Remove the panel support bracket.
3. Remove the weatherscreen carefully.
4. Remove the speaker.
5. Remove the mirror.
6. Remove the front beltline molding.
7. Raise the window to the point at which the glass retaining screws can be reached the access holes.
8. Carefully lift the glass from the channel.
9. With manual regulators, remove the attaching bolts and lift out the regulator.
10. With power regulators, remove the attaching bolts and drill out the rivets. Lift out the regulator.
11. Place the regulator in position and secure. Install then components in reverse order. Use new rivets on power regulators.

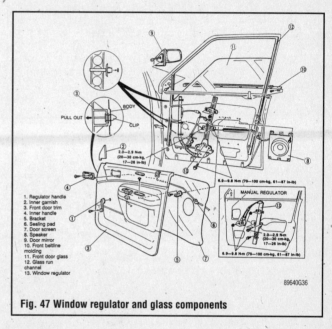

Fig. 47 Window regulator and glass components

Navajo

GLASS

♦ See Figure 48

1. Remove the door trim panel and speaker if applicable.
2. Remove the screw from the division bar. Remove the inside belt weatherstrip(s) if equipped.
3. Remove the 2 vent window attaching screws from the front edge of the door.
4. Lower the glass and pull the glass out of the run retainer near the vent window division bar, just enough to allow the removal of the vent window, if equipped.
5. Push the front edge of the glass downward and remove the rear glass run retainer from the door.
6. If equipped with retaining rivets, remove them carefully. Otherwise, remove the glass from the channel using Glass and Channel Removal Tool 2900, made by the Sommer and Mala Glass Machine Co. of Chicago, ILL., or its equivalent. Remove the glass through the belt opening if possible.

To install:

7. Install the glass spacer and retainer into the retention holes.
8. Install the glass into the door, position on the bracket and align the retaining holes.

BODY AND TRIM 10-15

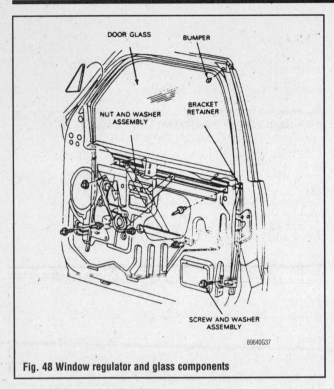

Fig. 48 Window regulator and glass components

9. Carefully install the retaining rivets or equivalent.
10. Raise the glass to the full closed position.
11. Install the rear glass run retainer and glass run. Install the inside belt weatherstrip(s).
12. Check for smooth operation before installing the trim panel.

REGULATOR

♦ See Figure 48

1. Remove the door trim panel and watershield.
2. Remove the inside door belt weatherstrip and glass stabilizer.
3. Remove the door glass.
4. Remove the 2 nuts attaching the equalizer bracket.
5. Remove the rivets attaching the regulator base plate to the door.
6. Remove the regulator and glass bracket as an assembly from the door and transfer to a workbench.
7. Carefully bend the tab flat in order to remove the air slides from the glass bracket C-channel.
8. Install new regulator arm plastic guides into the C-channel and bend the tab back 90°. If the tab is broken or cracked, replace the glass bracket assembly. Make sure the rubber bumper is installed properly on the new glass bracket, if applicable.

✱✱ CAUTION

If the regulator counterbalance spring is to be removed, make sure the regulator arms are in a fixed position prior to removal. This will prevent possible injury when the C-spring unwinds.

To install:

9. Assemble the glass bracket and regulator assembly.
10. Install the assembly in the door. Set the regulator base plate to the door using the base plate locator tab as a guide.
11. Attach the regulator to the door using new rivets. ¼ inch–20 x ½ inch bolts and nuts may be used in place of the rivets to attach the regulator.
12. Install the equalizer bracket, door belt weatherstrip and glass stabilizer.
13. Install the glass and check for smooth operation before installing the door trim panel.

Electric Window Motor

REMOVAL & INSTALLATION

1. Raise the window fully if possible. If not, you will have to support the window during this procedure. Disconnect the battery ground.
2. Remove the door trim panel.
3. Disconnect the window motor wiring harness.
4. There may be a drill dimple in the door panel, opposite the concealed motor retaining bolt. Drill out the dimple to gain access to the bolt. Be careful to avoid damage to the wires.
5. Remove the motor mounting bolts (front door) or rivets (rear door).
6. Push the motor towards the outside of the door to disengage it from the gears. You'll have to support the window glass once the motor is disengaged.
7. Remove the motor from the door.
8. Installation is the reverse of removal. To avoid rusting in the drilled areas, prime and paint the exposed metal, or, cover the holes with waterproof body tape. Make sure that the motor works properly before installing the trim panel.

Windshield and Fixed Glass

REMOVAL & INSTALLATION

If your windshield, or other fixed window, is cracked or chipped, you may decide to replace it with a new one yourself. However, there are two main reasons why replacement windshields and other window glass should be installed only by a professional automotive glass technician: safety and cost.

The most important reason a professional should install automotive glass is for safety. The glass in the vehicle, especially the windshield, is designed with safety in mind in case of a collision. The windshield is specially manufactured from two panes of specially-tempered glass with a thin layer of transparent plastic between them. This construction allows the glass to "give" in the event that a part of your body hits the windshield during the collision, and prevents the glass from shattering, which could cause lacerations, blinding and other harm to passengers of the vehicle. The other fixed windows are designed to be tempered so that if they break during a collision, they shatter in such a way that there are no large pointed glass pieces. The professional automotive glass technician knows how to install the glass in a vehicle so that it will function optimally during a collision. Without the proper experience, knowledge and tools, installing a piece of automotive glass yourself could lead to additional harm if an accident should ever occur.

Cost is also a factor when deciding to install automotive glass yourself. Performing this could cost you much more than a professional may charge for the same job. Since the windshield is designed to break under stress, an often life saving characteristic, windshields tend to break VERY easily when an inexperienced person attempts to install one. Do-it-yourselfers buying two, three or even four windshields from a salvage yard because they have broken them during installation are common stories. Also, since the automotive glass is designed to prevent the outside elements from entering your vehicle, improper installation can lead to water and air leaks. Annoying whining noises at highway speeds from air leaks or inside body panel rusting from water leaks can add to your stress level and subtract from your wallet. After buying two or three windshields, installing them and ending up with a leak that produces a noise while driving and water damage during rainstorms, the cost of having a professional do it correctly the first time may be much more alluring. We here at Chilton, therefore, advise that you have a professional automotive glass technician service any broken glass on your vehicle.

WINDSHIELD CHIP REPAIR

♦ See Figures 49 and 50

➥ Check with your state and local authorities on the laws for state safety inspection. Some states or municipalities may not allow chip repair as a viable option for correcting stone damage to your windshield.

10-16 BODY AND TRIM

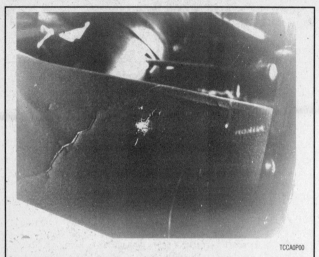

Fig. 49 Small chips on your windshield can be fixed with an aftermarket repair kit, such as the one from Loctite®

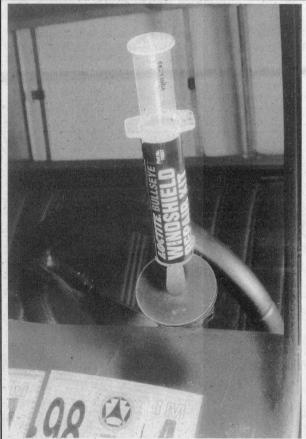

Fig. 50 Most kits us a self-stick applicator and syringe to inject the adhesive into the chip or crack

Although severely cracked or damaged windshields must be replaced, there is something that you can do to prolong or even prevent the need for replacement of a chipped windshield. There are many companies which offer windshield chip repair products, such as Loctite's® Bullseye™ windshield repair kit.

These kits usually consist of a syringe, pedestal and a sealing adhesive. The syringe is mounted on the pedestal and is used to create a vacuum which pulls the plastic layer against the glass. This helps make the chip transparent. The adhesive is then injected which seals the chip and helps to prevent further stress cracks from developing

➥Always follow the specific manufacturer's instructions.

Inside Rear View Mirror

The mirror is held in place with a single setscrew. Loosen the screw and lift the mirror off. Don't forget to unplug the electrical connector if the truck has an electric Day/Night mirror.

Repair kits for damaged mirrors are available and most auto parts stores. The most important part of the repair is the beginning. Mark the outside of the windshield to locate the pad, then scrape the old adhesive off with a razor blade. Clean the remaining adhesive off with chlorine-based window cleaner (not petroleum-based solvent) as thoroughly as possible. Follow the manufacturers instructions exactly to complete the repair..

Seats

REMOVAL & INSTALLATION

Pickups

♦ See Figure 51

1. Remove the seat anchor bolts.
2. Lift out the seat.

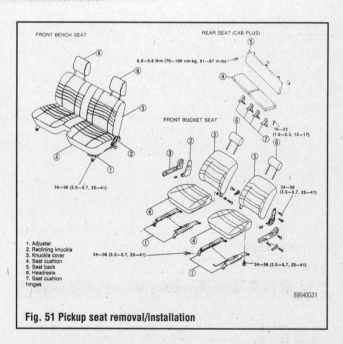

Fig. 51 Pickup seat removal/installation

3. Installation is the reverse of removal.

MPV

DRIVER'S SEAT

♦ See Figure 52

1. Remove the fuel filler lid opener.
2. Remove the parking brake lever.
3. Remove the seat bracket covers.
4. Remove the seat leg covers.

BODY AND TRIM 10-17

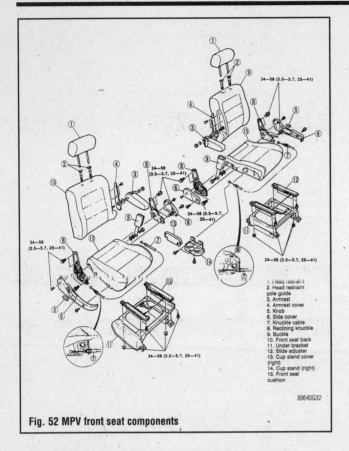

Fig. 52 MPV front seat components

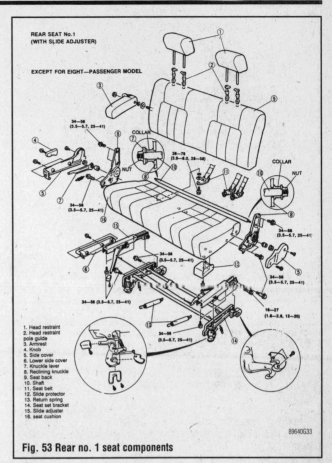

Fig. 53 Rear no. 1 seat components

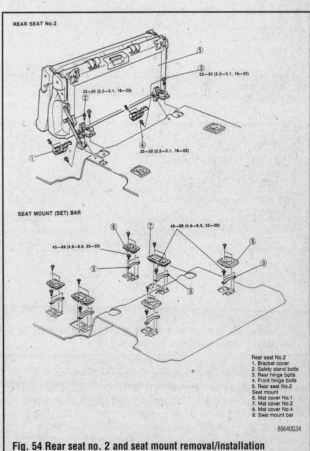

Fig. 54 Rear seat no. 2 and seat mount removal/installation

5. Remove the seat bracket-to-floor bolts and lift out the seat.
6. Installation is the reverse of removal. Torque the bracket-to-floor bolts to 38 ft. lbs. (52 Nm).

FRONT RIGHT SEAT

▶ See Figure 52

1. Remove the jack cover.
2. Remove the jack and jack handle.
3. Remove the seat bracket covers.
4. Remove the seat leg covers.
5. Remove the seat bracket-to-floor bolts and lift out the seat.
6. Installation is the reverse of removal. Torque the bracket-to-floor bolts to 38 ft. lbs. (52 Nm).

NO.1 REAR SEAT

▶ See Figure 53

The seat is removed by simply unlatching the release levers on the seat legs.

NO.2 REAR SEAT

▶ See Figure 54

1. Remove the seat bracket cover.
2. Remove the folding link bolts.
3. Remove the rear hinge bolts.
4. Remove the front hinge bolts.
5. Lift out the seat.
6. Installation is the reverse of removal. Torque all bolts to 22 ft. lbs. (30 Nm).

Navajo

▶ See Figure 55

1. Remove the 4 seat track-to-floorpan screws (2 each side) and lift the seat and track assembly from the vehicle.

10-18 BODY AND TRIM

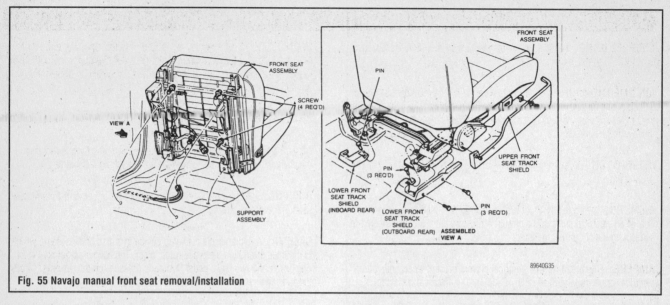

Fig. 55 Navajo manual front seat removal/installation

2. To remove the seat tracks from the seat cushion, position the seat upside down on a clean bench.
3. Disconnect the latch tie rod assembly and assist spring from the tracks.
4. Remove 4 track-to-seat cushion screws (2 each side) from the track assemblies. Remove the tracks from seat cushion.

To install:
5. Position the tracks to the seat cushion. Install the 4 track-to-seat cushion screws (2 each side) and tighten.
6. Connect the latch tie rod assembly and assist spring to the tracks.
7. Position the seat and track assembly in the vehicle.
8. Install 4 track-to-floorpan screws and tighten to specification.

Power Seat Motor

REMOVAL & INSTALLATION

The optional six-way power seat system on the Navajo uses a screw-type drive for seat adjustment.

The six-way power seat provides horizontal, vertical, and tilt adjustments. It consists of a reversible armature motor, switch and housing assembly, vertical screw, and horizontal screw drives.

The motor assembly, which contains three armatures, the flexible shafts, the screw drive and gear mechanism are not serviced individually and can be serviced only by replacing the complete track assembly.

TORQUE SPECIFICATIONS

Component	U.S.	Metric
Door hinge bolts		
MPV	10 ft. lbs.	13.6 Nm
Door striker bolts		
MPV	20 ft. lbs.	27.2 Nm
Seat bracket-to-floor bolts		
MPV front seats	38 ft. lbs.	51.68 Nm
MPV no.2 rear seat bolts	22 ft. lbs.	29.92 Nm
Strut-to-tailgate bolts		
MPV	104 inch lbs.	11.648 Nm
Tailgate hinge-to-tailgate bolts		
MPV	10 ft. lbs.	13.6 Nm

GLOSSARY

AIR/FUEL RATIO: The ratio of air-to-gasoline by weight in the fuel mixture drawn into the engine.

AIR INJECTION: One method of reducing harmful exhaust emissions by injecting air into each of the exhaust ports of an engine. The fresh air entering the hot exhaust manifold causes any remaining fuel to be burned before it can exit the tailpipe.

ALTERNATOR: A device used for converting mechanical energy into electrical energy.

AMMETER: An instrument, calibrated in amperes, used to measure the flow of an electrical current in a circuit. Ammeters are always connected in series with the circuit being tested.

AMPERE: The rate of flow of electrical current present when one volt of electrical pressure is applied against one ohm of electrical resistance.

ANALOG COMPUTER: Any microprocessor that uses similar (analogous) electrical signals to make its calculations.

ARMATURE: A laminated, soft iron core wrapped by a wire that converts electrical energy to mechanical energy as in a motor or relay. When rotated in a magnetic field, it changes mechanical energy into electrical energy as in a generator.

ATMOSPHERIC PRESSURE: The pressure on the Earth's surface caused by the weight of the air in the atmosphere. At sea level, this pressure is 14.7 psi at 32°F (101 kPa at 0°C).

ATOMIZATION: The breaking down of a liquid into a fine mist that can be suspended in air.

AXIAL PLAY: Movement parallel to a shaft or bearing bore.

BACKFIRE: The sudden combustion of gases in the intake or exhaust system that results in a loud explosion.

BACKLASH: The clearance or play between two parts, such as meshed gears.

BACKPRESSURE: Restrictions in the exhaust system that slow the exit of exhaust gases from the combustion chamber.

BAKELITE: A heat resistant, plastic insulator material commonly used in printed circuit boards and transistorized components.

BALL BEARING: A bearing made up of hardened inner and outer races between which hardened steel balls roll.

BALLAST RESISTOR: A resistor in the primary ignition circuit that lowers voltage after the engine is started to reduce wear on ignition components.

BEARING: A friction reducing, supportive device usually located between a stationary part and a moving part.

BIMETAL TEMPERATURE SENSOR: Any sensor or switch made of two dissimilar types of metal that bend when heated or cooled due to the different expansion rates of the alloys. These types of sensors usually function as an on/off switch.

BLOWBY: Combustion gases, composed of water vapor and unburned fuel, that leak past the piston rings into the crankcase during normal engine operation. These gases are removed by the PCV system to prevent the buildup of harmful acids in the crankcase.

BRAKE PAD: A brake shoe and lining assembly used with disc brakes.

BRAKE SHOE: The backing for the brake lining. The term is, however, usually applied to the assembly of the brake backing and lining.

BUSHING: A liner, usually removable, for a bearing; an anti-friction liner used in place of a bearing.

CALIPER: A hydraulically activated device in a disc brake system, which is mounted straddling the brake rotor (disc). The caliper contains at least one piston and two brake pads. Hydraulic pressure on the piston(s) forces the pads against the rotor.

CAMSHAFT: A shaft in the engine on which are the lobes (cams) which operate the valves. The camshaft is driven by the crankshaft, via a belt, chain or gears, at one half the crankshaft speed.

CAPACITOR: A device which stores an electrical charge.

CARBON MONOXIDE (CO): A colorless, odorless gas given off as a normal byproduct of combustion. It is poisonous and extremely dangerous in confined areas, building up slowly to toxic levels without warning if adequate ventilation is not available.

CARBURETOR: A device, usually mounted on the intake manifold of an engine, which mixes the air and fuel in the proper proportion to allow even combustion.

CATALYTIC CONVERTER: A device installed in the exhaust system, like a muffler, that converts harmful byproducts of combustion into carbon dioxide and water vapor by means of a heat-producing chemical reaction.

CENTRIFUGAL ADVANCE: A mechanical method of advancing the spark timing by using flyweights in the distributor that react to centrifugal force generated by the distributor shaft rotation.

CHECK VALVE: Any one-way valve installed to permit the flow of air, fuel or vacuum in one direction only.

CHOKE: A device, usually a moveable valve, placed in the intake path of a carburetor to restrict the flow of air.

CIRCUIT: Any unbroken path through which an electrical current can flow. Also used to describe fuel flow in some instances.

CIRCUIT BREAKER: A switch which protects an electrical circuit from overload by opening the circuit when the current flow exceeds a predetermined level. Some circuit breakers must be reset manually, while most reset automatically.

COIL (IGNITION): A transformer in the ignition circuit which steps up the voltage provided to the spark plugs.

COMBINATION MANIFOLD: An assembly which includes both the intake and exhaust manifolds in one casting.

GLOSSARY

COMBINATION VALVE: A device used in some fuel systems that routes fuel vapors to a charcoal storage canister instead of venting them into the atmosphere. The valve relieves fuel tank pressure and allows fresh air into the tank as the fuel level drops to prevent a vapor lock situation.

COMPRESSION RATIO: The comparison of the total volume of the cylinder and combustion chamber with the piston at BDC and the piston at TDC.

CONDENSER: 1. An electrical device which acts to store an electrical charge, preventing voltage surges. 2. A radiator-like device in the air conditioning system in which refrigerant gas condenses into a liquid, giving off heat.

CONDUCTOR: Any material through which an electrical current can be transmitted easily.

CONTINUITY: Continuous or complete circuit. Can be checked with an ohmmeter.

COUNTERSHAFT: An intermediate shaft which is rotated by a mainshaft and transmits, in turn, that rotation to a working part.

CRANKCASE: The lower part of an engine in which the crankshaft and related parts operate.

CRANKSHAFT: The main driving shaft of an engine which receives reciprocating motion from the pistons and converts it to rotary motion.

CYLINDER: In an engine, the round hole in the engine block in which the piston(s) ride.

CYLINDER BLOCK: The main structural member of an engine in which is found the cylinders, crankshaft and other principal parts.

CYLINDER HEAD: The detachable portion of the engine, usually fastened to the top of the cylinder block and containing all or most of the combustion chambers. On overhead valve engines, it contains the valves and their operating parts. On overhead cam engines, it contains the camshaft as well.

DEAD CENTER: The extreme top or bottom of the piston stroke.

DETONATION: An unwanted explosion of the air/fuel mixture in the combustion chamber caused by excess heat and compression, advanced timing, or an overly lean mixture. Also referred to as "ping".

DIAPHRAGM: A thin, flexible wall separating two cavities, such as in a vacuum advance unit.

DIESELING: A condition in which hot spots in the combustion chamber cause the engine to run on after the key is turned off.

DIFFERENTIAL: A geared assembly which allows the transmission of motion between drive axles, giving one axle the ability to turn faster than the other.

DIODE: An electrical device that will allow current to flow in one direction only.

DISC BRAKE: A hydraulic braking assembly consisting of a brake disc, or rotor, mounted on an axle, and a caliper assembly containing, usually two brake pads which are activated by hydraulic pressure. The pads are forced against the sides of the disc, creating friction which slows the vehicle.

DISTRIBUTOR: A mechanically driven device on an engine which is responsible for electrically firing the spark plug at a predetermined point of the piston stroke.

DOWEL PIN: A pin, inserted in mating holes in two different parts allowing those parts to maintain a fixed relationship.

DRUM BRAKE: A braking system which consists of two brake shoes and one or two wheel cylinders, mounted on a fixed backing plate, and a brake drum, mounted on an axle, which revolves around the assembly.

DWELL: The rate, measured in degrees of shaft rotation, at which an electrical circuit cycles on and off.

ELECTRONIC CONTROL UNIT (ECU): Ignition module, module, amplifier or igniter. See Module for definition.

ELECTRONIC IGNITION: A system in which the timing and firing of the spark plugs is controlled by an electronic control unit, usually called a module. These systems have no points or condenser.

END-PLAY: The measured amount of axial movement in a shaft.

ENGINE: A device that converts heat into mechanical energy.

EXHAUST MANIFOLD: A set of cast passages or pipes which conduct exhaust gases from the engine.

FEELER GAUGE: A blade, usually metal, or precisely predetermined thickness, used to measure the clearance between two parts.

FIRING ORDER: The order in which combustion occurs in the cylinders of an engine. Also the order in which spark is distributed to the plugs by the distributor.

FLOODING: The presence of too much fuel in the intake manifold and combustion chamber which prevents the air/fuel mixture from firing, thereby causing a no-start situation.

FLYWHEEL: A disc shaped part bolted to the rear end of the crankshaft. Around the outer perimeter is affixed the ring gear. The starter drive engages the ring gear, turning the flywheel, which rotates the crankshaft, imparting the initial starting motion to the engine.

FOOT POUND (ft. lbs. or sometimes, ft.lb.): The amount of energy or work needed to raise an item weighing one pound, a distance of one foot.

FUSE: A protective device in a circuit which prevents circuit overload by breaking the circuit when a specific amperage is present. The device is constructed around a strip or wire of a lower amperage rating than the circuit it is designed to protect. When an amperage higher than that stamped on the fuse is present in the circuit, the strip or wire melts, opening the circuit.

GEAR RATIO: The ratio between the number of teeth on meshing gears.

GENERATOR: A device which converts mechanical energy into electrical energy.

HEAT RANGE: The measure of a spark plug's ability to dissipate heat from its firing end. The higher the heat range, the hotter the plug fires.

GLOSSARY

HUB: The center part of a wheel or gear.

HYDROCARBON (HC): Any chemical compound made up of hydrogen and carbon. A major pollutant formed by the engine as a byproduct of combustion.

HYDROMETER: An instrument used to measure the specific gravity of a solution.

INCH POUND (inch lbs.; sometimes in.lb. or in. lbs.): One twelfth of a foot pound.

INDUCTION: A means of transferring electrical energy in the form of a magnetic field. Principle used in the ignition coil to increase voltage.

INJECTOR: A device which receives metered fuel under relatively low pressure and is activated to inject the fuel into the engine under relatively high pressure at a predetermined time.

INPUT SHAFT: The shaft to which torque is applied, usually carrying the driving gear or gears.

INTAKE MANIFOLD: A casting of passages or pipes used to conduct air or a fuel/air mixture to the cylinders.

JOURNAL: The bearing surface within which a shaft operates.

KEY: A small block usually fitted in a notch between a shaft and a hub to prevent slippage of the two parts.

MANIFOLD: A casting of passages or set of pipes which connect the cylinders to an inlet or outlet source.

MANIFOLD VACUUM: Low pressure in an engine intake manifold formed just below the throttle plates. Manifold vacuum is highest at idle and drops under acceleration.

MASTER CYLINDER: The primary fluid pressurizing device in a hydraulic system. In automotive use, it is found in brake and hydraulic clutch systems and is pedal activated, either directly or, in a power brake system, through the power booster.

MODULE: Electronic control unit, amplifier or igniter of solid state or integrated design which controls the current flow in the ignition primary circuit based on input from the pick-up coil. When the module opens the primary circuit, high secondary voltage is induced in the coil.

NEEDLE BEARING: A bearing which consists of a number (usually a large number) of long, thin rollers.

OHM: (Ω) The unit used to measure the resistance of conductor-to-electrical flow. One ohm is the amount of resistance that limits current flow to one ampere in a circuit with one volt of pressure.

OHMMETER: An instrument used for measuring the resistance, in ohms, in an electrical circuit.

OUTPUT SHAFT: The shaft which transmits torque from a device, such as a transmission.

OVERDRIVE: A gear assembly which produces more shaft revolutions than that transmitted to it.

OVERHEAD CAMSHAFT (OHC): An engine configuration in which the camshaft is mounted on top of the cylinder head and operates the valve either directly or by means of rocker arms.

OVERHEAD VALVE (OHV): An engine configuration in which all of the valves are located in the cylinder head and the camshaft is located in the cylinder block. The camshaft operates the valves via lifters and pushrods.

OXIDES OF NITROGEN (NOx): Chemical compounds of nitrogen produced as a byproduct of combustion. They combine with hydrocarbons to produce smog.

OXYGEN SENSOR: Use with the feedback system to sense the presence of oxygen in the exhaust gas and signal the computer which can reference the voltage signal to an air/fuel ratio.

PINION: The smaller of two meshing gears.

PISTON RING: An open-ended ring with fits into a groove on the outer diameter of the piston. Its chief function is to form a seal between the piston and cylinder wall. Most automotive pistons have three rings: two for compression sealing; one for oil sealing.

PRELOAD: A predetermined load placed on a bearing during assembly or by adjustment.

PRIMARY CIRCUIT: the low voltage side of the ignition system which consists of the ignition switch, ballast resistor or resistance wire, bypass, coil, electronic control unit and pick-up coil as well as the connecting wires and harnesses.

PRESS FIT: The mating of two parts under pressure, due to the inner diameter of one being smaller than the outer diameter of the other, or vice versa; an interference fit.

RACE: The surface on the inner or outer ring of a bearing on which the balls, needles or rollers move.

REGULATOR: A device which maintains the amperage and/or voltage levels of a circuit at predetermined values.

RELAY: A switch which automatically opens and/or closes a circuit.

RESISTANCE: The opposition to the flow of current through a circuit or electrical device, and is measured in ohms. Resistance is equal to the voltage divided by the amperage.

RESISTOR: A device, usually made of wire, which offers a preset amount of resistance in an electrical circuit.

RING GEAR: The name given to a ring-shaped gear attached to a differential case, or affixed to a flywheel or as part of a planetary gear set.

ROLLER BEARING: A bearing made up of hardened inner and outer races between which hardened steel rollers move.

ROTOR: 1. The disc-shaped part of a disc brake assembly, upon which the brake pads bear; also called, brake disc. 2. The device mounted atop the distributor shaft, which passes current to the distributor cap tower contacts.

10-22 GLOSSARY

SECONDARY CIRCUIT: The high voltage side of the ignition system, usually above 20,000 volts. The secondary includes the ignition coil, coil wire, distributor cap and rotor, spark plug wires and spark plugs.

SENDING UNIT: A mechanical, electrical, hydraulic or electro-magnetic device which transmits information to a gauge.

SENSOR: Any device designed to measure engine operating conditions or ambient pressures and temperatures. Usually electronic in nature and designed to send a voltage signal to an on-board computer, some sensors may operate as a simple on/off switch or they may provide a variable voltage signal (like a potentiometer) as conditions or measured parameters change.

SHIM: Spacers of precise, predetermined thickness used between parts to establish a proper working relationship.

SLAVE CYLINDER: In automotive use, a device in the hydraulic clutch system which is activated by hydraulic force, disengaging the clutch.

SOLENOID: A coil used to produce a magnetic field, the effect of which is to produce work.

SPARK PLUG: A device screwed into the combustion chamber of a spark ignition engine. The basic construction is a conductive core inside of a ceramic insulator, mounted in an outer conductive base. An electrical charge from the spark plug wire travels along the conductive core and jumps a preset air gap to a grounding point or points at the end of the conductive base. The resultant spark ignites the fuel/air mixture in the combustion chamber.

SPLINES: Ridges machined or cast onto the outer diameter of a shaft or inner diameter of a bore to enable parts to mate without rotation.

TACHOMETER: A device used to measure the rotary speed of an engine, shaft, gear, etc., usually in rotations per minute.

THERMOSTAT: A valve, located in the cooling system of an engine, which is closed when cold and opens gradually in response to engine heating, controlling the temperature of the coolant and rate of coolant flow.

TOP DEAD CENTER (TDC): The point at which the piston reaches the top of its travel on the compression stroke.

TORQUE: The twisting force applied to an object.

TORQUE CONVERTER: A turbine used to transmit power from a driving member to a driven member via hydraulic action, providing changes in drive ratio and torque. In automotive use, it links the driveplate at the rear of the engine to the automatic transmission.

TRANSDUCER: A device used to change a force into an electrical signal.

TRANSISTOR: A semi-conductor component which can be actuated by a small voltage to perform an electrical switching function.

TUNE-UP: A regular maintenance function, usually associated with the replacement and adjustment of parts and components in the electrical and fuel systems of a vehicle for the purpose of attaining optimum performance.

TURBOCHARGER: An exhaust driven pump which compresses intake air and forces it into the combustion chambers at higher than atmospheric pressures. The increased air pressure allows more fuel to be burned and results in increased horsepower being produced.

VACUUM ADVANCE: A device which advances the ignition timing in response to increased engine vacuum.

VACUUM GAUGE: An instrument used to measure the presence of vacuum in a chamber.

VALVE: A device which control the pressure, direction of flow or rate of flow of a liquid or gas.

VALVE CLEARANCE: The measured gap between the end of the valve stem and the rocker arm, cam lobe or follower that activates the valve.

VISCOSITY: The rating of a liquid's internal resistance to flow.

VOLTMETER: An instrument used for measuring electrical force in units called volts. Voltmeters are always connected parallel with the circuit being tested.

WHEEL CYLINDER: Found in the automotive drum brake assembly, it is a device, actuated by hydraulic pressure, which, through internal pistons, pushes the brake shoes outward against the drums.

MASTER INDEX

ADJUSTMENTS (AUTOMATIC TRANSMISSION) 7-10
   KICK-DOWN SWITCH 7-10
   KICKDOWN CABLE 7-11
   MANUAL LINKAGE 7-11
   SHIFT LINKAGE 7-10
ADJUSTMENTS (DISTRIBUTOR IGNITION SYSTEM) 2-4
   AIR GAP ADJUSTMENT 2-4
AIR CLEANER 1-12
   REMOVAL & INSTALLATION 1-12
AIR CONDITIONING 1-32
   PREVENTIVE MAINTENANCE 1-34
   SYSTEM INSPECTION 1-34
   SYSTEM SERVICE & REPAIR 1-32
AIR CONDITIONING COMPONENTS 6-12
   REMOVAL & INSTALLATION 6-12
AIR INJECTION SYSTEM 4-6
   OPERATION 4-6
   SERVICE 4-6
ALTERNATOR 2-9
   REMOVAL & INSTALLATION 2-10
   TESTING 2-9
ALTERNATOR PRECAUTIONS 2-9
ANTENNA 10-6
   REPLACEMENT 10-6
AUTOMATIC LOCKING HUBS 7-19
   REMOVAL & INSTALLATION 7-19
AUTOMATIC TRANSMISSION 7-7
AUTOMATIC TRANSMISSION (FLUIDS AND LUBRICANTS) 1-41
   DRAIN AND REFILL 1-42
   FLUID RECOMMENDATION 1-41
   LEVEL CHECK 1-41
AUTOMATIC TRANSMISSION ASSEMBLY 7-8
   REMOVAL & INSTALLATION 7-8
AVOIDING THE MOST COMMON MISTAKES 1-2
AVOIDING TROUBLE 1-2
AXLE SHAFT, BEARING AND SEAL 7-24
   REMOVAL & INSTALLATION 7-24
BACK-UP LIGHT SWITCH 7-2
   REMOVAL & INSTALLATION 7-2
BALANCE (COUNTERSHAFTS) SHAFTS 3-39
   REMOVAL & INSTALLATION 3-39
BASIC ELECTRICAL THEORY 6-2
   HOW DOES ELECTRICITY WORK: THE WATER
     ANALOGY 6-2
   OHM'S LAW 6-2
BASIC FUEL SYSTEM DIAGNOSIS 5-2
BASIC OPERATING PRINCIPLES 9-2
   DISC BRAKES 9-2
   DRUM BRAKES 9-2
BASIC STARTING SYSTEM PROBLEMS 2-16
BATTERY 1-15
   BATTERY FLUID 1-16
   CABLES 1-16
   CHARGING 1-17
   GENERAL MAINTENANCE 1-15
   PRECAUTIONS 1-15
   REPLACEMENT 1-17
BATTERY CABLES 6-8

10-24 INDEX

BELTS 1-17
    ADJUSTMENT 1-18
    INSPECTION 1-17
    REMOVAL & INSTALLATION 1-18
BLOWER MOTOR 6-8
    REMOVAL & INSTALLATION 6-8
BODY LUBRICATION AND MAINTENANCE 1-48
    BODY DRAIN HOLES 1-48
    DOOR HINGES AND HINGE CHECKS 1-48
    LOCK CYLINDERS 1-48
    TAILGATE 1-48
BOLTS, NUTS AND OTHER THREADED RETAINERS 1-6
BRAKE BLEEDING 9-6
    BLEEDING SEQUENCE 9-6
    FLUSHING HYDRAULIC BRAKE SYSTEMS 9-7
    MANUAL BLEEDING 9-7
BRAKE DRUMS 9-12
    INSPECTION 9-12
    REMOVAL & INSTALLATION 9-12
BRAKE HOSES AND LINES 9-5
    REMOVAL & INSTALLATION 9-6
BRAKE LIGHT SWITCH 9-3
    REMOVAL & INSTALLATION 9-3
BRAKE MASTER CYLINDER 1-47
    FLUID RECOMMENDATIONS 1-47
    LEVEL CHECK 1-47
BRAKE OPERATING SYSTEM 9-2
BRAKE PADS 9-7
    INSPECTION 9-7
    REMOVAL & INSTALLATION 9-7
BRAKE ROTOR 9-11
    INSPECTION 9-11
    REMOVAL & INSTALLATION 9-11
BRAKE SHOES 9-12
    ADJUSTMENTS 9-16
    INSPECTION 9-12
    REMOVAL & INSTALLATION 9-12
BRAKE SPECIFICATIONS 9-26
BUY OR REBUILD? 3-45
CABLES 9-18
    ADJUSTMENT 9-19
    REMOVAL & INSTALLATION 9-18
CALIPER 9-9
    OVERHAUL 9-10
    REMOVAL & INSTALLATION 9-9
CAMSHAFT 3-35
    INSPECTION 3-37
    REMOVAL & INSTALLATION 3-35
CAMSHAFT BEARINGS 3-38
    REMOVAL & INSTALLATION 3-38
CAMSHAFT POSITION (CMP) AND CRANKSHAFT POSITION (CKP) SENSORS 2-7
CAMSHAFT POSITION (CMP) SENSOR 4-16
    OPERATION 4-16
    REMOVAL & INSTALLATION 4-17
    TESTING 4-17
CAPACITIES 1-56
CARBURETED FUEL SYSTEM 5-3
CARBURETOR 5-3
    ADJUSTMENTS 5-3
    OVERHAUL 5-3
    REMOVAL & INSTALLATION 5-3

CENTER BEARING 7-17
    REPLACEMENT 7-17
CHARGING SYSTEM 2-8
CHASSIS AND CAB MOUNT BUSHINGS 10-6
    REMOVAL & INSTALLATION 10-6
CHASSIS MAINTENANCE 1-48
CIRCUIT BREAKERS 6-38
    REPLACEMENT 6-38
CIRCUIT PROTECTION 6-35
CLEARING CODES 4-22
CLUTCH 7-4
CLUTCH MASTER CYLINDER (CLUTCH) 7-6
    REMOVAL & INSTALLATION 7-6
CLUTCH MASTER CYLINDER (FLUIDS AND LUBRICANTS) 1-47
    FLUID RECOMMENDATIONS 1-47
    LEVEL CHECK 1-47
CLUTCH SLAVE CYLINDER 7-6
    BLEEDING THE HYDRAULIC SYSTEM 7-7
    REMOVAL & INSTALLATION 7-6
CODE DESCRIPTION 4-22
    NAVAJO 4-22
    PICKUP AND MPV 4-22
COIL SPRINGS (FRONT SUSPENSION) 8-4
    REMOVAL & INSTALLATION 8-4
COIL SPRINGS (REAR SUSPENSION) 8-25
    REMOVAL & INSTALLATION 8-25
COMBINATION SWITCH 8-30
    REMOVAL & INSTALLATION 8-30
COMPONENT LOCATIONS 4-23
    FRONT SUSPENSION COMPONENT LOCATIONS 8-2
    FRONT SUSPENSION COMPONENT LOCATIONS—MPV REAR WHEEL DRIVE SHOWN 8-3
    REAR SUSPENSION COMPONENT LOCATIONS—MPV SHOWN 8-24
COMPRESSION RODS 8-13
    REMOVAL & INSTALLATION 8-13
CONTROL ARMS/LINKS 8-27
    REMOVAL & INSTALLATION 8-27
CONTROL PANEL 6-13
    REMOVAL & INSTALLATION 6-13
COOLANT TEMPERATURE SENDER 2-13
    REMOVAL & INSTALLATION 2-13
    TESTING 2-13
COOLING SYSTEM 1-44
    CHECK THE RADIATOR CAP 1-46
    CLEAN RADIATOR OF DEBRIS 1-46
    DRAIN, REFILL & FLUSHING 1-46
    FLUID RECOMMENDATIONS 1-45
    LEVEL CHECK 1-45
CRANKCASE VENTILATION SYSTEM 4-2
    COMPONENT TESTING 4-2
    OPERATION 4-2
    REMOVAL & INSTALLATION 4-2
CRANKSHAFT POSITION (CKP) SENSOR 4-18
    OPERATION 4-18
    REMOVAL & INSTALLATION 4-19
    TESTING 4-18
CRANKSHAFT PULLEY (VIBRATION DAMPER) 3-26
    REMOVAL & INSTALLATION 3-26
CRUISE CONTROL 6-14
CRUISE CONTROL (CHASSIS ELECTRICAL) 6-16

INDEX

CV-BOOTS 1-25
   INSPECTION 1-25
CYLINDER HEAD (ENGINE MECHANICAL) 3-16
   REMOVAL & INSTALLATION 3-16
CYLINDER HEAD (ENGINE RECONDITIONING) 3-47
   ASSEMBLY 3-53
   DISASSEMBLY 3-48
   INSPECTION 3-51
   REFINISHING & REPAIRING 3-52
DECELERATION CONTROL SYSTEM 4-7
   OPERATION 4-7
   SERVICE 4-7
DESCRIPTION AND OPERATION (DISTRIBUTOR IGNITION SYSTEM) 2-2
DESCRIPTION AND OPERATION (REAR WHEEL ANTI-LOCK BRAKE SYSTEM) 9-20
DETERMINING ENGINE CONDITION 3-44
   COMPRESSION TEST 3-44
   OIL PRESSURE TEST 3-45
DIAGNOSIS AND TESTING (DISTRIBUTOR IGNITION SYSTEM) 2-3
   EXTERNAL RESISTOR 2-4
   IGNITION COIL SECONDARY VOLTAGE (SPARK) TEST 2-3
   PICKUP COIL RESISTANCE 2-3
   SERVICE PRECAUTIONS 2-3
   SPARK ADVANCE CONTROL 2-4
   TESTING PROCEDURES 2-3
DIAGNOSIS AND TESTING (DISTRIBUTORLESS IGNITION SYSTEM) 2-6
   PRELIMINARY CHECKS 2-6
   SECONDARY SPARK TEST 2-6
   SERVICE PRECAUTIONS 2-6
DIFFERENTIAL CARRIER 7-23
   REMOVAL & INSTALLATION 7-23
DISC BRAKES 9-7
DISCONNECTING THE CABLES 6-8
DISTRIBUTOR 2-5
   REMOVAL & INSTALLATION 2-5
DISTRIBUTOR CAP AND ROTOR 1-29
   INSPECTION 1-29
   REMOVAL & INSTALLATION 1-29
DISTRIBUTOR IGNITION SYSTEM 2-2
DISTRIBUTORLESS IGNITION SYSTEM 2-6
DO'S 1-4
DON'TS 1-5
DOOR AND LIFTGATE LOCKS 10-13
   REMOVAL & INSTALLATION 10-13
DOOR GLASS AND REGULATOR 10-14
   REMOVAL & INSTALLATION 10-14
DOOR PANELS 10-11
   REMOVAL & INSTALLATION 10-11
DOORS 10-2
   ADJUSTMENT 10-2
DRIVE AXLE 1-12
   FRONT 1-12
   REAR 1-12
DRIVELINE 7-13
DRIVEN DISC AND PRESSURE PLATE 7-4
   ADJUSTMENTS 7-6
   REMOVAL & INSTALLATION 7-4

DUCKBILL CLIP FITTING 5-2
   REMOVAL & INSTALLATION 5-2
ELECTRIC FUEL PUMP 5-3
   REMOVAL & INSTALLATION 5-3
   TESTING 5-3
ELECTRIC WINDOW MOTOR 10-15
   REMOVAL & INSTALLATION 10-15
ELECTRICAL COMPONENTS 6-2
   CONNECTORS 6-4
   GROUND 6-3
   LOAD 6-4
   POWER SOURCE 6-2
   PROTECTIVE DEVICES 6-3
   SWITCHES & RELAYS 6-3
   WIRING & HARNESSES 6-4
ELECTRO-HYDRAULIC VALVE 9-24
   REMOVAL & INSTALLATION 9-24
ELECTRONIC CONTROL UNIT (4-WHEEL ANTI-LOCK BRAKE SYSTEM) 9-25
   REMOVAL & INSTALLATION 9-25
ELECTRONIC CONTROL UNIT (REAR WHEEL ANTI-LOCK BRAKE SYSTEM) 9-23
   REMOVAL & INSTALLATION 9-23
ELECTRONIC ENGINE CONTROLS 4-9
EMISSION CONTROLS 4-2
EMISSION MAINTENANCE WARNING LIGHT 4-8
   RESETTING 4-8
ENGINE (ENGINE MECHANICAL) 3-2
   REMOVAL & INSTALLATION 3-2
ENGINE (SERIAL NUMBER IDENTIFICATION) 1-10
   NAVAJO 1-10
   PICK-UP & MPV 1-10
ENGINE (TRAILER TOWING) 1-50
ENGINE BLOCK 3-54
   ASSEMBLY 3-57
   DISASSEMBLY 3-54
   GENERAL INFORMATION 3-54
   INSPECTION 3-55
   REFINISHING 3-57
ENGINE COOLANT TEMPERATURE (ECT) SENSOR 4-12
   OPERATION 4-12
   REMOVAL & INSTALLATION 4-12
   TESTING 4-12
ENGINE FAN 3-13
   REMOVAL & INSTALLATION 3-13
ENGINE IDENTIFICATION 1-10
ENGINE MECHANICAL 3-2
ENGINE MECHANICAL SPECIFICATIONS 3-60
ENGINE OIL 1-39
   OIL & FILTER CHANGE 1-39
   OIL LEVEL CHECK 1-39
ENGINE OVERHAUL TIPS 3-45
   CLEANING 3-46
   OVERHAUL TIPS 3-46
   REPAIRING DAMAGED THREADS 3-46
   TOOLS 3-45
ENGINE PREPARATION 3-47
ENGINE RECONDITIONING 3-44
ENGINE START-UP AND BREAK-IN 3-60
   BREAKING IT IN 3-60
   KEEP IT MAINTAINED 3-60
   STARTING THE ENGINE 3-60

INDEX

ENTERTAINMENT SYSTEMS 6-16
EVAPORATIVE EMISSION CANISTER 1-15
   SERVICING 1-15
EVAPORATIVE EMISSION CONTROL SYSTEM 4-2
   COMPONENT TESTING 4-2
   OPERATION 4-2
   REMOVAL & INSTALLATION 4-5
EXHAUST GAS RECIRCULATION SYSTEM 4-5
   OPERATION 4-5
   REMOVAL & INSTALLATION 4-6
   SERVICE 4-5
EXHAUST MANIFOLD 3-10
   REMOVAL & INSTALLATION 3-10
EXHAUST SYSTEM 3-41
EXTENSION HOUSING SEAL (AUTOMATIC TRANSMISSION) 7-8
   REMOVAL & INSTALLATION 7-8
EXTENSION HOUSING SEAL (MANUAL TRANSMISSION) 7-2
   REMOVAL & INSTALLATION 7-2
EXTERIOR 10-2
FASTENERS, MEASUREMENTS AND CONVERSIONS 1-6
FENDERS 10-6
   REMOVAL & INSTALLATION 10-6
FIRING ORDERS 2-7
FLASHERS AND RELAYS 6-38
   REPLACEMENT 6-38
FLOOR CONSOLE 10-11
   REMOVAL & INSTALLATION 10-11
FLUID DISPOSAL 1-38
FLUIDS AND LUBRICANTS 1-38
4-WHEEL ANTI-LOCK BRAKE SYSTEM (4WABS) 9-25
FREEWHEEL MECHANISM 7-19
   REMOVAL & INSTALLATION 7-19
FRONT AND REAR DRIVE AXLE 1-43
   DRAIN AND REFILL 1-43
   FLUID RECOMMENDATIONS 1-43
   LEVEL CHECK 1-43
FRONT COVER OIL SEAL 3-29
   REMOVAL & INSTALLATION 3-29
FRONT DRIVE AXLE 7-18
FRONT DRIVESHAFT AND U-JOINTS 7-13
   REMOVAL & INSTALLATION 7-13
   U-JOINT REPLACEMENT 7-14
FRONT OUTPUT SHAFT SEAL 7-12
   REMOVAL & INSTALLATION 7-12
FRONT SUSPENSION 8-2
FRONT SUSPENSION COMPONENT LOCATIONS 8-2
FRONT SUSPENSION COMPONENT LOCATIONS—MPV REAR WHEEL DRIVE SHOWN 8-3
FRONT WHEEL BEARINGS 8-20
   ADJUSTMENT 8-22
   REMOVAL & INSTALLATION 8-20
FRONT WHEEL SPEED SENSOR 9-25
   REMOVAL & INSTALLATION 9-25
   TESTING 9-25
FUEL AND ENGINE OIL RECOMMENDATIONS 1-38
   ENGINE OIL 1-38
   FUEL 1-39
   OPERATION IN FOREIGN COUNTRIES 1-39
FUEL CHARGING ASSEMBLY 5-13
   REMOVAL & INSTALLATION 5-13

FUEL FILTER 1-14
   REMOVAL & INSTALLATION 1-14
FUEL INJECTION SYSTEM 5-7
FUEL INJECTORS 5-11
   REMOVAL & INSTALLATION 5-11
   TESTING 5-13
FUEL LINES AND FITTINGS 5-2
FUEL, OIL PRESSURE, VOLTAGE AND COOLANT TEMPERATURE GAUGES 6-26
FUEL PRESSURE REGULATOR 5-14
   REMOVAL & INSTALLATION 5-14
FUEL PUMP 5-8
   REMOVAL & INSTALLATION 5-8
   TESTING 5-9
FUEL SYSTEM PROBLEMS 5-16
FUEL TANK 5-15
FUSES 6-35
   REPLACEMENT 6-35
FUSIBLE LINKS 6-38
GASOLINE ENGINE TUNE-UP SPECIFICATIONS 1-33
GENERAL ENGINE SPECIFICATIONS 1-11
GENERAL INFORMATION (CHARGING SYSTEM) 2-8
GENERAL INFORMATION (DISTRIBUTORLESS IGNITION SYSTEM) 2-6
GENERAL INFORMATION (FUEL INJECTION SYSTEM) 5-7
   FUEL SYSTEM SERVICE PRECAUTIONS 5-7
GENERAL INFORMATION (FUEL LINES AND FITTINGS) 5-2
GENERAL INFORMATION (STARTING SYSTEM) 2-10
GENERAL INFORMATION (TROUBLE CODES) 4-20
   MALFUNCTION INDICATOR LAMP (MIL) 4-20
GENERAL RECOMMENDATIONS 1-50
GRILLE 10-5
   REMOVAL & INSTALLATION 10-5
HAIRPIN CLIP FITTING 5-2
   REMOVAL & INSTALLATION 5-2
HALFSHAFT 7-22
   REMOVAL & INSTALLATION 7-22
HANDLING A TRAILER 1-50
HEADLIGHT SWITCH 6-26
   REMOVAL & INSTALLATION 6-26
HEADLIGHTS 6-27
   AIMING THE HEADLIGHTS 6-28
   REMOVAL & INSTALLATION 6-27
HEATER CASE ASSEMBLY 6-12
   REMOVAL & INSTALLATION 6-12
HEATER CORE 6-10
   REMOVAL & INSTALLATION 6-10
HEATING AND AIR CONDITIONING 6-8
HITCH (TONGUE) WEIGHT 1-50
HOOD 10-2
   ALIGNMENT 10-2
HOSES 1-21
   INSPECTION 1-21
   REMOVAL & INSTALLATION 1-22
HOW TO USE THIS BOOK 1-2
HYDRAULIC CONTROL UNIT (HCU) 9-25
   PUMP TESTING 9-25
   REMOVAL & INSTALLATION 9-25

INDEX

I-BEAM AXLE 8-17
   REMOVAL & INSTALLATION 8-17
IAT/ECT SENSOR VOLTAGE AND RESISTANCE
  SPECIFICATIONS 4-12
IDLE AIR CONTROL (IAC) VALVE 4-11
   OPERATION 4-11
   REMOVAL & INSTALLATION 4-11
   TESTING 4-11
IDLE SPEED AND MIXTURE ADJUSTMENTS 1-31
   IDLE SPEED 1-31
   MIXTURE ADJUSTMENT 1-32
IGNITION COIL (DISTRIBUTOR IGNITION
  SYSTEM) 2-4
   REMOVAL & INSTALLATION 2-5
   TESTING 2-4
IGNITION COIL PACK (DISTRIBUTORLESS IGNITION
  SYSTEM) 2-6
   REMOVAL & INSTALLATION 2-7
   TESTING 2-6
IGNITION MODULE (IGNITER) (DISTRIBUTOR IGNITION
  SYSTEM) 2-5
   REMOVAL & INSTALLATION 2-5
IGNITION MODULE (DISTRIBUTORLESS IGNITION
  SYSTEM) 2-7
   REMOVAL & INSTALLATION 2-7
IGNITION SWITCH 8-31
   REMOVAL & INSTALLATION 8-31
IGNITION TIMING 1-29
   GENERAL INFORMATION 1-29
   INSPECTION & ADJUSTMENT 1-30
INSIDE REAR VIEW MIRROR 10-16
INSPECTION (EXHAUST SYSTEM) 3-42
   REPLACEMENT 3-42
INSTRUMENT CLUSTER 6-24
   REMOVAL & INSTALLATION 6-24
INSTRUMENT PANEL AND PAD 10-9
   REMOVAL & INSTALLATION 10-9
INSTRUMENTS AND SWITCHES 6-24
INTAKE AIR TEMPERATURE (IAT) SENSOR 4-13
   OPERATION 4-13
   REMOVAL & INSTALLATION 4-13
   TESTING 4-13
INTAKE MANIFOLD 3-8
   REMOVAL & INSTALLATION 3-8
INTERIOR 10-9
JACKING 1-52
JACKING PRECAUTIONS 1-53
JUMP STARTING A DEAD BATTERY 1-51
JUMP STARTING PRECAUTIONS 1-51
JUMP STARTING PROCEDURE 1-51
KNUCKLE AND SPINDLE 8-18
   REMOVAL & INSTALLATION 8-18
LEAF SPRINGS 8-25
   REMOVAL & INSTALLATION 8-25
LIFTGATE 10-3
   ALIGNMENT 10-5
   REMOVAL & INSTALLATION 10-3
LIGHT BULB APPLICATIONS 6-34
LIGHTING 6-27
LOAD SENSING G-VALVE 9-5
   REMOVAL & INSTALLATION 9-5

LOW OIL LEVEL SENSOR 2-14
   REMOVAL & INSTALLATION 2-15
   TESTING 2-14
LOWER BALL JOINTS 8-11
   INSPECTION 8-11
   REMOVAL & INSTALLATION 8-11
LOWER CONTROL ARM 8-15
   CONTROL ARM BUSHING REPLACEMENT 8-17
   REMOVAL & INSTALLATION 8-15
MAINTENANCE INTERVALS 1-53
MAINTENANCE OR REPAIR? 1-2
MANUAL LOCKING HUBS 7-18
   ADJUSTMENT 7-18
   REMOVAL & INSTALLATION 7-18
MANUAL STEERING GEAR (FLUIDS AND
  LUBRICANTS) 1-48
   FLUID RECOMMENDATIONS 1-48
   LEVEL CHECK 1-48
MANUAL STEERING GEAR (STEERING) 8-33
   REMOVAL & INSTALLATION 8-33
MANUAL TRANSMISSION 7-2
MANUAL TRANSMISSION (FLUIDS AND
  LUBRICANTS) 1-40
   DRAIN & REFILL 1-41
   FLUID RECOMMENDATIONS 1-40
   LEVEL CHECK 1-40
MANUAL TRANSMISSION ASSEMBLY 7-2
   REMOVAL & INSTALLATION 7-2
MASS AIR FLOW (MAF) SENSOR 4-13
   OPERATION 4-13
   REMOVAL & INSTALLATION 4-15
   TESTING 4-14
MASTER CYLINDER 9-3
   REMOVAL & INSTALLATION 9-3
MECHANICAL FUEL PUMP 5-3
   REMOVAL & INSTALLATION 5-3
   TESTING 5-3
NEUTRAL START SWITCH/BACK-UP SWITCH 7-7
   ADJUSTMENTS 7-8
   REMOVAL & INSTALLATION 7-7
OIL PAN 3-20
   REMOVAL & INSTALLATION 3-20
OIL PRESSURE SENDER SWITCH 2-14
   REMOVAL & INSTALLATION 2-14
   TESTING 2-14
OIL PUMP 3-22
   INSPECTION 3-25
   REMOVAL & INSTALLATION 3-22
OUTSIDE MIRRORS 10-5
   REMOVAL & INSTALLATION 10-5
OXYGEN SENSOR 4-10
   OPERATION 4-10
   REMOVAL & INSTALLATION 4-10
   TESTING 4-10
PARKING BRAKE 9-18
PCV VALVE 1-15
   REMOVAL & INSTALLATION 1-15
PINION SEAL (FRONT DRIVE AXLE) 7-23
   REMOVAL & INSTALLATION 7-23
PINION SEAL (REAR AXLE) 7-26
   REMOVAL & INSTALLATION 7-26

10-28 INDEX

POWER BOOSTER 9-3
  PUSHROD CLEARANCE 9-5
  REMOVAL & INSTALLATION 9-3
POWER SEAT MOTOR 10-18
  REMOVAL & INSTALLATION 10-18
POWER STEERING GEAR 8-34
  REMOVAL & INSTALLATION 8-34
POWER STEERING PUMP (FLUIDS AND LUBRICANTS) 1-48
  FLUID RECOMMENDATIONS 1-48
  LEVEL CHECK 1-48
POWER STEERING PUMP (STEERING) 8-36
  BLEEDING THE SYSTEM 8-38
  REMOVAL & INSTALLATION 8-36
POWER STEERING RACK AND PINION 8-35
  REMOVAL & INSTALLATION 8-35
POWER SUNROOF 10-8
  REMOVAL & INSTALLATION 10-8
POWERTRAIN CONTROL MODULE (PCM)/ELECTRONIC CONTROL MODULE (ECM) 4-9
  OPERATION 4-9
  REMOVAL & INSTALLATION 4-9
PRESSURE DIFFERENTIAL VALVE 9-5
  CENTRALIZING THE PRESSURE DIFFERENTIAL VALVE 9-5
  REMOVAL & INSTALLATION 9-5
PROPORTIONING VALVE 9-5
  REMOVAL & INSTALLATION 9-5
RADIATOR 3-12
  REMOVAL & INSTALLATION 3-12
RADIO RECIEVER/TAPE PLAYER 6-16
  REMOVAL & INSTALLATION 6-16
RADIUS ARM 8-14
  REMOVAL & INSTALLATION 8-14
READING CODES 4-20
  NAVAJO 4-21
  PICKUP AND MPV 4-20
REAR AXLE 7-24
REAR DRIVESHAFT 7-15
  DRIVESHAFT BALANCING 7-16
  REMOVAL & INSTALLATION 7-15
  U-JOINT REPLACEMENT 7-16
REAR DRUM BRAKES 9-12
REAR MAIN OIL SEAL 3-40
  REPLACEMENT 3-40
REAR OUTPUT SHAFT SEAL 7-12
  REMOVAL & INSTALLATION 7-12
REAR SPEED SENSOR 9-26
  REMOVAL & INSTALLATION 9-26
  TESTING 9-26
REAR SUSPENSION 8-24
REAR SUSPENSION COMPONENT LOCATIONS—MPV SHOWN 8-24
REAR WHEEL ANTI-LOCK BRAKE SYSTEM 9-20
REAR WHEEL BEARINGS 8-28
  REMOVAL AND INSTALLATION 8-28
REAR WIPER SWITCH 6-26
  REMOVAL & INSTALLATION 6-26
REGULATOR 2-10
  REMOVAL & INSTALLATION 2-10
RELIEVING FUEL SYSTEM PRESSURE 5-7
  NAVAJO 5-7
  PICKUP & MPV 5-7

ROCKER ARM (VALVE) COVER 3-2
  REMOVAL & INSTALLATION 3-2
ROCKER ARMS/SHAFTS 3-4
  REMOVAL & INSTALLATION 3-4
ROUTINE MAINTENANCE AND TUNE-UP 1-12
SAFETY PRECAUTIONS 3-41
SEATS 10-16
  REMOVAL & INSTALLATION 10-16
SENDING UNITS AND SENSORS 2-13
SERIAL NUMBER IDENTIFICATION 1-9
SERVICING YOUR VEHICLE SAFELY 1-4
SHOCK ABSORBERS (FRONT SUSPENSION) 8-7
  REMOVAL & INSTALLATION 8-7
  TESTING 8-8
SHOCK ABSORBERS (REAR SUSPENSION) 8-26
  REMOVAL & INSTALLATION 8-26
SIGNAL AND MARKER LIGHTS 6-29
  REMOVAL & INSTALLATION 6-29
SPARK PLUG WIRES 1-28
  REMOVAL & INSTALLATION 1-28
  TESTING 1-28
SPARK PLUGS 1-25
  INSPECTION & GAPPING 1-26
  REMOVAL & INSTALLATION 1-26
  SPARK PLUG HEAT RANGE 1-25
SPEAKERS 6-18
  REMOVAL & INSTALLATION 6-18
SPECIAL TOOLS 1-4
SPECIFICATIONS CHARTS
  BRAKE SPECIFICATIONS 9-26
  CAPACITIES 1-56
  ENGINE IDENTIFICATION 1-10
  ENGINE MECHANICAL SPECIFICATIONS 3-60
  GASOLINE ENGINE TUNE-UP SPECIFICATIONS 1-33
  GENERAL ENGINE SPECIFICATIONS 1-11
  IAT/ECT SENSOR VOLTAGE AND RESISTANCE SPECIFICATIONS 4-12
  MAINTENANCE INTERVALS 1-53
  TORQUE SPECIFICATIONS (BODY AND TRIM) 10-18
  TORQUE SPECIFICATIONS (DRIVE TRAIN) 7-28
  TORQUE SPECIFICATIONS (ENGINE AND ENGINE OVERHAUL) 3-69
  TORQUE SPECIFICATIONS (SUSPENSION AND STEERING) 8-39
  VEHICLE IDENTIFICATION CHART 1-9
SPEED SENSOR 9-24
  REMOVAL & INSTALLATION 9-24
  TESTING 9-24
SPINDLE BEARINGS 7-20
  REMOVAL & INSTALLATION 7-20
SPINDLE, SHAFT AND JOINT ASSEMBLY 7-22
  REMOVAL & INSTALLATION 7-22
SPRING LOCK COUPLING 5-2
  REMOVAL & INSTALLATION 5-2
STABILIZER BAR 8-27
  REMOVAL & INSTALLATION 8-27
STANDARD AND METRIC MEASUREMENTS 1-7
STARTER 2-11
  REMOVAL & INSTALLATION 2-12
  SOLENOID REPLACEMENT 2-13
  TESTING 2-11

INDEX 10-29

STARTING SYSTEM 2-10
STEERING 8-29
  STEERING LINKAGE 8-31
    REMOVAL & INSTALLATION 8-31
  STEERING WHEEL 8-29
    REMOVAL & INSTALLATION 8-29
  STRUTS 8-8
    OVERHAUL 8-8
    REMOVAL & INSTALLATION 8-8
  SWAY BAR 8-12
    REMOVAL & INSTALLATION 8-12
  SYSTEM SELF-TEST (4-WHEEL ANTI-LOCK BRAKE SYSTEM) 9-25
  SYSTEM SELF TEST (REAR WHEEL ANTI-LOCK BRAKE SYSTEM) 9-20
    PRECAUTIONS 9-23
  TAILGATE 10-3
    REMOVAL & INSTALLATION 10-3
  TANK ASSEMBLY 5-15
    REMOVAL & INSTALLATION 5-15
  TEST EQUIPMENT 6-5
    JUMPER WIRES 6-5
    MULTIMETERS 6-5
    TEST LIGHTS 6-5
  TESTING 6-6
    OPEN CIRCUITS 6-6
    RESISTANCE 6-7
    SHORT CIRCUITS 6-6
    VOLTAGE 6-7
    VOLTAGE DROP 6-7
  THERMOSTAT 3-6
    REMOVAL & INSTALLATION 3-6
  THROTTLE BODY 5-10
    REMOVAL & INSTALLATION 5-10
  THROTTLE POSITION (TP) SENSOR 4-15
    OPERATION 4-15
    REMOVAL & INSTALLATION 4-16
    TESTING 4-16
  TIMING BELT AND SPROCKETS 3-29
    REMOVAL & INSTALLATION 3-29
  TIMING BELT COVERS 3-26
    REMOVAL & INSTALLATION 3-26
  TIMING BELTS 1-21
    INSPECTION 1-21
  TIMING CHAIN AND TENSIONER 3-31
    REMOVAL & INSTALLATION 3-32
  TIMING CHAIN COVER 3-27
    REMOVAL & INSTALLATION 3-27
  TIRES AND WHEELS 1-35
    INFLATION & INSPECTION 1-36
    TIRE DESIGN 1-36
    TIRE ROTATION 1-35
    TIRE STORAGE 1-36
TOOLS AND EQUIPMENT 1-2
  TORQUE 1-6
    TORQUE ANGLE METERS 1-7
    TORQUE WRENCHES 1-6
  TORQUE SPECIFICATIONS (BODY AND TRIM) 10-18
  TORQUE SPECIFICATIONS (DRIVE TRAIN) 7-28
  TORQUE SPECIFICATIONS (ENGINE AND ENGINE OVERHAUL) 3-69

  TORQUE SPECIFICATIONS (SUSPENSION AND STEERING) 8-39
  TORSION BAR 8-5
    REMOVAL & INSTALLATION 8-5
TOWING THE VEHICLE 1-50
TRAILER TOWING 1-50
  TRAILER WEIGHT 1-50
TRAILER WIRING 6-34
TRANSFER CASE 7-12
  TRANSFER CASE (FLUIDS AND LUBRICANTS) 1-43
    DRAIN AND REFILL 1-43
    FLUID RECOMMENDATION 1-43
    LEVEL CHECK 1-43
  TRANSFER CASE (SERIAL NUMBER IDENTIFICATION) 1-12
  TRANSFER CASE ASSEMBLY 7-12
    ADJUSTMENTS 7-13
    REMOVAL & INSTALLATION 7-12
  TRANSMISSION (SERIAL NUMBER IDENTIFICATION) 1-12
    AUTOMATIC TRANSMISSION 1-12
    MANUAL TRANSMISSION 1-12
  TRANSMISSION (TRAILER TOWING) 1-50
TROUBLE CODES 4-20
TROUBLESHOOTING CHARTS
  BASIC STARTING SYSTEM PROBLEMS 2-16
  CRUISE CONTROL 6-16
  FUEL SYSTEM PROBLEMS 5-16
  TROUBLESHOOTING ELECTRICAL SYSTEMS 6-6
UNDERSTANDING AND TROUBLESHOOTING ELECTRICAL SYSTEMS 6-2
  UNDERSTANDING THE AUTOMATIC TRANSMISSION 7-7
  UNDERSTANDING THE CLUTCH 7-4
  UNDERSTANDING THE MANUAL TRANSMISSION 7-2
  UPPER BALL JOINTS 8-10
    INSPECTION 8-10
    REMOVAL & INSTALLATION 8-10
  UPPER CONTROL ARM 8-15
    CONTROL ARM BUSHING REPLACEMENT 8-15
    REMOVAL & INSTALLATION 8-15
VACUUM DIAGRAMS 4-30
  VALVE LASH 1-30
    ADJUSTMENT 1-30
  VALVE LIFTERS (TAPPETS) 3-39
    REMOVAL & INSTALLATION 3-39
  VEHICLE 1-9
    NAVAJO 1-9
    PICK-UP & MPV 1-9
  VEHICLE IDENTIFICATION CHART 1-9
  VEHICLE SPEED SENSOR (VSS) 4-19
    OPERATION 4-19
    REMOVAL & INSTALLATION 4-20
    TESTING 4-19
  WATER PUMP 3-14
    REMOVAL & INSTALLATION 3-14
  WHEEL ALIGNMENT 8-23
    CAMBER 8-23
    CASTER 8-23
    TOE 8-23
  WHEEL BEARINGS 1-48
    REPACKING 1-48
  WHEEL CYLINDER 9-17
    OVERHAUL 9-17
    REMOVAL & INSTALLATION 9-17

WHERE TO BEGIN 1-2
WINDSHIELD AND FIXED GLASS 10-15
    REMOVAL & INSTALLATION 10-15
    WINDSHIELD CHIP REPAIR 10-15
WINDSHIELD WASHER MOTOR 6-23
    REMOVAL & INSTALLATION 6-23
WINDSHIELD WIPER BLADE AND ARM 6-20
    REMOVAL & INSTALLATION 6-20
WINDSHIELD WIPER MOTOR 6-21
    REMOVAL & INSTALLATION 6-21
WINDSHIELD WIPER SWITCH 6-26
    REMOVAL & INSTALLATION 6-26
WINDSHIELD WIPERS 6-20
WINDSHIELD WIPERS (ROUTINE MAINTENANCE AND TUNE-UP) 1-35
    ELEMENT (REFILL) CARE & REPLACEMENT 1-35
WIRE AND CONNECTOR REPAIR 6-8
WIRING DIAGRAMS 6-39